9th International Conference on Tribology (BALKANTRIB'17)

IOP Conference Series: Materials Science and Engineering
Volume 295

Nevsehir, Turkey
13 - 15 September 2017

Editor:

Mehmet Baki Karamis

ISBN: 978-1-5108-5884-8
ISSN: 1757-8981

Printed from e-media with permission by:

Curran Associates, Inc.
57 Morehouse Lane
Red Hook, NY 12571

CURRAN ASSOCIATES INC. proceedings.com

Some format issues inherent in the e-media version may also appear in this print version.

Printed by Curran Associates, Inc. (2018)

Institute of Physics
Dirac House, Temple Back
Bristol BS1 6BE UK

Phone: 44 1 17 929 7481
Fax: 44 1 17 920 0979

techtracking@iop.org

Additional copies of this publication are available from:

Curran Associates, Inc.
57 Morehouse Lane
Red Hook, NY 12571 USA
Phone: 845-758-0400
Fax: 845-758-2633
Email: curran@proceedings.com
Web: www.proceedings.com

Table of contents

Volume 295

9th International Conference on Tribology (Balkantrib'17)

13–15 September 2017, Nevsehir, Turkey

Accepted papers received: 4 January 2018
Published online: 19 February 2018

Preface

Papers

IOP Conf. Series: Materials Science and Engineering **295** (2018) 011001 doi:10.1088/1757-899X/295/1/011001

Preface

The International Conferences on Tribology 'BALKANTRIB' is supported by the Balkan Tribological Association and is organized, every three years since 1993, in various Balkan countries and aim to stimulate collaboration between Balkan and other countries for developing research and development in tribology, as well as application aspects like innovative products and services.

The First Balkan Conference on Tribology BALKANTRIB '93 was held on October 1-3, 1993 in Sofia, Bulgaria. The next conferences BALKANTRIB were as follows: BALKANTRIB'96, Thessaloniki, Greece; BALKANTRIB'99, Sinaia, Romania; BALKANTRIB'02, Kayseri, Turkey; BALKANTRIB'05, Kragujevac, Serbia, BALKANTRIB'08, Sozopol, Bulgaria, BALKANTRIB'11, Thessaloniki, Greece.BALKANTRIB'14, Sinaia, Romania. The organizer of the 9th International Conference on Tribology (BalkanTRib'17) is TURKEY, will be held between 13-15 September 2017 in Cappadocia, Nevsehir.

BalkanTRib'17 provides an ideal academic platform for researchers to present the latest research findings and describe emerging technologies, and directions in Tribology. It also provides the premier interdisciplinary forum for scientists, engineers, and practitioners to present their latest research results, ideas, developments and applications of Tribology Science.

BalkanTRib'17 helded in Turkey by the supports of Erciyes University and Tribology Association (Kayseri). I hope the conference has been beneficial to all participants.

Editor

Prof. Dr. Mehmet Baki KARAMIŞ

Erciyes University, Kayseri, Turkey

BalkanTRib'17

9'th International Conference on Tribology

13-15 September 2017-Cappadocia

Supporters

We are deeply grateful for the generosity of our partners who support symposium that move science forward.

Sponsors

We are deeply grateful for the generosity of sponsors who support symposium that move science forward.

Editor

Prof. Dr. Mehmet Baki KARAMIŞ, Erciyes University, Turkey

Co-Editor

Fehmi NAİR, Soner SAVAS, Abdullah GÖÇER, Mustafa HAMAMCI, Erciyes University, Turkey

Publication layout

İmge Arda ÇATALTAS, İsmet Aydın YILMAZ, Doğukan HARMAN, Erciyes University, Turkey

Scientific Committee

Prof. Dr. A. A. Eker,Turkey
Prof. Dr. A. Erdemir, USA
Prof. Dr. A. Rac, Serbia
Prof. Dr. A. T. Alpas, Canada
Prof. Dr. A. Tudor, Romania
Prof. Dr. A. Vencl, Serbia
Prof. Dr. Anne Neville, UK
Prof. Dr. B. Bhushan, USA
Prof. Dr. B. Selçuk,Turkey
Prof. Dr. D. Olaru, Romania
Prof. Dr. E. Assenova, Bulgaria

IOP Conf. Series: Materials Science and Engineering 295 (2018) 011001 doi:10.1088/1757-899X/295/1/011001

Balkan

IOP Conf. Series: Materials Science and Engineering **295** (2018) 011002 doi:10.1088/1757-899X/295/1/011002

Peer review statement

All papers published in this volume of *IOP Conference Series: Materials Science and Engineering* have been peer reviewed through processes administered by the proceedings Editors. Reviews were conducted by expert referees to the professional and scientific standards expected of a proceedings journal published by IOP Publishing.

Dynamic Frictional Sliding Modes between Two Homogenous Interfaces

Taj W[2], Coker D[1]
[1]Department of Aerospace Engineering, Middle East Technical University, 06800, Ankara, Turkey
[2] Department of Aerospace Engineering, Middle East Technical University, Ankara, Turkey
[2] TUBITAK-UZAY Space Technologies Research Institute, 06800, Ankara, Turkey
coker@metu.edu.tr[1], waheedullah.taj@metu.edu.tr[2]

Abstract: Dynamic frictional sliding behaviour between two homogenous Homalite blocks is studied. The blocks are initially held together using a constant compressive stress. The sliding motion is initiated by impacting one of the blocks with an external object. The resulting sliding modes are inspected. Initiation of fretting is examined by studying the opening stresses in the interface.

Keywords: FEM, Dynamic friction, impact, crack

1. Introduction

Dynamic frictional sliding between two bodies is a fundamental problem encountered in many engineering and science problems in various scales such as fiber reinforced composite material failure, earthquake faults during rupture, machining processes and surface interactions of moving machinery.
The behavior of the frictional interface during sliding motion is of extreme importance as it could explain many important phenomena such as fretting, energy dissipation and fatigue failure of the surface. Previous studies have discovered many sliding modes such as crack-like mode, which resembles a shear crack, a train of pulses mode and a growing pulses mode [1-3]. In a crack-like sliding mode the speed at the crack tip rapidly grows to a moderate level and stays there until the end of the interface. On the other hand, train of pulses sliding mode consists of small but powerful pulses across the interface reaching relatively high speed. In this sliding mode, the pulses maintain a relatively constant peak velocity. Finally, in growing pulses sliding mode there is one leading pulse with extremely high speed followed by a large number of smaller pulses. The aforementioned sliding modes are seen in specific loading and impact velocity conditions, however the regions of transition between the modes has not yet been studied in detail.
In this study, we examine the internal stresses and openings in sliding bodies. Our model uses a rate- and state-dependent friction law developed by Ref. [4]. The model uses the same configuration as in Ref. [3]. The model simulates two thin Homalite blocks held together in space using a constant compressive stress. One of the blocks is then impacted on one side with an external object. The resulting interfacial shear and normal stresses, opening stresses in the material, sliding velocity distribution across the interface and coefficient of friction are then examined in detail.

2. Problem Formulation

2.1. Cohesive law

The finite element code has two constitutive relations consisting of the bulk material constitutive law relating the stresses and strains in the material along with the cohesive law that gives the relation along the interface.

$$\dot{T}_n = -C_n \Delta \dot{u}_n \,, \quad \dot{T}_s = C_s[\Delta \dot{u}_s - sgn(T_s)\dot{q}_{slip} \tag{1}$$

where slip $\dot{q}_{slip}$ is the sliding rate at the interface. The expression for the slip or sliding rate is the dependent relation shown below

$$\dot{q}_{slip} = \begin{cases} \dot{q}_0(\beta^m - 1) & for\, \beta > 1 \\ 0 & for\, \beta \leq 1 \end{cases} \tag{2}$$

Where β is defined as

$$\beta = \frac{|T_s|}{T_n g(\theta)} \tag{3}$$

The friction equations are then updated using a single step tangent modulus method for rate dependent solids [5].
The following equation is used to calculate the change in sliding (horizontal) velocity

$$\Delta u(t) = u_x^+ - u_x^- \tag{4}$$

where u_x is the displacement in the X-direction of an interfacial point such that u_x^+ belongs to the point in the upper plate and u_x^- belongs to the corresponding point in the lower plate. The relative slip velocity can be found using a derivative definition as

$$\dot{q}_{slip} = \frac{\Delta u(t+\Delta t) - \Delta u(t)}{\Delta t} \tag{5}$$

2.2. Rate- and state-dependent friction law

The rate- and state-dependent friction law proposed by Ref. [4] and used by Ref. [3] has the following form

$$\mu(\theta_{ss}, \dot{u}_{slip}) = \mu_{ss} = \mu_d + (\mu_s - \mu_d)\exp\left(-\left(\frac{\dot{u}_{slip}}{V_1}\right)^p\right) \tag{6}$$

where μ_s and μ_d are coefficients of static and dynamic friction, and V_1, V_0, m and p are constants obtained from experiments. The values of these parameters are given in Table 1. This model of friction predicts that coefficient of friction decreases with slip velocity, as shown in Figure 1 (left). If there is no motion in the model initially, but then a finite slip velocity is applied, the model predicts that the coefficient of friction first rises to a maximum value, before decreasing to a value lower than before. If on the other hand, the velocity is suddenly decreased, the coefficient of friction first drops to a floor value before reaching a steady state higher than before. This behavior is seen in Figure 1 (right).

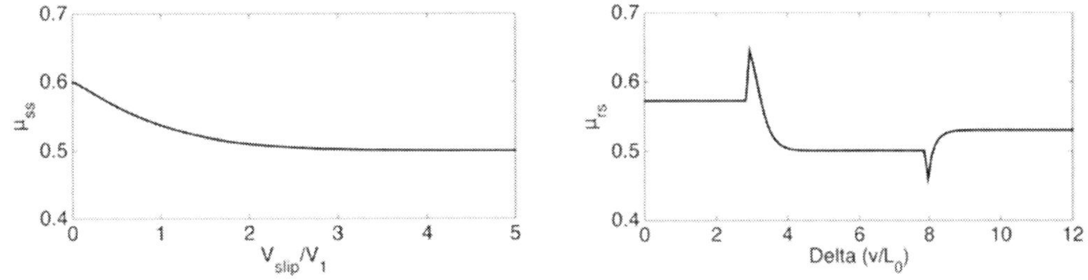

Figure 1. Coefficient of friction in steady state versus normalized velocity (left), and coefficient of friction versus delta with slip velocities: v_{slip} = 10 m/s for *delta* = 0 to *delta* = 3, v_{slip} = 100 m/s for *delta* = 3 to *delta* = 8 and v_{slip} = 30 m/s for *delta* = 80 to *delta* = 100 (right).

Table 1. Values of the parameters used in friction calculations in this study.

μ_s	μ_d	V_0 (μm/s)	V_1 (μm/s)	B	P	m	L_0 (μm)
0.6	0.5	100	26	1	1.2	5	20

2.3. *Finite element model*

The finite element model consists of two two dimensional blocks with dimensions of h = 75 mm and w = 200 mm each. The two blocks are pressed together with a uniform external loading of σ_0 until time t=0, at t=0 an impact velocity of $V(t)$ is given to the upper left side of the bottom block starting from X=0 up to Y=b=25mm. The dashed line in Figure 3 shows the cohesive interface between the two blocks. The X-Y coordinate system is defined such that X is parallel to the cohesive interface and Y is perpendicular to it. The origin of the coordinate system is selected as the left end of the cohesive interface. The mesh with rectangular elements used in finite element modelling is shown in Figure 4. A very fine uniform mesh was used at the left side of the model from X=0 to Y=80mm and from Y=0 to Y=±2mm. The mesh was then increased in coarseness starting from X=80mm to X=120mm at the left and Y=±2mm to Y=±20mm at the top. The rest of the mesh consisted of a uniform and relatively coarse mesh.

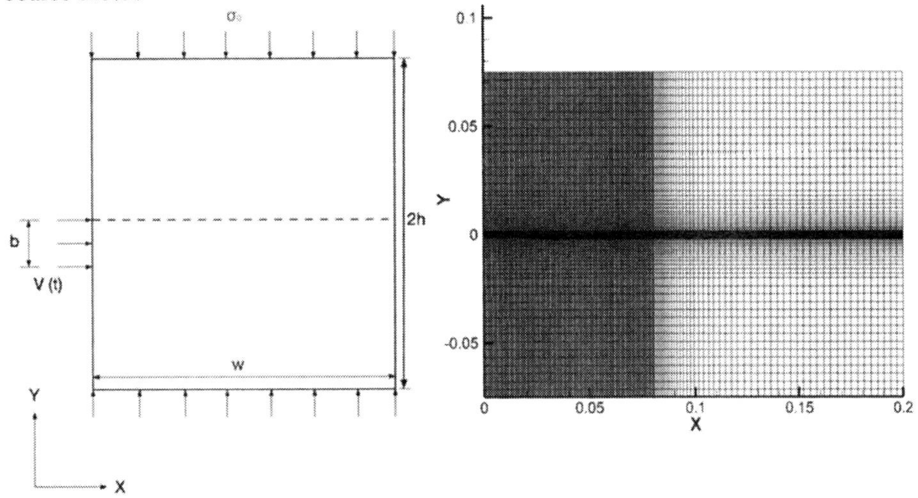

Figure 2. Finite element model (left) and finite element mesh (right).

The elastic properties of homalite used in this simulation are listed in Table 1.

Table 1. Elastic properties of Homalite used in the simulations.

E (GPa)	ν	ρ	c_l	c_s	c_R
5.3	0.35	1246	2201	1255	1170

Note that c_l, c_s, and c_R in Table 1 denote longitudinal wave speed, shear wave speed and Rayleigh wave speed of Homalite-100 respectively.

3. Results

Three different sliding modes were observed in the simulations. A crack-like mode, a train of pulses mode and a growing pulses mode. It has been observed that for a constant impact velocity of 10 m/s as compressive stress is increased the sliding mode transitions from crack-like to train of pulses to growing pulses mode.

3.1. Crack-like sliding mode

Figure 3 shows plots of sliding velocity and interfacial shear stress (left) and total slip displacement (right) vs. distance from the left side for the case σ_o = 5 MPa and V_{imp} = 10 m/s at time t = 34 μs. In this sliding mode sliding velocity suddenly jumps to a certain level, and then stays relatively constant until the end of the interface. Note that shear stress in the interface has progressed ahead of the sliding velocity. Shear stress builds up in the interface until it is enough to generate motion.
Figure 4 shows contour plots of maximum shear stress and normal stress in the horizontal direction, S_{XX}. The shape of the shear stress plot at X = 0.6 m closely resembles shear cracks, and hence the name of this sliding mode. S_{XX} is zero throughout the interface, except for the region of impact where it rises to very high levels, as is expected.
Figure 5 displays normal and shear stresses at the upper and lower surfaces of the interface and a close up of the interface. It is worth noting that S_{XX} is positive in the upper surface. This could lead to fretting and consequently fatigue failure of the interface. The close up of the interface suggests that there is no opening of the interface in this case. However there is an interpenetration. A small degree of interpenetration is allowed by the code to prevent numerical divergence.

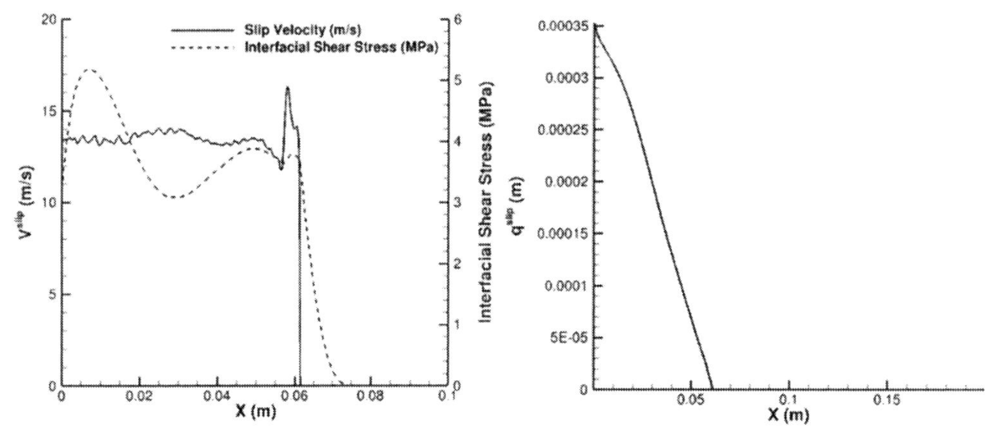

Figure 3. Sliding velocity and interfacial shear stress (left) and total slip displacement (right) vs. distance from the left side (t = 34 μs, σ_o = 5 MPa, V_{imp} = 10 m/s)

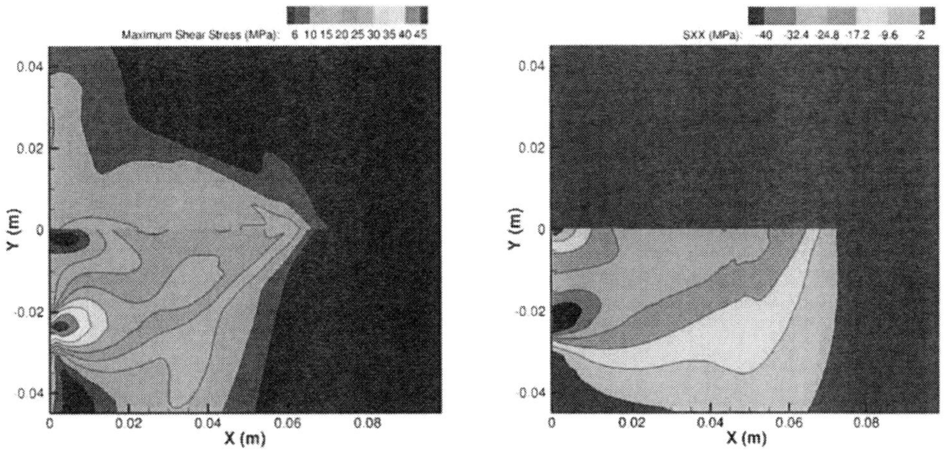

Figure 4. Contour plots of maximum shear stress (left) and horizontal normal stress, S_{XX} (right) throughout the specimen (t = 34 μs, σ_o = 5 MPa, V_{imp} = 10 m/s)

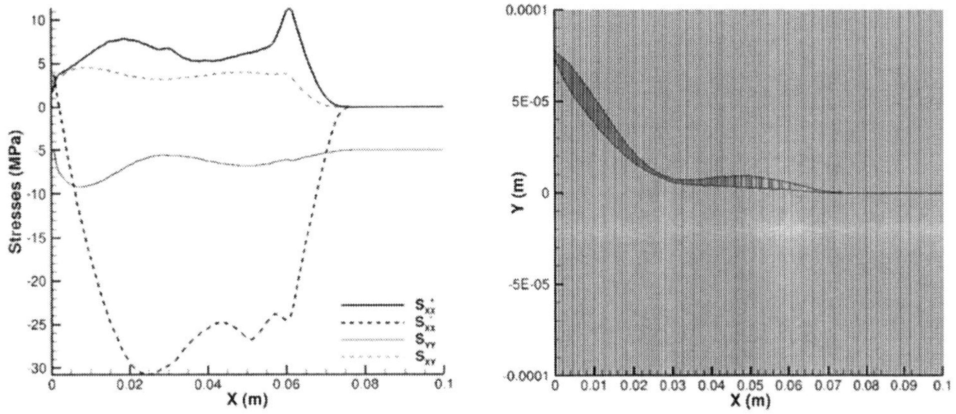

Figure 5. Plots of horizontal normal stress at the upper surface (S_{XX}^+), lower surface (S_{XX}^-), vertical normal stress (S_{YY}) and shear stress (S_{XY}) (left) and close-up of the mesh at the region of impact (right) (t = 34 μs, σ_o = 5 MPa, V_{imp} = 10 m/s)

3.2. Train of pulses sliding mode

Figure 6 shows plots of sliding velocity and interfacial shear stress (left) and total slip displacement (right) vs. distance from the left side for the case σ_o = 15 MPa and V_{imp} = 10 m/s at time t = 38 μs. In this sliding mode sliding velocity suddenly jumps to a moderately high level and then suddenly drops to zero. This pattern is repeated throughout the interface. Figure 6 (left) shows that shear stress in the interface builds up to relatively high levels before suddenly all the energy is released into a pulse. The plot of total displacement (Figure 6 (left)) has a stair-like shape which is a direct result of the velocity pulses in the interface.

Figure 7 plots contours of maximum shear stress and horizontal normal stress S_{XX} throughout the specimen. Notice that in this case the "crack tip" moves faster than the previous case. This is evident by the slope of the "crack tip" in the maximum shear stress plot.

Horizontal normal stress is again zero away from the region of impact, but close to region of impact it is positive in the upper surface and negative in the lower surface. This is further evident in the plot of Figure 8 (left). The stresses follow the pattern of the pulses just like interfacial shear stress. Figure 8 (right) is particularly interesting because it reveals that despite the heavier compressive loading there is an opening in the interface.

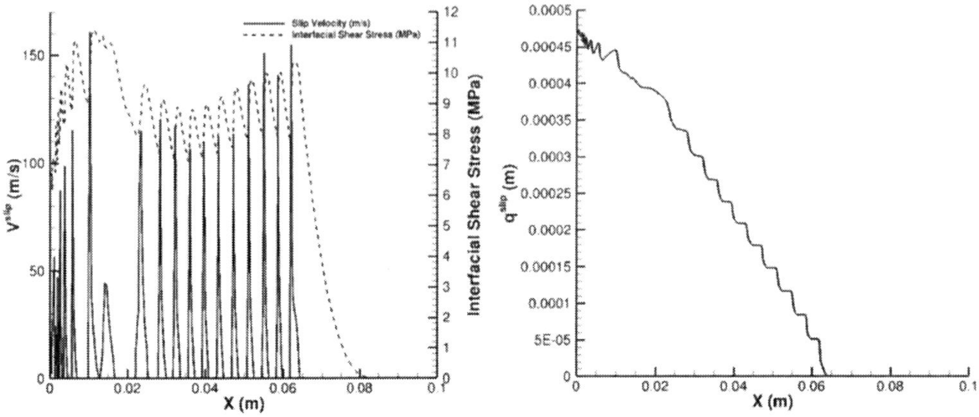

Figure 6. Sliding velocity and interfacial shear stress (left) and total slip displacement (right) vs. distance from the left side (t = 38 μs, σ_o = 15 MPa, V_{imp} = 10 m/s)

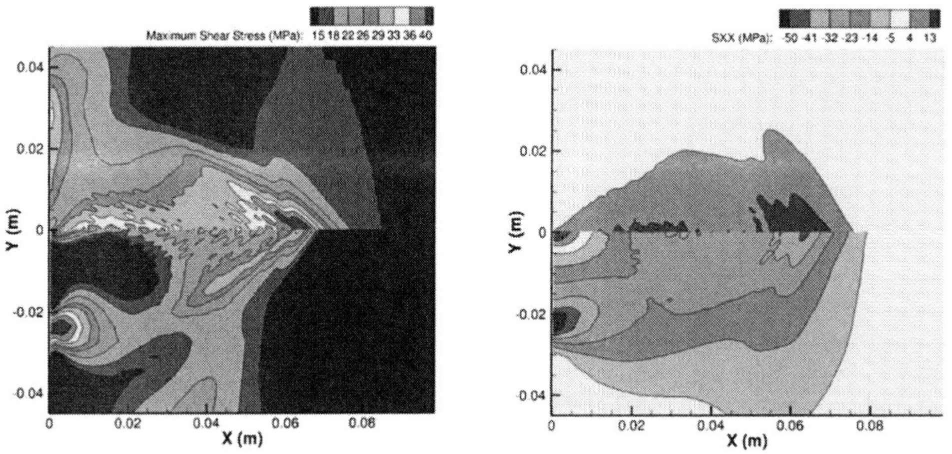

Figure 7. Contour plots of maximum shear stress (left) and horizontal normal stress, S_{XX} (right) throughout the specimen (t = 38 μs, σ_o = 15 MPa, V_{imp} = 10 m/s)

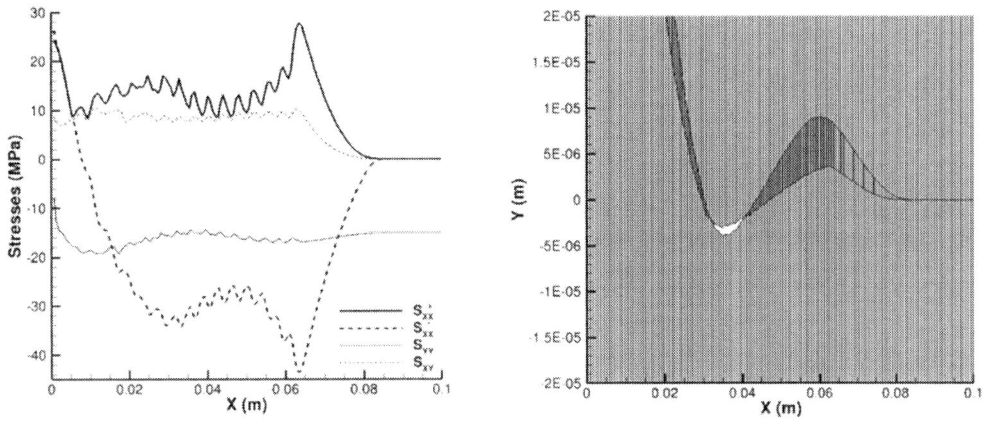

Figure 8. Plots of horizontal normal stress at the upper surface (S_{XX}^{+}), lower surface (S_{XX}^{-}), vertical normal stress (S_{YY}) and shear stress (S_{XY}) (left) and close-up of the mesh at the region of impact (right) (t = 38 μs, σ_o = 15 MPa, V_{imp} = 10 m/s)

3.3. Growing pulses sliding mode

We now look at the case where the compressive stress is increased to σ_o = 25 MPa, at the same impact speed of V_{imp} = 10 m/s. Figure 9 shows the sliding velocity and interfacial shear stress (left) and total slip displacement (right) vs. distance on the interface at time t = 20 μs. Multiple slip pulses with oscillating stresses corresponding to stick-slip motion are observed. At first sight this mode is similar to train of pulses sliding mode, however a closer look reveals the differences between the two cases. Although, pulses are observed in the sliding velocity profile, just like the previous sliding mode, unlike the previous mode the peak velocities of the pulses in this mode reach very high levels and the pulses are further apart and act as individual solitary pulses. Plot of total displacement is again stair-shaped, but the stairs are larger and less numerous as a direct consequence of the sliding velocity profile. Contour plot of maximum shear stress shown in Figure 10 (left) is even more inclined than the previous case meaning that the shear front is moving much faster. Plot of horizontal normal stress in Figure 10 shows traits very similar to the previous case, but in this case the stress level is lower. In this case the opening in the mesh vanishes.

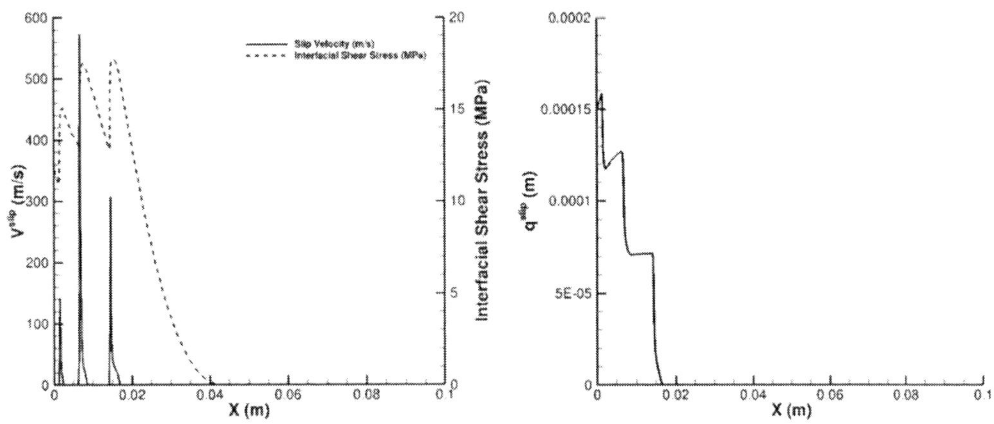

Figure 9. Sliding velocity and interfacial shear stress (left) and total slip displacement (right) vs. distance on the interface. (t = 20 μs, σ_o = 25 MPa, V_{imp} = 10 m/s)

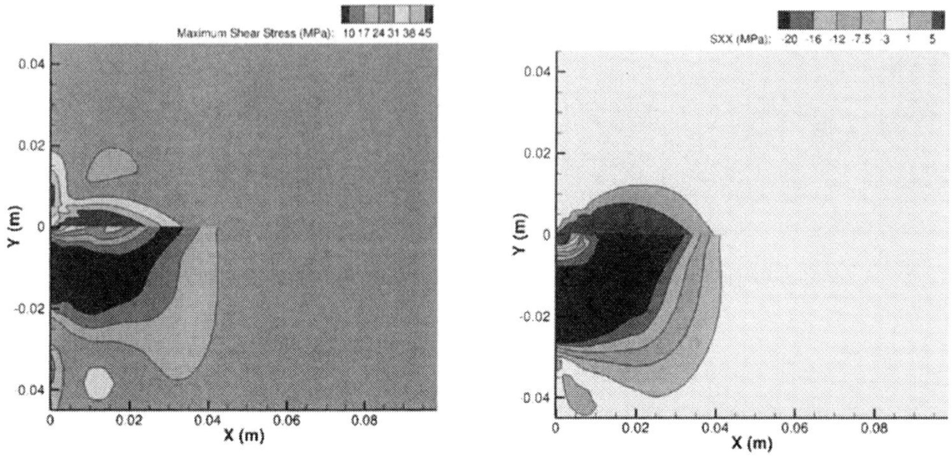

Figure 10. Contour plots of maximum shear stress (left) and horizontal normal stress, S_{XX} (right).
(t = 20 μs, σ_o = 25 MPa, V_{imp} = 10 m/s)

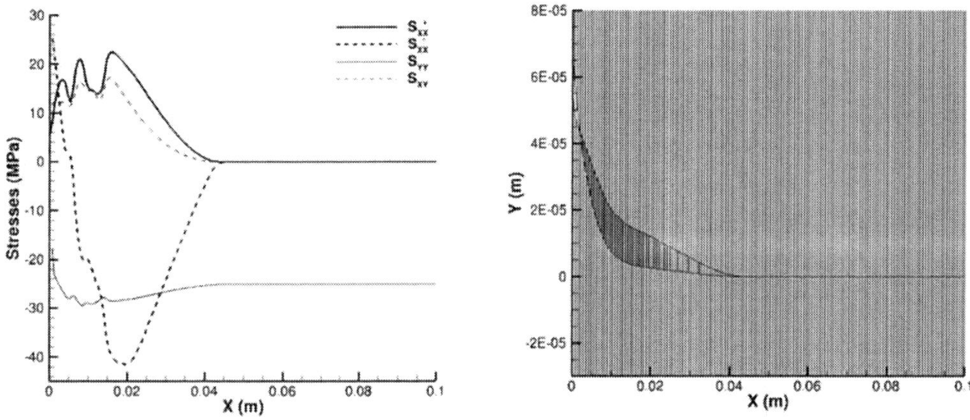

Figure 11. Plots of horizontal normal stress at the upper surface (S_{XX}^{+}), lower surface (S_{XX}^{-}), vertical normal stress (S_{YY}) and shear stress (S_{XY}) (left) and close-up of the mesh at the region of impact (right)
(t = 20 μs, σ_o = 25 MPa, V_{imp} = 10 m/s)

4. Discussion

As compressive stress increases initiation of frictional sliding is delayed. For the three cases examined in this study sliding motion starts at 8 μs for crack-like mode, 12 μs for train of pulses mode and 15 μs for growing pulses mode. Hence the total slip in the interface is lower and behind for the case of growing pulses sliding mode, as is evident from Figure 12 (left). However Figure 12 (right) shows that when slip displacement is plotted at a constant time after initiation (in this case 6 μs after initiation) growing pulses mode is ahead of the others and total slip displacement is much higher compared to the other two modes. Note that the three sliding modes in Figure 12 are the same instances as the ones discussed so far.

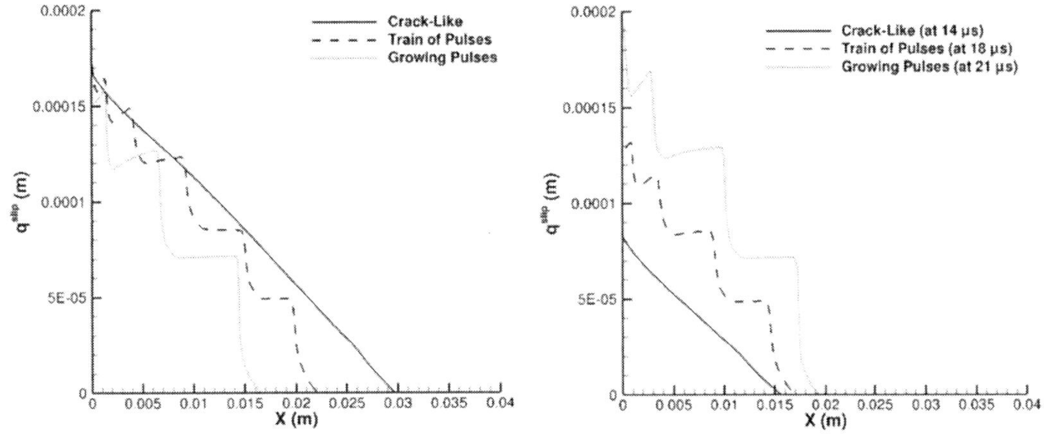

Figure 12. Total displacement in the three sliding modes 20 μs after the initiation of impact (left) and 6 μs after the initiation of sliding (right).

5. Conclusions

Dynamic frictional sliding modes between two plates under impact loading is numerically modeled using a rate- and state-dependent friction law. The three modes of sliding, expanding crack-like sliding, train of slip pulses, and solitary slip pulses are shown to exist depending on the impact speed and compressive load.

References

1. Heaton T.H., Evidence for and implications of self-healing pulses of slip in earthquake rupture, Physics of the Earth and Planetary Interiors, vol. 64, 1-20, 1990
2. Zheng G., Rice J.R., Conditions under which velocity-weakening friction allows a self-healing versus a cracklike mode of rupture, Bulletin of the Seismological Society of America, vol. 88, 1466-1483, 1998
3. Coker D., Lykotrafitis G., Needleman A., Rosakis A.J., Frictional sliding modes along an interface between identical elastic plates subject to shear impact loading, Journal of the Mechanics and Physics of Solids, vol. 53, 884-922, 2005.
4. Povirk G.L., Needleman A., Finite Element Simulations of Fiber Pull-Out, Journal of Engineering Materials and Technology, vol. 115, 286-291, 1993
5. Pierce D., Shih C.F., Needleman A., A tangent modulus method for rate dependent solids, Computers & Structures, vol. 18, 875-887, 1984

Wear characteristics of UHMW polyethylene by twist method

Chișiu G.[1], *Popescu A. M.*[1], *Tudor A.*[1], *Petrescu A. M.*[1], *Stoica G.F.*[1], *Subhi K.A.*[1, 2]
[1]*University "Politehnica" of Bucharest, Department of Machine Elements and Tribology, Spl. Independentei 313, Bucharest 060042, Romania*
[2]*AL-FURAT AL-AWSAT Technical University, Iraq*
georgiana.chisiu@upb.ro[1], andreitudor17@gmail.com[1]

Abstract: A wear test of the twist movement was performed as a new method to estimate the in vivo wear behavior of an acetabular cup material for total knee replacements. A series of UHMWPE samples was used to evaluate the dynamic coefficient of friction in twist movement in contact with steel. The experimental data were conducted to validate the related theoretical model developed in the present study.

Keywords: UHMWPE, friction, wear, twist, rolling joint

1. Introduction

It is well known that polyethylene ultra high molecular is the choice for hip and knee replacement arthroplasty. It is also used in industry. But, for medical using the failure of the artificial components usually is due to wear of acetabular component made of ultra high molecular weight polyethylene (UHMWPE) from total hip joint replacements and the wear of femoral component from total knee replacements. This material, in combination with metallic or ceramic biomaterials, provides low friction joint and generally excellent results for the short and medium term [1]. The motions occurring in an artificial joint, but also like in natural, are complex. The kinematic conditions in can include sliding, rolling and pivoting. In this work we focused on the twist movement.
The aim of this study was to evaluate the dynamic coefficient of friction in twist movement of UHMWPE in contact with steel.

2. Theoretical aspects

The contact we supposed to be studied in this paper is sphere on flat plane in pivoting motion. Such a sphere is shown in cross-section in figure 1. Pivoting occurs when a fixed point of the metallic component is rotating around a still axis. For simplify theoretical aspects we consider the hertzian contact between the bearing ball, which is elastic, and the flat surface as a rigid plan of UHMWPE.

The axi-symmetrical pressure distributions are given by [2]:

$$p_n = p_0\left(1-\frac{r^2}{a_H^2}\right)^{\frac{1}{2}} \tag{1}$$

We assume that pressure is exerted on a circle-shaped area with the radius a_H.

The resulting total force is:

$$F_n = \int_0^{a_H} p_n \cdot 2\pi r dr = \frac{2}{3} p_0 \cdot \pi a_H^2 \tag{2}$$

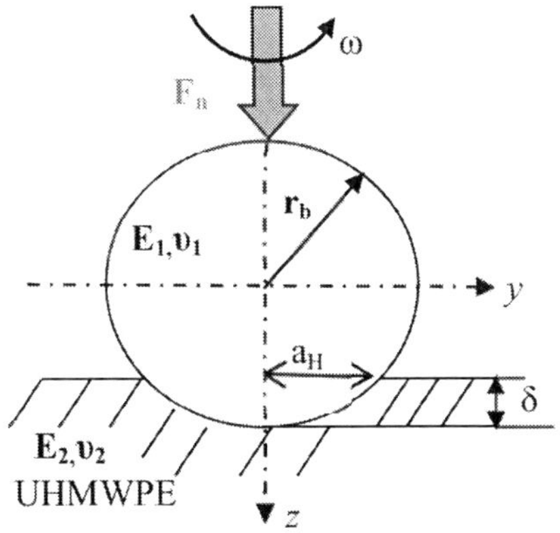

Figure 1. Hertzian contact sphere on flat plane, when is considered the bearing ball is the elastic sphere and the flat surface is a rigid plan of UHMWPE.

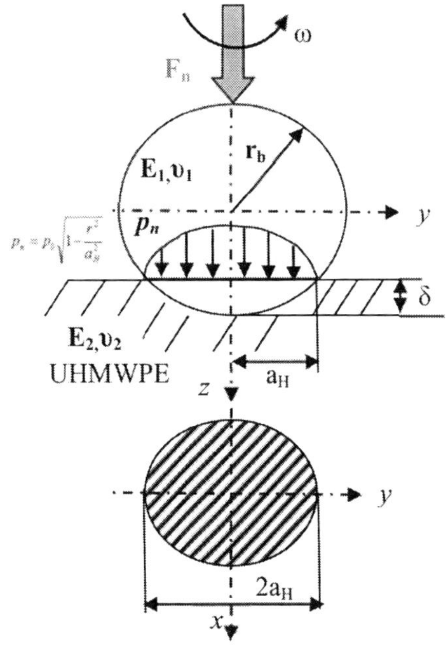

Figure 2. Contact stress has an elliptical distribution across contact over zone of diameter $2a_H$.

The displacement of the points on the surface in the contact area between an originally even surface and a sphere of radius r_b is equal to:

$$u_z = \delta - \frac{r^2}{2r_b} \tag{3}$$

We will try to find the parameters a_H and p_0 that cause exactly the displacement in Equation 3.

$$\frac{1}{E^*}\frac{\pi p_0}{4a_H}\left(2a_H^2 - r^2\right) = \delta - \frac{r^2}{2r_b} \tag{4}$$

The following expression for E* is:

$$\frac{2}{E^*} = \frac{1-\nu_1^2}{E_1} + \frac{1-\nu_2^2}{E_2} \tag{5}$$

Here, E_1 and E_2 are the moduli of elasticity and ν_1 and ν_2 the Poisson's ratios of bearing ball and polyethylene samples.

The pressure in the center of the contact area can be calculated as the contact radius as a function of the normal force:

IOP Conf. Series: Materials Science and Engineering **295** (2018) 012002 doi:10.1088/1757-899X/295/1/012002

$$p_0 = \left(\frac{6F_n E^{*2}}{\pi^3 \cdot r_b^2} \right)^{\frac{1}{3}} \tag{6}$$

And the contact radius:

$$a_H = \left(\frac{3F_n r_b}{4 \cdot E^*} \right)^{\frac{1}{3}} \tag{7}$$

The tangential tractions which act in a circumferential direction induce a state of torsion in the half-space. Johnson [2] consider the traction:

$$q_r = q_0 r_{red} \left(a_H{}^2 - r_{red}{}^2 \right)^{-1/2}, r_{red} \le a_H \tag{8}$$

Where r_{red} is radius reduced of contact area.

Thus the traction produced a rigid rotation of the loaded circle. The traction gives rise to a resultant twisting moment:

$$M_z = \int_0^{a_H} q(r)\, 2\pi r\, dr = \frac{4}{3} \pi a_H^4 \cdot q_0 \tag{9}$$

The twist moment could be written also:

$$M_z = \frac{3\pi}{16} \cdot \mu_{pivoting} \cdot F_n \cdot a_H \tag{10}$$

The coefficient of friction in pivoting movement is resulting from equation 10:

$$\mu_{pivoting} = \frac{16 \cdot M_z}{3 \cdot \pi \cdot F_n \cdot a_H} \tag{11}$$

The penetration depth is:

$$\delta = r_b - \sqrt{r_b^2 - a_H^2} \tag{12}$$

Table 1. Values for calculating Hertzian contact stress for a sphere on flat [3].

Item	Symbol	Value
Load	F_n	200, 300, 400, 600 *N*
Radius of rotating ball	r_b	6.375 *mm*
Poisson's ratio of steel ball	υ_1	0.3
Poisson's ratio of UHMWPE samples	υ_2	0.4
Young's modulus of steel ball	E_1	200 *GPa*
Young's modulus of UHMWPE samples	E_2	500 *MPa*

3. Experimental details

A four-ball apparatus specially adapted to pivoting movement, connected to a computer, was used to evaluate the twist moment of the UHMWPE against steel under dry and egg albumen lubricated conditions.

3.1. Test method

During the experiments, the polyethylene specimens were pressed against the metallic ball, at the constant load, as shown in figure 3. The ball rotates uniformly around its vertical axis, occurring the pivoting movement. For the experimental tests the load was obtained with this formula:

$$F_n = m_g \cdot g \cdot k \tag{13}$$

Where is m_g - weight; $g = 9.81\,m/s^2$ gravitational acceleration; $k = 10$ constant of balance.

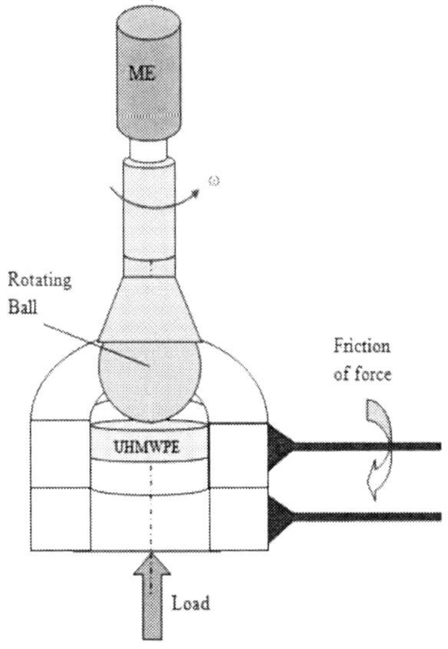

Figure 3. Illustration of the ball-on-flat assembly.

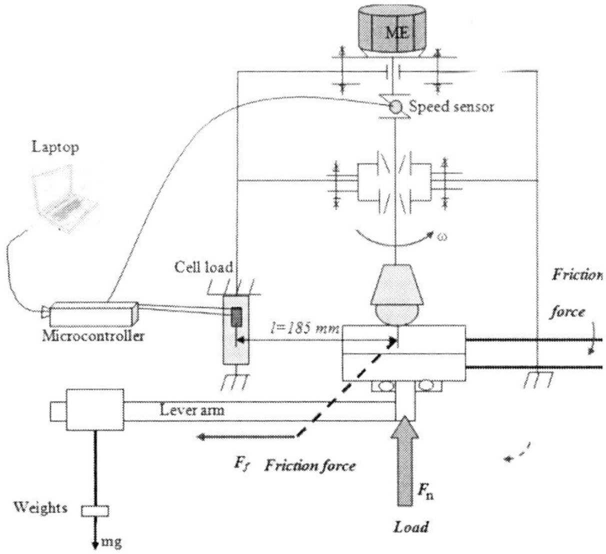

Figure 4. Schematic illustration of the four-ball tester specially adapted to pivoting movement.

3.2. Test equipment

The realization of the test method is possible using the modified four-ball tester, under dry conditions and at the controlled temperature of 23 ± 1^{0}C. The employed four-ball tester was controlled via special microprocessor-aided controller (ARDUINO), motor speed controller (tipul de ME) and PC with a special program installed. The motor speed controller has the role to changing the frequency of electric current that is enable to set the rotational speed within a wide range. A schematic representation of the arrangement used for testing the materials is shown in figure 4. A stationary polyethylene specimen was placed in twist contact with the surface of a steel ball of 12.27 diameter and 4 mm thick. The polyethylene specimens were cut from extruded sheets of UHMWPE GUR 1050. A series of pivoting tests were carried out on UHMWPE and the mechanical properties of the material were presented in previous works [4,5]. The mean roughness of the samples was 0.4± 0.6 μm. The wear tests were performed using a velocity of the rotation of the ball of 144 rpm, 173 rpm and 323 rpm. The applied load values were 200 N, 300N, 400 N and 600 N. For each load, the test was repeated three times. The polyethylene specimens were changed after each test. The steel ball and the polyethylene samples were cleaned thoroughly with propanol and then allowed to dry at open air, before each run.

The tests were carried out under dry pivoting and egg albumen lubricated conditions. The lubricant was distributed over the entire surface of the sample in an amount sufficient to ensure the presence of this during the test period. During the test, the force of friction was measured by a transducer mounted to the arm and recorded with a data acquisition system. The worn surface of tested samples were observed by optical microscopy using optical microscope Olympus back sight Gx51 by the magnification in the range 50× to 1000×.

IOP Conf. Series: Materials Science and Engineering **295** (2018) 012002 doi:10.1088/1757-899X/295/1/012002

4. Experimental results

4.1 Surface analyses

The optical microscopy examination of worn surfaces of UHMWPE samples against bearing ball at 200 N to 600 N applied load and at rotating speeds test for dry and lubricated conditions can be seen in table 2 and table 3.

Table 2. The optical microscopy examination of worn surfaces of UHMWPE samples in dry conditions.

	Load			
Rotational speed	**200 N**	**300 N**	**400 N**	**600 N**
144 rpm				
173 rpm	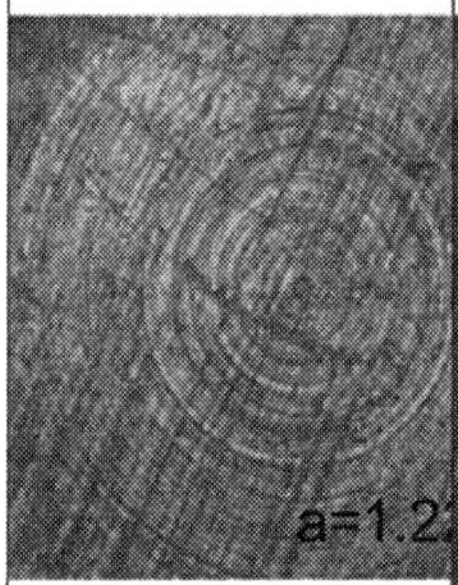			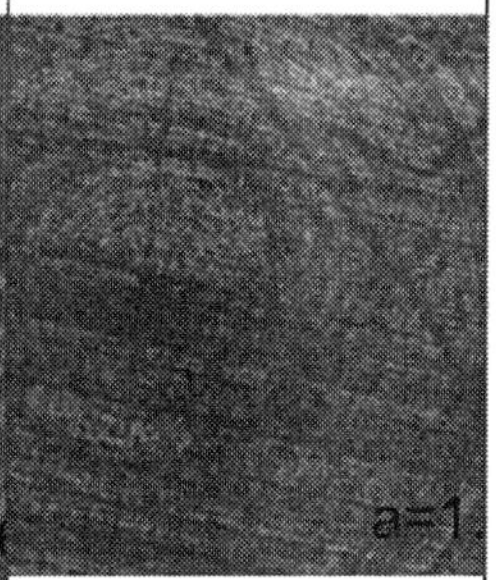
323 rpm	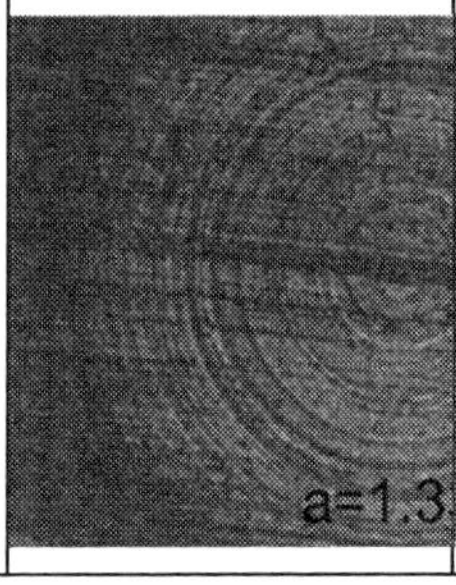		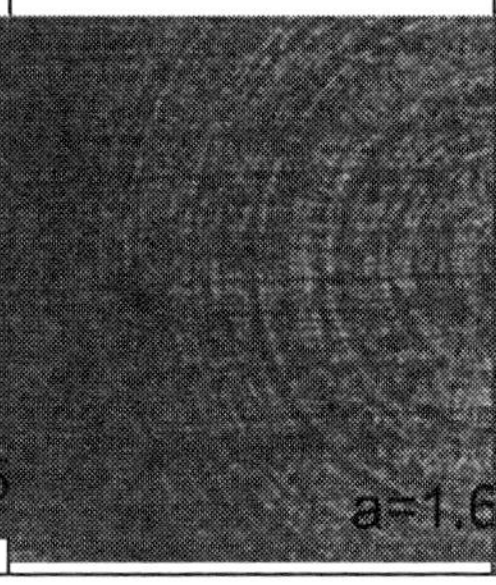	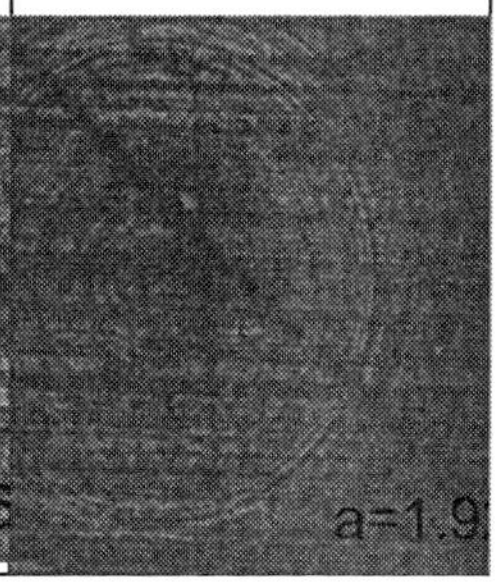

Table 3. The optical microscopy examination of worn surfaces of UHMWPE samples in lubricated conditions.

	Load			
Rotational speed	**200 N**	**300 N**	**400 N**	**600 N**
144 rpm	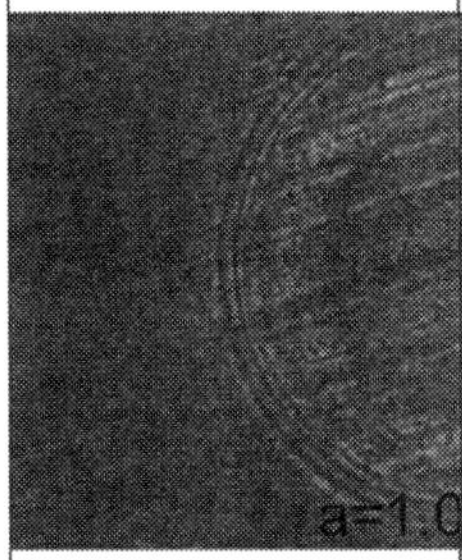	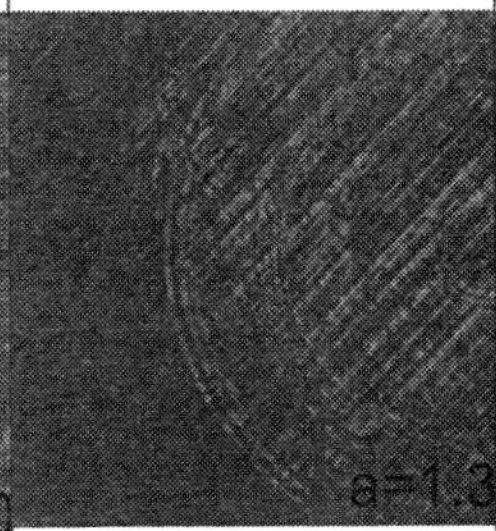	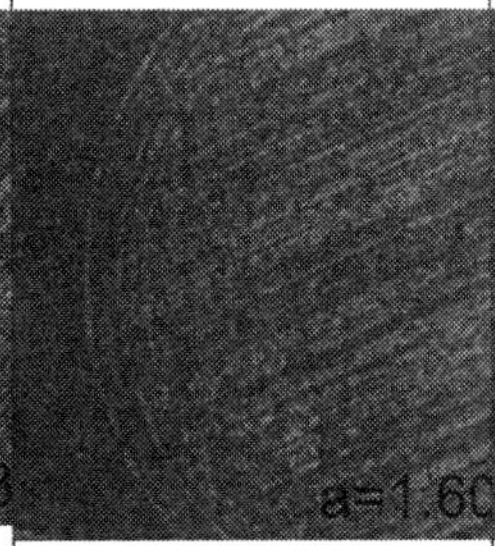	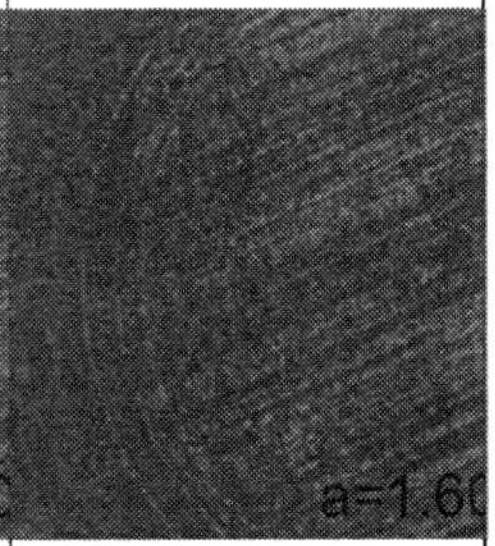
173 rpm	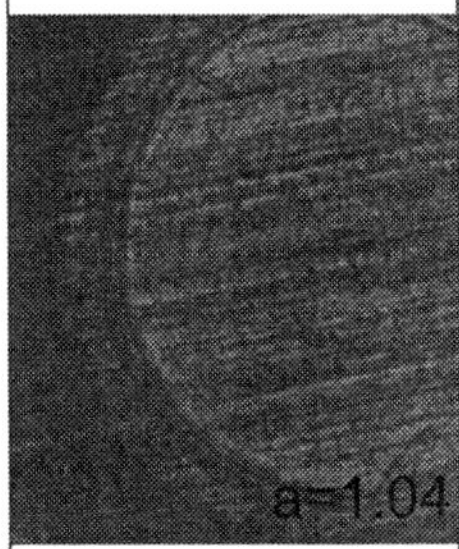	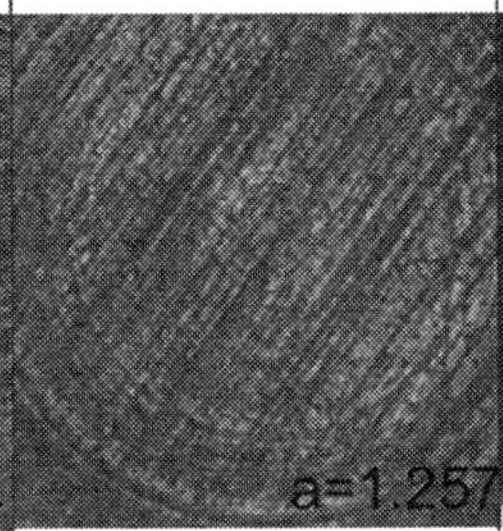	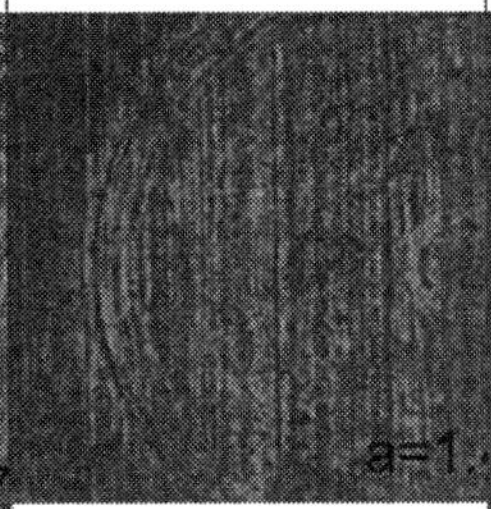	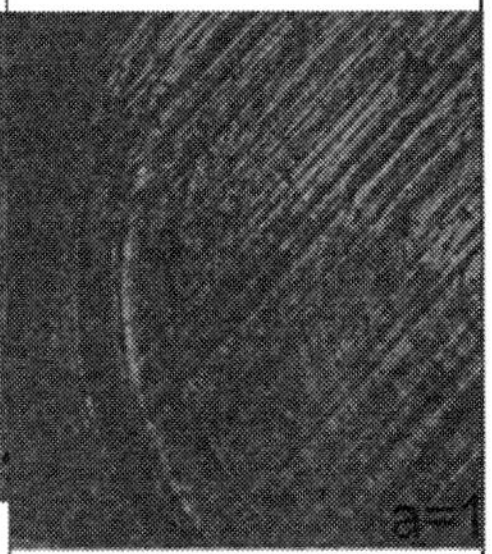
323 rpm		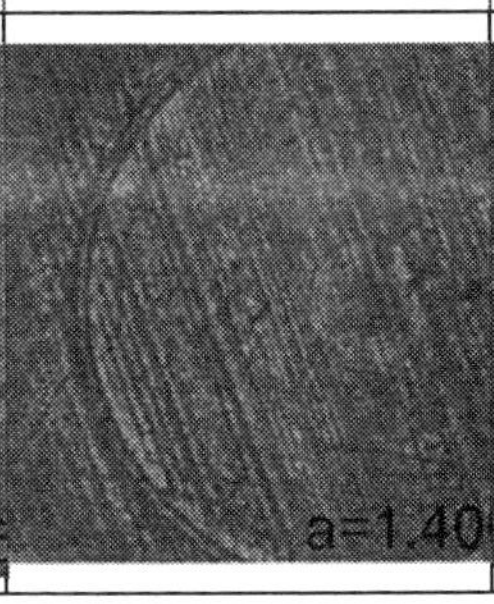	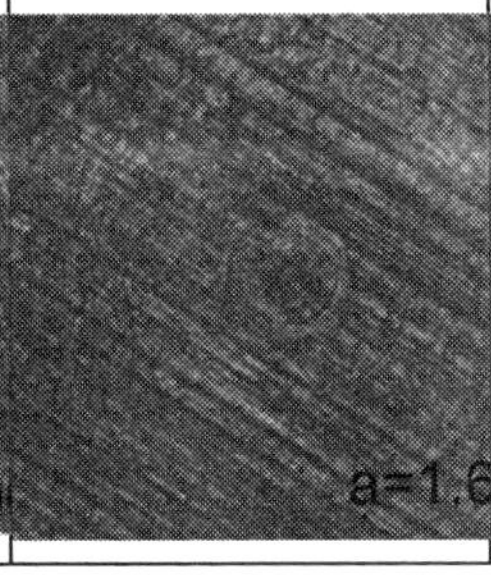	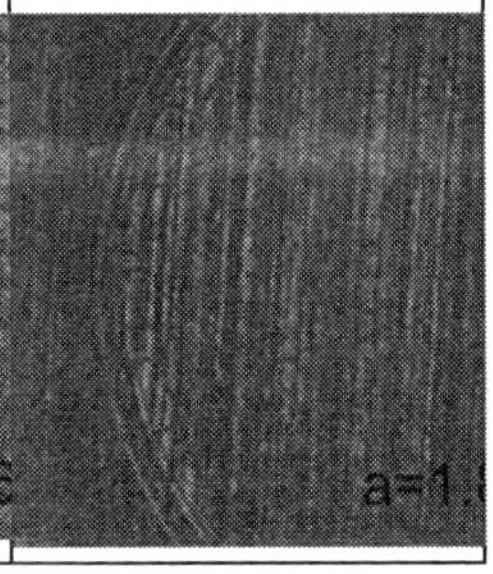

4.2 Tribological experiments

The results of tribological investigations are presented in figures 5-12. Figures 5 and 6 illustrate the comparison between theoretical and experimental results of contact radius of UHMWPE in contact with a steel ball under dry and lubricated conditions. The following notes used in this figures concerns: the experimental 1at the rotating speed 144 rpm, the experimental 2 at the rotating speed 173 rpm, the experimental 3 at the rotating speed 323 rpm. For each tested sample of UHMWPE the contact pressure was calculated with formula: $p_{Stribeck} = \frac{F_n}{\pi r_b^2}$

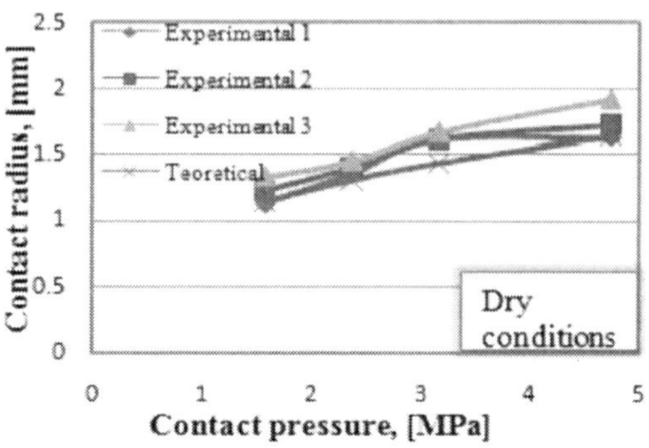

Figure 5. Experimental contact radius – contact pressure curves for UHMWPE under dry conditions.

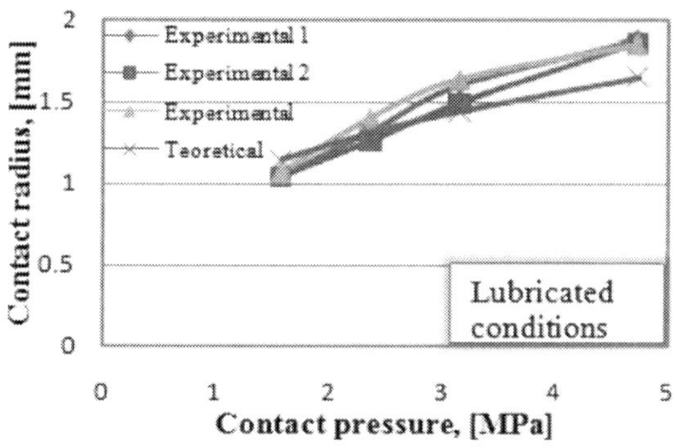

Figure 6. Experimental contact radius –contact pressure curves for UHMWPE under lubricated conditions.

Figures 7 and 8 illustrate the experimental results of penetration depth of UHMWPE in contact with a steel ball under dry and lubricated conditions.

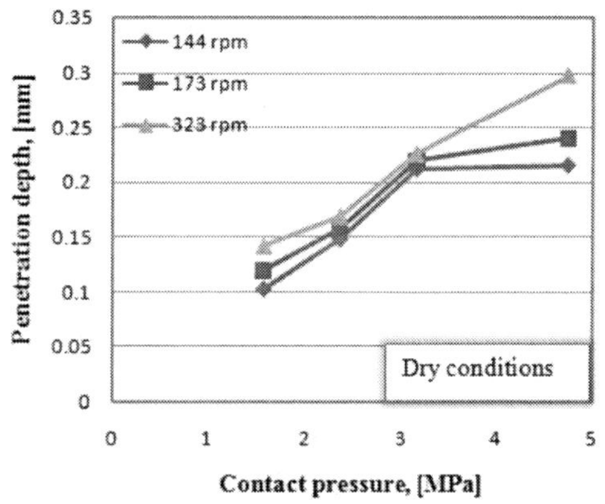

Figure 7. Experimental penetration depth– contact pressure curves for UHMWPE under dry conditions.

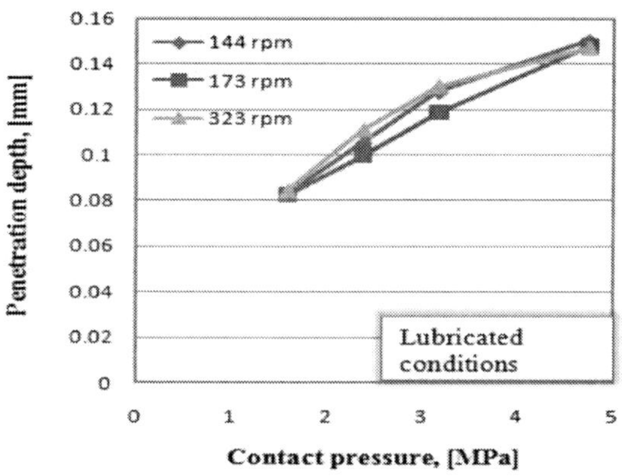

Figure 8. Experimental penetration depth – contact pressure curves for UHMWPE under lubricated conditions.

Figures 9 and 10 illustrate the experimental results of twist moment of UHMWPE in contact with a steel ball under dry and lubricated conditions.

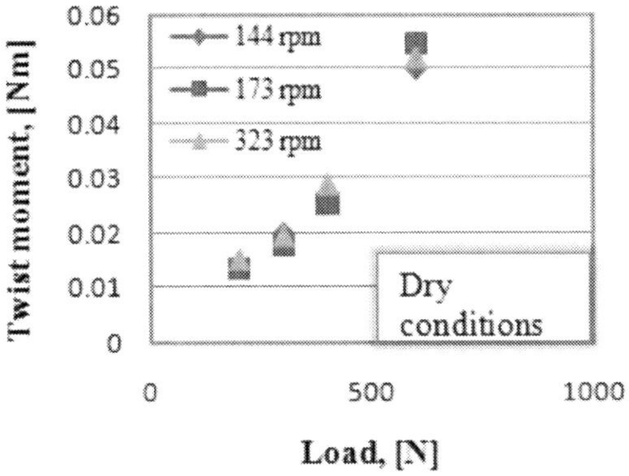

Figure 9. Experimental contact radius – contact pressure curves for UHMWPE under dry conditions.

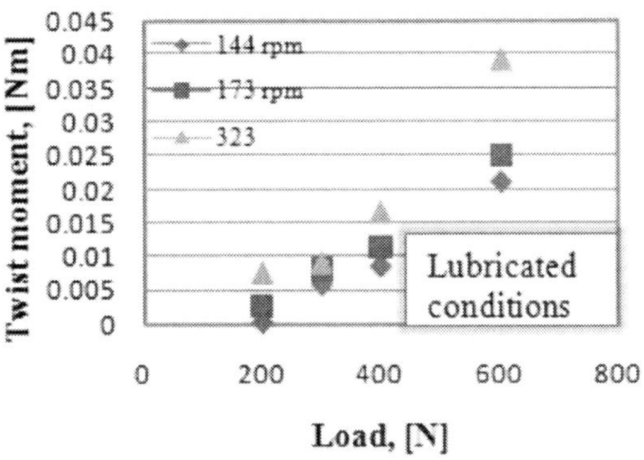

Figure 10. Experimental contact radius –contact pressure curves for UHMWPE under lubricated conditions.

Figure 11 shows the variation of the dynamic coefficient of friction of UHMWPE in contact with a steel ball and rotational speed. Figure 12 shows the variation of the dynamic coefficient of friction of UHMWPE in contact with a steel ball and load.

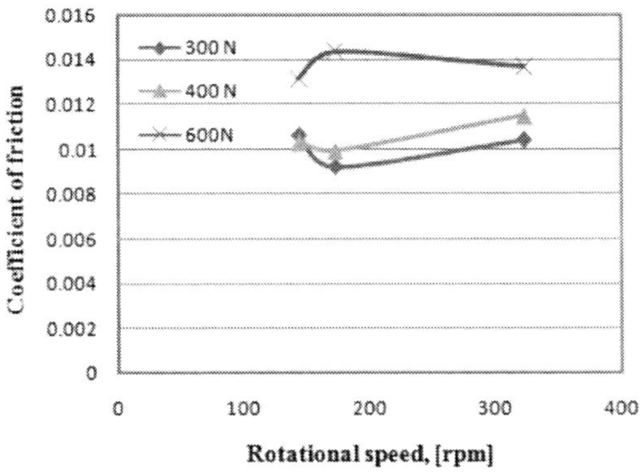

Figure 11. Experimental coefficient of friction-rotational speed curves under dry conditions.

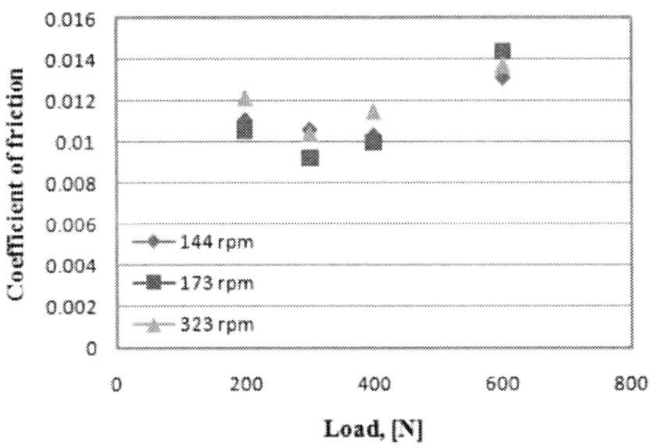

Figure 12. Experimental coefficient of friction-load curves under dry conditions.

Conclusions

A model for the contact between a ball bearing and an UHMWPE plane surface has been developed using the twist method. Also, an experimental study has been elaborated for the analysed problem.
Contact radius calculated has the appropriate values like at the contact radius measured from optical investigations. Penetration depths increase with the contact pressure. Twist moment calculated increases with the load. The coefficient of friction obtained is very small and increases with the rotational speed. Subha K.A. et al. [6] in their research also found that the coefficient of friction of the cow skin in contact with the steel increases with the rotational speed.

Acknowledgements

This work has been funded by University Politehnica of Bucharest, through the "Excellence Research Grants" Program, UPB – GEX. Identifier: UPB–EXCELENȚĂ–2016, Experimental study of biomaterials used in human joint prostheses, no. 344.

IOP Conf. Series: Materials Science and Engineering **295** (2018) 012002 doi:10.1088/1757-899X/295/1/012002

References

1. Sinha S.K., Chong W.L.M., Lim S.C., Scratching of polymers – Modelling abrasive wear, Wear, vol. 262, 1038-1047, 2007.
2. Johnson K.L., Contact mechanics, Cambridge University Press, 1985.
3. Bruck, Angela Lynette, Friction and wear behavior of ultra-high molecular weight polyethylene as a function of crystallinity and in the presence of the phospholipid DPPC (dipalmitoyl phosphatidylcholine), Retrospective Theses and Dissertations, Paper 14890, 2007.
4. Chisiu G., Tudor A., Wear characteristics of UHMW polyethylene by scratching method, Journal of the Balkan Tribological Association, vol. 19, no. 4, 661-672, 2013.
5. Chisiu G., Laurian T., Tudor A., Indentation tests to study the mechanical tribological properties of UHMWPE, U.P.B. Sci. Bull., Series D Mechanical Engineering, vol. 73, Iss. 4, 209-222, ISSN 1454-2358, 2011.
6. Subhi K. A., Tudor A., Hussein E. K., Chisiu G., Ex-Vivo Cow skin viscolestic Effect for tribological aspects in endoprotheses, 9 International Conference on Tribology (BalkanTRib'17), 13-15 September 2017, Cappadocia/TURKEY. (Paper submitted).

Numerical Investigation of the Effect of Radial Lip Seal Geometry on Sealing Performance

Tok G[1], Parlar Z[2], Temiz V[3]
[1]Işık University, Department of Mechanical Engineering, 34980, Istanbul Turkey
[2]Istanbul Technical University, Mechanical Engineering Faculty, 34437, Istanbul, Turkey
[3]Istanbul Technical University, Mechanical Engineering Faculty, 34437, Istanbul, Turkey
gorkem.tok@isikun.edu.tr[1], parlarze@itu.edu.tr[2], temizv@itu.edu.tr[3]

Abstract: Sealing elements are often needed in industry and especially in machine design. With the change and development of machine technology from day to day, sealing elements show continuous development and change in parallel with these developments. Many factors influence the performance of the sealing elements such as shaft surface roughness, radial force, lip geometry etc. In addition, the radial lip seals must have a certain pre-load and interference in order to provide a good sealing. This also affects the friction torque. Researchers are developing new seal designs to reduce friction losses in mechanical systems. In the presented study, the effect of the lip seal geometry on sealing performance will be examined numerically. The numerical model created for this purpose will be verified with experimental data firstly. In the numerical model, shaft and seal will be modeled as hyper-elastic in 2D and 3D. NBR (Nitrile Butadiene Rubber) as seal material will be analyzed for the rotating shaft state at constant speed by applying a uniform radial force.

Key Words: Lip Seal, Lip geometry, sealing performance, NBR (Nitrile Butadiene Rubber), Mooney Rivlin

1. Introduction

The use of nitrile butadiene rubber (NBR) in seals has increased significantly in recent years and is now the most used sealing material in the industry. In the majority of industry, lip seals are used as important sealing equipment also elastomer class material NBR with chemical name butadiene acrylonitrile has the ability to avoid of deformations after the applied force is removed or after large deformations. The NBR, which has high wear resistance for this reason, is considered to be a standard sealing element material.

Mechanical properties of NBR is very useful in many applications, but particularly non-linear mechanical properties of NBR make difficult to analyze the elastomer. Nitrile rubber is more resistant to oil and some additives are resistant to airborne degradation. But, NBR is not favorable to work in high temperature and cold environments. The working temperature is between -40 and 110°C [1]. In this study, analysis has been carried out at room temperature.

Except for the high amount of deformation, calculations must be done according to non linear mechanical properties and incompressibility of material. When the recent studies are examined, finite element applications are performed to material which has hyper-elastic behavior. The sealing performance of the new geometry designs was investigated by analysis with finite element method before prototyping. The results such as stress, deformation, contact pressure, etc. in the seal and shaft contact area were investigated during the studies.

It was found that the stress concentration in the contact area of the NBR radial lip seals were investigated in the study of Chung Kyun Kim and Woo Jeon Shim and it was concluded that the stress concertation were changed by the spring force and design parameters such as seal geometry, oil side angle, air side angle and spring position [2].

When the study of Xue-Guan SONG, Lin WANG and Young-Chul PARK is examined, 4 new seal models are formed for different loading conditions and different interferences in NBR radial lip seal. In these models, the maximum contact pressure change in the contact area and if it is greater than internal pressure it prevents leakage. The study of Xue-Guan SONG, Lin WANG, Young-Chul PARK 's the study of Chung Kyun Kim and Woo Jeon Shim showed that loading and geometry affect sealing performance of the seal [3].

How sealing performance is affected by α and β angles (see Fig. 1) is already studied, but in this study, how sealing performance is affected by lip geometry is investigated by using frictional torque values by using the stress values generated from the friction between seal and shaft.

2. Numerical Method and Verification

Hyperplastic modeling methods were used because of the hyperplastic behaviors of material when numerical model was formed. In this study, radial lip seal and shaft having 40 mm in diameter were modelled with two-dimensional asymmetric modeling instead of 3D modelling which is required more resource. As shown in Figure 1, the finite element model which was built in the ANSYS software and boundary conditions of seal and shaft were given. Then, the Mooney-Rivlin hyperelastic material model was chosen to characterize the seal. Mooney Rivlin is very popular in the elastomer modelling because Mooney Rivlin assumes that the material is homogeneous, isotropic and incompressible.

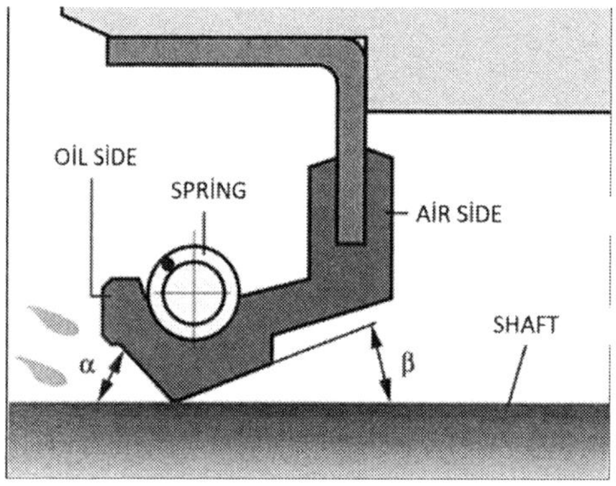

Fig. 1: 2D view of radial lip seal and shaft

After the FEM model (Fig. 2) is established, a uniformly distributed radial force which is caused by garter spring and shaft seal tightness, was applied. This force data was obtained by previous experiments which include NBR material [4]. In addition, the results are confirmed by the results of this previous experiment's data.

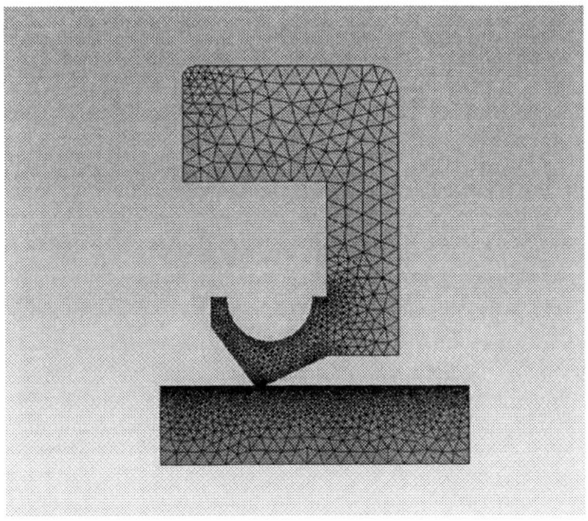

Fig. 2: FEM model of Lip seal and shaft

In order to verify model, some assumptions were made regarding previous experimental work [4]. The internal pressure which is applied by oil side, was modeled as 0.05 MPa and the external pressure was accepted as 0.1 MPa, provided that the angles α and β of the lip remain within the known valid ranges. Frictional contact is defined due to shaft rotation at the contact area and according to result of the researches, friction coefficient is defined as μ=0.1....0.01. As a result of the conducted analysis, stress values at the contact area which are caused by radial force, were examined and average stress values were obtained and then frictional torque was calculated using average stress at the contact area. Calculated frictional torque values were verified with experimental data of the "Experimental Study of Friction torque of Rotary Lip Seals" [4]. Also calculations are shown below;

$T_f = \sigma_{avg} \times L \times \pi \times D \times r$

σ_{avg} (average contact stress) = 7 MPa

L (Contact width)= 0.3 mm

D (diameter of the shaft)= 40 mm

r (radius of the shaft)= 0.02 m

μ (friction coefficient)= 0.05

T_f is found as 0.26 Nm and it was seen that the frictional torque value in the experimental study [4] was 0.24 Nm. When the analysis performed is compared with experimental data, the error rate is 8%.

In order to model The NBR radial lip seal whose boundary conditions was determined above, 2 parameters option was chosen for Mooney Rivlin equation [3, 5-7]. These two parameters, C10=1.87 and C01=0.47, were calculated by uniaxial tensile and uniaxial compression test data. Also the Possion ratio was assumed as 0.49.

3. Geometry

It is known that sealing performance of the lip seal is influenced by many parameters but researches show that the geometry of the contact area is deformed by applied force and it affects the strain values at contact zone. When the study of A. Tasora, E.Prati, T.Marin [8] is examined, the magnitude of the applied force appears to influence the stress values in the contact zone. It means that it affects friction torque at contact area. In this study, friction torque and stress values at contact area, were investigated based on the change in geometry of the seal lip. In the contact zone, the parameters that affect friction torque primarily are thickness of the lip (d) and eccentricity between garter spring and seal lip (e) (see. Fig. 3). Then models were created with different e and d values, as seen in the Table 1. In commercially available standard seal models values of e = 0.3 and d = 1.3, are taken as reference values.

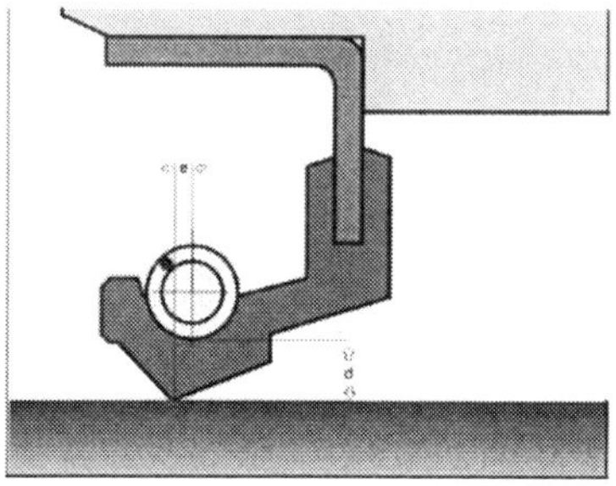

Fig. 3: Dimensional parameters of the lip which were studied; e: eccentricity between garter spring and seal lip d: thickness of lip

Table 1. Models and parameters are designed for analysis

Eccentricity (mm)	Model No
e=0.15	Model 2
e=0.3	Model 1
e=0.6	Model 3
d=1.3	Model 1
d=1.6	Model 4
d=1.9	Model 5
e=0.15 d=1.3	Model 6

Model 1 is the reference model and in other models, e and d values were changed. Also two parameters (e and d) were combined to form mixed model (model 7) and then in these models friction torque was obtained.

4. Results

The eccentricity and lip thickness values of reference model shown in the Fig. 4, were changed and new model was analyzed. Since the contact pressure is higher than the internal pressure, it can be said that sealing is provided. As seen in Table 2, when eccentricity was reduced from 0.3 to 0.15, friction torque caused by friction between seal and shaft, reduced. But when the eccentricity was increased, friction torque also increases. It means that increase in eccentricity also increases friction torque.

Fig. 4. Reference Model (model 1)

Fig. 5. Model 5 (Lip thickness d = 1.9 mm)

Lip thickness which is other geometrical parameter, is at first reduced from 1.3 to a smaller value, but it has seen that when the lip thickness reduced from reference size, α and β angles becomes out of optimum range so it affects the sealing performance negatively. Therefore models were created by increasing d value (lip thickness) (Fig. 5). Then as a result, it was decided increasing lip thickness from 1.3 to 1.9 mm. In the new model strain and stress values at the lip decreased, but friction torque increased.

Table 2. Friction torque results of the change in lip geometry

Eccentricity (mm)	T_f (Nm)
e=0.15	0.21
e=0.3	0.26
e=0.6	0.34
d=1.3	0.26
d=1.6	0.32
d=1.9	0.38
e=0.15 d=1.3	0.49

In the last model, analysis was performed in a mixed model in which the eccentricity was reduced (e=0.15 mm), and the lip thickness was increased (d = 1.6 mm). In this model, it is seen that the effect of decreasing friction torque is lost as a result of decreasing the eccentricity and the effect caused by the increase of lip thickness is dominant and the friction torque is increased in the model.

5. Conclusion

In this study, the effects of the lip geometry on the sealing performance were investigated and the effects of two parameters were emphasized; Garter spring and seal lip (e) and lip thickness (d). It was seen in the analyzes that when the eccentricity was reduced, the friction torque decreased, but an increase in the friction torque was observed with the increase of the eccentricity and lip thickness.

As a result, researchers have found that reducing the eccentricity improves the sealing performance, but it has been obtained that increasing the lip thickness affects the sealing performance negatively. In the mixed models, we have seen that the advantage of the eccentricity is lost due to the effect of lip thickness and then it becomes the worst model. For the future study, the value or range of values in which the eccentricity have optimum can be found, and new geometric models may be developed that may improve the sealing performance.

Referanses

1. D.BULUT, V.TEMIZ, Z. PARLAR, The Effect of Interference on the Friction Torque Characteristics of TPU Based Rotary Lip Seals, Tribology in Industry, 2015
2. CHUNG KYUN KİM AND WOO JEON SHİM, Analysis of contact force and thermal behavior of lip seals, Tribology International Vol 30, No 2, pp. 113-119. 1996
3. XUE-GUAN SONG, LİN WANG, YOUNG-CHUL PARK , Analysis And Optimization Of Nitrile Butadiene Rubber Sealing Mechanism Of Ball Valve, Trans. Non Ferrous Met. Soc. China, s220-s224, 2009
4. H. ÖZPERK, Experimental Study of Friction torque of Rotary Lip Seals, İstanbul Teknik Üniversitesi, January, 2010
5. ISMAİL J. ABUBAKAR, PETER MYLER, ERPİNG ZHOU, Constitutive Modelling of Elastomeric Seal Material under Compressive Loading, Modeling and Numerical Simulation of Material Science, 2016, 6, 28-40, 2016
6. M. ELGADARI, A. FATU, M. HAJJAM, Effect of Grooved Shaft On The rotary Lip Seal Performance İn Transient Condition: Elasto-Hydrodynamic Simulations, Tribology International, 2015
7. CHUN-YİNG LEEA, CHAO-SUNG LİNB, RUİ-QUAN JİANA, CHİH-YUNG WEN, Simulation and experimentation on the contact width and pressure distribution of lip seals, Tribology International 39 (2006) 915–920
8. A. TASORA , E.PRATİ, T.MARİN, A Method For The Characterization Of Static Elastomeric Lip Seal Deformation, Tribology International, Parma, Italy, 2012

Wear resistance of Polymethyl Methacrylate (PMMA) with the Addition of Bone Ash, Hydroxylapatite and Keratin

Emre G[1], Akkus A[2], Karamış M B[3]
[1]Cumhuriyet University, Department of Railway Systems, 58140, Sivas, Turkey
[2]Cumhuriyet University, Department of Mechanical Engineering, 58140, Sivas, Turkey
[3]Erciyes University, Department of Mechanical Engineering, 38039, Kayseri, Turkey
emregorgun@msn.com[1], aakkus@erciyes.edu.tr[2], karamisb@erciyes.edu.tr[3]

Abstract: In this study mechanichal and tribological properties of keratin, bone ash and hydroxylapatite by adding to PMMA (known as the main prosthesis material) were investigated. Hydroxylapatite, bone ash, and keratin materials were added as PMMA in to the content of PMMA, in the proportions of %1, %3 and %5, respectively. The resulting mixtures were put into the molds and solidified in order to obtain samples to be used in the wear experiments. Each experiment was conducted by preparing three experimental samples. The wear data were compared according to the average values of the experimental samples. In the wear test, the results were also evaluated according to the average values obtained from each group and the results of the control group. It was observed that, the wear resistance of the PMMA including 3%, 5% bone ash and PMMA including 5% keratin flour were higher than the values of the control group.

Keywords: Polymethyl Methacrylate (PMMA), Bone Ash, Hydroxylapatite, Keratin,Wear Resistance

1. Introduction

PMMA is an inevitable material because of its characteristics such as its widespread range of mechanical traits, impact maintenance, transparency, optical traits, and compliance with food regulations, bright colors, and its low aqueous absorption. Polymethyl methacrylate (PMMA), which is still in use in commercial bone cement formulations, is a shapeless polymer, polymerized on its own. In the commercially marketed Polybutlymethylmethacrylate PBMMA particles are implanted instead of PMMA- styrene copolymer particles. The relative proportions of the other added materials do not exhibit any changes. PMMA has various applications in industry and medicine. PMMA, which has been used in airplane glasses, spectacle lenses, lenses, magnifiers, flasher lamps of the automobiles, bathtubs, production of tubes and manufacturing of various ornaments in industry, has been used with biocompatible bone cement mixtures in medicine. In medicine, PMMA is added to the internal bone fractures with a local reaction. Moreover, it has been used as the base material of prosthesis in dentistry. Many studies have been conducted to improve the maintenance characteristics of PMMA, which has such a wide range of uses. Kinzl et al. (2011) pointed out the effect of PMMA in increasing the porous structure, while they were investigating the mechanical characteristics on PMMA-bone biopsies [1]. Jeremy et al. (1995) identified SRC-PMMA mixture as a composite material and they showed that higher molecular orientation is attained with the thermodynamic pull applied on the composite. While the fracture maintenance was measured as 170MPA and tension fault as 5% in the studied fibers with 40 m diameter, the fracture maintenance was found to be as 100MPa and the tension faults as 15% in the fibers with 120 m diameters [2]. Weam et al. (1999) investigated the effect of PMMA powder, which is the essential ingredient of PMMA based bone cement, the conditions related to its different characteristics, its size and molecular weight on the methods of processing, mechanical characteristics and permeability of the bone. By adding different bioactive cements, they showed that while the compression maintenance, elasticity module, the mechanical characteristics such as fracture, hardness improved significantly, bending force very slightly decreased and pulling force significantly decreased [3]. In a study conducted by B. Pascual et al. (1998), modified HPMA cement with a low T/S proportion as 1,86 was used. In order to comprehensively understand the effect of lower T/S proportion, studies were conducted on the exothermic, characteristics of polymerization: the mechanical characteristics were tested by means of compression and fracture tests. Polymerization pull and porosity were calculated from the density measurements [4]. In a study conducted by Rodford (1990), polybutadienestyrene copolymers (Macromere) with low molecular weight were used in order to strengthen polymethyl methacrylate. Moreover, butadiene styrene block copolymer with 30% styrene was polymerized by mixing it with methyl methacrylate monomer. The impact maintenance and elasticity module values of the obtained copolymer structures were compared [5].

As can be seen in the relevant studies in the literature, it has been known that the mechanical, chemical and thermal characteristics of the obtained copolymer structure can be improved by means of mixing the copolymers with different characteristics in various proportions, and that this improvement is related to the synergist behaviors of the used polymers [6]. In the present study, by means of mixing bio-cement materials in different proportions, we aimed to make PMMA structure more durable and to decrease the amount of contraction seen in the drying process in the base materials [7].

PMMA has the 18-20 Knoop hardness score and it is relatively hard. It is its contraction resistance is 60 MPa, density is 1,19 gr/cm3 and elasticity module is 2400 MPa [8]. As the organic structures such as keratin, bone ash and hydroxylapatite added within it improves improve the characteristics of PMMA, it is thought that it is use in medicine and dentistry has become more convenient.

Keratin is a structurally strong material, which is the base material of nail and horn, insoluble in aqueous. Bone ash has been used in implant treatment in people with bone loss in their mouths, as bone ash is a very similar tissue to the living tissue. As hydroxylapatite present in dentin and bone without getting in the interaction with oxygen, it is expected to exhibit high compatibility. As a result, it is thought that the produced composites might constitute facilitations especially in the field of medicine.

It is aimed to compare the abrasion resistance of acrylic materials by examining the probable erosions of the teeth due to the influence of the jaws moving relative to each other in the mouth.

2. Material and Method

2.1 Bone ash

The bone ash required in the present study was obtained from the grinding of the cow foot bones as a whole. The big particles in the obtained bone ash were separated by means of a sieve, and later they were dehumidified by means of storing them in a dry environment.

2.2 Hydroxylapatite

Hydroxylapatite was produced in the laboratory environment by using reaction method. It is stored in powder form.

2.3 Keratin

Keratin, obtained in liquid form, was stored in 10 ml ampule bottles.

IOP Conf. Series: Materials Science and Engineering **295** (2018) 012004 doi:10.1088/1757-899X/295/1/012004

2.4 The preparation of the Mixtures

By means of analytical balance, Bone ash, hydroxylapatite and keratin, 1%, 3% and 5% in weight respectively, were mixed with PMMA, which will be placed in the molds of the experimental samples. Through these preparations, bone ash and hydroxylapatite both in powder form were mixed with PMMA, which was also in powder form, and the final process of the preparations was conducted by means of adding MMA in liquid form to the PMMA mixture. And then the mixture was stored in the molds [9].

Table 1. Hardness values of samples

Group No:	*Group Name:*	*Brinell Hardness (Br5)*
1	Kontrol Grubu	20,6
2	%1Bone Ash	21,38
3	%3 Bone Ash	21,55
4	%5Kemik Tozu	22,29
5	%1 Keratin	19,16
6	%3 Keratin	17,71
7	%5 Keratin	13,50
8	%1 Hidroksiapatit	13,26
9	%3 Hidroksiapatit	17,17
10	%5 Hidroksiapatit	20,27

2.5 Preparation of the Test Samples

The metal molds were prepared for the compression test in 16 mm diameter and 18 mm height according to ASTM D695-02a standards, for the charpy impact text in dimensions of 55*10*10 square prisms according to TS EN ISO 179-1 standards and for the hardness tests circular cross section prisms in 16 mm diameter and 16 mm height. The modeling wax poured into these molds in melted form. After the wax solidified, the obtained models were extracted from the molds. For obtaining acrylic resin polymerized with conventional heat, 250 g plaster powder for 100 ml water was mixed and then the resulting mixture was poured the lower part of the muffle. The wax samples were placed into the lower part of the muffle before the plaster was solidified. After the plaster is solidified, varnish was applied. After placing the upper lid of the muffle, plaster was poured. The muffle was placed into the compression with 150 kg/cm^2 pressure. The plaster was left to solidify for a 30-minute period. In the end of this period, the muffle was opened and the wax models were carefully extracted from the plaster by means of model separator. Isolator was applied to the muffles [2, 10].

2.6 Test Methods

2.6.1 Compression Test. The experimental samples of PMMA- hydroxylapatite, PMMA-bone ash and PMMA-Keratin preparations mixed in 1%, 3% and 5%, respectively, were placed on the compression surface in the given order. After the dimensions of the experimental samples were installed to the compression device, the experiment was initiated by determining the process speed as 2mm./min. The evaluations were made with reference to the peak points at which cracks were appeared on the surface area of the experimental samples.

2.6.2 Charpy Impact Test. The experimental samples in dimensions of 10*10*55 prepared according to TS EN ISO 179-1 standards were placed on the charpy experimental module. The values after the fractures were recorded in Joule unit.

2.6.3 Hardness Test. The hardness conditions of the samples were found by measuring the resulting trace diameters on the Brinell hardness module by using bits with 10mm diameter under 250 kg load.

2.6.4 Wear Tests. The experiments made are based on the unit obtained by applying abrasive discs 3 and 6, which rotate at 337 rpm, at two different loads of 47 N and 87 N, respectively, to the charging power. Initial weights of the test specimens were measured by means of precision scales and comparisons were made by calculating the evaluation result differences as a percentage.
For the abrasion test, the test specimens were prepared with the appropriate dimensions of D: 12.7mm, L: 12.7mm. ASLE (American Society of Lubrication Engineers) has specified from 100 test systems as abrasion tests and measurement methods, the most commonly used abrasion amount, thickness difference and trace change methods. The "weight difference method", which is the most widely used method, is preferred in our experiments.

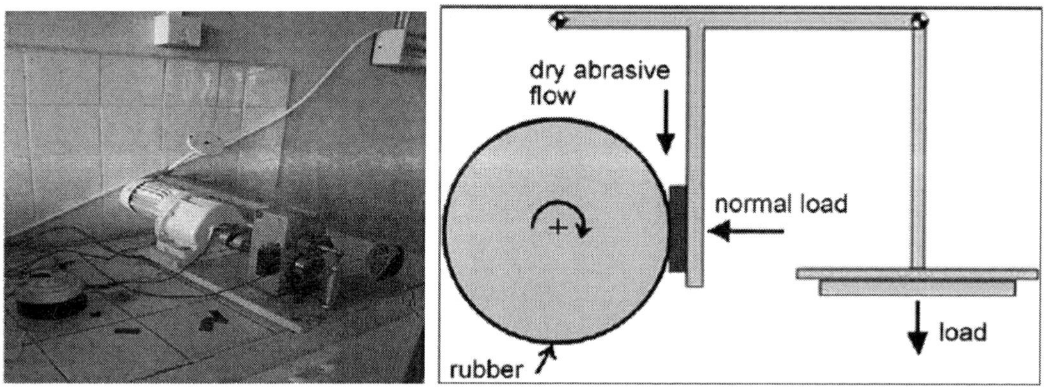

Figure 1. Wear test bench and test configuration

Measurement of wear loss 1/1000 gr precision scales. Wear can be expressed in general (gr / km), (mg / m) and expressed as weight loss (g / cm2) for the unit area. The following formula is used for the specific erosion of the material.Before the test, the surface of the test samples prepared was polished with sandpaper No. 6 in appropriate dimensions and then the friction surface to do smoothly. The polished samples were weighed before the experiment on the sensitive scale and the initial weight was recorded. The test sample was placed in the test setup and the load, temperature and time values to be the test are determined during apply. The operation is performed by running the test setup [11, 12].

$$W_s = \frac{\Delta V}{F_n L} = \frac{\Delta_m}{\rho F_n L} \quad [4]$$

W_s: Amount of specific wear of the material

ΔV: Volumetric material loss

Δ_m: Weight loss

P: Density

F_n: Normal force applied

L: Total distance travelled

3. Results and Discussions

3.1 Wear Properties

The results obtained from the wear tests conducted in this study are given below for all samples.It can be seen from the figures that samples behave different according to applied load and sliding duration.

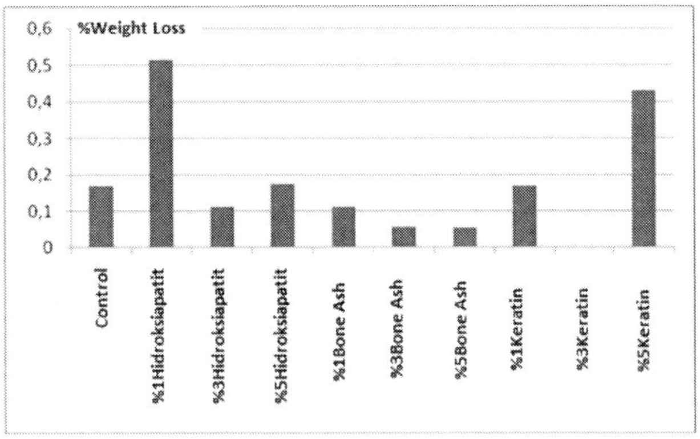

Figure 2.a. The weight loss of the samples ınder load of 47N and 3mins.

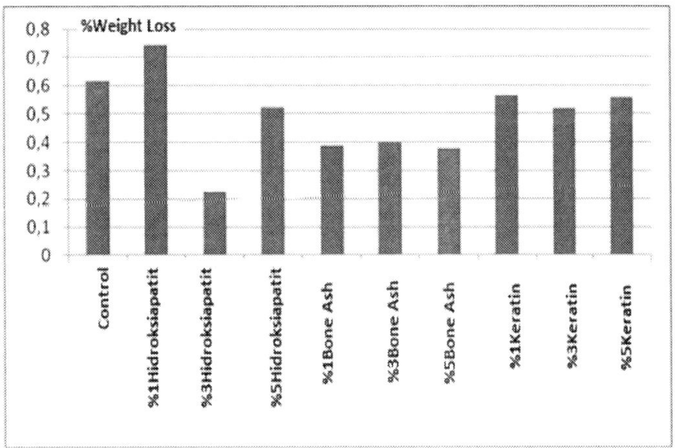

Figure 2.b. The weight loss of the samples ınder load of 47N and 6mins

The wear test results are given in Fig.2 and Fig.3 under 47N and 87N respectively. It can be seen from the figures that the weight loss results have smilar tendency except %3Hidroksiapatit.When the comperation of the relation between compression and wear result it can be seen that, altought %3Hidroksiapatit group has the lowest hardness values, their wear resistance are the highest among the

others.It can be derived from here that the wear performances of the hard samples are reduced by the microstructures after samples. On the other hand, the homogenity and porous of the powdered samples have an important role on the wear performance of the material.Hard materials wear perform could be bad affected.

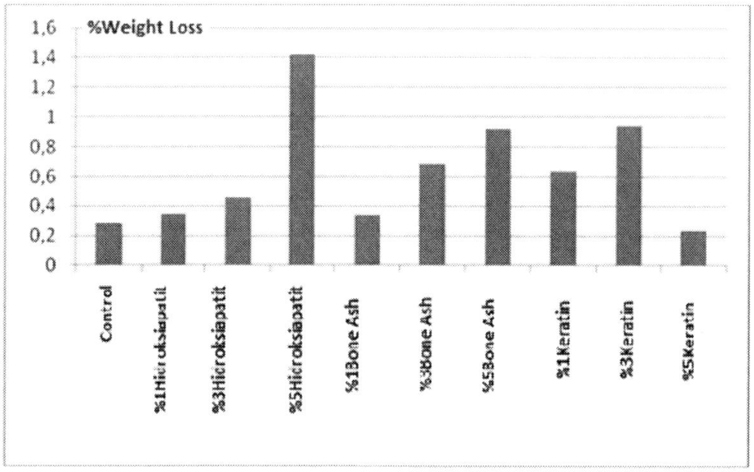

Figure 3.a. The weight loss of the samples ınder load of 87N and 3mins

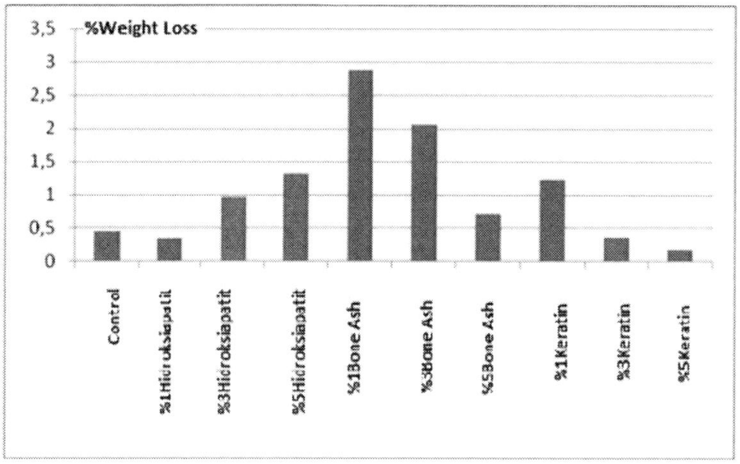

Figure 3.b. The weight loss of the samples ınder load of 87N and 6mins

On the other hand when the load increased to 87N. Wear resistance values also changed. When the hardness values of the samples are examined, it can be seen that %5Bone Ash and %5Hidroksiapatit have best wear resistance values.

All experimental samples were applied the same experimental compression test under the same conditions, same force and same speed and in the conclusion of the experiments.

Table 2. Compression Test Results

Mixture	Compression (N)	Charpy Impact (J)	Brinell Hardness (HB)
Kontrol (PMMA)	27891,90	3,33	20,6
%1 Hidroksiapatit	23359,15	2,63	13,26
%3 Hidroksiapatit	16473,95	2,57	17,17
%5 Hidroksiapatit	18998,20	2,47	20,27
%1 Bone Ash	23315,45	2,47	21,38
%3 Bone Ash	29911,80	2,90	21,55
%5 Bone Ash	26234,40	3,13	22,29
%1Keratin	24446,60	2,40	19,16
%3Keratin	21679,90	2,67	17,71
%5Keratin	19973,90	2,67	13,5

In this study we finded smilar values under 87N.Specially %5Keratin added group flourished.This group had high wear resistance in addition to toughness according to Table1.İt could be shown us this PMMA could be use with keratin.

It is so important like as toughness wear resistance values specially in dental industry. Othervise this group of compression resistance can improve.In dental using PMMA which is dental implant material always expose compression forces.

One of the biggest advantage %5 Keratin PMMA group it is the cheap material.

4. Conclusion

The conclusions derived from this study can be follow as

1. In the 18-20 age group, the average of the bite force is 176N. If considered to be, the 87N group was examined as the closest value, and in these groups, the mass loss of the 5% keratin group decreased significantly compared to the control group
2. Variable findings were found as a result of the abrasion tests conducted and it was observed that the 5% keratin-PMMA and hydroxyapatite-PMMA groups had higher wear resistance under high loadings as a general wear resistance and especially the 5% keratin-PMMA mixture gave better results than the control group.
3. Considering an acrylic exposed to continuous wear in the mouth, it was observed that the 5% hydroxyapatite group could give better results.
4. it was seen that the sample with 3% bone ash mixture yielded better results in comparison to the control group.

References

1. Phoenix, R.D., *Denture base materials.* Dent Clin North Am, 1996. **40**(1): p. 113-20.
2. Polat, T., *Polimetil Metakrilat (PMMA)'ların cam fiberle güçlendirilmesi.* Cumhuriyet Üniversitesi Dişhekimliği Fakültesi Dergisi, 2002. **5**(1): p. 41-44.
3. Gur, A.K., *Metalik malzemeler ve biyouyum.* Doğu Anadolu Bölgesi Araştırmaları, 2004: p. 106-113.
4. Frederick, A., *From vulcanite to vinly, a history of resins in restorative dentistry.* Rueggeberg, 2002. **87**(J. Prosthet. Dent.): p. 364-379.
5. Özdemir, A.K., *Hareketli Protezler Kliniğe Hazırlık.* Cumhuriyet Üniversitesi Yayınları, 2002. **1**.
6. www.polimernedir.com. *Polimer nedir.* 2013.
7. Şahin, O., *Farklı Polimerizasyon Teknikleri Uygulanan Kopolimer Yapıdaki Akrilik Rezinlerin Bazı Mekanik Özellikleri ve Sitotoksitesinin İncelenmesi*, in *Protetik Diş Tedavisi Anabilim Dalı.* 2010, Cumhuriyet Üniversitesi: Sivas.
8. Çalıkkocaoğlu, S., *Tam Protezler.* Protez Akademisi ve Gnatoloji Derneği Yayınları, 1998. **2**(3).
9. *New American Dental Association specification no. 30 for dental zinc oxide-eugenol type restorative materials. Council on Dental Materials and Devices.* J Am Dent Assoc, 1977. **95**(5): p. 991-5.
10. Sarsılmaz, F., *Ortopedide kullanılan polimer esaslı kompozit malzemeler.* Doğu Anadolu Bölgesi Araştırmaları, 2003. **3**: p. 117-113.
11. Crosa, G. (1993).Tribology of Polymer Composites Used as Friction Materials, Elsevier,
12. Yılmaz, F. (1997).Sürtünme ve Aşınma, Uluslararası Metalurji ve malzeme Kongresi, 9, 229-256.

Tribological Behaviour of Ti:Ta-DLC Films Under Different Tribo-Test Conditions

İhsan Efeoglu[a], Ayşenur Keleş[a], Yaşar Totik[a], Hikmet Çiçek[b], Ebru Emine Süküroglu[c],
[a]*Atatürk University, Faculty of Engineering, Erzurum, Turkey*
[b]*Erzurum Technical University, Faculty of Engineering, Erzurum, Turkey*
[c]*Gümüşhane University, Faculty of Engineering, Gümüşhane, Turkey*

Abstract

Diamond-like carbon (DLC) films are suitable applicants for cutting tools due to their high hardness, low friction coefficient and wear rate. Doping metals in DLC films have been improved its tribological properties. In this study, titanium and tantalum doped hydrogenated DLC films were deposited by closed-field unbalanced magnetron sputtering system onto M2 high speed steels in $Ar/N_2/C_2H_2$ atmosphere. The friction and wear properties of Ti:Ta-DLC film were investigated under different tribo-test conditions including in atmospheric pressure, distilled water, commercial oil and Ar atmosphere. The coated specimens were characterized by SEM and X-ray diffraction techniques. The bonding state of C-C (sp^3) and C=C (sp^2) were obtained with XPS. The tribological properties of Ti:Ta-DLC were investigated with pin-on-disc wear test. Hardness measurements performed by micro-indentation. Our results suggest that Ti:Ta-doped DLC film shows very dense columnar microstructure, high hardness (38.2 GPa) with low CoF ($\mu \approx 0.02$) and high wear resistance (0.5E-6 mm^3/Nm).

Keywords: Magnetron sputtering, Ti:Ta-DLC, friction, wear

1.Introduction

Lubrication of machine elements provides low coefficient of friction and wear rate so that energy losses are avoided and operating life is increased. But under some of the condition of operating (high and cryogenic temperatures, vacuum, etc.). Liquid lubrication is not convenient. In similar conditions, liquid lubricants are replaced by solid lubricants [1]. The most commonly used solid lubricants are Diamond Like Carbon (DLC), transition metal dichalcogenides (TMDC) and polymeric composite coatings [2]. DLC coatings provide low friction and high wear resistance in oil-lubricated conditions. DLC is a metastable form of amorphous carbon consists of sp^2 and sp^3 bonds. Having high hardness, chemically inert and semiconductor DLC coatings can be used some many applications including machine parts, cutting tools, parts of automotive [3], Li-ion baterry[4]. But the low adhesion properties are the major disadvantage of the DLC films [5]. For this reason elements or compounds doped the DLC films to improve some properties such as adhesion between the substrate, wear resistance, etc. According to the literature, there are some study about the doped DLC films. For example; Ag doped in DLC films by unbalanced magnetron sputtering reduce coefficient of friction (CoF.) [6], Mo doped DLC coatings by CFUBMS (Closed Field Unbalanced Magnetron Sputtering) have good adhesion and low CoF. [7], Zr doped DLC coatings decreased too small values coefficient of friction [8]. Ti doped DLC coatings by using CFUBMS is decreased CoF. [9] and also the reduced CoF was affected TiC interlayer on a DLC coatings by using CFUBMS [10]. Deposition method, substrate material, chemical and structural nature, working temperature, contact pressure and test environment are the significant on the friction and wear properties of DLC coating films [11]. For example, DLC films that contain hydrogen have ultra-low friction in inert or vacuum environments but DLC films that not contain hydrogen have ultra-low friction and wear in the presence of oxygen, hydrogen or water molecules [12-13]. Searching literature, it was shown that tribological properties of Ta doped DLC coatings have not been investigated upto now. Ta has attractive properties like corrosion resistance, thermal stability, high resistivity, fracture toughness. Ta affects material's performance through microstructural modifications. Through carbide formation, Ta provides strength, malleable and thermal stability [14].

In the present research; Ti and Ta doped hydrogenated DLC films were deposited by closed field unbalanced magnetron sputtering (CFUBMS) system onto M2 high speed steels in $Ar/N_2/C_2H_2$ atmosphere. The friction and wear properties of Ti:Ta-DLC film were investigated under different tribo-test conditons including in atmoshperic pressure, distilled water, commercial oil and the Ar atmosphere. The results showed that Ti:Ta-doped DLC film shows very dense microstructure, high hardness with low CoF and high wear resistance.

2. Experimental Methods

Ti and Ta doped DLC was coated on M2 high speed steel, in size 25x25x2 mm^3, as a substrate metal and glass wafer by CFUBMS. The chemical composition of steel is 0.3 wt% Si, 4.2 wt% Cr, 0.3 wt% Mn, 5.1 wt% Mo, 6.3 wt% W, 2 wt% S and C balance. The substrates were polished to a roughness value of $R_a < 0.1$ µm, ultrasonically cleaned with ethyl alchol and then etched in the 2% Nital solution and dried. The Ti:Ta doped DLC was perfomed by CFUBMS system by produced Teer Coating Ltd [15]. Schematic picture of magnetron configuration system is shown Figure 1.

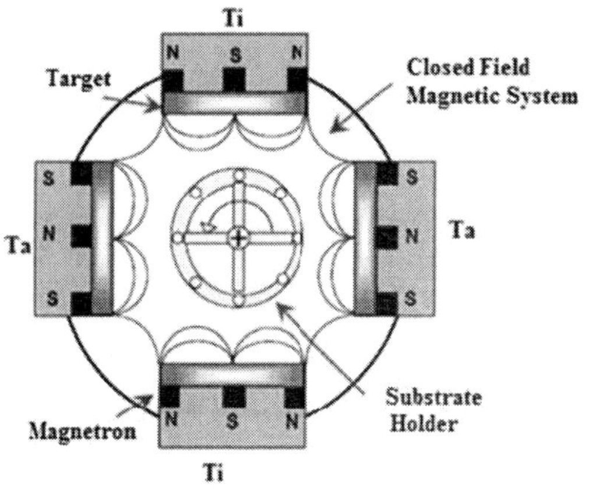

Figure 1. Schematic picture of magnetron configuration CFUBMS system [16]

Depositing the Ti:Ta DLC films; 2 Ta and 2 Ti targets and N_2, C_2H_2 reactive gases were used and for ionization Ar gase was used. Before deposition process to eliminate contamination and improve adhesion on the substrate metals, ion cleaning was realized for 20 minutes. The parameters of coating process were given in Table 1. Current of Ti targets were kept constant at 5 A in 3 runs during process and the current values of Ta targets were 5A, 4A and 3A, respectively. The pulsed DC bias was applied to the substrates rotated in the system. For deposition Ti-Ta DLC were bias voltage -60 V, working pressure 0.33 Pa. The deposition time was 75 min and to improve adhesion between substrate and DLC coating, Ti/TiN/TiTaN/TiTaCN graded layer was coated on substrate. The schematic picture of coating was shown Figure 2.

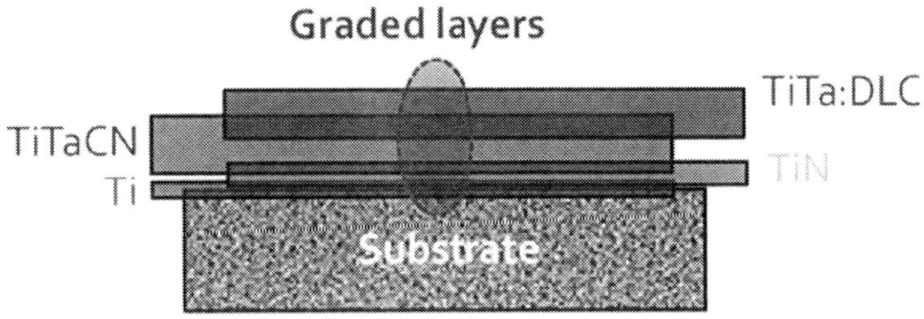

Figure 2. The schematic picture of DLC coating

The thickness of coating, the microstructure, the stoichiometry and the wear tracks were analyzed with JEOL-6400-SEM and energy dispersive spectrometry (EDS). The microhardness of the coatings was measured by Buehler Micromet 2001 microhardness tester (using Knoop indenter, 10gf load). The crystallographic orientation was characterized using Rigaku DMax-2000 XRD with a Cu-K_α radiation source (at 5° to 100° scan range). The bonding state of C-C and C=C were obtained by PHI 5000 VersaProbe XPS. The wear tests under different conditions were carried out CSM high temperatuere tribo-tester (3N load, 25 °C, %45-55RH, 10cm/s velocity, Si_3N_4 counterpart), respectively.

Table 1. The Deposition Parameters

Coating Film Parameters		
Number of Coating	Target Currents (A)	
	Ta	Ti
R1	5	5
R2	4	
R3	3	
Constant Parameters		
Substrate Bias Voltage (-V)	60	
Working Pressure (Pa)	0.33	
Coating Time	75	

3. Results And Discussion

3.1. Microstructure of Coating

Cross-section and surface SEM images of the Ti:Ta-DLC film coatings are given in Figure 3.

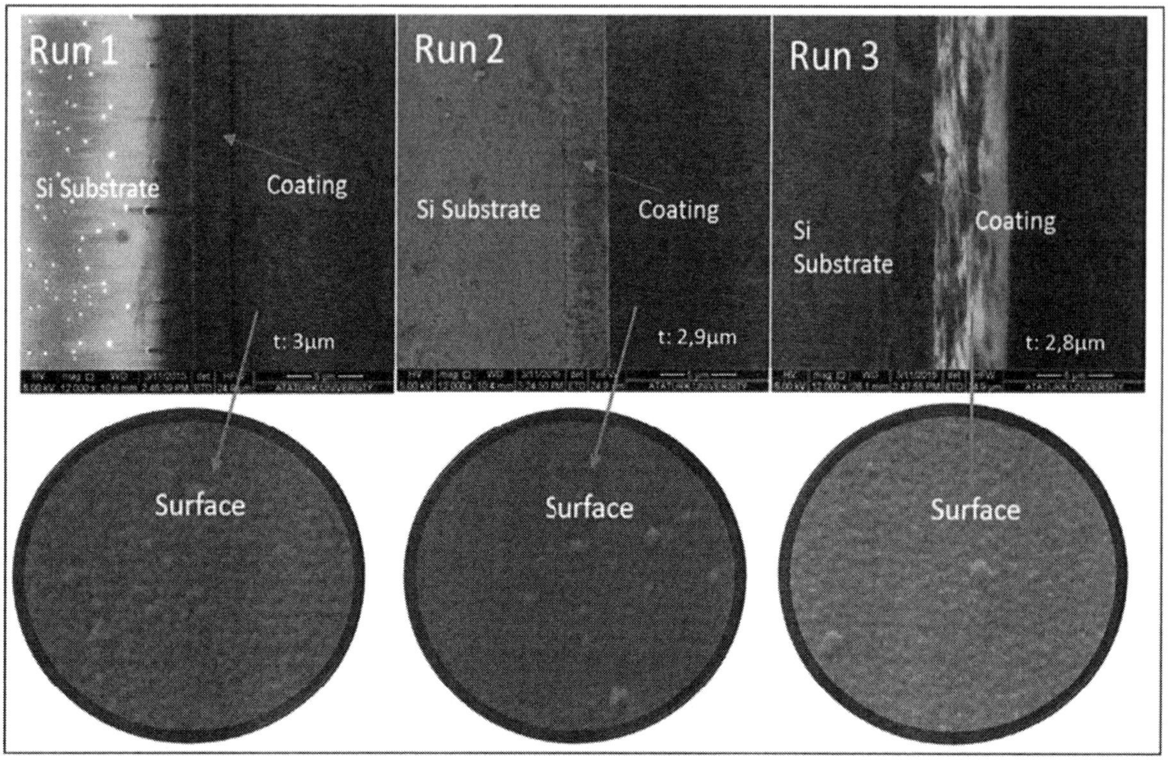

Figure 3. Crossection and surface SEM images of Ti:Ta-DLC film coatings.

The cross-section and the surface SEM images showed that all Ti:Ta-DLC films grown as dense columnar structures. It seems that the density of the columns decreased with decreasing target current values of Ta. The interfaces between the substrate and the Ti:Ta-DLC films were well defined. Run 3 (R3) films showed brittle type fracture and the other Run 1 (R1) and Run 2 (R2) films showed ductile fractures. According to these results, it can be said that when the Ta target current value increase, it caused the change of fracture type of the film from brittle to ductile.

The total coating thickness for the Ti:Ta-DLC films were 3.0 µm, 2.9 µm, 2.8 µm for R1 , R2 and R3 respectively. It was observed that thickness of the films was affected directly from the Ta target current value increasing in a linear manner.

Chemical compositions of the films determined by EDS. The results of the films are given in Table 2.

Table 2: Chemical composition of the Ti:Ta-DLC films

Elements	R1 (at.%)	R2 (at.%)	R3 (at.%)
C	44,5	48,2	64,8
N	19,1	18,3	10,7
O	4,8	4,2	3,4
Ta	14,7	16,5	17,2
Ti	16,9	12,8	3,9

According to EDS analyzes, when the Ta target current increased chemical composition ratio (at.%) of Ta in films increased and Ti ratio decreased as expected. When looking at the carbon content in the films, the highest carbon ratio (R3) obtained from the highest Ta target current (5A) and the lowest (R1) obtained the lowest Ta current (3A). It is clearly seen that carbon content directly affected the Ta target current with parallel behavior. The chemical ratio of nitrogen showed an inverse status with carbon. Reducing the carbon content of the film was increased the nitrogen content. It appeared that oxygen ration in the films showed the same behavior with the nitrogen.

XRD analysis was performed to investigate the crystal structure of the Ti:Ta-DLC films. XRD patterns of the films were given in Figure 4 comparatively.

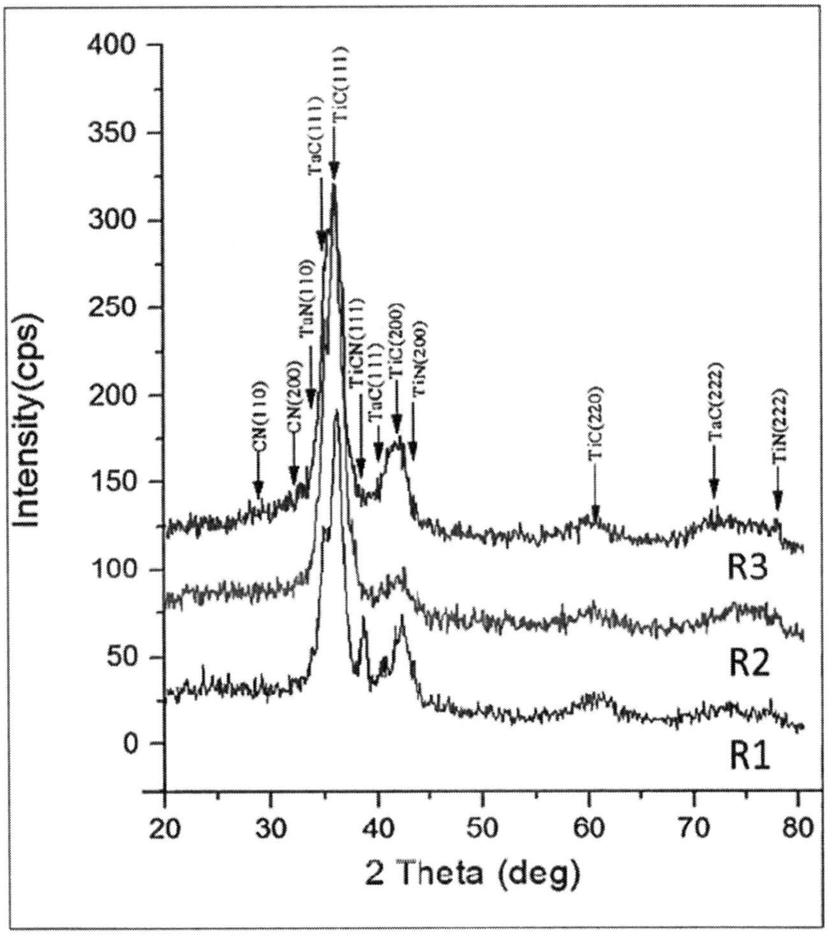

Figure 4. XRD patterns of Ti:Ta-DLC deposited films.

The XRD patterns showed that lots of different crystal structure such as TiC, TaC, CN, TaC, TiN formed by used elements (Ti, Ta, C, N). It was observed that TiC (111) and TaC (111) phases predominantly placed in the films. The TiC (111) peak reached the highest value at the R3 film with increasing the Ti ratio (at. %) and decreasing the Ta ratio (at.%) in film. It was found that crystallographic orientation depends on the Ta target current. TiC (200) phase is second dominant phase in the films although some reduction at R2. TiC (220), TaC (222), TiN (222), CN (110), CN (220) and TaN (110) peaks do not show significant change at the films. On the other hand, although the TiCN (111) peak clearly appears at R1, but it disappears at R2 and R3 films with the decreasing of carbon content ratio from %64 to approximate %44,5.

The XPS patterns of bonding state of C-C (C-sp^3) and C=C (C-sp^2) were shown Figure 5. Some researchers reported that the peaks at around 284.3 ±0.1 eV and 285.3±0.1 eV were attributed to C-sp^2 and C-sp^3 bondings [17, 18]. The figure 5 showed that C-1s, C-sp^2 and C-sp^3 bondings were measured

282.6 eV, 284.4 eV and 285.2 eV, respectively. The ratio of $sp^3/(sp^3+sp^2)$ was found 0.45 for R1, 0.48 for R2 and 0.52 for R3. It is evident that Ta facilitates the growth of C=C (sp^2) bonding in DLC coating.

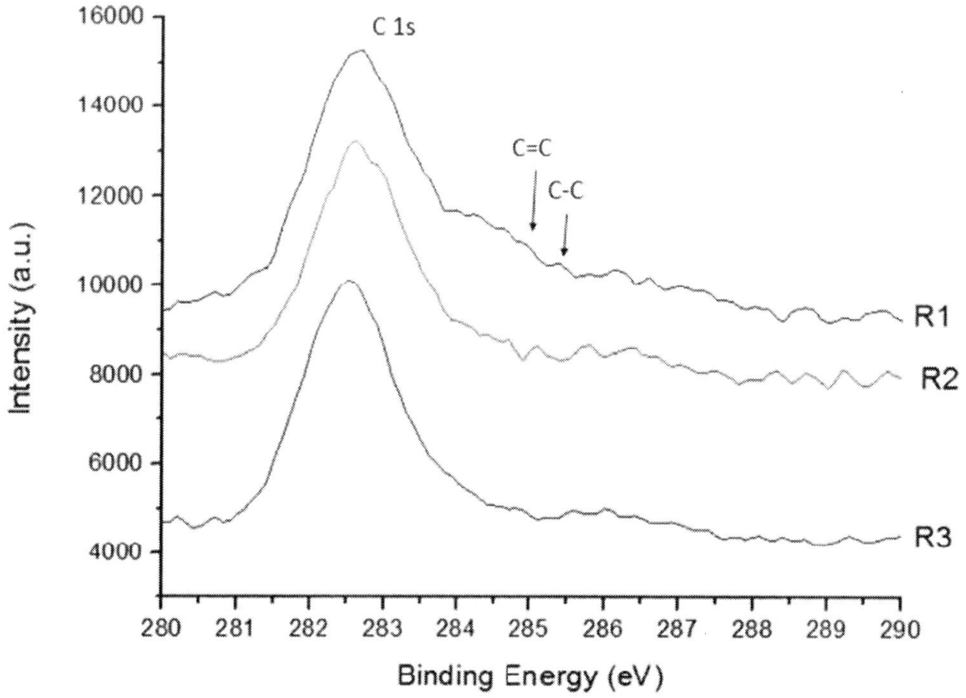

Figure 5. The XPS spectra of Carbon Element of The Coatings

3.2. Microhardness of films

The microhardness of Ti-Ta DLC films at a 10 gf load was R1, R2 and R3, from low to high, respectively. The value of hardness with ascending order was 13.9 GPa, 17.8 GPa and 38.2 GPa. It was observed that decreasing the Ta current values, the microhardness and the carbon content increased. According to EDS analyze, R3 has the highest value of the carbon content. It seems that large crystallite disperses in high C-C (C-sp^3) content amorphous carbon shows the highest hardness among the films [19]. The hardness of the DLC films depends on the C=C (C-sp^2) / C-C (C-sp^3) ratio of the films [20]. Results show that the hardness of the DLC films increase with decreasing C=C (C-sp^2) / C-C (C-sp^3) ratio of the DLC films.

3.3. Coefficient of Friction (CoF) and Wear Rate of films

Four different environments were used to investigation for tribological properties of the Ti:Ta-DLC films. The coefficient of friction of films under atmosphere, argon, commercial oil and distilled water conditions were given in Figure 6. It is showen that the friction coefficient at the begin of the test (approximate 0-500 s) exhibited a high value (for atmosphere condition:R1≈0.35μ, R2≈0.30μ, R3≈0.32μ, for Argon atmosphere condition: R1≈0.34μ, R2≈0.32μ, R3≈0.25μ, for Distilled water contion:R1≈0.25μ, R2≈0.15μ, R3≈0.12μ, for Commercial Oil condition: R1≈0.12μ, R2≈0.04μ, R3≈0.03μ). After then the friction returned again a stable value by showing a decreasing tendency. This case was related to a transfer film formation on the counterpart and dense film structure result.

It was shown that the hardest R3 was the lowest coefficient of friction and wear rate in all conditions. It is realized that the R1 has the highest CoF and wear rate in all conditions and also it has the lowest hardness value. Compared with different conditions, it is reported that the CoF was changed in commercial oil condition, distiled water condition, argon condition and atmosphere condition, from low to high, respectively. It was obtained that the friction coefficient of R3 films were 0.03 - 0.1 - 0.16 and 0.25 and the wear rate were 0.45×10^{-6} mm^3/Nm, 11.5×10^{-6} mm^3/Nm, 15.7×10^{-6} mm^3/Nm ve 18.3×10^{-6} mm^3/Nm commercial oil condition, distiled water condition, argon condition and atmosphere condition, from low to high, respectively. According to classical theories of wear, hardness affects the wear resistance of a film and higher hardness led to lower wear rate [19, 21]. According to the results, the R3 that was deposited at low Ta currents shows the highest hardness among DLC films. The friction of coefficient of R3 has stable all of the conditions. This cause of ability of R3 has ability to soften small debris of wear from the counterpart. It is very important for stable coefficient of friction [22].

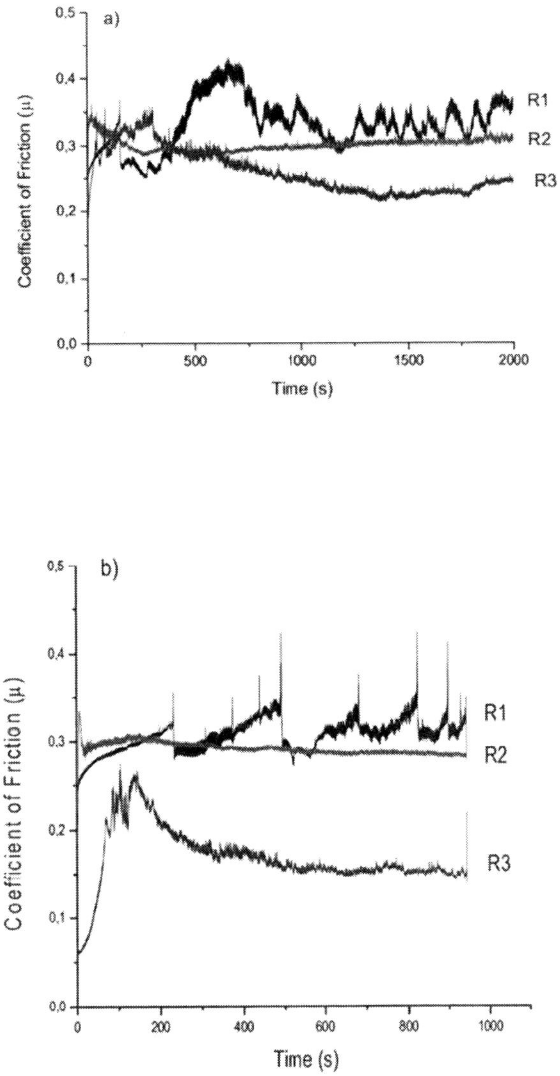

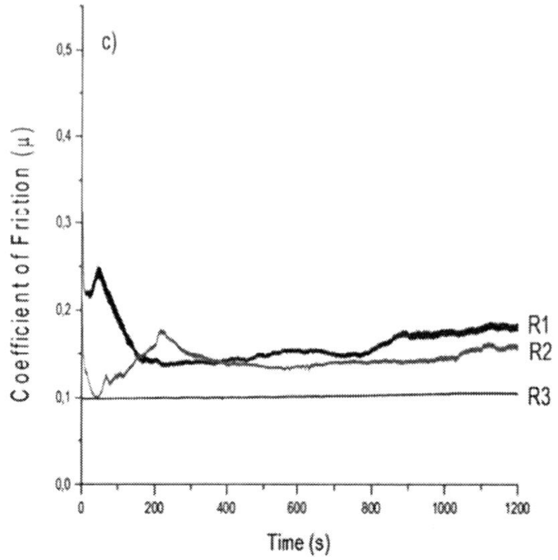

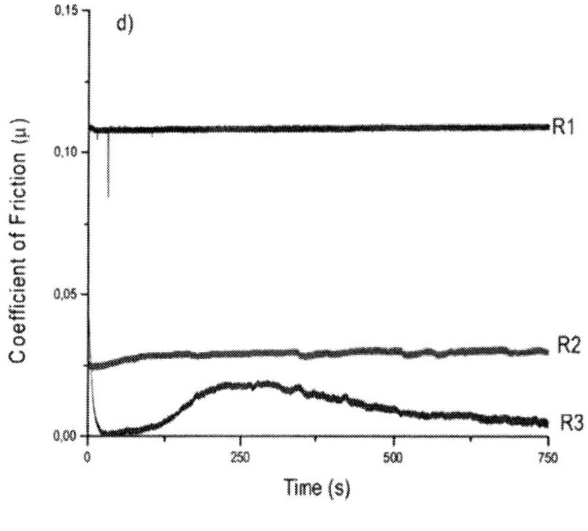

Figure 6. Coefficient of Films a) Atmosphere condition b)Argon Atmosphere condition c) Distilled water contion d) Commercial Oil condition

SEM images of wear tracks of the film and Optical images of abrasive balls at four different environments are given in Figure 7.

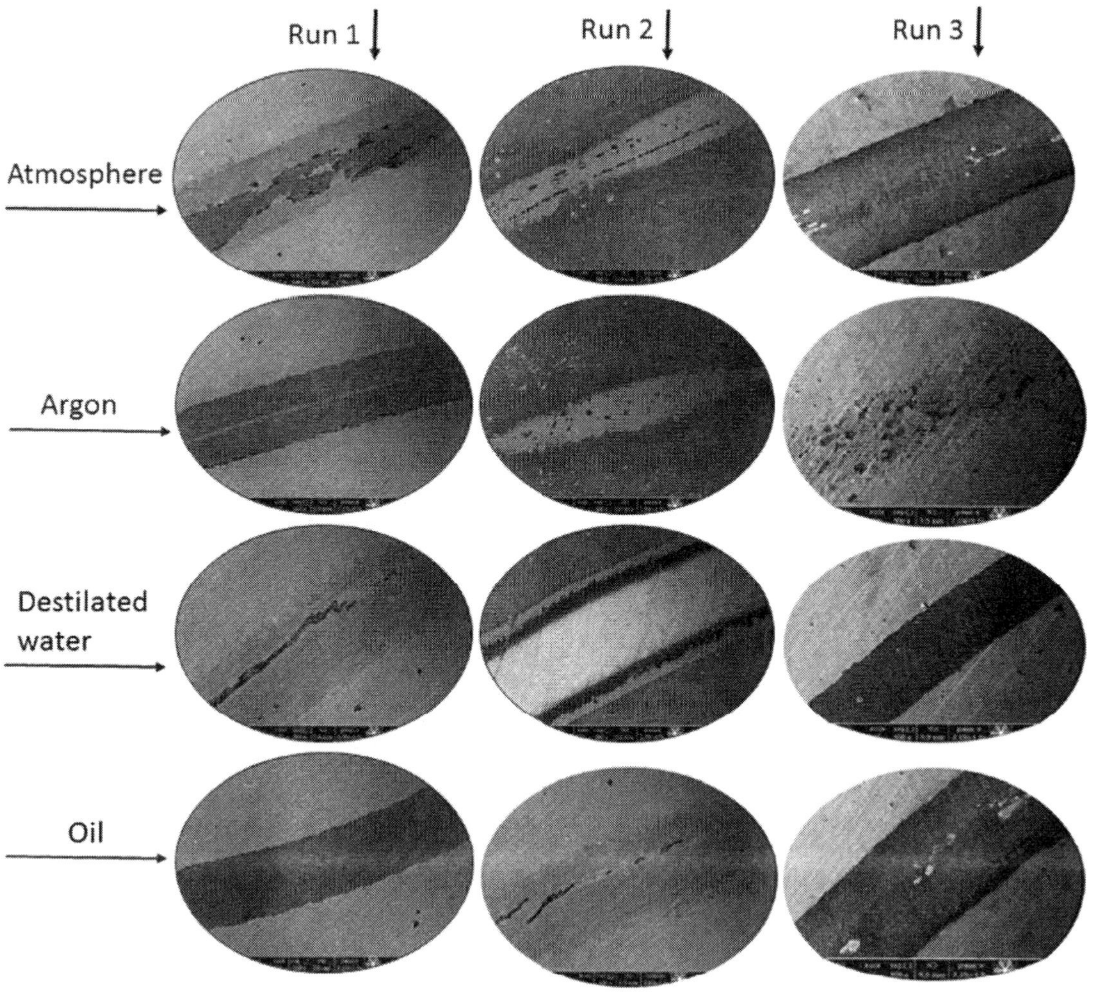

Figure 7. SEM images of wear tracks of the films and Optical images of abrasive balls

It was seen that the wear track obtained for the all wear conditions, the R3 film is very smooth and steady, but it did not have deep abrasive scratches on track as that of the films obtained under the same wear conditions. While plenty of small particles dragged on the sides of wear track obtained from the film grown by R1 and R2 were observed, less particles and some flakes on track existed also in that of grown by R3 at the same condition. The wear track obtained from the R3 includes plenty of wear debris as a result of abrasion due to the high Ta content (Table 3). But wear debris amount is more than that of the film obtained by R1 and R2 under the same wear conditions. It can be said that the wear mechanism of the Ti:Ta-DLC films produced by lower Ta content is more abrasive according to that of R3 because of higher Ta content.

When evaluated in terms of EDS analyzes after wear test (Table 3), R3 has the highest value of the carbon and Ta content. The wear track and the counterpart ball images show that the highest value of the carbon and Ta content in R3 films caused not showing an abrasive behavior.

IOP Conf. Series: Materials Science and Engineering **295** (2018) 012005 doi:10.1088/1757-899X/295/1/012005

Table 3. Chemical composition of wear tracks

% Atomic	Elements	R1	R2	R3
Atmosphere	*C*	49	57,8	62,4
	N	20,9	12,8	10,2
	O	12,3	18,6	15,7
	Fe	0,1	0,1	0,5
	Ta	4,4	3,4	6,1
	Ti	13,3	7,3	5,1
Argon	*C*	52,4	56,1	69,1
	N	20,7	12,9	10,5
	O	11,8	18,5	10,3
	Fe	0,1	0,2	0,4
	Ta	4,6	3,6	5,8
	Ti	10,4	8,7	3,9
Distilated Water	*C*	46,5	50,5	63,6
	N	23,6	20,6	17,16
	O	10,8	12,8	9,15
	Fe	0	0,4	0,2
	Ta	4,6	5,4	5,5
	Ti	14,5	10,3	4,3
Oil	*C*	56,5	56,7	62,9
	N	16,4	15,1	12,5
	O	11,2	19,1	14,1
	Fe	0,1	0,1	0,1
	Ta	4,5	3,2	6,3
	Ti	11,3	5,8	4,1

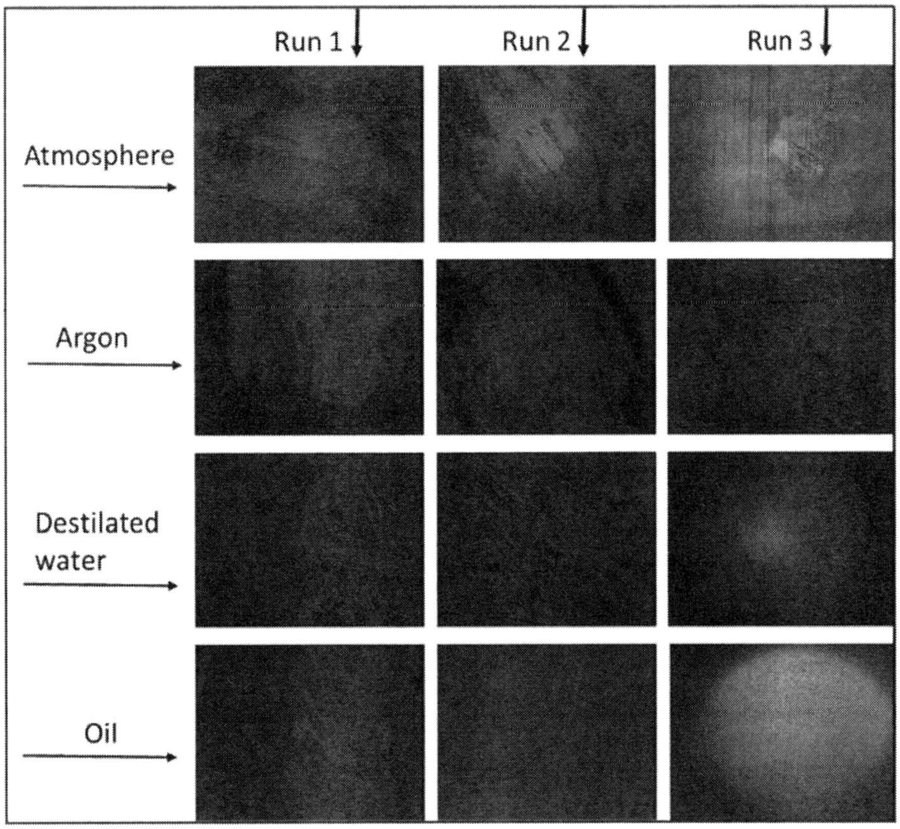

Figure 8. Optical images of abrasive balls

4. Conclusions

The following conclusions can be derived from the above results and discussion:

- All three films grown as dense columnar structures. It was seen that the density of the columns decreased with decreasing target current values of Ta.
- With decreasing Ta current, carbon content was increased and therefore the hardness of film was increased. The hardness value of the hardest coated sample was 38.2 MPa.
- The reason of low friction coefficient was Ta current. With decreasing current of Ta, hardness value was increased and this reason CoF ($\mu \approx 0.02$) and wear rate (0.5E-6 mm^3/Nm).were decreased.

References

1. A. Erdemir, C. Donnet, Tribology of diamond-like carbon films: recent progress and future prospects, J. Phys. D Appl. Phys. 39 (2006) 311–327. doi:10.1088/0022-3727/39/18/R01.

2. T.W. Scharf, S. V. Prasad, Solid lubricants: A review, J. Mater. Sci. 48 (2013) 511–531. doi:10.1007/s10853-012-7038-2.

3. J. Vetter, 60years of DLC coatings: Historical highlights and technical review of cathodic arc processes to synthesize various DLC types, and their evolution for industrial applications, Surf. Coatings Technol. 257 (2014) 213–240. doi:10.1016/j.surfcoat.2014.08.017.

4. J.K. Lee, J.H. Lee, B.K. Kim, W.Y. Yoon, Electrochemical characteristics of diamond-like carbon/Cr double-layer coating on silicon monoxide-graphite composite anode for Li-Ion batteries, Electrochim. Acta. 127 (2014) 1–6. doi:10.1016/j.electacta.2014.01.166.

5. J. Robertson, Diamond-like amorphous carbon, Mater. Sci. Eng. R Reports. 37 (2002) 129–281. doi:10.1016/S0927-796X(02)00005-0

6. Y. Wu, J. Chen, H. Li, L. Ji, Y. Ye, H. Zhou, Preparation and properties of Ag/DLC nanocomposite films fabricated by unbalanced magnetron sputtering, Appl. Surf. Sci. 284 (2013) 165–170. doi:10.1016/j.apsusc.2013.07.074.

7. X.S. Tang, H.J. Wang, L. Feng, L.X. Shao, C.W. Zou, Mo doped DLC nanocomposite coatings with improved mechanical and blood compatibility properties, Appl. Surf. Sci. 311 (2014) 758–762. doi:10.1016/j.apsusc.2014.05.155.

8. T. Vitu, A. Escudeiro, T. Polcar, A. Cavaleiro, Sliding properties of Zr-DLC coatings: The effect of tribolayer formation, Surf. Coatings Technol. 258 (2014) 734–745. doi:10.1016/j.surfcoat.2014.08.003.

9. W.Q. Bai, L.L. Li, X.L. Wang, F.F. He, D.G. Liu, G. Jin, et al., Effects of Ti content on microstructure, mechanical and tribological properties of Ti-doped amorphous carbon multilayer films, Surf. Coatings Technol. 266 (2015) 70–78. doi:10.1016/j.surfcoat.2015.02.024.

10. F. Bülbül, I. Efeoğlu, E. Arslan, The effect of bias voltage and working pressure on S/Mo ratio at MoS2-Ti composite films, Appl. Surf. Sci. 253 (2007) 4415–4419. doi:10.1016/j.apsusc.2006.09.059.

11. A. Erdemir, C. Donnet Tribology of diamond-like carbon films: recent progress and future prospects Journal of Physics D—Applied Physics, 39 (2006), pp. R311–27

12. J. Andersson, R.A. Erck, A. Erdemir, Friction of diamond-like carbon films in different atmospheres, Wear, 254 (2003), pp. 1070–1075

13. J. Robertson, Diamond-like amorphous carbon, Materials Science and Engineering R-Reports, 37 (2002), pp. 129–281

14. S.M. Cardonne, P. Kumar, C.A. Michaluk, H.D. Schwartz, Tantalum and its Alloys, Journal of Refractory Metals and Hard Materials, 13 (1995), pp. 187-194

15. D.G. Teer, UK patent, GB 2 258 343 B (1991).

16. F. Bulbul, I. Efeoglu, Characterization of Ti-CN-a:C films prepared by pulsed-dc bias sputtering, Proc. Inst. Mech. Eng. Part J-Journal Eng. Tribol. 224 (2010) 677–684. doi:10.1243/13506501jet702.

17. M. Jelinek, T. Kocourek, J. Zemek, J. Miksovsky, S. Kubinova, J. Remsa, et al., Chromium-doped DLC for implants prepared by laser-magnetron deposition, Mater. Sci. Eng. C. 46 (2015) 381–386. doi:10.1016/j.msec.2014.10.035.

18. J.C. Pu, S.F. Wang, C.L. Lin, J.C. Sung, Characterization of boron-doped diamond-like carbon prepared by radio frequency sputtering, Thin Solid Films. 519 (2010) 521–526. doi:10.1016/j.tsf.2010.07.018.

19. Mansoureh Ebrahimi, Farzad Mahboubi, M. Reza Naimi-Jamal., RSM base study of the effect of deposition temperature and hydrogen flow on the wear behavior of DLC films, Tribology International, Volume 91, November 2015, Pages 23–31

IOP Conf. Series: Materials Science and Engineering **295** (2018) 012005 doi:10.1088/1757-899X/295/1/012005

20. A Gangopadhyay, Mechanical and tribological properties of amorphous carbon films, Tribol Lett, 5 (1998), pp. 25–39

21. A Leyland, A Matthews, On the significance of the H/E ratio in wear control: a nanocomposite coating approach to optimised tribological behaviour, Wear, 246 (2000), pp. 1–11

22. J.M. Martin, T. Onodera, C. Minfray, F. Dassenoy, A. Miyamoto, The origin of anti-wear chemistry of ZDDP, Faraday Discuss., 156 (2012), pp. 311–323

Finite element analysis of fretting contact for nonhomogenous materials

Korkmaz Y M[1,2]***, Coker D***[1]
[1]*Middle East Technical University, Department of Aerospace Engineering, 06800, Ankara, Turkey*
[2]*Helicopter group, Turkish Aerospace Industries, 06800, Ankara, Turkey*
yedzdanmedet.korkmaz@tai.com.tr[1], *coker@metu.edu.tr*[2]

Abstract: *Fretting problem arises in the case of relatively small sliding motion between contacting surfaces. Fatigue life of the components that are in contact with each other, especially in rotorcraft may be significantly reduced due to fretting. The purpose of this study is to investigate material inhomogeneity near the contact region on the fretting problem in a cylindrical on flat contact configuration. A finite element (FE) model was constructed by using commercial finite element package ABAQUS*^{TM}*to study partial sliding and stress concentrations. In order to investigate the effect of material inhomogeneity, the fretting contact is analyzed by introducing voids near the contact region. The void size and an array of voids is introduced into the substrate. The results are compared in terms of pressure, shear traction, tangential stress magnitudes and relative slip between the contacting materials.*

Keywords: fretting; finite element modelling; contact modelling

1. Introduction

When relative small sliding takes in place between two contacting surfaces, failure due to fretting can occur, such as bolted joints and lug attachments of aircrafts. Waterhouse [1] mentioned that bolted and riveted structures which are subjected to fluctuating loads are possible sources of fretting fatigue failures in his works. In order to investigate the fretting phenomena, there are numerous studies which were done with using both analytical and finite element (FE) methods can be found in literature. To observe the pressure distribution in the frictionless contact region between two elastic materials, the analytical approach is developed by Hertz [2] which is called Hertzian contact theory. Later, to observe the effect of the both tangential loading and corresponding friction at the contact interface Mindlin and Deresiewicz [3] further improved Hertz contact theory. Mindlin theory divides the contact interface in two main regions, which are called as stick and slip, with using shear traction distribution. Furthermore, the effect of the bulk stress in calculations of the shear traction distribution taken in consideration by Hills and Nowell [4]. In addition to that Nowell [5] reported that Mindlin theory is inadequate to calculate shear traction distribution for contact between elastically dissimilar materials. Finite element analysis provides has an opportunity to discover and simulate different contact conditions by taking the geometrical non-linearity into account for various geometries. Numerous works were done in the literature in order to model fretting contact by using finite element methods. For instance, Ruiz et al. [6] and Stower et al. [7] simulated fretting contact using FEA. The effect of dissimilar materials on fretting fatigue behavior of Ti-6Al-4V was shown by Lee and Mall [8]. Simulation of the crack propagation were performed in the contact region with using extended finite element method by Giner et al. [9]. The differences between 2D and 3D finite element analysis of the fretting fatigue were investigated Kim et al. [10]. Korkmaz and Coker [11] simulated the contact and friction behavior of dissimilar materials using finite element method. In this study, we investigate the effect of material inhomogeneity in cylindrical on flat fretting contact configuration by introduction of voids in the substrate near the contact region.

2. Theoretical Background

Hertzian contact theory is used to calculate the contact pressure and to determine the contact area between the two different elastic bodies without friction (Figure 1). This theory is only applicable when the effective radius of curvature is much greater than the semi contact width. In accordance with Hertzian theory, the contact pressure becomes maximum at the center of the contact region and

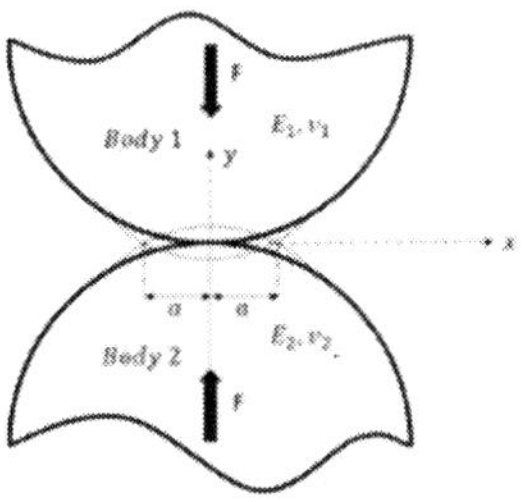

Figure 1. Representative view of Hertzian contact.

becomes zero at the edges when two cylindrical bodies are in contact,

$$p(x) = -\frac{2F}{\pi a l}\sqrt{a^2 - x^2} \tag{1}$$

where $p(x)$ is the contact pressure, F is the normal load, l is the thickness, and a is the contact radius given by,

$$a = \sqrt{\frac{2F}{\pi}\left(\frac{1-\nu_1^2}{E_1} + \frac{1-\nu_2^2}{E_2}\right)\left(\frac{1}{R_1} + \frac{1}{R_2}\right)} \tag{2}$$

where E_i, ν_i, R_i are the elastic modulus, Poisson's ratio and radius of the top and bottom surfaces for $i=1,2$.

The contact region is divided into two regions consisting of stick and slip regions (Figure 2). Part of the contact region where there is no relative displacement between the pad and the specimen is called the stick region and the other part where there is relative displacement is called the slip region [12].

The shear traction distribution, $q(x)$, is calculated by,

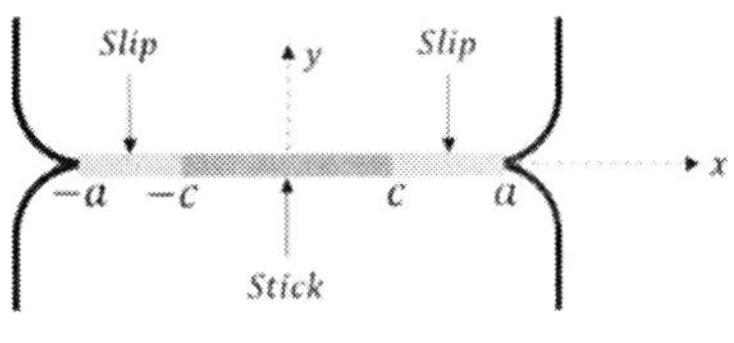

Figure 2. Representative view of the stick slip relation in the contact region.

$$q(x)=\begin{cases} -fp_{\max}\sqrt{1-(x/a)^2} & c\le |x|\le a \\ -fp_{\max}\left[\sqrt{1-(x/a)^2}-2\dfrac{c}{a}\sqrt{1-(x/c)^2}\right] & |x|<c \end{cases} \tag{3}$$

Where f is the friction coefficient and c is the length of the stick region given by,

$$\frac{c}{a}=\sqrt{1-\frac{Q}{2fF}} \tag{4}$$

However, when the shear loading is generated by an axial bulk stress applied to one component, the stick region is shifted by the eccentricity value e which causes a change in the shear traction distribution given by the following equation,

$$q(x)=\begin{cases} -fp_{\max}\sqrt{1-(x/a)^2} & c\le |x|\le a \\ -fp_{\max}\left[\sqrt{1-(x/a)^2}-2\dfrac{c}{a}\sqrt{1-\left(\dfrac{x+e}{c}\right)^2}\right] & |x+e|<c \end{cases} \tag{5}$$

3. Finite element modelling

Finite element model of a cylinder on flat contact configuration is simulated with ABAQUS™. Half of the specimen is modelled due to symmetry conditions in the y-direction and the vertical movement along the bottom symmetry axis is restricted. A specimen with length $L=40$ *mm*, width $w=5mm$ and thickness t=4mm was used [13]. The radius of the pad, R, was chosen as 10mm. In FE model, where a uniform axial stress is applied to the right boundary, the cylinder pad is restricted in the horizontal direction along its side surfaces and is free to move in the vertical direction (Figure 3). The loading is applied to the model in two steps. In the first step, normal load is applied. In step 2 axial stress and a reaction is applied while keeping the normal stress constant as shown in Table 1.

Table 1. The loading conditions of fretting fatigue contact [14].

	Step 1	Step 2
Normal Load [N]	543	543
Axial stress [MPa]	0	100
Reaction stress [MPa]	0	92.2

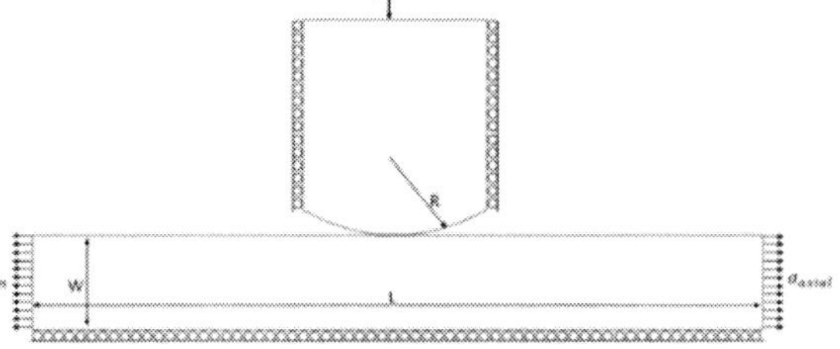

Figure 3 Geometric and loading conditions of fretting contact.

The model is discretized using a 2D quadrilateral, 4-node, plane strain, reduced integration element (CPE4R). For accurate results, the characteristic mesh was selected as 5 μm (Figure 4). The contact between the pad and the specimen was described by utilizing master slave algorithm in ABAQUS [15]. The simulation of contact friction was achieved by using Lagrange Multiplier formulation for the tangential behavior and Augmented Lagrange Algorithm for the normal behavior. Multi-Point Constraint (MPC) was used at the top of the pad in model in order to prevent pad from rotating because of the applied loads. The material of the specimen and pad was selected as Al2024 T3. The modulus of elasticity is 72.1 GPa and the Poisson's ratio is 0.33 for Al2024 T3.

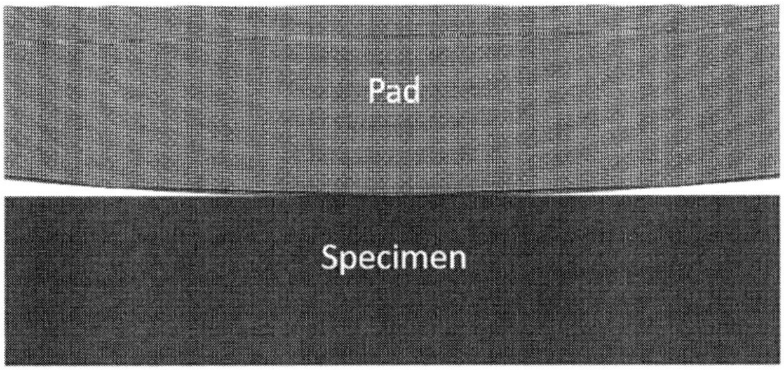

Figure 4. A close-up view of the contact region of the finite element model.

4. Results and Discussion

Materials are not homogeneous and voids may exist in materials. Murakami and Endo [16] reported the effect of these voids in engineering alloys were crucial for fatigue crack formation. To observe the effect of a void, a circular hole was generated under the contact region in the substrate [17]. The position and size of the void was changed, in Figure 5a and 5b, respectively, to investigate the effect on the stresses at the contact surface.

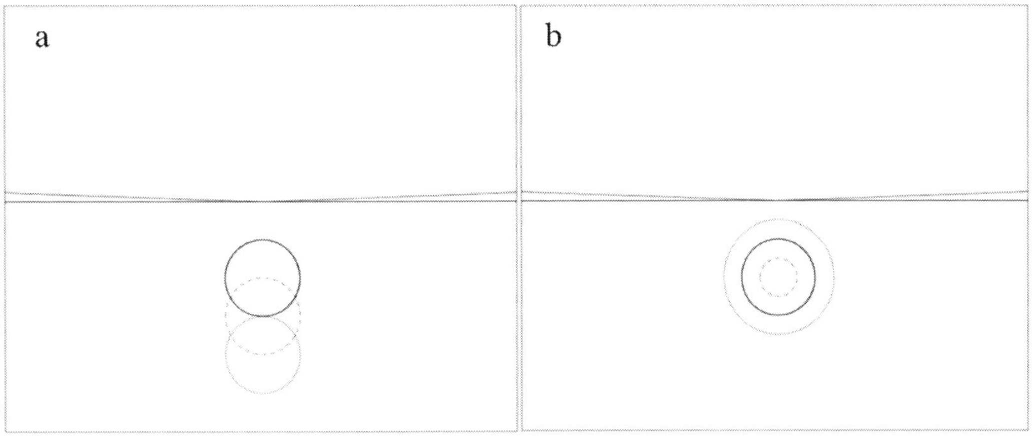

Figure 5. Voids in the substrate (a) at different distance of the centroid from the contacting edge in the finite element model, (b) with different void radius

In the first case, Figure 5a, the distance of the center of circular void to the contact surface is selected as h=0.2, 0.3 and 0.4mm. The von Mises stress distribution for no hole case and hole distance of 0.2, 0.3 and 0.4 are shown figure 6a, 6b, 6c and 6d respectively. It is seen that when the void is closest to the surface, the stresses at the contact region actually increases. However, as the void is moved away by a distance of r or 1.5r, the von Misses stresses at the sliding surface is decreased. It is observed that hertz contact theory cannot predict the results approximately in terms of contact pressure when voids inserted to the material. The von Mises stress distribution for no hole case and hole distance of 0.2, 0.3 and 0.4 are shown figure 6a, 6b, 6c and 6d respectively.

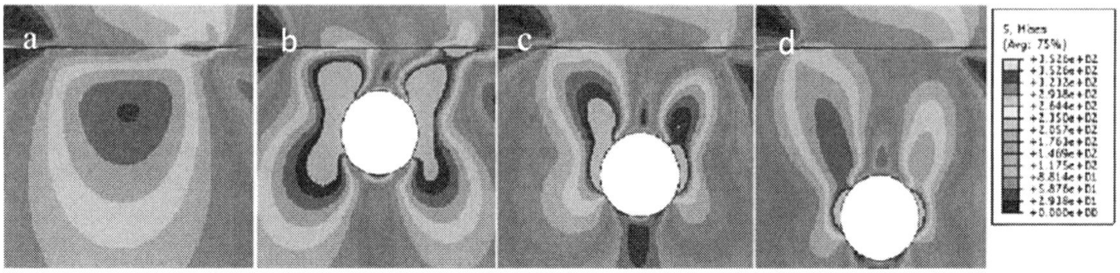

Figure 6. Von Mises stress contours (a) no hole (b) hole distance equal to 0.2 (c) hole distance equal to 0.3 (d) hole distance equal to 0.4

Figure 7 shows the normal traction (Figure 7a) and shear traction (Figure 7b) on the interface, the axial stress in the substrate (Figure 7c) and the relative slip amplitude (Figure 7d) at the contact surface. We will focus on the results for the case when the hole is one radius away from the surface. The peak value of the contact pressure decreases when the void is close to the surface (Figure 7a). The maximum value of the shear traction distribution decreases at the right edge of the contact area while increasing at the left contact region (Figure 7b). The maximum tangential stress tends to increase when the hole becomes closer to the surface. In contrast, the introduction of the void increases the maximum axial stress for the closest void (Figure 7c). Relative slip amplitude decreases when the void is inserted (Figure 7d). Thus, the insertion of a void decreases the frictional shear stresses and relative slip while increasing the normal axial stresses.

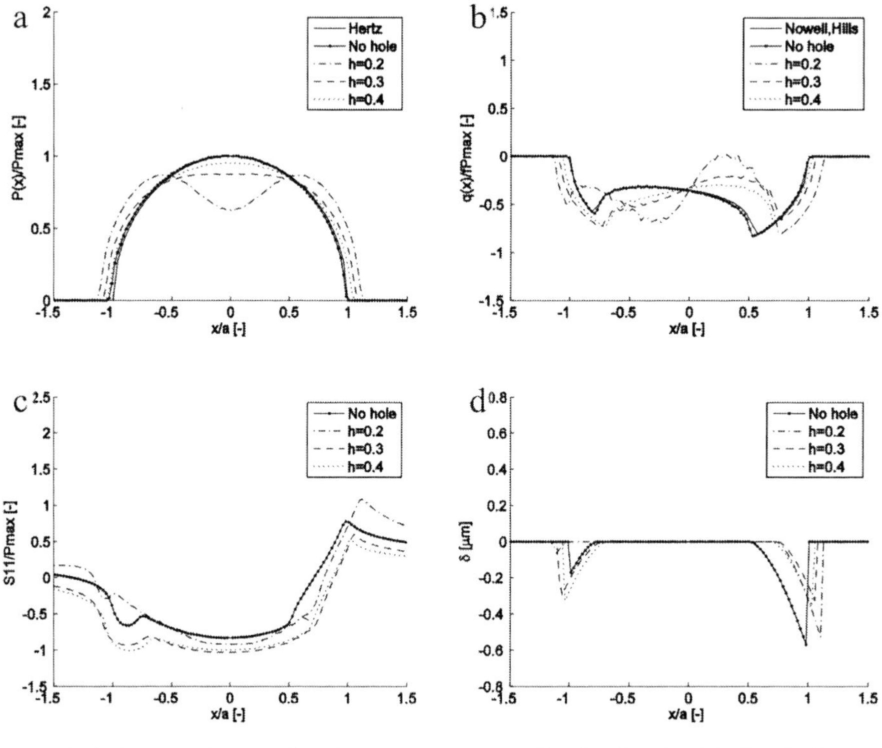

Figure 7. Variation of (a) normalized contact pressure distribution (b) frictional shear stress (c) normalized axial stress (d) relative slip amplitude along the contact surface between the pad and the specimen.

The second comparison is done by the fixing distance between center of void and contact surface and changing its radii which are chosen as 0.05, 0.1 and 0.15mm. The von Mises stress distribution for no hole case and hole radii of 0.05, 0.1 and 0.15mm are shown in figure 8a, 8b, 8c and 8d respectively.

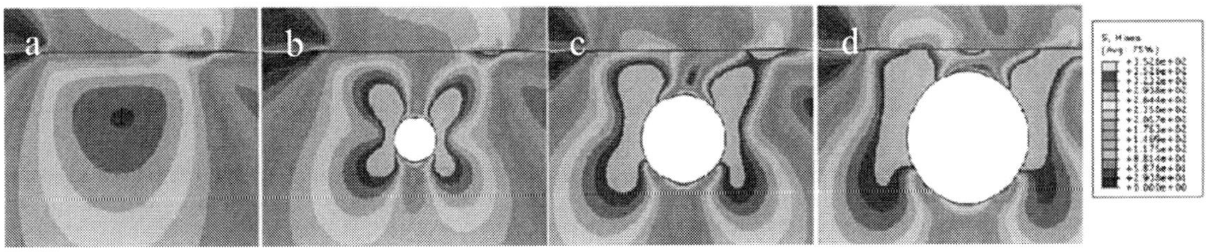

Figure 8. Von Mises stress contours (a) no hole (b) hole radius is equal to 0.05 (c) hole radius is equal to 0.1 (d) hole radius is equal to 0.15

It is seen that contact pressure drops suddenly at the center of the contact region when the void radii increases (Figure 9a). Furthermore, shear traction distribution changes dramatically by changing the void radii (Figure 9b). The maximum value of the tangential stress is also increases (Figure 9c). The stick/slip relation in the contact area depends on both the contact pressure and shear traction distribution. The decreases in the contact pressure at the center of the contact area may leads to slip in this region. When the void radius takes its maximum value relative slip occurs at the center of the contact are (Figure 9d).

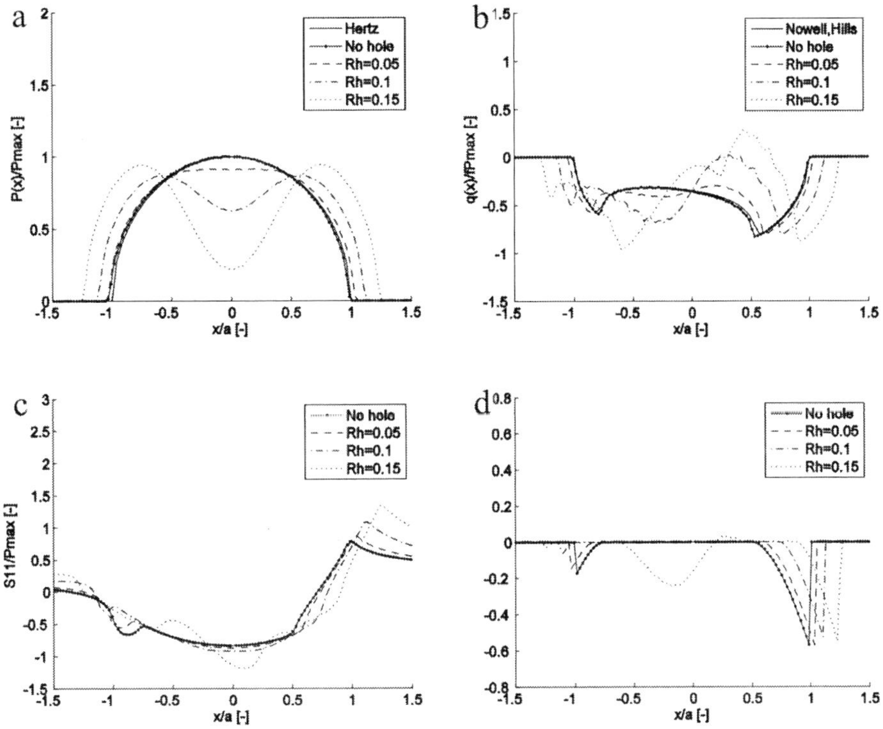

Figure 9. Variation of (a) normalized contact pressure distribution (b) frictional shear stress (c) normalized tangential stress (d) relative slip amplitude along the contact surface between the pad and the specimen.

In the next case, the effect of the multiple voids are considered. An array of 3 by 5 circular voids with radius 0.1 mm is located next to the contact plane in the substrate as shown in Figure 10. The von

Mises stress distribution for no hole case, single hole and multiple holes cases are shown in the figure 11a, 11b, 11c and 11d respectively.

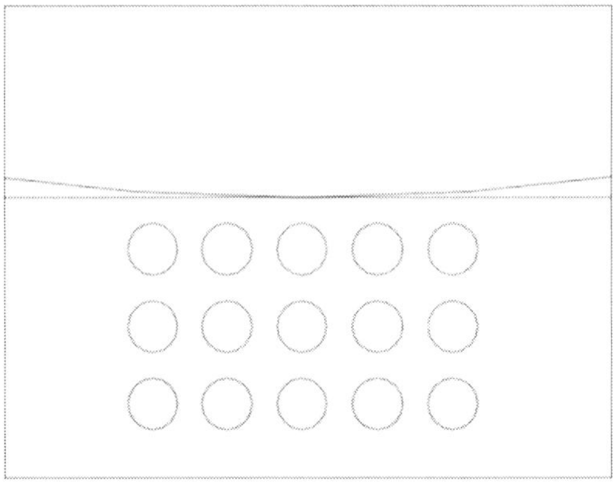

Figure 10. Substrate with an array of voids.

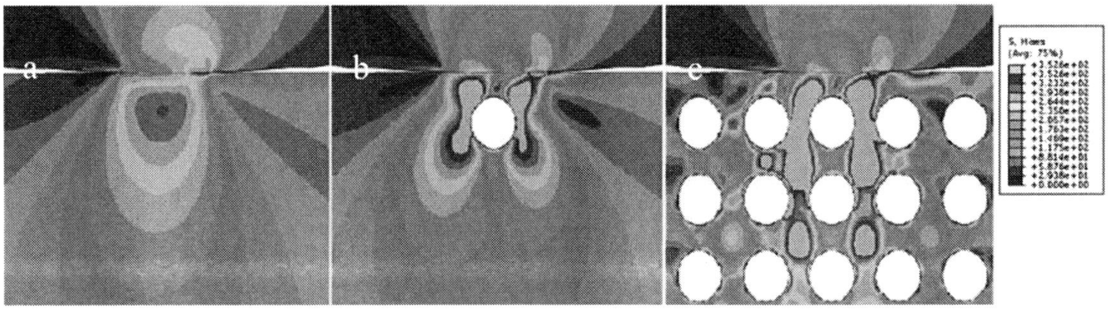

Figure 11. Von Mises stress contours (a) no hole (b) single hole (c) multiple holes

Figures 12a, 12b show that the maximum value of the contact pressure and the shear traction distributions are similar when the single and multi-void results are compared. However, it is seen that contact region increases in the multi-void case (Figure 12a). The effects of the multi voids becomes apparent in terms of maximum tangential stress values and relative slip magnitudes. The maximum value of the tangential stress decreases in the trailing edge of the contact area in the multi voids case (Figure 9c). In contrast, the relative slip amplitude increases (Figure 9d).

5. Conclusions

In this work, fretting contact conditions between a cylindrical pad and a flat specimen are examined for nonhomogeneous materials using finite element analysis. The summary of this work is as follows,

- The effect of the distance between contact region and the center of void is examined. According to the results, both contact pressure and relative slip amplitudes, which is important for the crack nucleation, decreases with increasing the diameter of hole size.

- The maximum value of the tangential stress rises with increasing with hole size. Furthermore, slip takes place at the contact center when these increments are severe.

- The multiple hole case is also examined. The general behaviour of the contact in terms of pressure, shear traction tangential stress distribution are similar for specimens consist of single hole and multiple holes. On the other hand, maximum value of the relative slip amplitudes of the specimen with multiple holes is higher than single hole case.

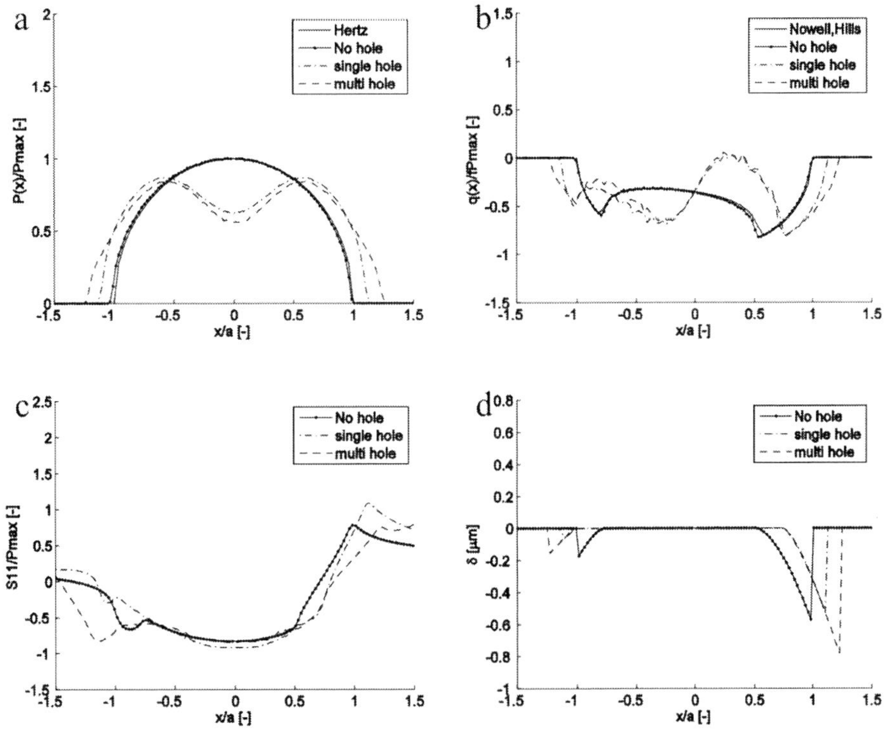

Figure 12. Variation of (a) normalized contact pressure distribution (b) frictional shear stress (c) normalized tangential stress (d) relative slip amplitude along the contact surface between the pad and the specimen.

References

1. Waterhouse, R. B., 1992. Fretting fatigue. The Institute of Materials and ASM International, 77-97.
2. Hertz., H., 1881. On the contact of elastic solids. 156-171.
3. Mindlin, R. and Deresiewica, H., 1953. Elastic spheres in contact under varying oblique forces. Journal of applied mechanics, 20.
4. Nowell, D., and Hills, D. A., 1986. Mechanics of fretting fatigue tests. International Journal of Mechanical Sciences 29(355-365).
5. Nowell, D., 1988. An analysis of fretting fatigue. Thesis (Ph. D.).
6. Ruiz, C., Boddington, P., and Chen, K., 1984. An investigation of fatigue and fretting in a dovetail joint. Experimental Mechanics, 24(3):208 217.
7. Stover, R. J., Mabie, H. H., and Furey, M. J., 1985. A Finite Element Investigation of a Bearing/Cartridge Interface for a Fretting Corrosion Study. Journal of Tribology,107(2), 157.
8. Lee, H., Jin, O., and Mall, S., 2003. Fretting fatigue behavior of Ti-6Al-4V with dissimilar mating materials. International Journal of Fatigue, 393-402.
9. Giner, E., Sukumar, N., Denia, F., & Fuenmayor, F., 2008. Extended finite element method for fretting fatigue crack propagation. International Journal of Solids and Structures, 45(22-23), 5675-5687.
10. Kim, S. H., 2011. Two dimensional and three dimensional finite element analysis of finite contact width on fretting fatigue. Materials Transactions, 147-154.

11. Korkmaz, Y., and Coker, D., 2017. Finite element analysis of fretting contact for dissimilar and nonhomogeneous materials. International Conference of Structural Integrity.
12. Nicholas, T., 2006. High Cycle Fatigue: A Mechanics of Materials Perspective. Elsevier.
13. Hojjati Talemi, R., 2014. Numerical Modelling Techniques for Fretting Fatigue Crack Initiation and Propagation (Unpublished master's thesis). Ghent University.
14. Hojjati-Talemi, R., Wahab, M. A., Pauw, J. D., and Baets, P. D., 2014. Prediction of fretting fatigue crack initiation and propagation lifetime for cylindrical contact configuration. Tribology International, 76, 73-91.
15. ABAQUS (2014) `ABAQUS Documentation', Dassault Systèmes, Providence, RI, USA.
16. Murakami, Y. and Endo, M., 1994. Effects of defects, inclusions and inhomogeneities on fatigue strength. International Journal of Fatigue, 16(3):163–182.
17. Kumar, D., Biswas, R., Hien Poh, L., and Wahab, M. A., 2017. Fretting fatigue stress analysis in heterogeneous material using direct simulations in solid mechanics. Tribology International, 109, 124-132.

Multidisciplinary Investigations Regarding the Wear of Machine Tools Operating Into the Soil

Cardei P.[1]***, Vladutoiu L. C.***[2]***, Gheorghe G.***[3]***, Fechete T.L.V.***[4]***, Chisiu G.***[5]
[1]*INMA, IT Department, 71592, Bucharest, Romania*
[2] *INMA, Testing Department, 71592, Bucharest, Romania*
[3]*INMA, Testing Department, 71592, Bucharest, Romania*
[4]*Technical University of Cluj-Napoca, Cluj-Napoca, Romania*
[5]*University POLITEHNICA of Bucharest, Faculty of Mechanical Engineering and Mechatronics, Bucharest, Romania*
petru_cardei@yahoo.com[1],
laurentiuvladutoiu82@gmail.com[2],gabrielvalentinghe@yahoo.com[3],
lucian.fechete@auto.utcluj.ro,[4]georgiana_bosoi@yahoo.com[5]

Abstract: The paper presents the results obtained by the authors in investigating the problem of wear of work organs of machines working in continuous interaction with the soil. The phenomenon of the interaction of the tools of agricultural machinery for ploughing, and the soil, is a complex of phenomena, one of the most difficult to model. Among the phenomena involved in this interaction, friction and wear (of many types) are the most important. We did not take into account the chemical wear, and by the wear caused by weather conditions. Research has focused on formulating a theory that has more than a descriptive character, for it be used for application purposes. For this we used classical theoretical models, mathematical models based on the theory of continuous bodies, theory of flow of fluids around the profiles, as well as other theories, approached or not, in an attempt to solve as satisfactorily the issue of the wear, for the tools of the agricultural machines for the tillage. We also sought to highlight the fact that wear is a phenomenon on a micro and macro-scale scale, and its generating causes must ultimately be related to observable effects, on the macro-structural scale.

Keywords: wear, friction, soil, tillage

1. Introduction

The physical phenomenon called wear is a very complex phenomenon, therefore difficult to define.In fact, a definition in a strict sense is impossible to give for any notion, because sooner or later it leads to a vicious circle.However, we accept definitions as many words that intuitively suggest the range of phenomena we refer to.For example, in [1], [2], [3], [4], wear is associated with notions: damage, eroding, scrape, scrape, scratch, friction, fret, wear. An engineering definition of the wear phenomenon we find in [5]: "Removal of material due to mechanical process under conditions of sliding, rolling, or repeated impact. Included are abrasive wear, fatigue wear and adhesive wear, but not the corrosive and thermal wear."In [8], wear is defined as the detachment of superficial layers of a solid body due to mechanical actions (friction, erosion, etc.) or chemical (oxidative corrosion, etc.) or to both kinds of actions.

There is no need to emphasize once again the special importance of studying the phenomenon of material wear. We limit ourselves to making some clarifications relative to the limited scope that we are referring to in this article. For agricultural machines for soil works, the wear of work organs is of particular importance.First of all, they must be designed to meet agro-technical requirements, but at the same time to reduce friction with soil and, implicitly, wear of organs.Secondly, the user of such a machine has to control the process parameters, having among the main objectives to reduce or maintain the value as close as possible to the minimum, for the forward-pull force. The draft (whose growth also involves increasing energy consumption and eventual qualitative losses in soil processing) depends heavily on the width of the attack board of the tools, and on the settlement angle of the work organ, [6]. All the investigations that we will make next, relate to a chisel plow or scarifiers. A typical example is drawn in figure 1.

Figure 1. Typical tools for chisel plow or scarifier: geometrical model (a, b), new chisel head (c), chisel head worn and loaded by welding.

Obviously, in order to estimate wear, we need to quantify this magnitude. According to [8], the wear intensity is expressed by the ratio between a parameter that characterizes the wear (e.g. the weight of the loose layer) and a parameter that characterizes the conditions in which the wear was produced.

Models and methods for approach with the wear problem of agricultural machinery used for soil tillage

The five subchapters of this chapter are all theoretical or theoretical-empirical methods involved in solving the problem of the wear of the tools of the agricultural machinery for soil tillage. Not all of them lead to a final estimate of wear, some of them just help others methods to get some information. This information is then used to develop the final wear estimation formula.

1.1 Wear of tools of agricultural machinery for soil tillage

The tools of agricultural machinery for soil tillage are: plow mouldboards, shares and knives, chisels, arrows and other types of tillage tools. These tools-are made up of a solid body in the category of knives, caught on the supports of the tools. These work bodies wear out by friction with soil particles and with the harsh components of the soil.

The main wear of the chisel body is manifested on the blade, especially at the top. The top of the blade evolves from a width of between 0.5 and 1 mm (at the fabrication, see figure 1, c), [9] up to a width of 2-3 mm or more, (see figure 1, d) until it is diagnosed as worn and is sharpened or replaced. Accordingly, the mass of the removed material increases. A more complex measure of wear is obtained by dividing the mass of the loss material to a product of sizes involved in describing the aggressiveness of the phenomenon: the length of the road travelled, the aggressiveness of the soil, the working speed, etc. In [9] it is also specified for such work organs, besides the condition on the width of the blade, the condition of not having cracks (especially in the blade top area). It is known that cracks are the causes of local or global damage, favouring rapid organ wear.

The users of agricultural machines equipped with tools of this type have a fairly simple way to estimate wear. They can be based on measuring the traction force in a standard polygon, where the plow was also tested on the first use of the work organs. A wear criterion is, for example, that a set of tools is considered to be worn when it increases by more than a certain percentage the resistance force opposite the tractor. The test can be done for the entire machine or separately for each work piece.

1.2 A list of parameters that characterize the wear phenomenon

In this paper, the wear phenomenon intervenes only through the action of the soil on the tools. So the wear is produced by rubbing a solid (work organ) with a material that can be considered solid, or, if necessary, another type of material (a special fluid, as will be seen in the following chapters).In these circumstances, among the main parameters that are considered for investigating this phenomenon, we considered:

Table 1. List of parameters involved in the wear process.

Parameter	Notation	Unit
The roughness of the metal surface of the tools	R_a, R_t	μm
The hardness of the tools	Br	
The yield strength of the tools material	σ_Y	Pa
Ultimate tensile strength of the tools material	σ_R	Pa
Fatigue strength of the tools material	σ_{Nf}	Pa
Shear strength of the tools material	τ_Y	Pa
Ultimate shear strength of the tools material	τ_R	Pa
Density of the tools material	ρ_w	$Kg \cdot m^{-3}$
Tools width	b	m
The width of the cutting edge of the workpiece	h	m
Working depth	a	m
Settlement angle	α	degree
Soil density	ρ_s	$Kg \cdot m^{-3}$
Working speed	v	$m \cdot s^{-1}$
The working width of the tillage machine	B	m

1.3 Wear and erosion phenomenon

Among the words that appear in the definition of notion of wear are erosion. This mutual positioning of the two notions is not accidental. It refers to the phenomenon of erosion of the slopes under the action of rain.

In order to open a new investigation path, it is enough to recall Wischmeier's formula [7] for estimating pluvial erosion. This equation is called Universal Soil Loss Equation and have the next form:

$$A = RKLSCP \tag{1}$$

where A the long-term average annual soil loss, R is the rainfall erosivity, K is the soil erodibility, L and S are topographic factors and C and P are cropping management factors.

By similarity we could replace formula (1) with formula (2):

$$U = R_s K_o L_o S_o C_o P_o \tag{2}$$

where U is a measure of wear, $R_s$is a measure of soil abrasion, K_o is a measure of the vulnerability to wear of the tools material, L_o and $S_o$are factors that represent the influence of the tools geometry, C_o and P_o may be factors indicating the processing, possibly initial treatments of the body.

1.4 Simulation of wear of the tools using structural models

Structural patterns of material wear phenomena have been used for some time.You can imagine in immense amount the structural models of the wear phenomenon.The selection of these is made by the extent to which they are applicative and not merely descriptive.This means that it is possible to simulate a phenomenon with the help of the structural model, in order to achieve concrete goals: behaviour prediction in difficult experimental conditions or optimal calculus.We used structural models to simulate the interaction between the tools of tillage machines and the soil.In all the elaborated structural models, the tools were modelled as a linear-elastic or rigid material.The soil, however, to be consistent with the most elemental experimental findings, should be considered as a material, at least elastic-plastic, even viscous-elastic-plastic.We have created a structural model in which we assumed the soil as a high density fluid, and the tools, is a rigid body around which the fluid that simulate the soil moves.

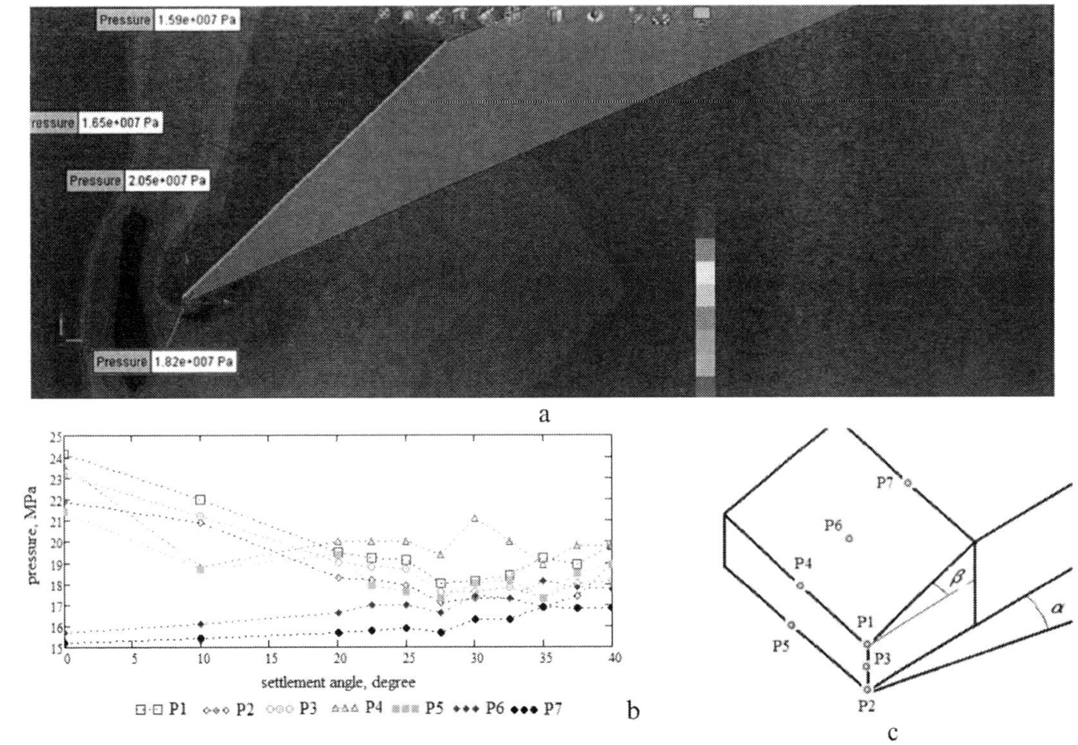

Figure 2. Structural model of motion of the tools in the soil modelled as rigid solid in the high density current flow and constant speed (2,7 m / s).

From the point of view of wear theory, the structural models are used to extract some necessary information for other mathematical models of the same phenomenon.The structural model in which soil is modelled as a fluid material is used to estimate soil pressures on the tools.This information is then used to model the reducing of the degree of roughness of the surface of these work organs, especially in the neighbourhood of edges and thin surfaces.The structural model in which the soil is considered to be a solid material, based on information from the fluid soil model (Figure 2, a), attempts to estimate what characteristic dimensions have those formations called "peaks" that can be irreversibly deformed or broken by pressure loads determined on the surface of the tools using the model with the soil modelled as fluid.

From the structural model in which the soil is modelled as a fluid, we take the pressure in the neighbourhood of the cutting edge of tools, assuming that the hydrostatic soil pressure as the fluid is given by the modulus of the soil elasticity, and the dynamic pressure is given by the product of the fluid density and the square of the movement speed of the organ or fluid.The pressure variation in seven points on the tools, according to the angle of settlement is shown in Figure 2, b.The location of the seven points on the tools is given in Figure 2, c. The pressure values so found are used to estimate the gauge of the metallic formations that can be irreversibly broken or deformed by the action of pressure.So, it is also be possible to approximate the intensity of the action of the abrasive bodies, which continues the wear beyond the appreciable reduction in roughness.

1.5 Roughness as a measure of wear

The initial roughness of a tools of a tillage machine is characterized by a series of statistical sizes. The roughness of the work surface changes according to the length of the trajectory travelled, the working speed and the aggressiveness of the soil (properly characterized). The acceptable wear limit in this particular case is related to the cutting edge of the tools, but also to the length of the inclined plane in the cutting area, measured along the organ.

In manufacturing, a roughness class is not required for the work organ.Through rapid wear, the tools reaches small values of the roughness class.For the wear of the tools of the agricultural machines for soil tillage, it is possible to use the roughness, drawing a curve that has as abscissa a measure of the performed work (e.g. the number of hectares worked), and on the ordinate, the average value of the roughness, in several points , or fixed measurement areas.It can be considered that the chisel body has, initially, the roughness class (Roughness ISO, [10]) N12 (R_a= 50 µm, R_t= 200 µm).For work organs studied in operation, on roughly 310 hectares worked in hard and sandy soil, the roughness reached class N6 (R_a= 0.8 µm, R_t= 4 µm). The shortening in length of the tools was approximately 70 mm (from 300 to 230 mm).Starting from these experimental findings, elementary calculations and simple considerations lead to the following form of wear dependence, R_a, on the worked surface, S:

$$R_a(S) = R_a^0 \exp(\frac{S}{S_1} \ln \frac{R_a^1}{R_a^0}) \quad (3)$$

where R_a^0 the initial roughness characteristic, and S_1 is the area for which the measured roughness indicates the value R_a^1. The wear function expressed in the roughness variation of the working surface can be expressed also in the mass of the current or lost material, using the density of the material and the geometric approximation of the material portion at the top of the organ that is lost by eroding (estimation is made only for the top of the tools). In the Figure 3 shows variations in the roughness and mass of the body as well as the mass of material lost through wear, depending on the worked surface.

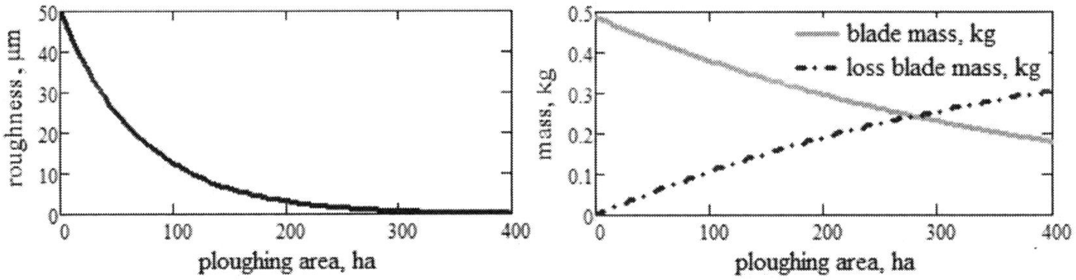

Figure 3. Graphical representations of the wear of the tools in roughness and themass (actual and lost), depending on the number of hectares worked.

If we annotate with s the length of the trajectory travelled by each tools, for surface processing S, we can accept the approximation:

$$s = \frac{S}{B} \quad (4)$$

then the roughness and mass functions of a tools can be written using as argument the length of the trajectory travelled:

$$R_a(s) = R_a^0 \exp(\frac{s}{s_1} \ln \frac{R_a^1}{R_a^0}) \quad (5)$$

and similar formula for current and lost weight.

The distance travelled by the machine and implicitly by the tools may be related to the mechanical work necessary to overcome the resistance to soil deformation and thereby the soil characteristics and the working speed. In this way can be obtained, also other interesting relationship for wear.To increase the accuracy of the empirical formula (1), we recommend as many experimental measurements as possible and the estimating the coefficients of this function by the least squares method.Also, on either the formula (1) or the more complicated forms, a dimensional analysis of the wear phenomenon (in the set of parameters in Table 1) is recommended. This analysis can lead to the determination of dimensional combinations with which new wear approximation functions can be formed.

2. Results

The results presented in this paper are included in the second part. Synthetically, these are:
-practical wear criteria for tools of tillage machinery,
-a list of parameters involved in various mathematical models of the surface wear process,
-the alternative of surface wear modelling using a pattern similar to the USLE model of slope erosion due to precipitation,
-information and explanations about the possibilities of using structural models in the partial or total description of the wear phenomenon,
-a description of the surface wear of tools of tillage machines, based on the variation of roughness and the connection with the formulation in terms of variation of the mass of the object subjected to wear.

3. Conclusions

In view of the results presented and the broad description, some important conclusions can be drawn for the continuation of the investigations and for the orientation, possibly preferential, to some of the directions described in the second chapter.
A first important conclusion is that the phenomenon of wear of surfaces (metallic or not) is a complex phenomenon, with factors of influence on both the micro scale and the macro scale.Moreover, the implication of statistical features is obvious. Even describing the surfaces in terms of roughness is a statistical one. Also the action of some of the factors that produce wear, has random character.
As a result of the first conclusion, an approach of the wear problem, using the methods of statistical physics, would be necessary. However, such an approach is very difficult and does not guarantee the solution of the practical problems of the wear phenomenon.
The results from subchapter 2.1 and 2.5 show that there are enough practical resources to formulate some phenomenological criteria (in terms of macrostructure) of wear.These criteria are used in practical applications. Many of these can be mathematically modelled after a reasonable amount of experiences. Subchapter 2.5 proves this statement.
The table with the factors involved in the surface wear process, from subchapter 2.2, shows that there are many parameters whose influence in this process is still waiting for a profound assessment.Generally, this investigation involves a great deal of experience. These are often very expensive. For this reason, we appreciate that important results in this direction will come at a slow pace.
The alternative to surface wear modelling through a formula of the type used in the classical theory of slope erosion due to rainfall is tempting. However, before performing ample experimental activity, it is necessary to give a complete definition to the variables involved in the equation (2).
The use of structural models in the surface wear study is useful for verifying hypotheses or for estimating the effects of some parameters in the process. However, the prospects of using these models are related to their experimental validation. Experimental validation is costly and this involves checking not only the results of these models but, to the extent possible, verifying working hypotheses.
Finally, in Chapter 2.5 we have shown that the roughness of the tools surface can give a measure of the wear, especially if it is supplemented with the variation in the length of the organ or the variation of the geometry of the cutting edge.
In essence, it can be said that applications that aim at optimizing or improving the wear process are only possible through a very extensive experimental activity. Even if we try to optimize some aspects of the wear problem by theoretical procedures, the models used need experimental validation.

Acknowledgements

This work was supported by Grant of the Romanian National Authority for Scientific Research, CNCS, UEFISCDI, PN-III-P2-2.1-BG-2016, Project number: 78BG/2016, SC Artecom SRL upgraded competences through deep soil tillage machinery optimization.

References

1. https://www.google.ro/?gws_rd=cr&ei=ysplWeX1LonA6QTchKG4Cg#q=wear+definition
2. Merriam –Webster Dictionary, https://www.merriam-webster.com/dictionary/wear
3. Cambridge Dictionary, http://dictionary.cambridge.org/dictionary/english/wear#translations
4. https://en.wikipedia.org/wiki/Wear
5. http://www.engineering-dictionary.org/MECHANICAL_WEAR
6. E.Y.H Bobobee, G. Gebresenbet, Effect of cutting edge thickness and state of wear of ploughshare on drought force and heart rates of Sanga oxen in Ghana, Soil & Tillage Research 95 (2007), 298-307, ELSEVIER.
7. W. H. Wischmeier, D.D. Smith, Predicting Rainfall-Erosion Losses from Cropland East of the Rocky Mountains, Guide for Selection of Practices for Soil and Water Conservation, Agricultural handbook, no. 282, 1965.
8. C. Iacob, S.I. Gheorghita, M. Soare, L. Dragos, Dictionar de mecanica, Editurastiintificasienciclopedica, Bucuresti, 1980.
9. V. Scripnic, P. Babiciu, Masiniagricole, Editura CERES, Bucuresti 1979.
10. https://en.wikipedia.org/wiki/Surface_roughness

Effects of Load and Speed on Wear Rate of Abrasive Wear for 2014 Al Alloy

Odabas D.
Erciyes University, Department of Mechanical Engineering, 38039, Kayseri, Turkey
odabasd@erciyes.edu.tr

Abstract: In this paper, the effects of the normal load and sliding speed on wear rate of two-body abrasive wear for 2014 Al Alloy were investigated in detail. In order to understand the variation in wear behaviour with load and speed, wear tests were carried out at a sliding distance of 11 m, a speed of 0.36 m/s, a duration of 30 s and loads in the range 3-11 N using 220 grit abrasive paper, and at a speed range 0.09-0.90 m/s, a load of 5 N and an average sliding distance of 11 m using abrasive papers of 150 grit size under dry friction conditions. Before the wear tests, solution treatment of the 2014 Al alloy was carried out at temperatures of 505 and 520 °C for 1 h in a muffle furnace and then quenched in cold water at 15 °C. Later, the ageing treatment was carried out at 185 °C for 8 h in the furnace. Generally, wear rate due to time increased linearly and linear wear resistance decreased with increasing loads. However, the wear rate was directly proportional to the load up to a critical load of 7 N. After this load, the slope of the curves decreased because the excessive deformation of the worn surface and the instability of the abrasive grains began to increase. When the load on an abrasive grain reaches a critical value, the groove width is about 0.17 of the abrasive grain diameter, and the abrasive grains begin to fail. The wear rate due to time increased slightly as the sliding speed increased in the range 0.09-0.90 m/s. The reason for this is that changes arising from strain rate and friction heating are expected with increasing sliding speeds.

Keywords: Abrasive wear, solution treatment, wear rate.

1.Introduction

Wear occurs in many different situations, e.g. on earth-moving equipment, agricultural ploughs, slurry pumps, crushers, ball bearings, plain bearings, piston rings, seals, gears, brakes or in daily life at knives, shoes, clothes, furniture, stairways or human joints at the hip, knee or elbow. Materials are worn most of all by abrasive wear mechanism. Abrasive wear is the displacement of material caused by the presence of hard particles, between or embedded in one or both of two surfaces,in relatively motion, or by the presence of hard asperities on one or both of the relatively maving surfaces. Therefore, abrasive wear classified as two-body and three-body abrasive wear mechanism [1]. It has been estimated that 50 % of all wear problems in industry are due to abrasion, and as such, much laboratory work has examined and sought to rationalise the abrasive wear behaviour of a wide range of materials [2]. Abrasive wear is experienced such as in feeders, belt conveyors, vibratory conveyors and screen conveyors. Any mechanical devices that involve the motion of bulk solids relative to surfaces will experience wear. The abrasive damage is occurred in aluminium alloy sheet used in truck's buckets under the action of sand friction resulting when transporting the sand in these trucks [3]. Aluminium alloys are known for their suitable mechanical properties, mainly due to their high strength-to-weight ratio [4]. Therefore, multiple experimental studies on abrasive wear of aluminium alloys have been conducted in recent decades due to the interest in the ever increasing use of light aluminium alloys and composites in engineering applications [5]. Aluminium is the most accepted matrix for metal matrix composites because of its low density, ability to be strengthened by precipitation hardening, very good corrosion resistance, high thermal conductivity and electrical conductivity, and its high damping capacity [6]. With hard particles dispersed in a relatively ductile matrix, aluminium-alloy matrix composites possess an ideal structure for wear-resistant materials; recent studies have revealed this concept. Wang et al. [7] reported that both SiC particulate and whisker-reinforced composites exhibited greater wear resistance than the unreinforced aluminium alloy. The whisker reinforcement was more effective than the particulate reinforcement in improving the wear resistance of the matrix alloy. Song et al. [8] studied changes in the abrasive wear resistance of aluminium-based composites heat treated for different ageing conditions. They found that the hardness and abrasive wear resistance increase with increasing the ageing temperature to 200 °C, but at 250 °C decrease due to the coarsening of the intermetallic precipitates. The composites with 20 μm SiC particles were slightly more wear resistant than those with 3 μm SiC. In another study, it was shown that most composites exhibit better abrasive wear resistance than the unreinforced matrix alloys, and the impact-abrasive resistance increases with decreasing particle size and increasing volume fraction [9]. In all the investigations discussed, abrasive wear behaviour of aluminium alloys has been carried out using pin-on disc or pin-on drum test machines under continuous sliding motion. Previous investigations were mostly focused on a single factor (Sliding distance, load or speed etc), and combined multiple factors have rarely been studied. In order to have a better understanding of the wear mechanism, studies on multiple factors need emphasis, for there may be synergistic effect among these factors, which was often neglected in previous studies. In this paper, the effect of the pv factor (product of contact pressure and sliding speed) on the abrasive wear of solution-treated and age-hardened 2014 Al alloy was investigated under dry frictional conditions during continuous wear in detail. The pv factor is a well-known index in the field of mechanical wear. It can be used to characterize the service life and wear rate of machine elements [10].

2. Determination of wear rate as a function of time

In a mechanical friction contact, where two-body abrasive wear occurs, the worn volume V (m^3) is proportional to the load F_N (N) and the sliding distance L (m), but inversely proportional to the hardness H (Nm^{-2}) of the softer surface. Thus leads to the well-known expression for abrasive wear volume [11-15]

$$V = \frac{k_a F_N L}{H_s} = \frac{G}{\gamma} \qquad (1)$$

where G (kg) is weight loss and γ (kgm^{-3}) is the density of material. V/L (m^3m^{-1}) is defined as the volumetric wear intensity. The wear resistance ε is calculated as ε=L/G (m/kg) or ε=L/V (m/m^3). The ratio k_a/H of the wear coefficient k_a and the hardness H_s (Nm^{-2}), of the softer contact surface is used here as a constant and the worn volume is related to the softer of the two material surfaces. Where k_a is called the wear coefficient for abrasive wear. The wear coefficient k_a is dimensionless. It is a principal value for a friction pair to describe its wear rate. The physical meaning of k_a is the wear volume fraction at the plastic contact zone, and it is strongly affected by the material properties and the geometry of the zone in compression and shearing. Where there is a frictional contact between a very hard and a much softer surface, the harder surface is always subject to wear, but the worn volume cannot be predicted quantitatively. For all possible tribo-conditions the ratio k_a/H_s gives no more practical information to describe wear than the simple K value which is more frequently used. K (m^2N^{-1}) is called the wear factor or specific wear rate. From equation (1), the wear factor of materials can be written as

$$K = \frac{k_a}{H_s} = \frac{V}{F_N L} \qquad (2)$$

Since the normal contact pressure with plastic deformation is almost equal to the hardness value H_s of the wearing material, the total real contact area can be expressed by

$$A_r = \frac{F_N}{H_s} \tag{3}$$

From equation (1) and equation (3), the abrasive wear volume of materials can be written as

$$V = k_a A_r L \tag{4}$$

The sliding distance is expressed as L=vxt. Where v (m s^{-1}) is sliding velocity and t (s) is duration of the wear process. The normal load F_N on the contact surface is expressed as F_N=pxA. Where p (Nm^{-2}) is the average contact pressure and A (m^2) is the nominal (apparent) contact area.

Equation (1) becomes

$$V = \frac{k_a F_N L}{H_s} = \frac{k_a (vt) F_N}{H_s} = \frac{k_a (vt) pA}{H_s} = \frac{k_a (pA) tv}{H_s} \tag{5}$$

The thickness h (m) of the wear linear amount can be expressed as

$$h = \frac{V}{A} = \frac{k_a (pv) t}{H_s} = \frac{G}{A\gamma} \tag{6}$$

h/L (mm^{-1}) is defined as the linear wear rate (intensity). If the wear depth h is taken as the basis upon which wear rate is defined, the variation in wear with the time can be expressed as λ=dh/dt [16]. In most calculations it is possible to assume a linear relationship between wear time t and the amount of wear h. In other word, it can be assumed that during the normal operating period, the wear rate remains constant. The variation in wear with the time can be expressed as λ=h/t. Thus, wear rate due to time can be expressed as

$$\lambda = \frac{h}{t} = \frac{k_a (pv)}{H_s} \tag{7}$$

where pv ($Nmm^{-2}s^{-1}$) is referred to a factor.

3. Experimental details

3.1. Material

In this study, 2014 Al all oy with higher Cu content was chosen as a test material. For this test material, chemical analysis (wt.%) was: Cu 4.18; Mn 0.67; Mg 0.72; Si 0.81; Fe 0.26; Zn 0.02 and Ti 0.019 [5]. In general, this alloy is used for heavy-duty forgings, aircraft fittings and truck frames. 2014 Al alloy was fabricated by easting in a metal mould at Seydişehir Aluminium Co. in Konya, Turkey. After casting, the homogenization heat treatment was applied to the alloy at 500 °C for 2 h and the hot-pressed billets with a 45 mm diameter were extruded at 450 °C into rods with a 16 mm diameter with an extruded ratio of 3:1. After the rods were left for cooling at the room temperature, they were straightened by a 3 % plastic deformation. The specimens were prepared from the rods for the hardness and abrasive wear tests. The specimens to be used for the wear tests were cylinders with a 16 mm diameter and a length of 20 mm while those for the hardness tests were blocks 2.5x8x5 mm in size. Before the wear tests, the specimens were solution treated in a muffte furnace and then quenched in cold water at 15 °C. Later, the ageing treatment was carried out at 185 °C for 8 h in the furnace. T6 heat-treatment conditions (solution heat treated, artificially aged) and the hardness values are given in Table 1. Figure 1 shows microstructure of solution treated 2014 Al alloy after ageing treatment.

Table 1. Solution heat treatment of 2014 Al alloy and final hardness values.

Solution heat treatment		Ageing treatment		After age hardening
Temperature (°C)	Time (h)	Temperature (°C)	Time (h)	Final hardness HB_{30} ($Kg.mm^{-2}$)
505	1	185	8	150
520	1	185	8	161

3.2. Test methods

Abrasive wear tests in continuous motion were carried out using a grinding-polishing machine with the pin-on disc configuration under dry sliding conditions [Fig.2]. The disc with 200 mm diameter was driven by an electric motor. The speed of the electric motor is changed by means of an A.C. speed-control converter. The waterproof SiC papers on the rotating disc were used as abrasive materials. The pin specimen (stationary) was mounted in a holder so that it could be removed for measurement and was rubbed against the surface of a SiC paper fixed to the fiat surface of a revolving disc. Friction occurred continuously along a single track over which the abrasive ability gradually deteriorated. The radius of a single track was 35 mm. Thus, the average periphery of this track is equal to $2\pi r=2\pi 35=220$ mm and nominal contact area is equal to 2.01×10^{-4} m^2. Wear was determined over a friction path of 11 m during the wear test.

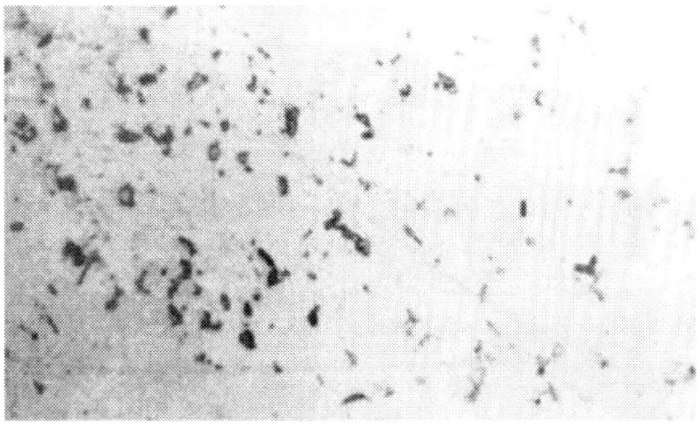

(a)Solution-treating temperature of 505 °C (x500)

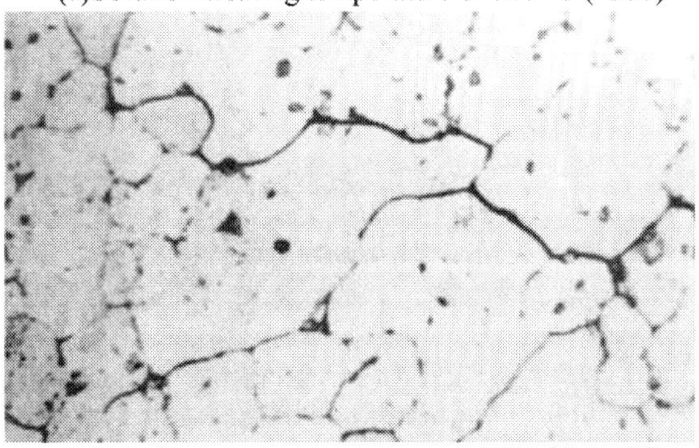

(b)Solution-treating temperature of 520 °C (x500)

Figure 1. Microstructure of solution treated 2014 Al alloy after ageing treatment (Etched in Keller's reagent)

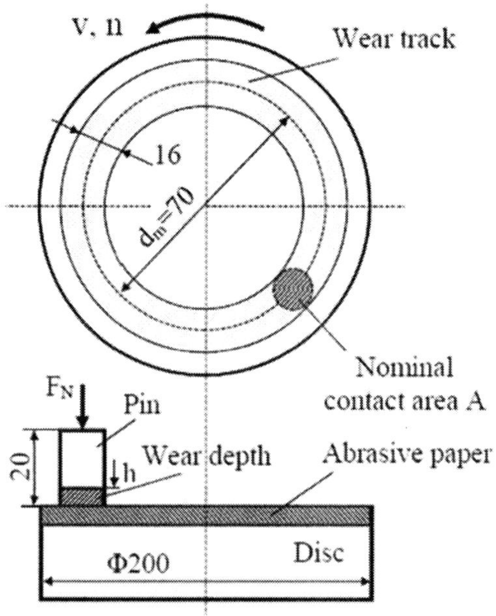

Figure 2. Schematic test rig of pin-on disc

Before the wear tests, the contact surfaces of the pins were polished with 500 grit size SiC paper and the initial average roughness was approximately R_a= 0.06 um. Prior to testing, all the specimens were cleaned by rinsing in ethanol and dried in hat air. During the tests, the pin specimens were loaded with dead weights and tests were conducted dry, at room temperature and room humidity. The wear lass was measured by means of an analytical balance with an accuracy of 0.1 mg. The wear surfaces, wear debris and subsurface deformation were also studied in optical and scanning electron microscopes. The density γ of test material for a 2014 Al alloy is 2800 kgm^{-3}.

4. Results

4.1. The effects of load on the abrasive wear behaviour of 2014 Al alloy

In order to understand the variation in wear behaviour with load for the continuous motion, wear tests were carried out at a sliding distance of 11 m, a speed of 0.36 m/s, a duration of 30 s and loads in the range 3-11 N using 220 grit abrasive paper under dry friction conditions. The specimen solution treated at 520 °C for 1 h were tested. The results were given in Table 2.

Table 2. The influence of the load on abrasive wear of 2014 Al alloy.

Load F_N (N)	p (Nm^{-2}) $x10^{-4}$	pv ($N.m^{-1}s^{-1}$) $x10^{-5}$	Weight loss G (kg) $x10^{-6}$	Wear volume V (m^3) $x10^{-9}$	Wear depth h (m) $x10^{-6}$	Wear rate λ (ms^{-1}) $x10^{-7}$	Linear wear rate h/L (mm^{-1}) $x10^{-7}$	Wear resistance ε (mm^{-1}) $x10^{9}$
3	1.492	5.371	3.5	1.25	6.218	2.072	5.652	17.692
5	2.487	8.953	6.7	2.392	11.90	3.966	10.818	9.243
7	3.482	12.535	11.2	4.00	19.90	6.633	18.090	5.527
9	4.477	16.117	13	4.642	23.094	7.698	20.994	4.763
11	5.472	19.699	16.2	5.785	28.781	9.593	26.164	3.822

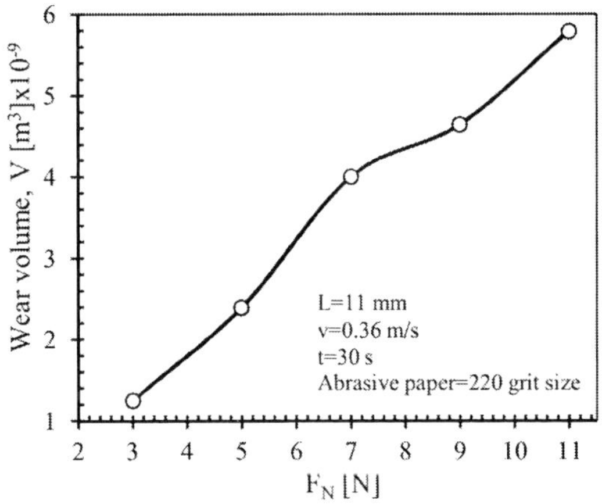

Figure 3. The variation in wear volume of 2014 Al alloy with load.

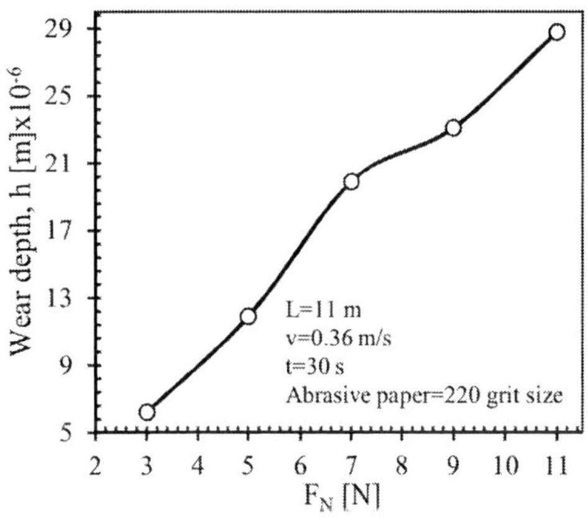

Figure 4. The variation in wear depth with load.

Figure 3 shows the relationship between the wear volume and the load. The wear volume increases linearly with an increase in load. The polynomial function can be used to propose the relationships between wear volume and load. The rough regression formula can be expressed as wear volume $V=(0.0172(F_N)^2+0.807F_N-1.054)x10^{-9}$ m^3. The correlation coefficient is defined as R % 99. Generally, wear depth increases linearly with increasing loads as shown in Figure 4. The regression formula of wear depth can be expressed as $h=[(0.085(F_N)^2 + 4.0157F_N - 5.244)]x10^{-6}$ (m). This polynomial function has a correlation coefficient R of % 99. However, the wear volume and wear depth was directly proportional to the load up to a critical load of 7 N. After this load, the slope of the curves decreases because the excessive deformation of the worn surface and the instability of the abrasive grains began to increase. When the load on an abrasive grain reaches a critical value, the groove width is about 0.17 of the abrasive grain diameter, and the abrasive grains begin to fail. Thus, groove widths are fairly constant with load but the number of contact points increases linearly with load. If grit wear can be eliminated, the wear volume may be directly proportion al to the applied load.

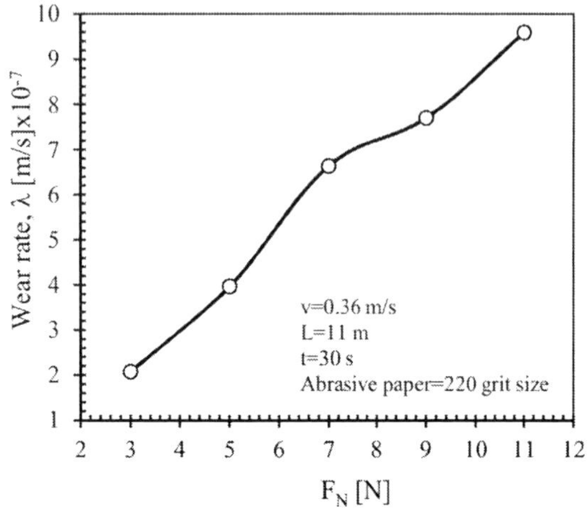

Figure 5. The variation in wear rate with load

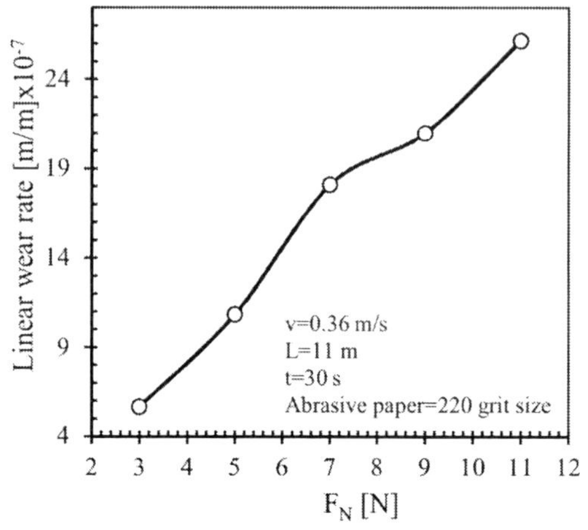

Figure 6. The variation in linear wear rate with load.

Figure 5 and 6 show the dependence of wear rate due to time and linear wear rate on the load. Generally, wear rate due to time and linear wear rate increase linearly with increasing loads as shown in these figures. The polynomial function can be used to propose the relationships between wear rate due to time and load. The regression formula of wear rate due time is expressed as $\lambda=(0.028(F_N)^2+1.338F_N1.749)x10^{-7}$ (ms^{-1}). This function has a correlation coefficient R of % 99. The polynomial function can be used to propose the relationships between linear wear rate and load. The regression formula of linear wear rate is expressed as $h/L=(-0.077(F_N)^2+3.65F_N-4.768)x10^{-7}$ (mm^1). This function has a correlation coefficient R of % 99.

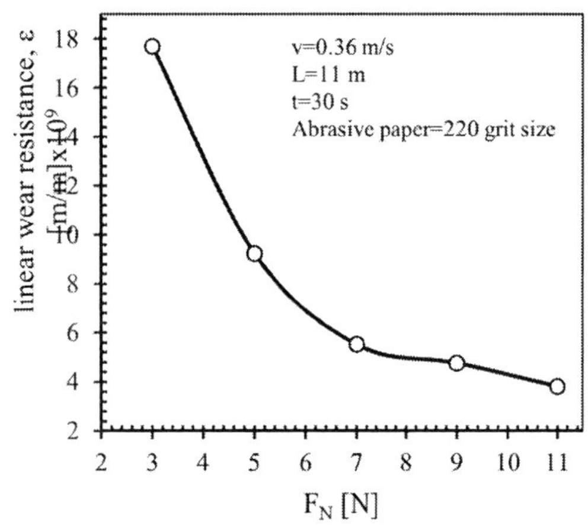

Figure 7. The variation in linear wear resistance with load.

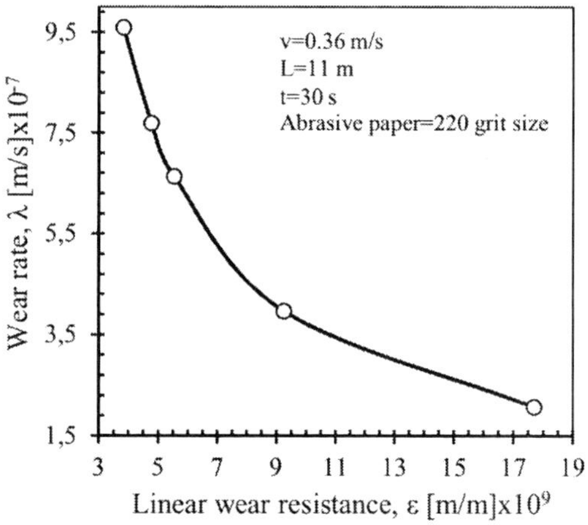

Figure 8. The variation in linear wear rate with linear wear resistance.

Figure 7 shows the relationship between the linear wear resistance and the load. The linear wear resistance decreases with an increase in load. The exponential function can be used to propose the relationships between linear wear resistance and load. The regression formula of linear wear resistance rate is expressed as $\varepsilon=(62.88xF_N^{-1.189})x10^9$ mm^{-1}.This function has a correlation coefficient R of % 98. Figure 8 shows the relationship between the wear rate due time and the load. The wear rate due to time decreases with an increase in linear wear resistance. The regression formula of wear rate due time is expressed as $\lambda=[36.67x\varepsilon^{-1}]x10^{-7}$ ms^{-1}. This exponential function has a correlation coefficient R of % 100. Figure 9 shows the relationship between the wear rate due time and linear wear rate. The wear rate due to time increases linearly with an increase in linear wear rate. The regression formula of wear rate due time is expressed as λ=[0.366(h/L)0.0005)]x10^{-7} ms^{-1}. This linear function has a correlation coefficient R of % 98.

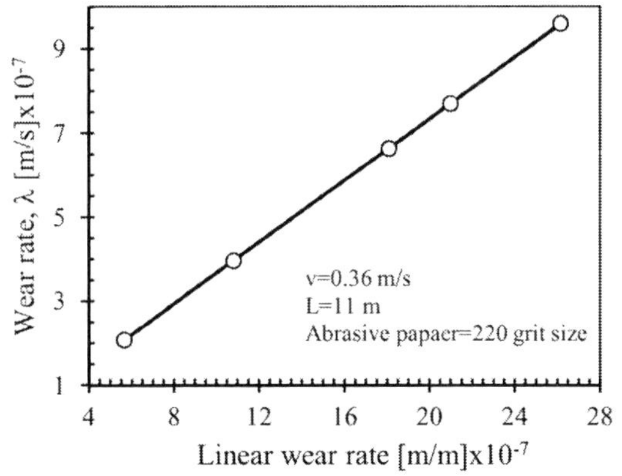

Figure 9. The variation in wear rate with linear wear rate.

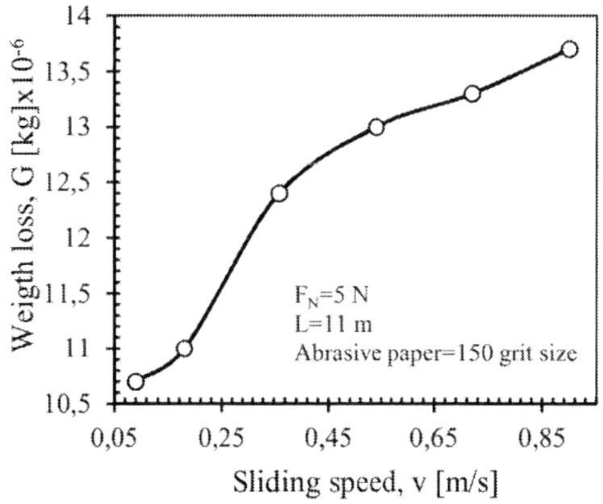

Figure 10. The variation in weigth loss with speed.

4.2. The effect of the sliding speed on the abrasive wear behaviour of 2014 Al alloy

In this section, in order to compare the variation in the wear depth of the 2014 Al alloy with sliding speed for the continuous motions, and the abrasive wear tests were performed at a speed range 0.09-0.90 m/s, a load of 5 N and an average sliding distance of 11 m using abrasive papers of 150 grit size. A test material solution treated at 505 °C for 1 h was used during the wear tests. The results are given in Table 3.

Table 3. The influence of the speed on abrasive wear of 2014 Al alloy.

Speed v (ms^{-1})	Time t (s)	pv [Nm/m^2 s] x10^{-5}	Weight loss G (kg) x10^{-6}	Wear volume V (m^3) x10^{-9}	Wear depth h (m) x10^{-6}	Wear rate λ (ms^{-1}) x10^{-7}	Linear wear rate h/L (mm^{-1}) x10^{-7}	Wear resistance ε (mm^{-1}) x10^{9}
0.09	122	2.238	10.7	3.821	19	1.557	17.272	5.789
0.18	61	4.476	11	3.928	19.542	3.203	17.765	5.629
0.36	30	8.953	12.4	4.428	22.029	7.343	20.026	4.993
0.54	20	13.429	13	4.642	23.039	11.519	20.944	4.774
0.72	15	17.906	13.3	4.750	23.631	15.519	21.482	4.655
0.90	12	22.383	13.7	4.892	24.338	20.281	22.125	4.519

Figure 10 shows the relationship between the weight loss and the speed. The weight loss volume increases slightly with an increase in speed. The reason for this is that changes arising from strain rate and friction heating are expected with increasing sliding speeds. The polynomial function can be used to propose the relationships between weight loss and speed. The rough regression formula can be expressed as wear volume G=(-4.188xv^2 +7.893xv +9.929)x10^{-6} (kg). The correlation coefficient is defined as R % 98. The wear volume increases slightly with an increase in speed as shown in Figure 11. The polynomial function can be used to propose the relationships between wear volume and speed. The regression formula is expressed as wear volume V=(-1.496xv^2+2.819xv+3.545)x10^{-9} (m^3). This polynomial function has a correlation coefficient R of % 98. Generally, wear depth increases linearly with increasing speeds as shown in Figure 12. The regression formula of wear depth can be expressed as h=[-7.295v^2 +13.879v +17.653]x10^{-6} (m). This polynomial function has a correlation coefficient R of % 98. In addition, the general pattern of the variation graphs in Figures 10, 11, 12 and 14 are similar.

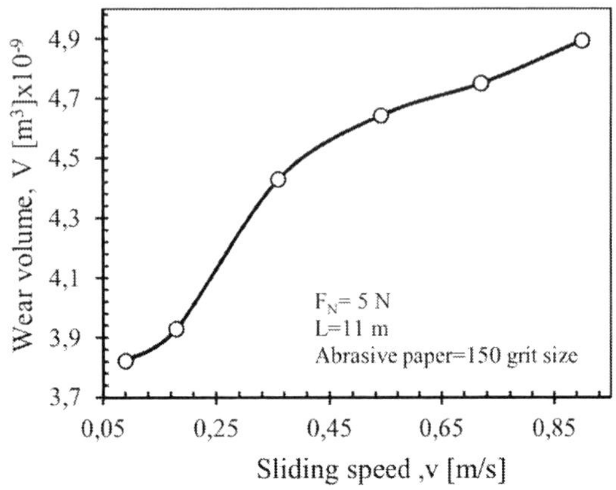

Figure 11. The variation in wear volume loss with speed.

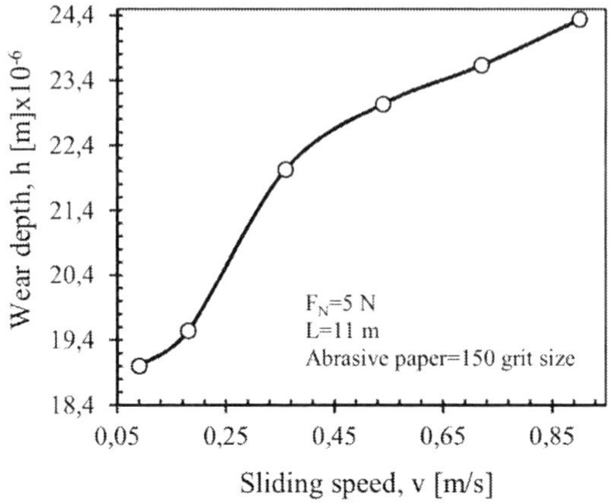

Figure 12. The variation in wear depth with speed.

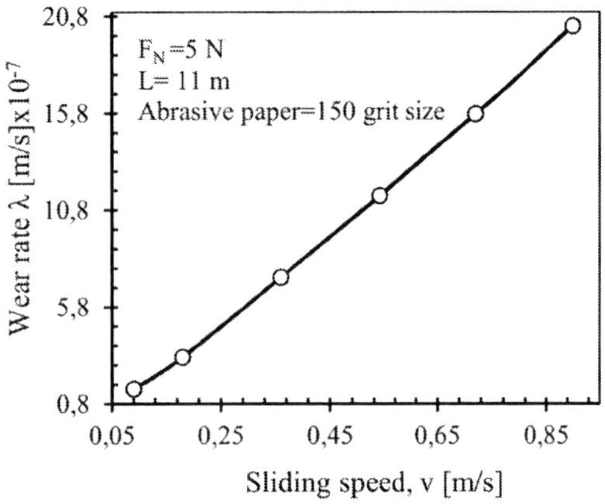

Figure 13. The variation in wear rate with speed.

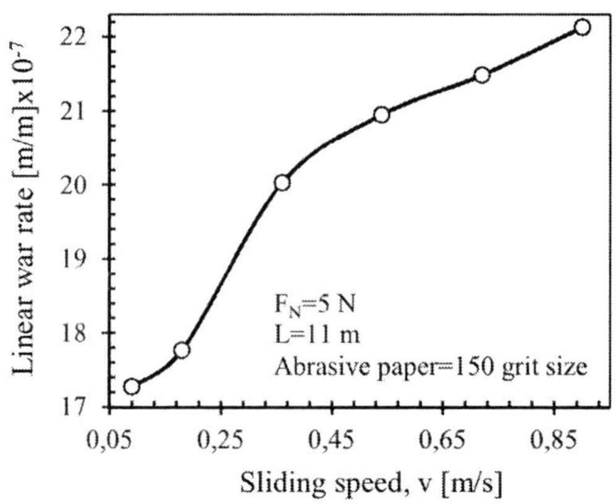

Figure 14. The variation in linear wear rate with speed

Figure 13 and 14 show the dependence of wear rate due to time and linear wear rate on the speed. Generally, wear rate due to time and linear wear rate increase linearly with increasing speeds as shown in these figures. The linear function can be used to propose the relationships between wear rate due to time and speed. The regression formula of wear rate due time is expressed as λ= (23.208v^2-0.848) x10^7 (ms^{-1}). This function has a correlation coefficient R of % 99. The polynomial function can be used to propose the relationships between linear wear rate and speed. The regression formula of linear wear rate is expressed as h/L=(-6.632$(v)^2$+12.618v+16.047) x10^{-7} (mm^1). This function has a correlation coefficient R of % 98. Figure 15 shows the relationship between the linear wear resistance and the speed. The linear wear resistance decreases with an increase in speed. The polynomial function can be used to propose the relationships between linear wear resistance and speed. The regression formula of linear wear resistance rate is expressed as ε=(2.129xv^2-3.676v+6.134) x10^9 (mm^{-1}). This function has a correlation coefficient R of % 98. Figure 16 shows the relationship between the wear rate due time and the wear resistance. The wear rate due to time decreases with an increase in linear wear resistance. The regression formula of wear rate due time is expressed as λ=[84331x$e^{-1.853\varepsilon}$]x10^{-7} (ms^{-1}). This exponential function has a correlation coefficient R of % 97. The general pattern of the variation graphs in Figures 15 and 16 are similar.

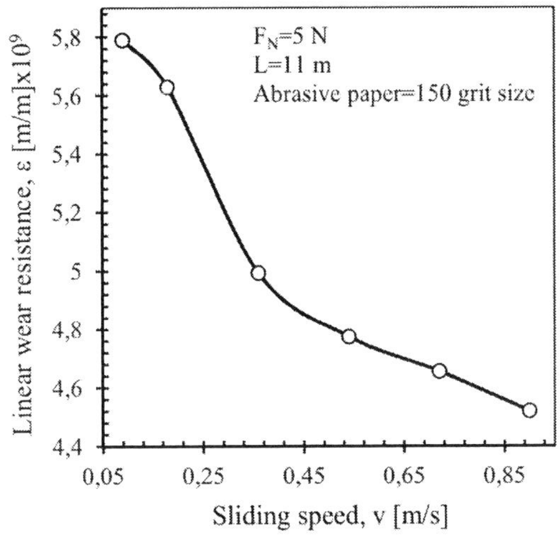

Figure 15. The variation in linear wear resistance with speed.

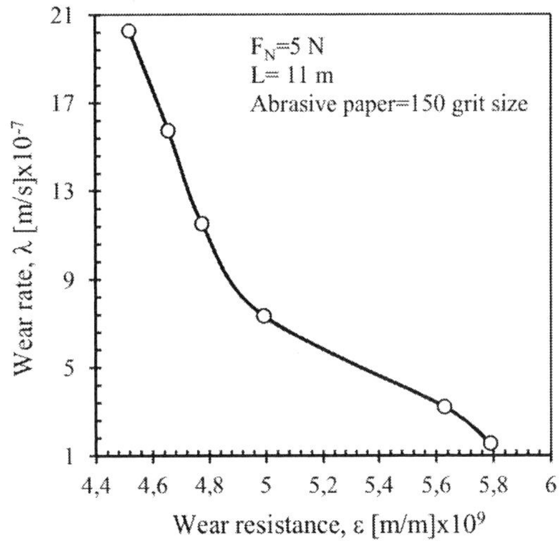

Figure 16. The variation in wear rate with wear resistance.

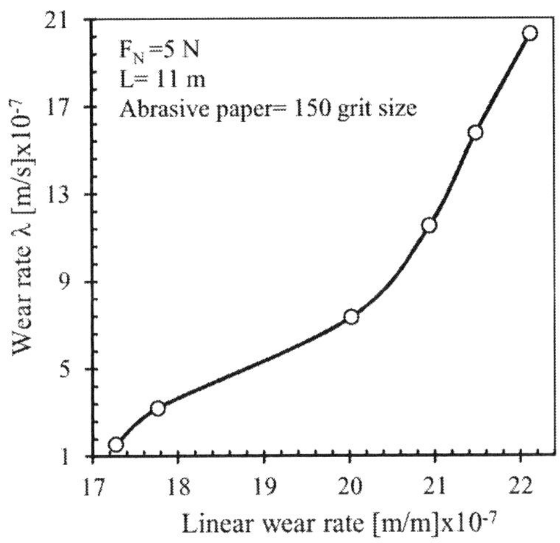

Figure 17. The variation in wear rate with linear wear rate.

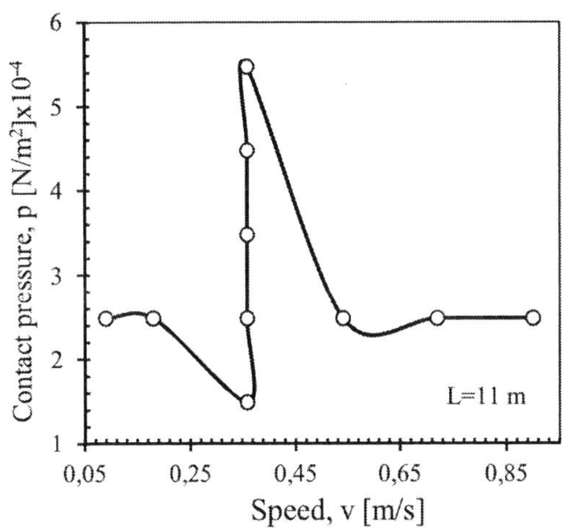

Figure 18. The relationships between contact pressure and speed during tests.

Figure 17 shows the relationship between the wear rate due time and the linear wear rate. The wear rate due to time decreases with an increase in linear wear rate. The regression formula of wear rate due time is expressed as $\lambda = [0.866(h/L)^2 - 30.448(h/L) + 269]x10^{-7}$ (ms^{-1}). This polynomial function has a correlation coefficient R of % 99.

4.3. The effects of the pv factor on the abrasive wear behaviour of 2014 Al alloy

The factor of pv is a well-known index in the field of mechanical wear. It can be used to characterize the service life and wear rate of machine. Hence, further study on it will be helpful to understand the wear mechanism of material. The rate of wear of materials is obviously a function of load and speed and the pv factor can be used as a design criterion [16]. The derivation of the pv factor is based on the reasonable assumption that the rate of wear will be proportional to the rate of energy dissipation at the sliding interface [15]. Starting from this assumption, it can be derived the relationship between wear rate and the pv factor for the pin-on disc configuration as shown in Figure 1. The amount of energy dissipated on a sliding distance L is given by $F_f L = \mu F_N L$. The rate of energy dissipation or frictional heat power is therefore given by $q=\mu F_N L/t=\mu pv$ ($Nm.m^{-2}s^{-1}$). Where F_N is normal load, v is the sliding speed at the interface and μ is the coefficient of sliding friction. With increasing pv factor, the frictional heat increases accordingly, if the friction coefficient may be fairly assumed to be constant, then pv factor can be a direct indicator of the frictional power dissipated at the interface, that is to say, frictional heat will be directly proportional to the pv factor.

Table 3. The influence of the pv on abrasive wear of 2014 Al alloy.

pv ($Nmm^{-2}s^{-1}$)x10^{-5}	Wear depth h (m)x10^{-6}	Wear rate λ (ms^{-1})x10^{-7}	Linear wear rate h/L (mm^{-1})x10^{-7}	Wear resistance ε=L/h (mm^{-1})x10^{9}
2.238	19.00	1.557	17.272	5.789
4.476	19.542	3.203	17.765	5.629
5.371	6.218	2.072	5.652	17.692
8.953	11.90	3.966	10.818	9.243
8.953	22.542	7.443	20.026	4.993
12.535	19.90	6.633	18.090	5.527
13.429	23.039	11.519	20.944	4.774
16.117	23.094	7.698	20.994	4.763
17.906	23.631	15.754	21.482	4.655
19.699	28.781	9.593	26.164	3.822
22.383	24.338	20.281	22.125	4.519

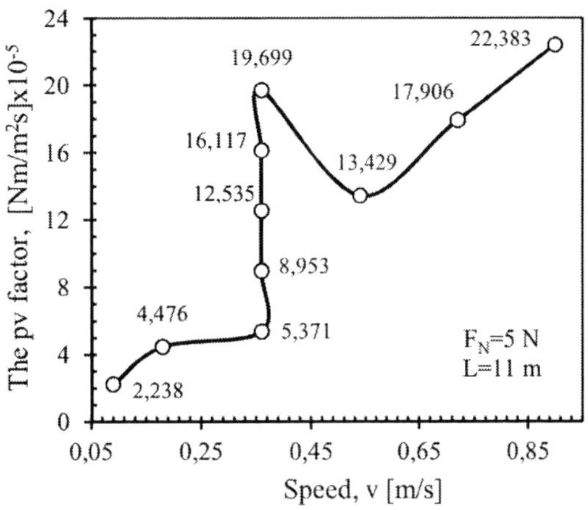

Figure 19. The variation in with the pv factor with speed.

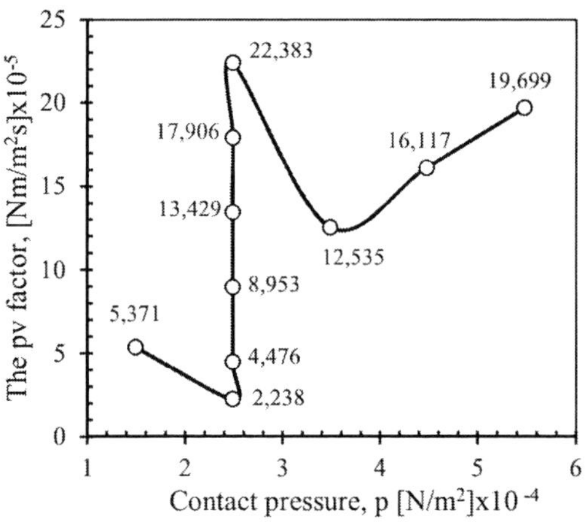

Figure 20. The variation in with the pv factor with contact pressure.

Figure 21 shows the variation in wear depth with the pv factor. In generally, the wear depth increases roughly with the pv factor as shown in this figure. But the wear depth reaches a minimum of 6.218x10^{-6} m at a pv of 5.371 $Nmm^{-2}s^{-1}$ because the contact pressure is a minimum of 1.492 Nm^{-2} as shown in Figure 19. Friction is the process of conversion of energy. The conversion of external mechanical energy into the energy of internal processes occurs in the process of friction. As shown in Figure 22, wear rate due to time increases roughly with increasing pv factor. The wear rate values of (1.557, 3.203,7.343, 11.519, 15.519) and 20.281) x10^{-7} ms^{-1} belong to a contact pressure of 2.487 speeds in the range 0.09-0.90 ms^{-1}. The wear rate values of (2.072, 3.966, 6.633, 7.698 and 9.593) x10^{-7} ms^{-1} belong to a speed of 0.36 contact pressures in the range (1.492-5.472)x10^{-4} Nm^{-2}. It is evident that the wear rate values for a contact pressure of 2.487 and speeds in the range 0.09-0.90 ms^{-1} are smaller than for a speed of 0.36 contact pressures in the range (1.492-5.472) x10^{-4} Nm^{-2}. Because of the conversion of external mechanical energy into the energy of internal processes occurs in the process of friction. Thus, the effects of contact pressure on wear rate is higher than the effect of sliding speed.

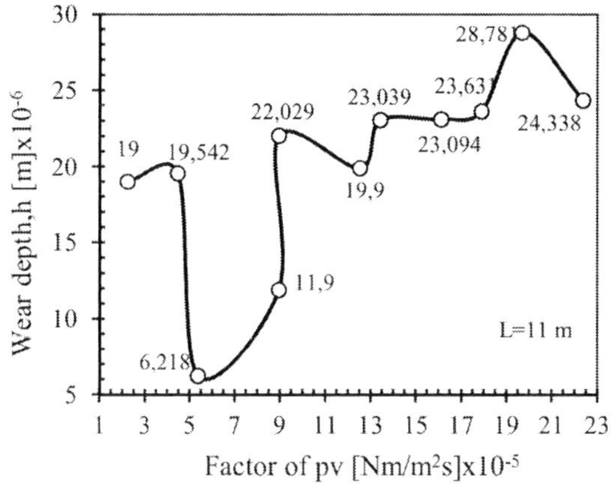

Figure 21. The variation in wear depth with the pv factor.

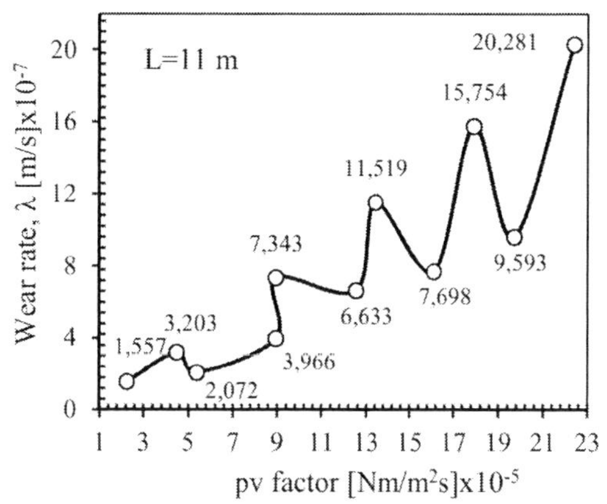

Figure 22. The variation in wear rate with the pv factor.

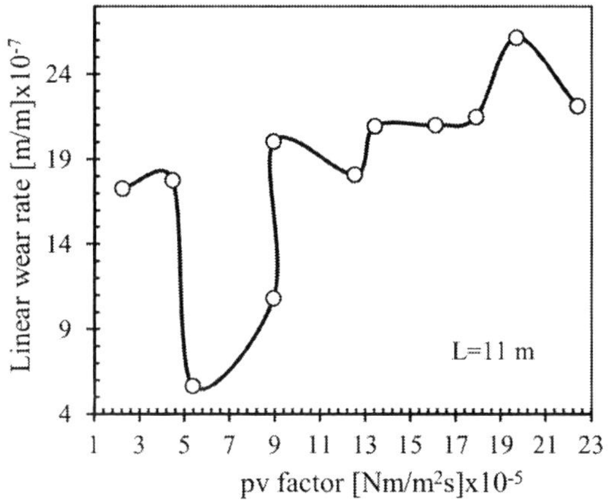

Figure 23. The variation in linear wear rate with the pv.

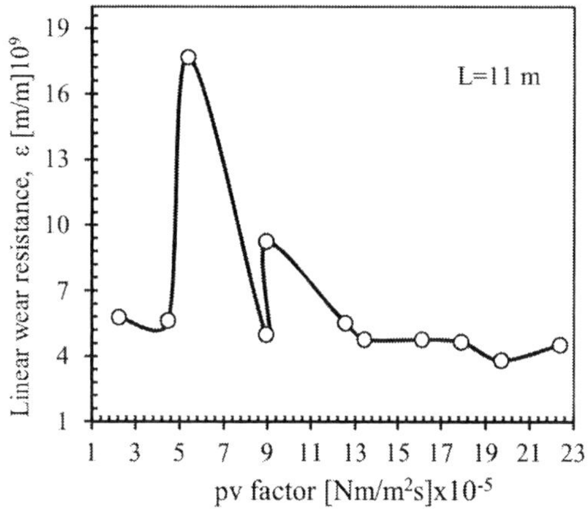

Figure 24. The variation in linear wear resistance with the factor of pv.

Figure 23 shows the variation in linear wear rate with the pv factor. In generally, the wear depth increases roughly with the pv factor as shown in this figure. But the wear depth reaches a minimum of $6.218x10^{-6}$ m at a pv of 5.371 $Nmm^{-2}s^{-1}$ because the contact pressure is a minimum of 1.492 Nm^{-2} as shown in Figure 19. Friction is the process of conversion of energy. The conversion of external mechanical energy into the energy of internal processes occurs in the process of friction. Figure 24 shows the relationship between the linear wear resistance and the pv factor. The linear wear resistance decreases roughly in the range $(2.238\text{-}12.535)10^{-5}$ $Nmm^{-2}s^{-1}$ of pv and then it decreases linearly in the range $(13.429\text{-}22.383)10^{-5}$ $Nmm^{-2}s^{-1}$ of pv. But the linear wear resistance reaches a maximum of $17.692x10^{9}$ mm^{-1} at a pv of 5.371 $Nmm^{-2}s^{-1}$ because the contact pressure is a minimum of 1.492 Nm^{-2} as shown in Figure 19. Friction is the process of conversion of energy. The conversion of external mechanical energy into the energy of internal processes occurs in the process of friction.

5. Conclusions

The conclusion derived from the present study may be summarised as follows:

1) Generally, the wear volume and wear depth increase linearly with an increase in load.

2)Generally, wear rate due to time and linear wear rate increase linearly with increasing loads. The linear wear resistance decreases with an increase in load and the wear rate due to time decreases with an increase in linear wear resistance.

3)The wear rate due to time increases linearly with an increase in linear wear rate. The wear volume increases slightly with an increase in speed. Generally, wear rate due to time and linear wear rate increase linearly with increasing speeds.

4)The linear wear resistance decreases with an increase in speed. The wear rate due to time decreases with an increase in linear wear resistance. The wear rate due to time decreases with an increase in linear wear rate.

5)In generally, the wear depth increases roughly with the pv factor. Wear rate due to time increases roughly with increasing pv factor. Thus, the effects of contact pressure on wear rate is higher than the effect of sliding speed. In generally, the wear depth increases roughly with the pv factor. The linear wear resistance decreases roughly with increasing pv factor.

References

1. Gahr K. H. Z., Microstructure and wear of materials, Elsevier Science Publishing Company Inc, New York, 1987
2. Elleuch K., Mezlini S., Guermazi N., Kapsa Ph., Abrasive wear of aluminium alloys rubbed against sand, Wear, 261, 1316-1321, 2006.
3. Mezlini S., Kapsa Ph., Henon C., Guillemenet J., Abrasion of aluminium alloy: effect of subsurface hardness and scratch interaction simulation, Wear, 257, 892–900,2004
4. Ruiz-Anders M., Conde E. A., Damborenta J. D. and Garcia I., Wear behaviour of aluminium alloys at slow sliding speeds, Tribology Transactions, 58, 955-962, 2015
5. Odabaş D., Su Ş., A comparison of the reciprocating and continuous two-body abrasive wear behaviour of solution-treated and age-hardened 2014 Al alloy, Wear, 208, 25-35, 1997.
6. Shiri N. D., Bhat T., Bhat S., Kumar G., Effect of heat treatment variables on the abrasive wear behaviour of Al-6061 Alloy and Al 6061-SiC_P composites using Taguchi Technique, American Journal of Materials Science, 151-156, 2015.
7. Wang A., Rack H.J., Abrasive wear of silicon carbide particulate- and whisker-reinforced 7091 aluminium matrix composite, Wear,146, 337-348, 1991
8. Song W.Q., Krauklis P., Mouritz A.P., Bandyopadhyay S., The effect of thermal ageing on the abrasive wear behaviour of age-hardened 2014 Al/SiC and 6061 Al/SiC composites, Wear, 185, 125-130, 1995.
9. Liang Y.N., Ma Z.Y., Li S.Z., Li S., Bi J., Impact-abrasive behaviour of SiC_p/2024 Al composites in single pendulum scratch testing, Wear, 178, 9-15, 1994.
10. Zhongliang H., Zhenhua C., Jintong X., Guoyun D. Effect of pv factor on the wear of carbon brushes for micromotors, Wear, 265, 336–340, 2008.
11. Ludema K. C., Classes of Wear, Ind. Eng. Chem. Prod. Res. Dev, 19, 335-337, 1980.
12. Zum Gahr K.H., Microstructure and Wear of Materials, Elsevier Science Publisher B.V. 1987
13. Verbeek H. J., Tribological system and wear factors, Wear, 56, 81-92, 1979.
14. Archard J. F., Contact and rubbing of flat surfaces, J. Appl. Phys., 24, 981-998, 1953.
15. Halling J., Principles of Tribology, Macmillan, London, p.101, 1983
16. Akkurt M., The reliability of machine elements due to wear, Journal of İTÜ, 30-2 ,111-127, 1975.

IOP Conf. Series: Materials Science and Engineering **295** (2018) 012009 doi:10.1088/1757-899X/295/1/012009

Tribological characterization of the drill collars and casing friction couples

RIPEANU R G[1], *BADICIOIU M*[2], *CALTARU M*[3], *DINITA A*[4], *LAUDACESCU E*[5]
[1]*University Petroleum-Gas of Ploiesti, Department of Mechanical Engineering, 100680, Ploiesti, Romania*
[2]*University Petroleum-Gas of Ploiesti, Department of Mechanical Engineering, 100680, Ploiesti, Romania*
[3]*University Petroleum-Gas of Ploiesti, Department of Mechanical Engineering, 100680, Ploiesti, Romania*
[4]*University Petroleum-Gas of Ploiesti, Department of Mechanical Engineering, 100680, Ploiesti, Romania*
[5]*University Petroleum-Gas of Ploiesti, Department of Mechanical Engineering, 100680, Ploiesti, Romania*
rrapeanu@upg-ploiesti.ro[1], *mbadicioiu@upg-ploiesti.ro*[2], *miki@upg-ploiesti.ro*[3], *adinita@upg-ploiesti.ro*[4], *leugen@upg-ploiesti.ro*[5]

Abstract: Drill collars are special pipes used in the drilling of wells for weighting the drill bit, enabling it to drill through the rock. In the drilling process, the drill collars are exposed to an intensive wear due to friction on inner surface of casing wall. In order to evaluate the tribological behaviour of this friction couple, paper presents the drill collars parent material, reconditioned and casing pipe chemical composition, microstructures, hardness and friction tests. For friction tests were prepared samples extracted from new and reconditioned drill collars and from casing pipes and tested on a universal tribometer. Were used plane-on-disk surface friction couples and tests were conducted at two sliding speeds and three normal loads for each materials couple. Plane static partner samples were extracted from casing pipes and disks samples were extracted from new and reconditioned drill collars. Were obtained friction coefficients values and also the temperatures increasing values due to friction working tests parameters. The temperature increasing values were obtained by measuring it with an infrared thermographic camera.

Keywords: drill collar, casing, friction coefficients, temperature.

1. Introduction

Drill collars used in petroleum industry works in heavy conditions. A drill collar is a device used in the drilling of oil wells for weighting the drill bit, enabling it to drill through rock. The drill collars it is manufactured from a bar made of non-magnetic steel alloy, drilled on entire length to permit the flow of drilling fluids. The drill collars are the most expensive elements of the drill string and their durability is essential for the economical drill work efficiency. These devices are typically about 9450 mm long and threaded at both ends (pin and box) to allow multiple drill collars to be joined above the bit assembly. The outside diameter of the drill collars may vary from about 76 mm to 279 mm and greater [1, 2, 3, 4].

In the exploitation time, the drill collars are subjected to torsion, tensile/ compression, internal pressure, external pressure and bending. Two of the degradation processes that reduce durability are abrasive wear of the external surfaces (due to the friction between the drill collars with the inner surface of the casing or the open hole wall) and internal corrosion (due to the passage of drilling fluids through drill collars), [2, 3, 4, 5, 6, 7]. It is important to evaluate the friction behaviour of drill collars because if the wear is advanced, the drill string can slip from the elevator shoulder (device used for handling the drill string) and fall into the borehole [2, 3, 4, 7]. Usually surfaces exposed to friction with casing are reconditioned or hardbanded by applying different alloys that contains carbides. [1, 2, 3, 7, 8].

The aim of this work is to evaluate the tribological behaviour of drill collars- casings friction couple. Thus, paper presents the drill collars parent material, reconditioned and casing pipe chemical composition, microstructures, hardness and friction tests. For friction tests were prepared samples extracted from new and reconditioned drill collars and from casing pipes and tested on a universal tribometer.

2. Experimental tests

2.1. Samples preparations

Samples were extracted from real drill collars and casing. In order to not modify the material proprieties during extraction process a water jet cutting machine was used. In figure 1 it is presented the pictures of water jet cutting.

(a) Water jet sample cutting. (b) Drill collar.

Figure 1. Extraction the samples from drill collars.

After extraction samples were manufactured by milling and then by grinding with some non-intensive cutting regimes and in the presence of cooling fluids.

2.2. Determination the chemical composition

The chemical composition of the parent material and the deposited layer for tested reconditioned drill collars, were obtained with FONDRY MASTER PRO type laboratory spectrometer. The chemical composition results are presented in table 1, [7].

Table 1. Chemical composition of the parent metal and deposited layer.

Chemical element, (wt.%)														
Fe	*C*	*Si*	*Mn*	*P*	*S*	*Cr*	*Mo*	*Ni*	*Al*	*Co*	*Cu*	*Nb*	*Ti*	*V*
Parent material														
96.1	0.488	0.277	1.086	0.018	0.014	1.160	0.323	0.182	0.017	0.011	0.220	0.0005	0.0079	0.0054
Deposited layer														
95.4	0.479	0.319	1.289	0.0178	0.014	1.116	0.347	0.555	0.022	0.009	0.183	0.0005	0.0088	0.0053

In table 2 it is shown, according to API and ISO standards, [4, 5, 6], the chemical composition of drill collars made of AISI 4145H and for casing, made of API 5CT N80-1 spectrometer measured values.

Table 2. Chemical composition of the steel AISI 4145H and N 80-1.

Material	*Chemical element,* (wt.%)							
AISI 4145H	*C*	*Si*	*Mn*	*P*	*S*	*Cr*	*Mo*	*Ni*
	0.43-0.49	0.10-0.35	0.85-1.10	max. 0.035	max. 0.040	0.80-1.10	0.15-0.25	max. 0.25
Casing N 80-1	*C*	*Si*	*Mn*	*P*	*S*	*Cr*	V	
	0.37	0.31	1.64	0.015	0.015	0.14	0.12	

The chemical composition obtained for parent material and deposition layer presented in table 1 are quite similar with the imposed standard values [4, 5].

2.3. Hardness measurements

The hardness tests were performed with an EMCO – DURASCAN 20 type micro hardness machines at a load 0.5 kgf. At the drill collars the microhardness HV0.5 variation with the distance from the deposited layer to the parent metal is presented in figure 2, [7], and for N80-1 casing in figure 3.

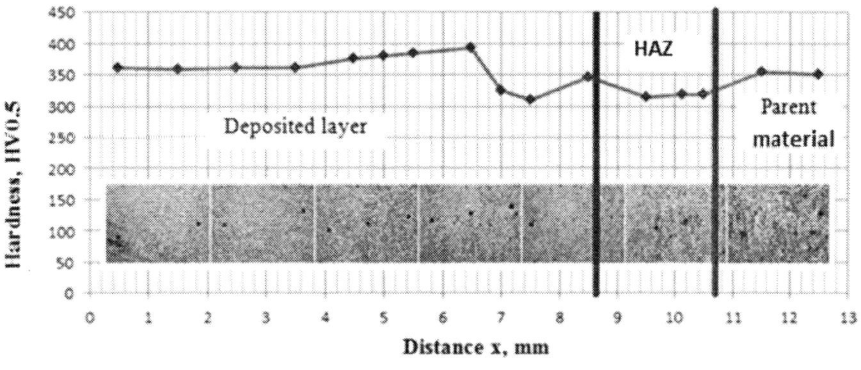

Figure 2. Microhardness HV0.5 variation at hardbanded drill collars samples, [7].

Figure 2 show that at hardbanded (reconditioned) drill collars samples the hardness is almost the same in deposited layer, heat affected zone (HAZ) and parent material AISI 4145H, respecting standard specifications [4, 5, 6].

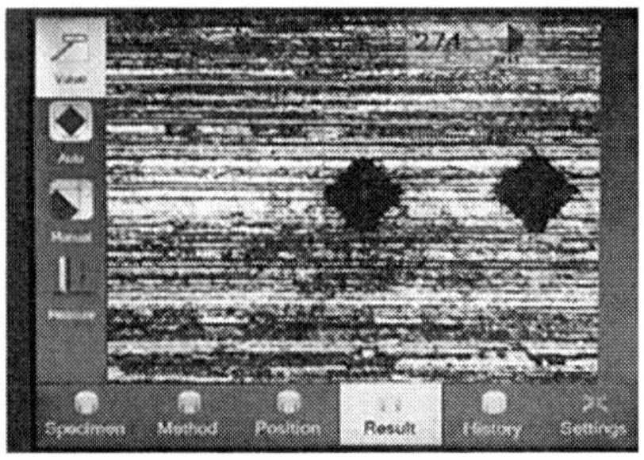

Figure 3. Image of hardness measuring at casing made of N80-1.

Based on the average of 5 measured values the hardness obtained for the N80-1 samples material was 270 HV 0.5.

2.4. Analysis by optical microscopy

The analyses by optical microscopy were performed on metallographic prepared samples drill collars material AISI 4145H, hardbanded drill collars and from casing N80-1. Specimens were prepared by polishing and metallographic etching with chemical reactive (NITAL), and analyses were performed with OLYMPUS BX60M type metallographic microscope. In figure 4 are presented the specimens microstructure images obtained.

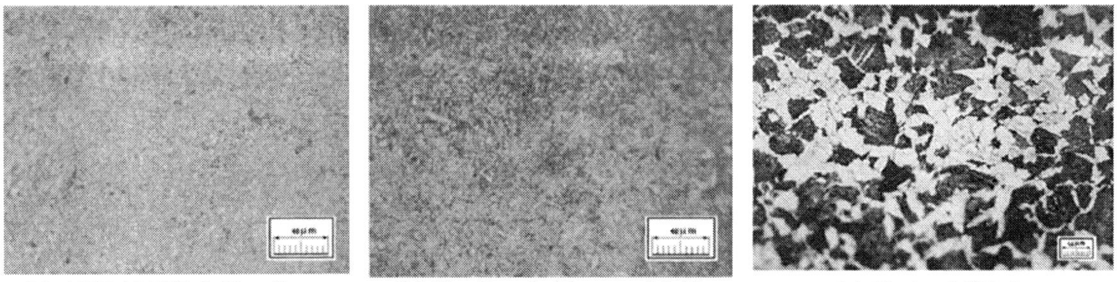

(a) AISI 4145H drill collars. (b) Hardbanded layer of drill collars. (c) Casing N80-1.

Figure 4. Microstructure images of the AISI 4145H drill collars and hardbanded and of casing.

The investigations have found the following:

- the AISI 4145H material drill collar specimen has a sorbitic structure;
- the hardbanded drill collar in deposited layer material specimen shows a typical dendritic structure for weld deposits with alloys that contains chemical elements which form carbides;
- the casing N80-1 specimen material has a ferrite-pearlite structure.

2.5. Friction tests

Friction tests were performed on plane-on-disk couples on a CSM type tribometer [8, 9, 10], disk samples with dimensions Ø 40x3 mm were made of drill collars 4145H material, and of hardbanded drill collars material. Plane static partner in tested friction couples has a cubic shape with dimensions 4x4x4 mm, and were made of casing N80-1 material. In figure 5 it is shown the picture of friction couple positioning during the tests on CSM tribometer.

IOP Conf. Series: Materials Science and Engineering **295** (2018) 012009 doi:10.1088/1757-899X/295/1/012009

Figure 5. Plane-on-disk friction couple.

The working conditions were:
- normal load of 1N, 2N and 4N;
- disks of AISI 4145H and of reconditioned by hardbanding drill collars materials;
- cubes of N80-1 casings material;
- sliding speed of 0.100 $m{\cdot}s^{-1}$ and 0.200 $m{\cdot}s^{-1}$;
- friction length 100 m;
- dry friction at temperature of air of 20^0C and RH=55%.

At the start and at the end of each test the temperature was measured and recorded by using an infrared thermographic camera, type FLIR E50 which was focused to the friction surface. In figure 6 it is shown the friction coefficients values results obtained on plane-on-disk tribometer for disks of 4145H drill collars-N 80-1 materials couple at sliding speed of 0.100 $m{\cdot}s^{-1}$ and in figure 7 at 0.200 $m{\cdot}s^{-1}$.

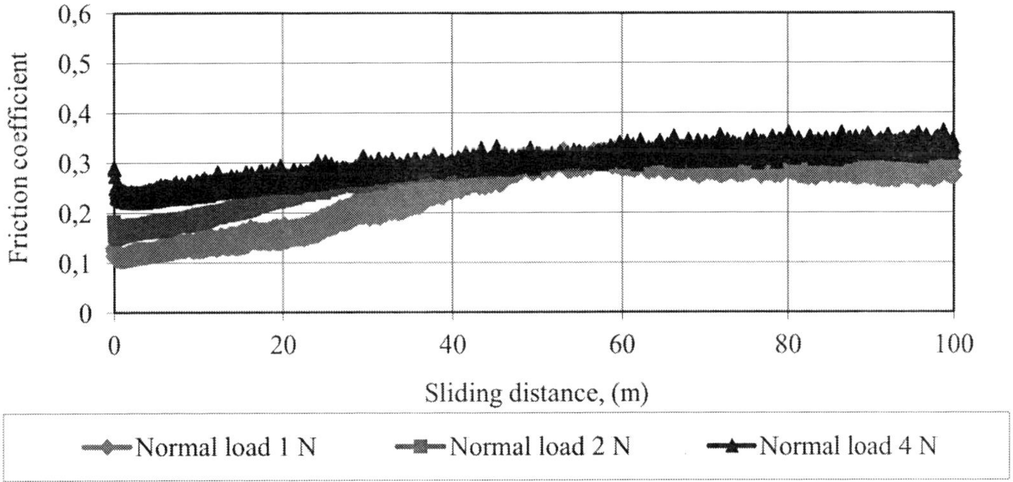

Figure 6. Friction coefficients vs. sliding distance for couple materials AISI 4145H drill collars-N80-1 casing at sliding speed of 0.100 $m{\cdot}s^{-1}$.

IOP Conf. Series: Materials Science and Engineering 295 (2018) 012009 doi:10.1088/1757-899X/295/1/012009

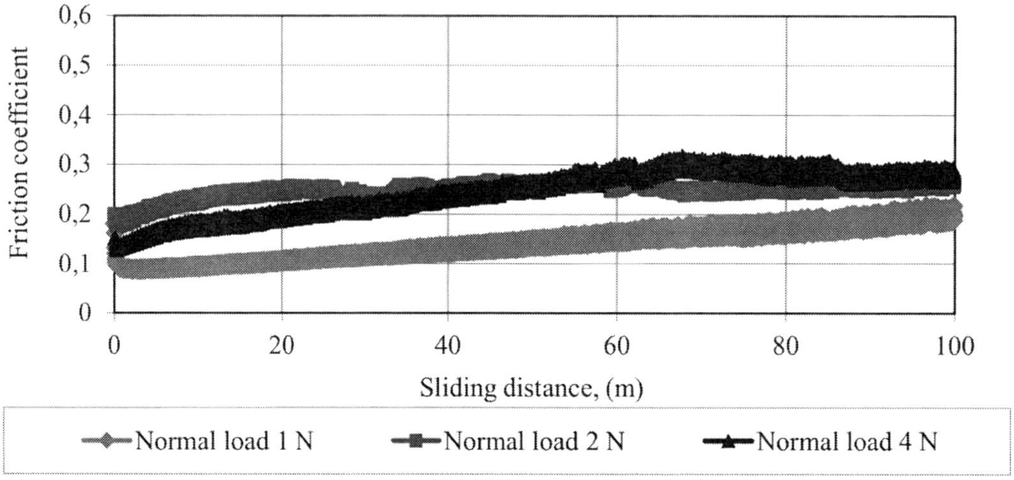

Figure 7. Friction coefficients vs. sliding distance for couple materials AISI 4145H drill collars-N80-1 casing at sliding speed of 0.200 m·s^{-1}.

In figure 8 are the friction coefficients values for disks of hardbanded drill collars material at sliding speed of 0.100 m·s^{-1} and in figure 9 at 0.200 m·s^{-1}.

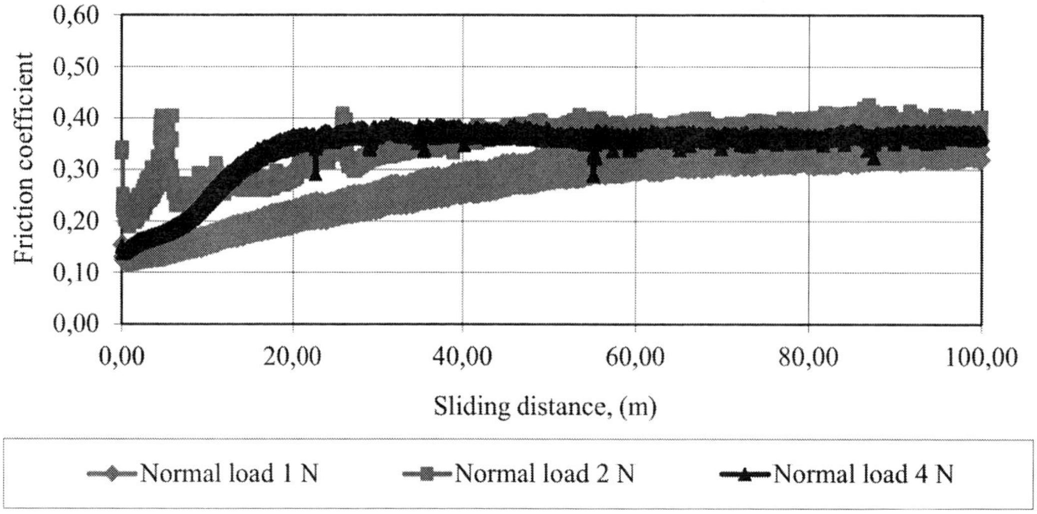

Figure 8. Friction coefficients vs. sliding distance for couple materials hardbanded drill collars- N80-1 casing at sliding speed of 0.100 m·s^{-1}.

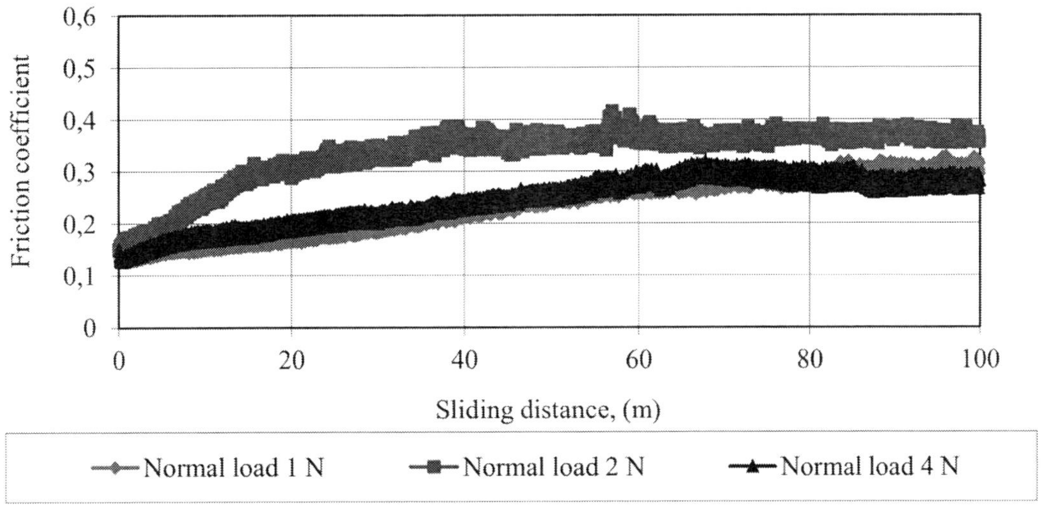

Figure 9. Friction coefficients vs. sliding distance for couple materials hardbanded drill collars- N80-1 casing at sliding speed of 0.200 m·s^{-1}.

Friction coefficients curves presented in figures 6, 7, 8 and 9 are quite similar for tested friction couples with a small increase of obtained values for couple materials hardbanded drill collars- N80-1 casing (0.27-0.39) instead of (0.19-0.34) for couple materials AISI 4145H drill collars- N80-1 casing. The sliding speed has a very small influence on both tested friction couples with a very small reduction at 0.2 m·s^{-1}. The normal load influence at couple materials AISI 4145H drill collars- N80-1 show that friction coefficients values has a very small increase with load increasing meanwhile at couple materials hardbanded drill collars- N80-1 casing were a very small increase of friction coefficients were obtained al 2N load instead of 1N and 2N loads. These behaviors are due the small differences of chemical compositions of the AISI 4145 drill collars and hardbanded drill collars materials disk samples with the presence especially of cupper in the second one material disk samples.

In table 3 are presented the images of temperatures measured values at the start at the end of each friction tests. The spot of infrared thermographic camera, as shown in images, was focused on friction surface.

Table 3. Temperature values at the start and at the end of friction tests.

Working parameters		Friction couple materials			
Normal load (N)	Sliding speed (m·s^{-1})	Disk of AISI 4145H- Cube of N80		Disk of hardbanded drill collars-Cube of N80	
		Spot temperature at friction surface, (^{0}C)			
		Initial	Final	Initial	Final
1	0.100	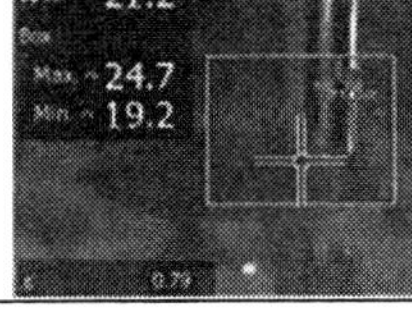	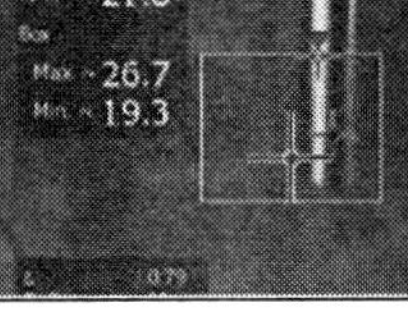	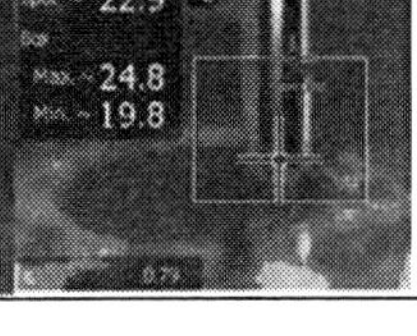	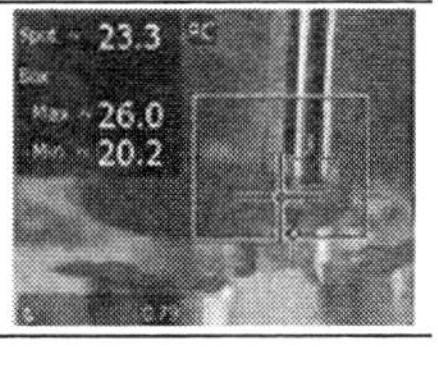

	0.200	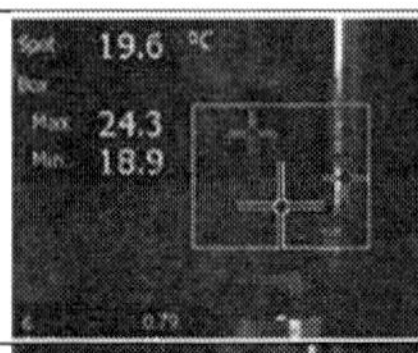	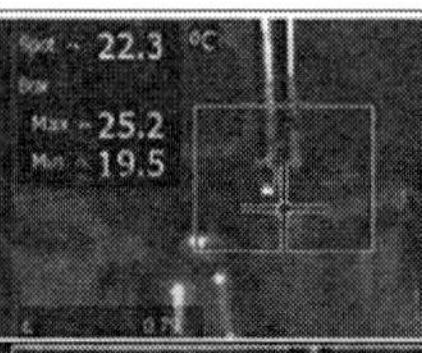	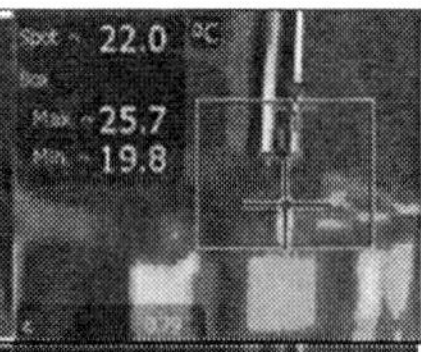	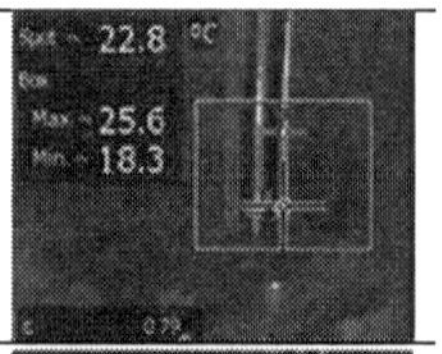
2	0.100	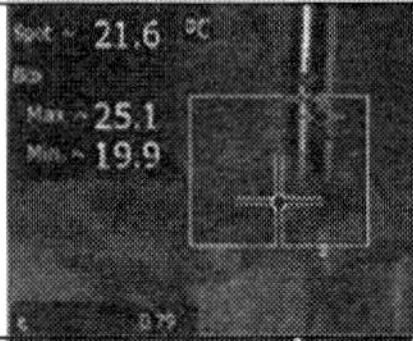	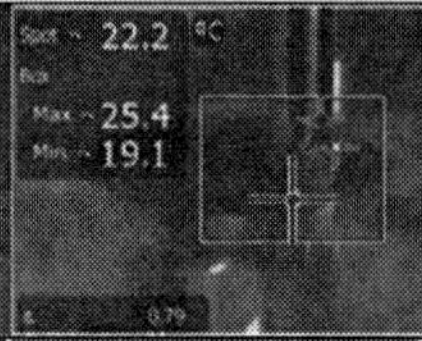	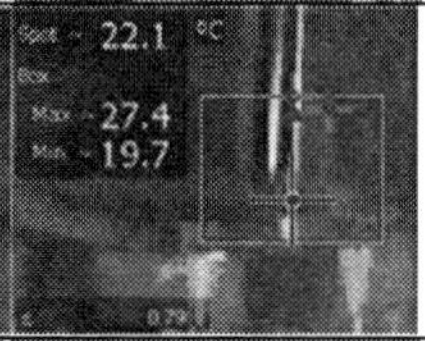	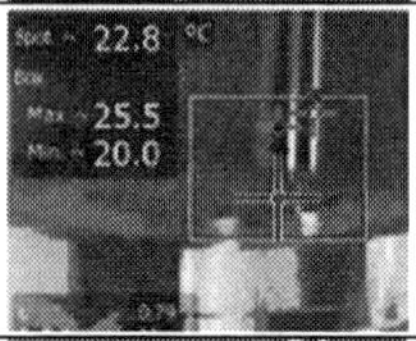
	0.200	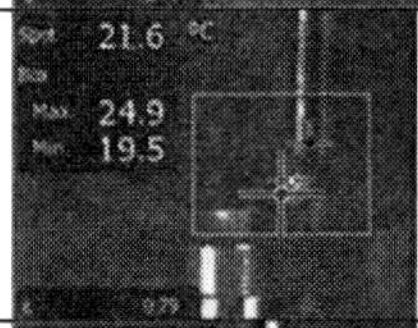	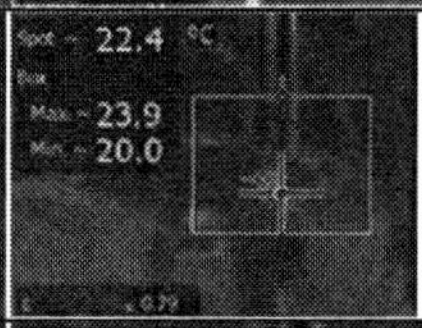	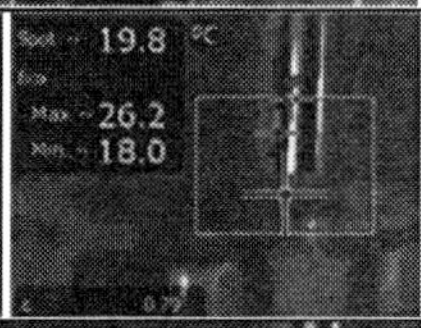	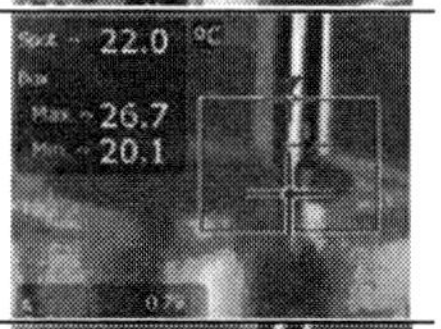
4	0.100		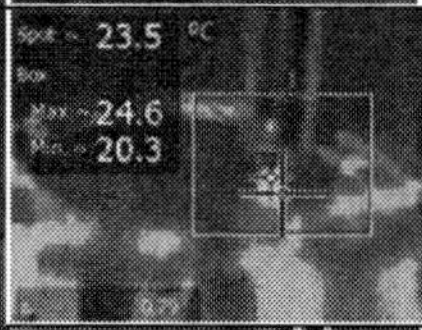	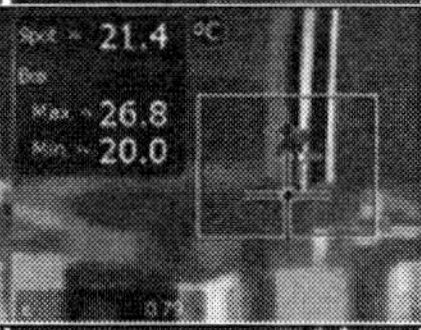	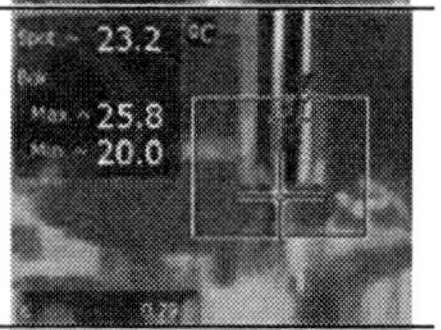
	0.200	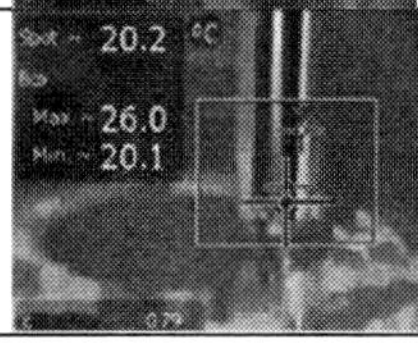	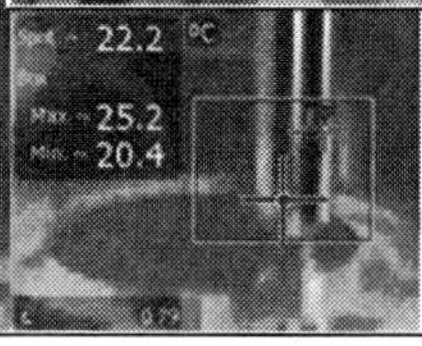	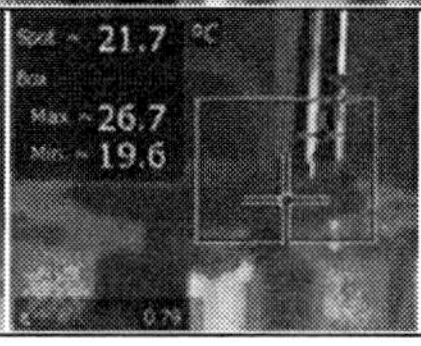	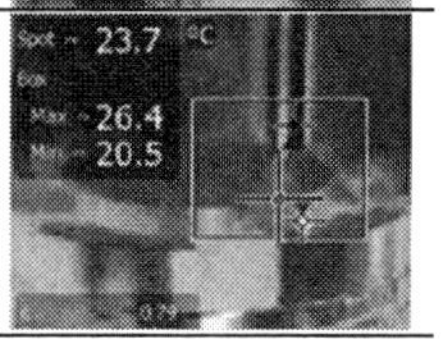

Analyzing the spot temperatures values presented in table 3 we could observe that both tested couple materials, at the same friction tests parameters, presents similar increase of the temperatures between the start and the end of tests. For friction tests conditions, these temperatures differences increase with sliding speed and with normal load.

3. Conclusions

Tribological behaviour of AISI 4145H and reconditioned by hardbanding drill collars materials in couple with N80-1 casing material is almost the same. This conclusion was confirmed by the results obtained after chemical composition determination, hardness measurements, optical microscopy analysis and friction tests.

The friction coefficients values curves obtained presents a small increase of obtained values for couple materials hardbanded drill collars- N80-1 casing (0.27-0.39) instead of (0.19-0.34) for couple materials AISI 4145H drill collars- N80-1 casing. The sliding speed has a very small influence on both tested friction couples with a very small reduction at 0.2 $m{\cdot}s^{-1}$. The normal load influence at couple materials AISI 4145H drill collars- N80-1 show that friction coefficients values has a very small increase with load increasing meanwhile at couple materials reconditioned by hardbanding drill collars- N80-1 casing were a very small increase of friction coefficients were obtained al 2N load instead of 1N and 2N loads. These behaviours are due the small differences of chemical compositions of the AISI 4145 drill collars and hardbanded drill collars materials disk samples with the presence especially of cupper in the second one material disk samples.

The spot temperatures values at both tested couple materials, at the same friction tests parameters, presents similar increase of the temperatures between the start and the end of tests. For friction tests conditions, these temperatures differences increase with sliding speed and with normal load.

References

1. Ulmanu V., Petroleum Tubular Materials (in Romanian), Technical Publ. House, ROU, 1992.
2. Ulmanu V., Petroleum Equipment Manufacturing and Repairing Technology (in Romanian), ILEX Publ. House, ROU, 2002.
3. Ulmanu V., Badicioiu M., Caltaru M., Dinita A., Researches regarding the hardbanding technology of drill pipes tool joints using hardfacing with welding electrode wire (Research project - in Romanian), Petroleum-Gas University, ROU, 2015.
4. Standard by American Petroleum Institute, API Specification 5D, Specification for Drill Pipe, Fifth Edition, October 2001.
5. International Standard Organization, ISO 11961:2008, Petroleum and natural gas industries - Steel drill pipe.
6. Standard by American Petroleum Institute, API Specification 5CT, Specification for Casing and Tubing, 9th Edition, June 2011.
7. Ripeanu R.G., Badicioiu, M., Caltaru, M., Reconditioning of drill collars by using welding technologies, Journal of the Balkan Tribological Association, vol. 21(2), 314-328, 2015.
8. Caltaru M., Badicioiu M., Ripeanu R.G., Establishing the tribological behavior of the HVOF hardfacing applied at petroleum gate valves, Journal of the Balkan Tribological Association, vol.19(3), 448-460, 2013.
9. Ripeanu R.G., Ulmanu V., Minescu M., Tribological characterisation of HVOF hardfacing applied at parallel slide gate valves, OilDoc Conference 2013, Lubricants, Maintenance, Tribology, 22-24 January, Rosenheim, Germany, 1-6, 2013.
10. Cursaru D.L., Andronescu C., Pirvu C., Ripeanu R.G., The efficiency of Co-based single wall carbon nanotubes (SWNTs) in comparison with a commercial additive as AW and EP additives to mineral base oil, Wear, vol. 290-291, 133-139, 2012.

Tribological characterization of the drill pipe tool joints reconditioned by using welding technologies

Caltaru M[1], Badicioiu M[2], Ripeanu R G[3], Dinita A[4], Minescu M[5], Laudacescu E[6]
[1]University Petroleum-Gas of Ploiesti, Department of Mechanical Engineering, 100680, Ploiesti, Romania
[2]University Petroleum-Gas of Ploiesti, Department of Mechanical Engineering, 100680, Ploiesti, Romania
[3]University Petroleum-Gas of Ploiesti, Department of Mechanical Engineering, 100680, Ploiesti, Romania
[4]University Petroleum-Gas of Ploiesti, Department of Mechanical Engineering, 100680, Ploiesti, Romania
[5]University Petroleum-Gas of Ploiesti, Department of Mechanical Engineering, 100680, Ploiesti, Romania
[6]University Petroleum-Gas of Ploiesti, Department of Mechanical Engineering, 100680, Ploiesti, Romania
miki@upg-ploiesti.ro[1], mbadicioiu@upg-ploiesti.ro[2], rrapeanu@upg-ploiesti.ro[3], adinita@upg-ploiesti.ro[4], mminescu@upg-ploiesti.ro[5], leugen@upg-ploiesti.ro[6]

Abstract: Drill pipe is a seamless steel pipe with upset ends fitted with special threaded ends that are known as tool joints. During drilling operations, the wall thickness of the drill pipe and the outside diameter of the tool joints will be gradually reduced due to wear. The present research work investigate the possibility of reconditioning the drill pipe tool joints by hardbanding with a new metal-cored coppered flux cored wire, Cr-Mo alloyed, using the gas metal active welding process, taking into considerations two different hardbanding technologies, consisting in: hardbanding drill pipe tool joints after removing the old hardbanding material and surface reconstruction with a compensation material (case A), and hardbanding tool joint drill pipe, without removing the old hardbanding material (case B). The present paper brings forward the experimental researches regarding the tribological characterization of the reconditioned drill pipe tool joint by performing macroscopic analyses, metallographic analyses, Vickers hardness measurement, chemical composition measurement and wear tests conducted on ball on disk friction couples, in order to certify the quality of the hardbanding obtained by different technological approaches, to validate the optimum technology.

Keywords: tool joints, reconditioning, welding, wear rate.

1. Introduction

Drill pipes represent about 90 % ... 95 % of the total length of the drill string, being the major component of the entire drill string, which makes their durability to be essential for the economical drill work efficiency [1].

Drill pipe is a seamless steel pipe with upset ends fitted with special threaded ends that are known as tool joints. The drill pipe has two tool joints, one female called a box and the other male called pin, these joints allows to connect of each drill pipe segment to the next segment [2, 3, 4, 5, 6, 7, 8]. The drill pipe configuration and the main elements (tool joints, pipe body, upset and welding zone) are specified in figure 1.

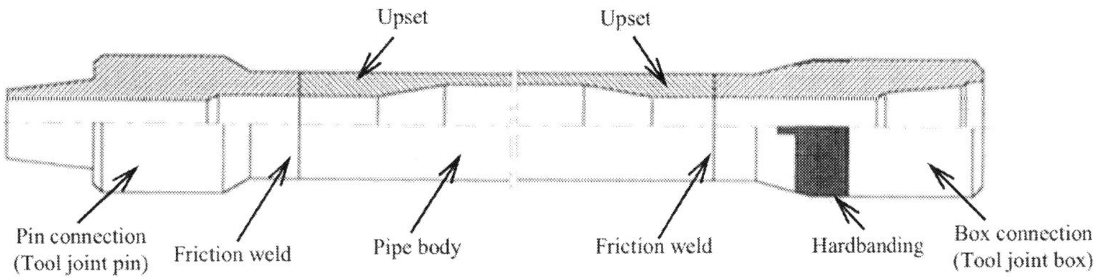

Figure 1. Drill pipe, [2].

The drill pipes are used to connect the rig surface equipment with the bottomhole assembly and the bit, to transmit the rotary motion from surface to the bit, and conduct the drilling fluid to the bit, thus it plays a vital part in the successful drilling of oil and gas wells.

During drilling operations, the drill pipes are subjected to torsion, tensile/compression, internal pressure, external pressure and bending. At the same time, due to the different environments, the drill pipes are subject to degradation processes resulting in their rejection. Two of the degradation processes that lead to the rapid decline in lifetime are abrasive wear of the external surfaces (due to the friction between the tool joints drill pipes with the inner surface of the casing – see figure 2) and internal corrosion (due to the passage of drilling fluids through drill pipe), [6, 7, 9, 10, 11, 13]. To improve the wear resistance, optional, the drill pipes are delivered with hardbanded tool joints, as shown in figure 3, [12].

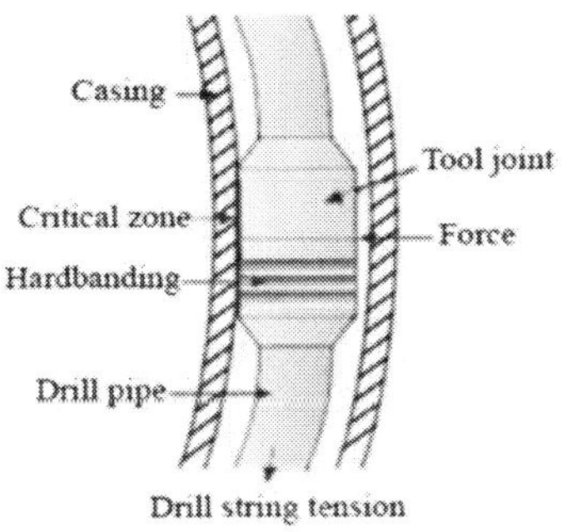

Figure 2. Schematic representation of the wear sources for drill pipes, [11].

Figure 3. Hardbanded drill pipe tool joint, [12].

In the exploitation period, the wall thickness of the drill pipe and the outside diameter of the tool joint will be gradually reduced by wear. This decrease the strength properties of drill pipe due to pipe cross sectional area reduction or tool joint shoulder area reduction. For this reason, API classifies drill pipe according to degree of wear based on dimensions, surface damage, and corrosion, [2, 3, 4, 6, 7]. Depending on the existing wear at some point, in order to avoid the downgrade of the drill pipe to a degree of past (class) lower resistance, frequently, are applying the reconditioning of the drill pipe tool joint.

Based on this, the present research work investigates the possibility of reconditioning the drill pipe tool joint by hardbanding with a new metal-cored coppered flux cored wire, Cr-Mo alloyed, using the gas metal active welding process, taking into considerations two different hardbanding technologies, consisting in:

- hardbanding drill pipe tool joint after removing the old hardbanding material and surface reconstruction with a compensation material;
- hardbanding tool joint drill pipe, without removing the old hardbanding material.

The present paper brings forward the experimental researches regarding the tribological characterization of the reconditioned drill pipe tool joint by performing macroscopic analyses, metallographic analyses, Vickers hardness measurement, chemical composition measurement and wear tests, to certify the quality hardbanding obtained by different technological approaches, to validate the optimum technology.

2. Reconditioning, results and discussions

2.1. Hardbanding specification

The tool joints drill pipe subjected to reconditioning process were manufactured by steel AISI 4145H, with the chemical composition, according to API standards, presented in table 1, [6].

Table 1. Chemical composition of the steel AISI 4145H.

Chemical element, (wt.%)							
C	*Si*	*Mn*	*P*	*S*	*Cr*	*Mo*	*Ni*
0.43-0.49	0.10-0.35	0.85-1.10	max. 0.035	max. 0.040	0.80-1.10	0.15-0.25	max. 0.25

The metal active gas welding process (GMAW) was used for drill pipe tool joint hardbanding. The shielding gas used was 82 % Ar and 18 % CO_2. The welding equipment had the devices necessary to grip and rotate the drill pipe, to grip and oscillate the welding torch parallel to the drill pipe tool joint surface, and to incline the welding torch, in the direction of the rotation. The drill pipe tool joints were reconditioned in two different technological approaches, as follows, [5]:

a) hardbanding drill pipe tool joint with a metal-cored coppered flux cored wire, Cr-Mo alloyed, after removing the old hardbanding material and surface reconstruction with a copper coated basic flux cored wire, like compensation material;
b) hardbanding drill pipe tool joint with a metal-cored coppered flux cored wire, Cr-Mo alloyed, without removing the old hardbanding material.

The chemical composition and the deposit mechanical characteristics for the hardbanding material (metal-cored coppered flux cored wire, Cr-Mo alloyed, of 1.6 mm diameter) and also for the compensation material (copper coated basic flux cored wire of 1,6 mm diameter) used are presented in table 2 and table 3, according to the manufacturer specifications.

Table 2. Chemical composition of the deposit.

Type of filler material	*Chemical element,* (wt.%)						
	C	*Mn*	*Si*	*S*	*P*	*Cr*	*Mo*
Compensation material	0.05	1.2	0.3	≤0.01	≤0.01	-	-
Hardbanding material	0.45-0.85	1.60-2.20	0.50-1.00	≤0.02	≤0.02	5.0-7.0	0.45-0.85

Table 3. Mechanical characteristics of the deposit.

Type of filler material	*Yield strength* $Rp_{0.2}$ (MPa)	*Tensile strength* Rm (MPa)	*Elongation* A5 (%)	*Impact energy* KV at -40°C (J)	*Hardness* HRC (-)
Compensation material	≥ 420	500-640	≥ 25	≥ 80	-
Hardbanding material	-	-	-	-	57-62

2.2. Results and discussions

The reconditioned tool joints drill pipe was tribological characterized in order to verify the deposition quality obtained by different technological approaches. The deposition quality was verified by: macroscopic analyses, metallographic analyses, Vickers hardness measurement, chemical composition measurement and wear tests. The experimental researches were made on two samples (A and B -as presented in figure 4), cut from the tool joints drill pipe hardbanded and prepared by polishing and metallographic etching with chemical reactive. The samples A and B were cut from tool joints drill pipe hardbanded in different technological approaches, as follows, [5]:

a) sample A – cut from drill pipe tool joint reconditioned by hardbanding with a metal-cored coppered flux cored wire, Cr-Mo alloyed, after removing the old hardbanding material and surface reconstruction with a copper coated basic flux cored wire, like compensation material;

b) sample B – cut from drill pipe tool joint reconditioned with a metal-cored coppered flux cored wire, Cr-Mo alloyed, without removing the old hardbanding material.

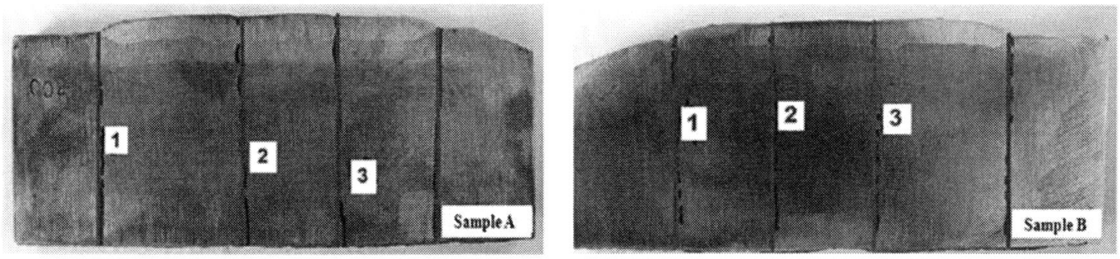

Figure 4. Metallographic samples for macroscopic analyses:
1, 2, 3 - locations for metallographic samples used in optical microscopy analyses.

2.2.1. Macroscopic analysis. The macroscopic analyses consisted in investigating with the naked eye (standard visual examination) and magnifying glass (magnifications of up to 10X) of the drill pipe tool joint hardbanded. The investigations, also made on the metallographic samples cut from the drill pipe tool joint hardbanded, have found the following:

- there were no cracks visible in the hardbanded layers;
- the adhesion between the substrate (parent metal) and the hardbanded layers is good (without discontinuities, porosities etc.);
- the adhesion between two adjacent rows and between two successive layers is good (without discontinuities, porosities etc.);
- the compensation layer applied to the drill pipe tool joint (Sample A) is executed correctly from the dimensional point of view, without cracks or pores in the transition to the substrate (parent metal).

2.2.2. Metallographic analyses by optical microscopy. The analyses by optical microscopy were performed on metallographic samples cut from the sampling locations (1, 2, 3) as indicated in figure 4. The metallographic samples, presented in figure 5 were embedded in polymeric material (resin) and special prepared by polishing and metallographic etching with chemical reactive - NITAL 5%, for examination by using the optical microscope OLYMPUS BX 60M.

In figure 6 are presented the microstructure images of the hardbanded layer, compensation layer, heat affected zone and parent metal, obtained after metallographic analysis of the Sample A (sample cut from the drill pipe tool joint reconditioned by hardbanding with a metal-cored coppered flux cored wire, Cr-Mo alloyed, after removing the old hardbanding material and surface reconstruction with a copper coated basic flux cored wire, like compensation material).

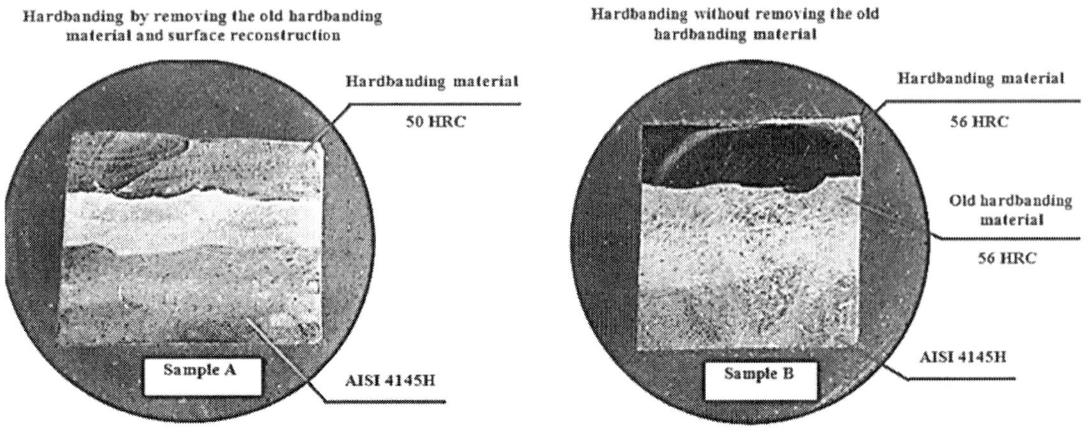

Figure 5. Samples for metallographic analysis using optical microscopy.

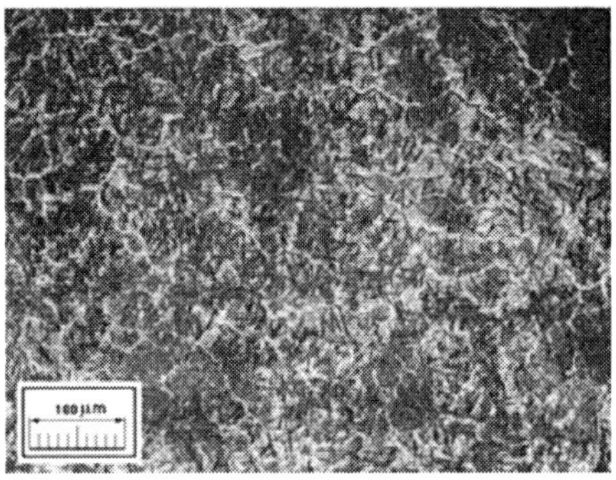

(a) Hardbanded layer.

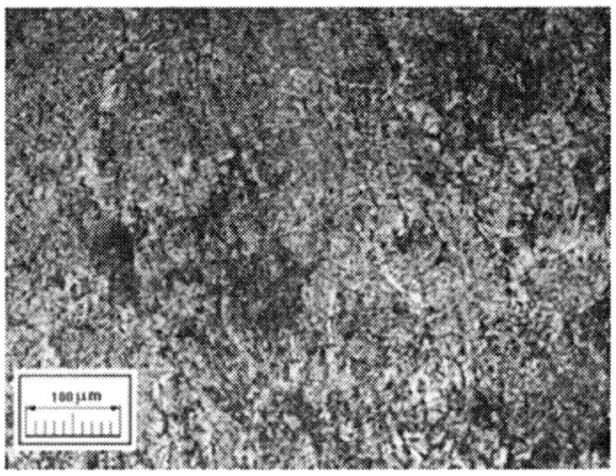

(b) Compensation layer.

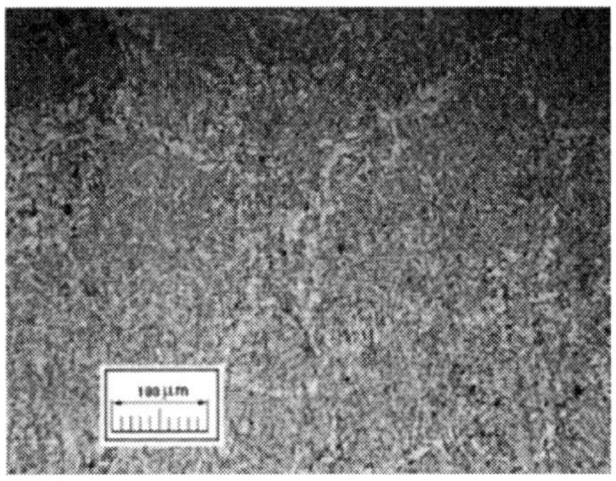

(c) Heat affected zone.

(d) Parent metal.

Figure 6. Microstructure images of the Sample A, starting from the surfaces of the hardbanded layer until the parent metal.

In figure 7 are presented the microstructure images of the hardbanded layer, heat affected zone and parent metal, obtained after metallographic analysis of the Sample B (sample cut from the drill pipe tool joint reconditioned by hardbanding with a metal-cored coppered flux cored wire, Cr-Mo alloyed, without removing the old hardbanding material).

The metallographic analyses by optical microscopy of the hardbanded samples A and B (figure 6 and figure 7) shows the following:

- the parent metal has a very fine sorbitic structure outside the heat affected zone;

IOP Conf. Series: Materials Science and Engineering **295** (2018) 012010 doi:10.1088/1757-899X/295/1/012010

- the parent metal in the adjacent weld bead area (transition area to the parent metal) has an acicular martensitic structure and there were no cracks and discontinuities visible;
- the fusion line has a small thickness and the nearly area has an inferior bainitic structure;
- the adhesion between the substrate (parent metal) and the hardbanded layer is good (no discontinuities are visible);
- the hardbanded layer shows a typical structure for weld deposits with alloys that contains chemical elements which form carbides.

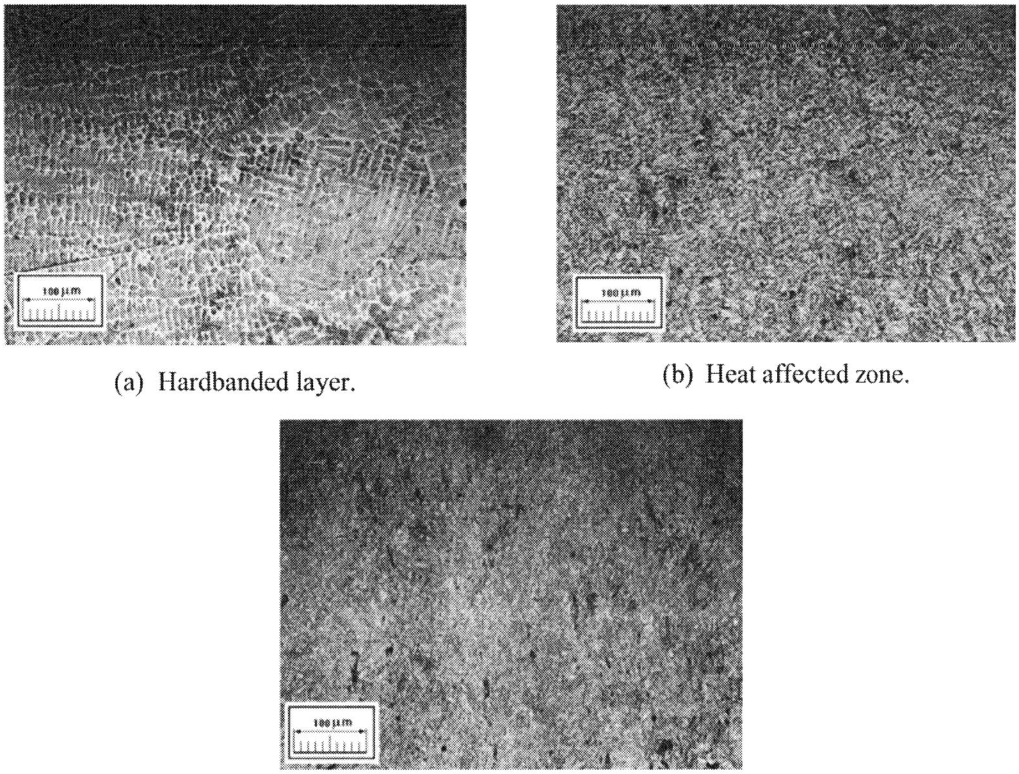

(a) Hardbanded layer.

(b) Heat affected zone.

(c) Parent metal.

Figure 7. Microstructure images of the Sample B, starting from the surfaces of the deposited layer until the parent metal.

2.2.3. Determination of the chemical compositions for parent metal and hardbanded layer. In order to establish the chemical composition of the parent metal and the hardbanded layer, the investigations were performed with FONDRY MASTER PRO type laboratory spectrometer. The results obtained are shown in table 4.

Table 4. Chemical composition of the parent metal and hardbanded layer.

Sample	*Chemical element*, (wt.%)								
	C	*Si*	*Mn*	*P*	*S*	*Cr*	*Mo*	*Ni*	*Al*
	Parent metal								
A	0.3979	0.3132	1.168	0.0074	0.0072	1.163	0.3903	0.1064	0.031
B	0.3937	0.2956	1.178	0.0076	0.00474	1.145	0.3659	0.1147	0.0266
	Deposited layer								
A	0.3042	0.6899	1.771	0.0114	0.0085	4.359	0.3554	0.0233	0.0321
B	0.3198	0.5834	1.539	0.0116	0.0048	4.811	0.4757	0.0581	0.0324

The quantitative chemical composition analysis by optical emission spectrometry of the parent metal and the deposited layer revealed the following:

- the chemical composition of the parent metal corresponds to the steel utilized for manufacturing of tool joint drill pipe, with the observation that, the carbon content is below the lower limit recommended for steel 4145H as presented in table 1;
- the chemical composition of the hardbanded layer is similar for sample A and B, being closer to the specifications indicated by the manufacturer of the hardbanding material regarding the chemical composition of the deposit; it is noted that the percentages of C, Cr, Mo are equal or less than the lower limit reported by the manufacturer (see table 2). This difference may be due to the fact that in the present researches the hardbanding were performed in a single layer, and the manufacturer of the hardbanding material does not specify after how many layers the chemical composition was determined.

2.2.4. Vickers hardness measurement. The hardness tests were performed with EMCO – DURASCAN 20 type hardness machines at a load of 49.05 N (HV5). For each sample (A and B), three rows of hardness were measured, each of them starting from the hardbanded layer, on a perpendicular direction, to the parent metal (Row 1, Row 2, Row 3), maintaining a measurement indentation distance of 0.2 mm. The hardness values obtained are presented in table 5.

Table 5. Hardness measurements.

Zone of measurement	Sample A HV5 ($kgf{\cdot}mm^{-2}$)			Sample B HV5($kgf{\cdot}mm^{-2}$)		
	Row 1	Row 2	Row 3	Row 1	Row 2	Row 3
Hardbanded layer	512	520	510	580	603	591
	501	511	509	591	623	602
	504	503	520	593	665	631
	500	499	517	601	654	645
	498	543	516	623	643	651
	523	523	512	641	638	647
	531	531	515	660	632	629
	539	-	-	637	629	631
	-	-	-	629	-	625
	-	-	-	625	-	-
Heat Affected Zone	217	220	225	356	395	394
	212	198	207	351	379	385
	207	200	210	362	382	363
	209	505	207	369	373	365
	203	212	209	353	376	362
	212	199	211	360	364	359
	-	-	-	365	-	-
Parent metal	303	328	305	315	289	317
	298	277	307	323	295	322
	300	294	310	317	312	309
	310	285	300	312	320	313
	307	279	298	-	-	-
Rockwell hardness of the hardbanded layer (obtained by Vickers hardness conversion of the average value)	Sample A			Sample B		
	50HRC			56HRC		

The hardness measurements showed the following:

- the hardness of the hardbanded layer is relatively constant on thickness layer;
- the hardness value for Sample A (50HRC) has a lower value than the one usually recommended, perhaps due to the reconstruction of the tool joint drill pipe, after removing the old hardbanding, with a compensation material having a low carbon content (0.05% C), which resulted in a dilution of the hardbanding layer.
- the hardness value for Sample B (56HRC) corresponds, to the requirements applied worldwide for the protection of the drill pipe tool joint hardbanding by welding, and to the manufacturer's specifications for hardbanding material of 57-62 HRC;
- it can be noted a gradual passage from the hardness of the deposited layer to the hardness of the heat affected zone and parent metal.

2.2.5. Wear tests. Wear tests were performed on ball-on-disk couples on a CSM type tribometer [13, 14, 15], disk samples were made also from drill pipe tool joint parent material AISI 4145H (not reconditioned) and samples with hardbanded layers (Sample A and Sample B). Samples A were cut from the drill pipe tool joint reconditioned by hardbanding with a metal-cored coppered flux cored wire, Cr-Mo alloyed, after removing the old hardbanding material and surface reconstruction with a copper coated basic flux cored wire, like compensation material. Samples B were cut from the drill pipe tool joint reconditioned by hardbanding with a metal-cored coppered flux cored wire, Cr-Mo alloyed, without removing the old hardbanding material.

The working conditions were:

- normal load of 2N;
- disks of AISI 4145H and of reconditioned by welding (hardbanded layer - samples A and B);
- ∅ 6 mm ball of steel 100Cr6;
- sliding speed of 0.200 m/sec.;
- friction length 200 m;
- dry friction at temperature of air of 20^{0}C and RH=57%.

In figure 8 it is shown the friction coefficients values results obtained on ball-on-disk tribometer.

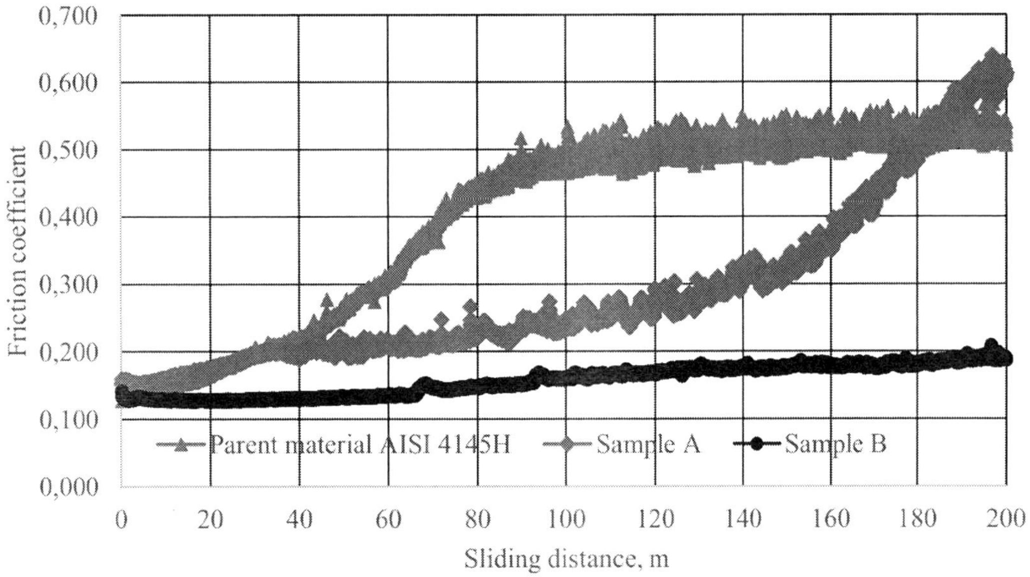

Figure 8. Friction coefficients vs. sliding distance.

In table 6 it is presented the wear rates obtained for tested materials on CSM microtribometer.

Table 6. Wear rate coefficients.

Material type	Parent material AISI 4145H	Sample A-Reconditioned by welding	Sample B-Reconditioned by welding
Disk wear rate coefficient, ($mm^3 \cdot N^{-1} \cdot m^{-1}$)	5.357E-005	2.689E-005	3.695E-006
Ball wear rate coefficient, ($mm^3 \cdot N^{-1} \cdot m^{-1}$)	1.358E-007	1.030E-007	2.681E-008

As we can see from figure 8 and table 6 the results obtained show that recondition by hardbanding with a metal-cored coppered flux cored wire, Cr-Mo alloyed, after removing the old hardbanding material and surface reconstruction with a copper coated basic flux cored wire, like compensation material deposition layer behavior at wear is quite similar (a little bit smaller) with the parent material and friction coefficients are smaller almost on entire sliding length, only at the last 20 m the friction coefficients becomes greater. For sample B reconditioned by hardbanding with a metal-cored coppered flux cored wire, Cr-Mo alloyed, without removing the old hardbanding material, friction coefficients in dry friction conditions and wear rate coefficients are significative smaller.

3. Conclusions

Reconditioning of the drill pipe tool joint by hardbanding with a metal-cored coppered flux cored wire, Cr-Mo alloyed, after removing the old hardbanding material and surface reconstruction with a copper coated basic flux cored wire (compensation material) and by hardbanding with a metal-cored coppered flux cored wire, Cr-Mo alloyed, without removing the old hardbanding material, using the gas metal active process (GMAW), allows to obtain of relatively smooth layers, without cracks or pores.

The hardness of the hardbanded layer is relatively constant on the thickness layer, with a gradual passage from the hardbanded layer to the heat affected zone and parent metal. The hardness of the drill pipe tool joint hardbanded, after removing the old hardbanding material and surface reconstruction with a compensation material, has a lower value (50HRC) than the one usually recommended (57-62HRC), possible when the compensation material used has a low carbon content (0.05% C), which resulted in a dilution of the hardbanding layer. The hardness of the drill pipe tool joint hardbanded (56HRC), without removing the old hardbanding material, corresponds to the requirements applied worldwide for the protection of the drill pipe tool joint hardbanding by welding, and to the manufacturer's specifications for hardbanding material of 57-62 HRC.

The microstructure of the hardbanded layer is similar for all two different technological approaches regarding the reconditioning process of the drill pipe tool joints, and consists in fine Cr and Mo carbides uniformly distributed in a fine tempered martensitic network. All heat affected zones have an identical inferior bainitic microstructure, caused by annealing treatment produced by hardbanded layers. The heat affected zone is narrow and provides a gradual transition from the parental metal to hardbanded layer. The parent metal has a very fine sorbitic structure outside the heat affected zone, being non-affected by the hardbanding processes, using the welding technologies. The hardness of the parental metal is 277 - 328 HV5, which corresponds to the standard regulations for drill pipe tool joints (30-36 HRC), [6, 7].

The result of experimental researches regarding the tribological characterization of the reconditioned drill pipe tool joints by macroscopic analyses, metallographic analyses, Vickers hardness measurement, chemical composition measurement and wear tests, certify the quality hardbanding obtained by two different approaches regarding the hardbanding technologies, as follow:

- both approaches regarding the hardbanding technologies are valid for reconditioning of the tool joints drill pipes with a metal-cored coppered flux cored wire, Cr-Mo alloyed, using the gas metal active process (GMAW), only if for the surface reconstruction is chosen another compensation material to avoid the dilution of the hardbanded layer;

- hardbanding drill pipe tool joint without removing the old hardbanding material increase drill pipe tool joints and casings durability and efficiency because wear and friction coefficients obtained were significative smaller than new drill pipe tool joints and reconditioned drill pipe tool joins by method A.

References

1. The drillingformulas.com website, 2017, Available: http://www.drillingformulas.com.
2. Ulmanu V., Drillpipe & BHA Design and Components, OMV Petrom Well Academy, OMV. ROU, 2011.
3. Ulmanu V., Petroleum Tubular Materials (in Romanian), Technical Publ. House, ROU, 1992.
4. Ulmanu V., Petroleum Equipment Manufacturing and Repairing Technology (in Romanian), ILEX Publ. House, ROU, 2002.
5. Ulmanu V., Badicioiu M., Caltaru M., Dinita A., Researches regarding the hardbanding technology of drill pipes tool joints using hardfacing with welding electrode wire (Research project - in Romanian), Petroleum-Gas University, ROU, 2015.
6. Standard by American Petroleum Institute, API Specification 5D, Specification for Drill Pipe, Fifth Edition, October 2001.
7. International Standard Organization, ISO 11961:2008, Petroleum and natural gas industries - Steel drill pipe.
8. The Bestar Steel Company website, 2017, Available: http://www.apicasing.com.
9. Ripeanu R.G., Badicioiu, M., Caltaru, M., Reconditioning of drill collars by using welding technologies, Journal of the Balkan Tribological Association, vol. 21(2), 314-328, 2015.
10. Caltaru M., Badicioiu M., Ripeanu R.G., Establishing the tribological behavior of the HVOF hardfacing applied at petroleum gate valves, Journal of the Balkan Tribological Association, vol.19(3), 448-460, 2013.
11. The Arnco Company website, 2017, Available: http://www.arncotech.com.
12. The Oil&Gas Equipment Sales Company website, 2017, Available: https://www.oilandgasequipmentsales.com.
13. Ripeanu R.G., Ulmanu V., Minescu M., Tribological characterisation of HVOF hardfacing applied at parallel slide gate valves, OilDoc Conference 2013, Lubricants, Maintenance, Tribology, 22-24 January, Rosenheim, Germany, 1-6, 2013.
14. Balaceanu M., Braic V., Kiss A., Zoita C.N., Vladescu A., Braic M., Tudor I., Popescu A., Ripeanu R.G., Logofatu C. and Negrila C.C., Characteristics of arc plasma deposited TiAlZrCN coatings, Surface and Coatings Technology, vol.202, 3981-3987, 2008.
15. Cursaru D.L., Andronescu C., Pirvu C., Ripeanu R.G., The efficiency of Co-based single wall carbon nanotubes (SWNTs) in comparison with a commercial additive as AW and EP additives to mineral base oil, Wear, vol. 290-291, 133-139, 2012.

Experimental Research and Mathematical Modeling of Parameters Effecting on Cutting Force and SurfaceRoughness in CNC Turning Process

Zeqiri F[1], Alkan M[2], Kaya B[3], Toros S[2]
[1]*MitrovicaUniversity, Department of Mechanical Engineering, 40000Mitrovica, Kosovo*
[2]*Nigde Omer Halisdemir University, Department of Mechanical Engineering, 51240Nigde, Turkey*
[3]*Erciyes University, Department of Mechanical Engineering, 38039, Kayseri, Turkey*
fitim.zeqiri@uni-pr.edu[1], malkan@ohu.edu.tr[2], bulentkaya@erciyes.edu.tr[3], serkantoros@ohu.edu.tr[2]

Abstract: In this paper, the effects of cutting parameters on cutting forces and surface roughness based on Taguchi experimental design method are determined. Taguchi L9 orthogonal array is used to investigate the effects of machining parameters. Optimal cutting conditions are determined using the signal/noise (S/N) ratio which is calculated by average surface roughness and cutting force. Using results of analysis, effects of parameters on both average surface roughness and cutting forces are calculated on Minitab 17 using ANOVA method. The material that was investigated is Inconel 625 steel for two cases with heat treatment and without heat treatment. The predicted and calculated values with measurement are very close to each other. Confirmation test of results showed that the Taguchi method was very successful in the optimization of machining parameters for maximum surface roughness and cutting forces in the CNC turning process.

Keywords: Inconel 625, Taguchi design of experiments, Surface roughness, Cutting force, vibrations and temperature.

1. Introduction

Inconel-625 is one material that has high mechanical properties, and high resistance to oxidation and corrosion. This material is widely used in the gas turbine blades, aerospace components and nuclear plants. Mostly industrial and engineering applications include gas turbine components, space vehicles, submarines, steam power plants and rockets/missiles components may be used. Cutting forces and surface roughness are two important during the machining process of Super alloys materials. Therefore, experimental measurement of the cutting forces and surface roughness became unavoidable. According this in different laboratories are made research to understand the principles of chip formation, developing cutting force models, cutting process control, tool geometry optimization, tool condition monitoring and for detection and suppression of chatter vibrations.
According the values of cutting parameters, the experiment is designed by Taguchi L9 in Minitab 17 software. The regression model of surface roughness is done with ANOVA and was presented the optimization of parameter impact, [2, 4].Nowadays a lot of works have been done to improve the capability of machine tools. The continuing development and trends in manufacturing and operations

engineering cannot be sustained using current methods and processes . Some set up parameters or machining processes are affected on main parameters such as tool life. One way to reduce work hardening effect on tool life is to conduct end-milling operations at high speed rates. The study of criteria for evaluating the surface roughness represents, today, one of the most important problem for the production of some specific and functional characteristics. For this reason many authors consider the roughness as the fourth dimension of the design. Increasing productivity, decreasing costs, and maintaining high product quality at the same time are the main challenges manufacturing face today. The importance of monitoring and controlling the cutting force in CNC turning process has been well recognized in machine tool communities, [4]. A considerable amount of investigations has been directed for the purpose of the prediction and measurement of cutting forces. That is because the cutting forces generated during metal cutting have a direct influence on the generation of heat because the experiment is taking without unleaded, tool wear, vibration, quality of machined surface and accuracy of the work piece. This research paper presents the use of coated tool for dry turning of Inconel 625 with heat treatment and without heat treatment investigating the cutting force and surface roughness. The specimens were annealed at 1350 ^{0}C for 1 hour and were cooled in room temperature, [5]

2. Experimental work - work piece material

Round bars with ∅62mm diameter and 180mm length were used as work piece materials. The work material used for the present investigation is Inconel 625. The chemical composition of the work piece materials is given in table 1.

Table 1: Chemical composition of the Inconel 625

Ni	Cr	Mo	Nb	Fe	S	Al
59.64*	22.34	9.20	3.57	4.29	0.002	0.16
Co	Cu	C	Mn	P	Ta	Si
0.032	0.091	0.020	0.057	<0.005	0.014	0.28

*Calculated as the difference between 100% and the sum of the measured elements.

2.1 Material and method

The experiments were conducted under dry cutting conditions on champions made all geared CNC lathe machine. The cutting tool used was CNMG 120408NN with grade coating of TiN/TiCN PVD coated. Based on the research Inconel 625 is hard processed, for turning it is decided to use the tool holder PCLNR 2525 12A product of renowned company ISCAR.

2.2 Cutting force and surface roughness measurement.

For the experiment is used Force Measurement aperture *Kistler Type 5070A*.The cutting force Measurement, dynamometers which were mounted on the CNC lathe. The measurement equipment for the temperature measure was used *Pyrometer Cella Temp PQ 38* which is specially designed for measuring the temperature. The measures of vibration are made with equipment *Kistler Type 5134B*, and for surface measurement is used machine for Surface Measurement Talsurf, *Taylor Hobson*. In this paper we will use two basic methods of determination of cutting forces and their characteristics:

- Analytical by Taguchi method and ANOVA :
- Experimental methods: direct measurement characteristicsFx, Fy and Fz, vibration, temperature and surface roughness.

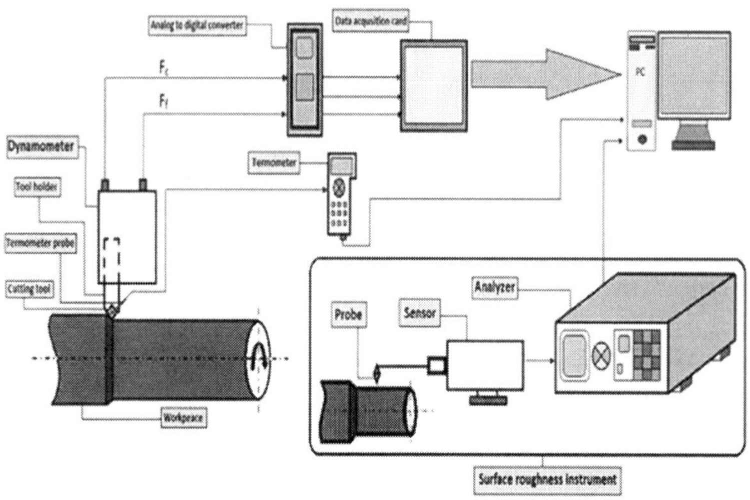

Figure 1. Diagram of experimental setup.

3. Design of the experiment

In our study paper was use Taguchi standard orthogonal array L9 with 9 DOF as the experimental design. Taguchi's orthogonal arrays are the most suitable techniques for conducting results with a less time and cost, [1, 3]. Compatible results can be achived with these design of experiment as compare to full factorial design. Experiments were planned using 3 levels for each input parameter as shown in table 1.

Table 2. Experimental results using L_9 orthogonal array.

Cutting parameters		Exp.	Speed	Feed	Depth ofcutting
Speed (v)	50	1	50	0,04	0,4
	65	2	50	0,06	0,7
	80	3	50	0,08	1,0
Feed (f)	0.04	4	65	0,04	0,7
	0.06	5	65	0,06	1,0
	0.08	6	65	0,08	0,4
Depth of cutting (d)	0.4	7	80	0,04	1,0
	0.7	8	80	0,06	0,4
	1.0	9	80	0,08	1,0

Table 3: Experimental results for cutting force.

Exp.	**Experimental results**					**Experimental results after annealed.**				
	T ^{0}C	Amplitude	Fx1	Fy1	Fz1	T ^{0}C	Amplitude	Fx2	Fy2	Fz2
1	329,05	0,4930	134,98	179.28	260,14	399.63	0.5988	150.73	214.81	298.24
2	361,21	0,3999	276,04	272.05	478,31	407.81	0.4514	305.02	299.12	517.90
3	374,29	0,1083	442,98	433.18	687,93	467.70	0.1354	512.76	490.68	729.58
4	333,22	0,3108	262,06	230.25	383,37	404.28	0.3771	299.74	266.25	428.25
5	286,44	0,4281	282,97	206.21	622,56	343.72	0.5137	325.78	243.12	660.19
6	299,44	0,4617	174,01	236.20	419,92	376.24	0.5800	198.21	273.59	461.24
7	443,66	0,4322	262,1	164.92	460,55	532.61	0.5189	308.56	200.24	515.52
8	333,75	0,5217	145,71	251.37	342,28	416.19	0.6506	184.84	369.06	449.12
9	372,71	0,6544	371,15	348.75	642,74	438.30	0.7695	425.23	406.66	682.39

Table 4: Experimental results for surface roughness.

Exp.	Experimental results					Experimental results after annealed.				
	T °C	Amplitude	Ra1	Rz1	Rt1	T °C	Amplitude	Ra2	Rz2	Rt2
1	329.05	0.4930	0.9111	7.7501	5.1382	399.63	0.5988	0.9974	8.4336	5.5721
2	361.21	0.3999	1.5957	16.4656	7.7612	407.81	0.4514	1.6015	11.6238	7.4998
3	374.29	0.1083	1.5733	14.5312	8.7791	467.70	0.1354	1.6281	10.4363	8.3869
4	333.22	0.3108	1.1743	10.3691	6.4573	404.28	0.3771	1.2317	8.6635	6.8242
5	286.44	0.4281	1.5142	9.7913	7.842	343.72	0.5137	1.6202	10.7137	8.5033
6	299.44	0.4617	2.0923	15.9306	10.3342	376.24	0.5800	2.2018	19.3488	10.6679
7	443.66	0.4322	0.8837	6.8065	5.411	532.61	0.5189	0.9604	8.2655	5.381
8	333.75	0.5217	1.3552	8.7036	7.1143	416.19	0.6506	1.2709	17.8425	7.6771
9	372.71	0.6544	2.3455	11.6071	10.2512	438.30	0.7695	2.3214	11.8395	10.2814

3.1 Main effect plots of S/N ratio for output responses

S/N ratio is the Signal-to-noise ratio, measuring the variation of the observation in a run. It is generally calculated to select the optimal conditions of input parameters for getting the best output results under a particular working conditions, [2-4]. The value of S/N ratio should be always maximum to reduce the effect of noise. There are three different formulae for calculating the S/N ratio depending on the type of responses. Here, smaller-is-better methodology has been implemented for each output responses which can be calculated as below: $SN_S = -10 \cdot \log\left(\frac{1}{n}\sum_{i=1}^{n} y_i^2\right)$

3. 2 Analysis of Variance (ANOVA)

ANOVA table is generally prepared to find the significant input parameters which will mostly effect the output responses. Analysis of variance table was prepared for each output response with a significant level of α = 0.05 and confidence level of 95%. The sources present in the ANOVA table having P-Value less than 0.05 had been treated as the significant parameter for the respective output response.In the table given the analysis with ANOVA, for cutting force and surface roughness without heat treatment.

Table 5: Analysis of Variance for cutting force F_{x1}.

	df	SS	MS	F	Significance F	
Regression	5	90751.12	18150.22	3.23	0.181	
Residual	3	16852.82	5617.60			
Total	8	107603.95				
		Coefficients	**Standard Error**	**P-value**	**Regression Statistics**	
Intercept		-214.39	274.88	0.492	Multiple R	0.91
Speed		-0.74	3.21	0.832	R Square	0.84
Feed		3336.74	1554.82	0.121	Adj R Square	0.58
Depth of cutting		288.39	137.62	0.127	Standard Error	74.95
Temperature		0.435	0.596	0.518	Observations	9
Amplitude		-42.28	257.01	0.87		

Table 6: Analysis of Variance for cutting force F_{y1}.

	df	SS	MS	F	Significance F	
Regression	5	44988.729	8997.746	2.082	0.290	
Residual	3	12964.480	4321.493			
Total	8	57953.209				
		Coefficients	**Standard Error**	**P-value**	**Regression Statistics**	
Intercept		-42.221	226.125	0.469	Multiple R	0.881
Speed		-0.655	2.486	0.660	R Square	0.776
Feed		3955.160	1374.106	0.118	Adj R Square	0.403
Depth of cutting		-17.608	119.349	0.126	Standard Error	65.738
Temperature		0.555	0.592	0.450	Observations	9
Amplitude		-177.843	236.637	0.948		

Table 7: Analysis of Variance for cutting force F_{z1}.

	df	SS	MS	F	Significance F
Regression	5	166360.228	33272.046	23.426	0.013
Residual	3	4260.898	1420.299		
Total	8	170621.126			

	Coefficients	Standard Error	P-value	Regression Statistics	
Intercept	-146.115	129.635	0.342	Multiple R	0.987
Speed	-0.920	1.425	0.564	R Square	0.975
Feed	5329.360	787.759	0.007	Adj R Square	0.933
Depth of cutting	481.723	68.422	0.006	Standard Error	37.687
Temperature	-0.155	0.339	0.679	Observations	9.000
Amplitude	189.862	135.661	0.256		

Table 8: Analysis of Variance for cutting force R_{a1}.

	df	SS	MS	F	Significance F
Regression	5	1.728	0.346	5.030	0.107
Residual	3	0.206	0.069		
Total	8	1.934			

	Coefficients	Standard Error	P-value	Regression Statistics	
Intercept	-0.203	0.902	0.836	Multiple R	0.945
Speed	-0.001	0.010	0.911	R Square	0.893
Feed	24.979	5.479	0.020	Adj R Square	0.716
Depth of cutting	0.185	0.476	0.723	Standard Error	0.262
Temperature	-0.001	0.002	0.708	Observations	9
Amplitude	1.147	0.943	0.311		

Table 9: Analysis of Variance for cutting force R_{z1}.

	df	SS	MS	F	Significance F
Regression	4	72.336	18.084	2.495	0.199
Residual	4	28.995	7.249		
Total	8	101.331			

	Coefficients	Standard Error	P-value	Regression Statistics	
Intercept	9.905	8.989	0.332	Multiple R	0.845
Speed	-0.136	0.077	0.152	R Square	0.714
Feed	146.165	56.252	0.060	Adj R Square	0.428
Depth of cutting	-1.224	4.132	0.782	Standard Error	2.692
Temperature	0.007	0.024	0.797	Observations	9
Amplitude	9.905	8.989	0.332		

Table 10: Analysis of Variance for cutting force R_{t1}.

	df	SS	MS	F	Significance F
Regression	5	26.181	5.236	7.341	0.066
Residual	3	2.140	0.713		
Total	8	28.321			

	Coefficients	Standard Error	P-value	Regression Statistics	
Intercept	2.092	2.905	0.523	Multiple R	0.961
Speed	0.010	0.032	0.784	R Square	0.924
Feed	100.458	17.654	0.011	Adj R Square	0.799
Depth of cutting	0.434	1.533	0.796	Standard Error	0.845
Temperature	-0.005	0.008	0.535	Observations	9
Amplitude	1.143	3.040	0.732		

In the table: 11-16 given the analysis with ANOVA, for cutting force and surface roughness after heat treatment (annealed).

Table 11: Analysis of Variance for cutting force F_{x2}.

	df	*SS*	*MS*	*F*	*Significance F*
Regression	5	69713.56	13942.71	3.220	0.182
Residual	3	12986.47	4328.82		
Total	8	82700.04			

	Coefficients	**Standard Error**	**P-value**	**Regression Statistics**	
Intercept	-187.274	226.316	0.469	Multiple R	0.91
Speed	-1.209	2.488	0.660	R Square	0.84
Feed	2994.792	1375.271	0.118	Adj R Square	0.58
Depth of cutting	250.989	119.450	0.126	Standard Error	65.79
Temperature	0.514	0.593	0.450	Observations	9.00
Amplitude	-16.785	236.838	0.948		

Table 12: Analysis of Variance for cutting force F_{y2}.

	df	*SS*	*MS*	*F*	*Significance F*
Regression	5	51805.03	10361	1.38	0.420
Residual	3	22533.07	7511		
Total	8	57953.20			

	Coefficients	**Standard Error**	**P-value**	**Regression Statistics**	
Intercept	-24.284	317.850	0.943	Multiple R	0.83
Speed	1.108	3.71476	0.784	R Square	0.69
Feed	4253.220	1797.85	0.098	Adj R Square	0.19
Depth of cutting	-79.192	159.136	0.652	Standard Error	86.66
Temperature	0.4250	0.6899	0.581	Observations	9
Amplitude	-233.652	297.184	0.489		

Table 13: Analysis of Variance for cutting force F_{z2}.

	df	SS	MS	F	Significance F
Regression	5	149968.07	29993.62	15.94	0.022
Residual	3	5643.14	1881.04		
Total	8	155611.22			

	Coefficients	**Standard Error**	**P-value**	**Regression Statistics**	
Intercept	-103.352	159.064	0.562	Multiple R	0.98
Speed	0.771	1.859	0.706	R Square	0.96
Feed	5200.68	899.717	0.010	Adj R Square	0.90
Depth of cutting	422.454	79.638	0.013	Standard Error	43.37
Temperature	-0.1427	0.345	0.707	Observations	9
Amplitude	63.558	148.722	0.697		

Table 14: Analysis of Variance for cutting force R_{a2}.

	df	SS	MS	F	Significance F
Regression	5	1.634	0.327	4.254	0.132
Residual	3	0.230	0.077		
Total	8	1.864			

	Coefficients	**Standard Error**	**P-value**	**Regression Statistics**	
Intercept	0.229	1.016	0.836	Multiple R	0.936
Speed	-0.001	0.012	0.919	R Square	0.876
Feed	24.022	5.750	0.025	Adj R Square	0.670
Depth of cutting	0.295	0.509	0.603	Standard Error	0.277
Temperature	-0.002	0.002	0.519	Observations	9
Amplitude	0.829	0.950	0.447		

Table 15: Analysis of Variance for cutting force R_{z2}.

	df	SS	MS	F	Significance F	
Regression	5	107.674	21.535	2.877	0.207	
Residual	3	22.454	7.485			
Total	8	130.128				
		Coefficients	**Standard Error**	**P-value**	**Regression Statistics**	
Intercept		12.586	10.034	0.299	Multiple R	0.910
Speed		0.189	0.117	0.206	R Square	0.827
Feed		126.441	56.753	0.112	Adj R Square	0.540
Depth of cutting		-11.108	5.023	0.114	Standard Error	2.736
Temperature		-0.018	0.022	0.463	Observations	9
Amplitude		-9.918	9.381	0.368		

Table 16: Analysis of Variance for cutting force R_{t2}.

	df	SS	MS	F	Significance F	
Regression	5	25.665	5.133	11.084	0.038	
Residual	3	1.389	0.463			
Total	8	27.055				
		Coefficients	**Standard Error**	**P-value**	**Regression Statistics**	
Intercept		4.976	2.496	0.523	Multiple R	0.974
Speed		0.036	0.029	0.784	R Square	0.949
Feed		91.039	14.117	0.011	Adj R Square	0.863
Depth of cutting		0.049	1.250	0.796	Standard Error	0.681
Temperature		-0.012	0.005	0.535	Observations	9
Amplitude		-0.060	2.334	0.732		

4. Regression analysis equations

Regression analysis is a statistical process for estimating the relationships among variables. It includes many techniques for modelling and analysing several variables, when the focus is on the relationship between a dependent variable and one or more independent variables. Regression analysis was employed to derive the predictive equations of the cutting forces, and roundness error. The predictive equations generated for cutting force and surface roughness in Minitab 17 by ANOVA analysis.
Regression analysis equations for cutting force:

$Fx_1 = -187-1.21\cdot v + 2995\cdot f + 251\cdot d + 0.514\cdot t - 17\cdot a,$ $R^2=84\%$
$Fy_1 = -42-0.66\cdot v + 3955\cdot f - 18\cdot d + 0.555\cdot t - 178\cdot a,$ $R^2=77\%$
$Fz_1 = -146 - 0.92\cdot v + 5329\cdot f + 481.7\cdot d - 0.155\cdot t + 190\cdot a,$ $R^2=97\%$
$Fx_2 = -187-1.21\cdot v + 2995\cdot f + 251\cdot d + 0.514\cdot t - 17\cdot a,$ $R^2=84\%$
$Fy_2 = -42-0.66\cdot v + 3955\cdot f - 18\cdot d + 0.555\cdot t - 178\cdot a,$ $R^2=77\%$
$Fz_2 = -146 - 0.92\cdot v + 5329\cdot f + 481.7\cdot d - 0.155\cdot t + 190\cdot a,$ $R^2=97\%$

Regression analysis equations for surface roughness:

$Ra_1 = -0.203 - 0.00121\cdot v + 24.98\cdot f + 0.185\cdot d - 0.00097\cdot t + 1.147\cdot a,$ $R^2=89\%$
$Rt_1 = 10.6 - 0.116\cdot v + 145.7\cdot f - 1.99\cdot d + 0.0062\cdot t - 2.8\cdot a,$ $R^2=71\%$
$Rz = 2.09 + 0.0095\cdot v + 100.5\cdot f + 0.43\cdot d - 0.00532\cdot t + 1.14\cdot a,$ $R^2=92\%$
$Ra_2 = -0.23 - 0.0013\cdot v + 24.02\cdot f + 0.295\cdot d - 0.00161\cdot t + 0.829\cdot a,$ $R^2=87\%$
$Rt_2 = 4.98 + 0.0359\cdot v + 91.0\cdot f - 0.05\cdot d - 0.01167\cdot t - 0.06\cdot a,$ $R^2=97\%$
$Rz_2 = 12.6 + 0.189\cdot v + 126.4\cdot f + 11.11\cdot d - 0.0183\cdot t - 9.92\cdot a,$ $R^2=82\%$

4.1 Normal probability plots

To check the validation or the authenticity of the above statistical model normal probability plot. There will be no inadequacy in the model and there will no unusual structure in the model. It will predict better result in future. So from the graphs as shown in figure 2 and 3, it has been cleared that our statistical model is an authenticate model.

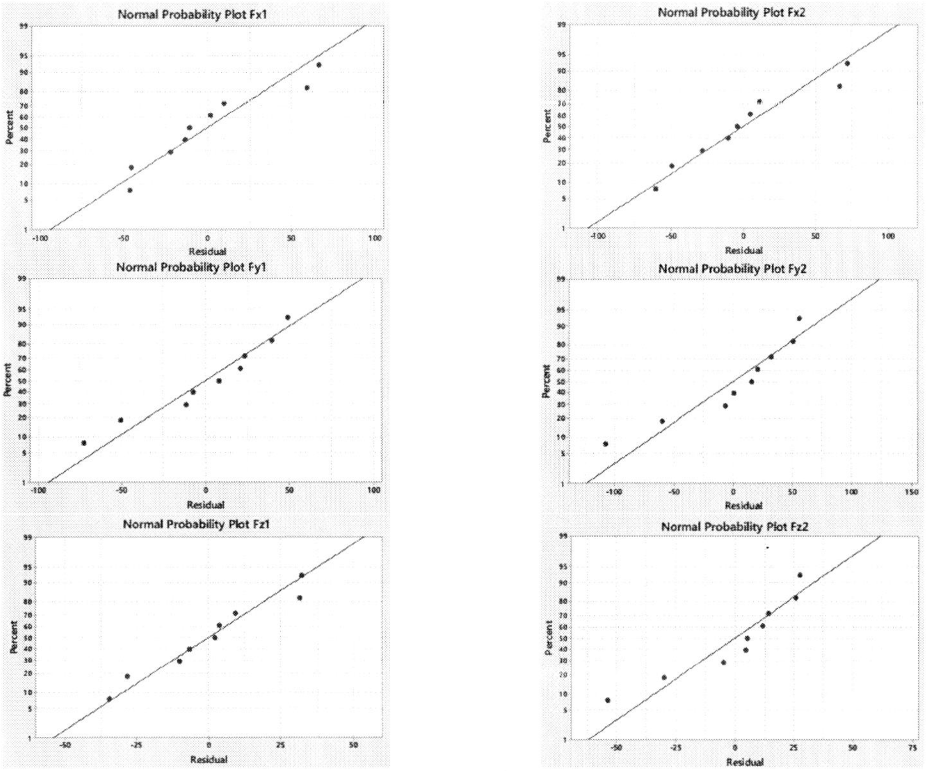

Figure 2. Normal Probability plots for cutting force.

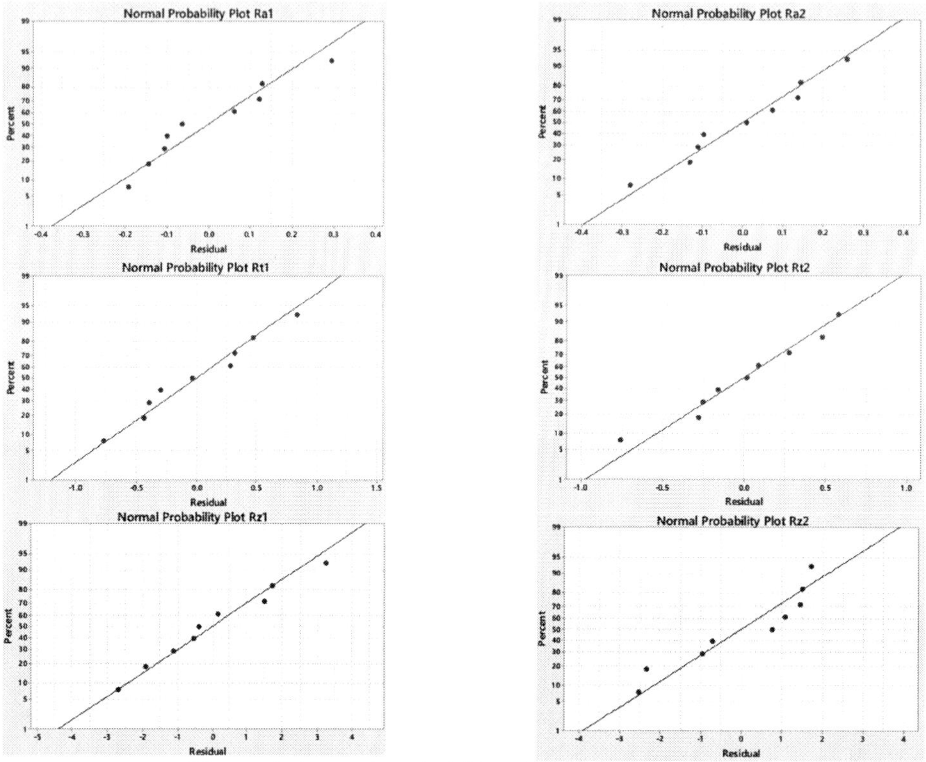

Figure 3. Normal Probability plots for surface roughness.

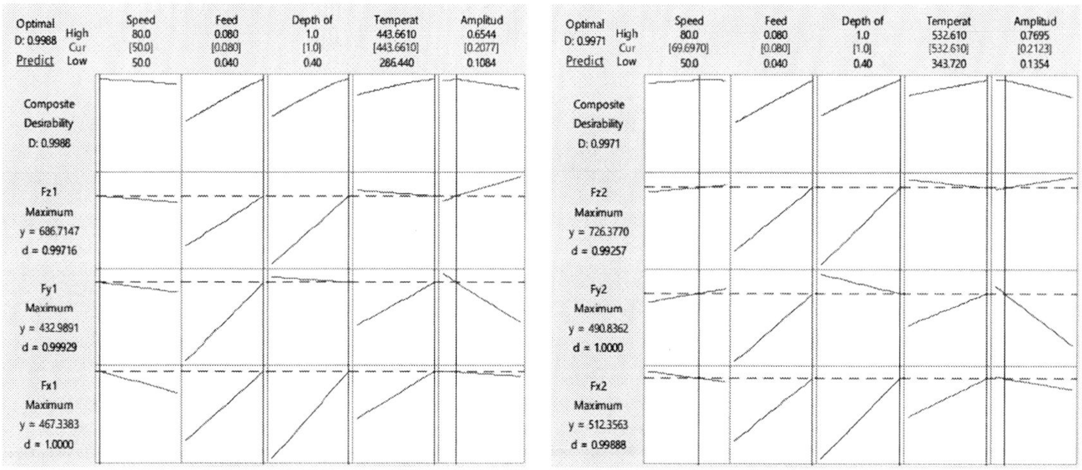

Figure 4. Multi objective optimization plot for cutting force.

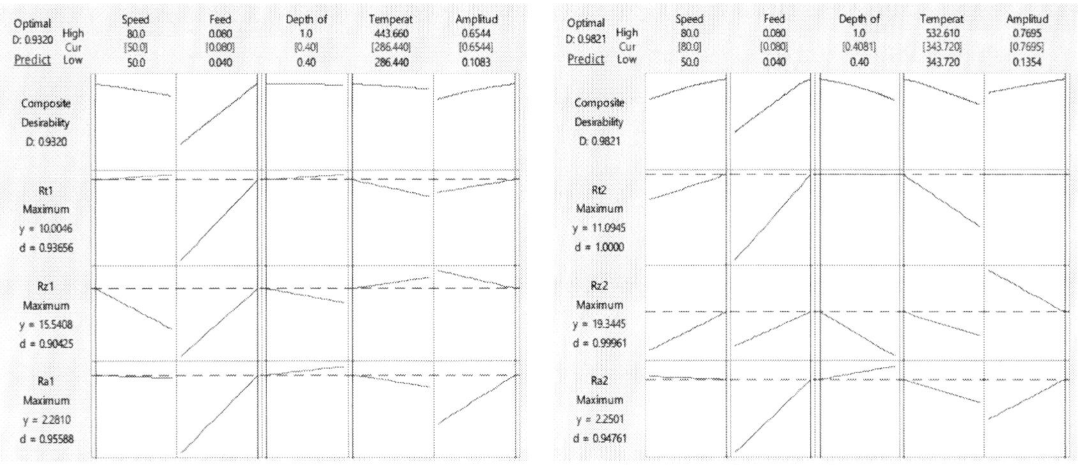

Figure 5. Multi objective optimization plot for surface roughness.

From the analysis of the obtained values in table 3, and from the analysis of figure 4 we can see that after heat treatment (annealed) the values of cutting forces increase. But optimization of maximal cutting forces must be done with: v_{min}, f_{max}, d_{max} and with t_{max}, a_{mes}.

In other hand form the analysis of experimental values on table 4 and analysis of diagrams in figure 5 we cannot see big difference between values of surface roughness. But the optimization of maximal values of surface roughness must be done with: v_{min}, f_{max}, d_{min} and with t_{min}, a_{max}.

Heat treatment more influence has in the cutting forces, which influence in the increase of the tool wear. While in the surface roughness make the structure adjustment so we can reach better surface roughness.

5. Conclusion

The realization of the experiment and analytical researches, as well as complete analysis of so far literature, enables us to come up with conclusions that due the complexity and the conditions under which the cutting occurs, cutting force, temperature, vibration and surface roughness defining is necessary to be presented by mathematical model, by which physical phenomenon changes and technological effects of the treated surface roughness is possible to be foreseen. From results of mathematical models for cutting forces and from their graphic shows we can conclude:

- ✓ With the increase of cutting speed, cutting forces decrease,
- ✓ With the increase of cutting feed, cutting forces increase,
- ✓ With the increase of depth of cutting, cutting force increase,
- ✓ With the increase of temperature and vibration, cutting force increase,

From the obtained results in experimental way, mathematical models and graphics interpretation we can see that with the increase of cutting speed, the roughness parameters Ra, Rt and Rz decrease. While with the increase of cutting depth also increase the roughness parameters. The temperature has e small influence in the surface roughness parameters, while the vibrations have much more influence on the surface roughness parameters.

6. References

1. Kaya B, Oysu C, Ertunc M. H, Force – torque based on-line wear estimation system for CNC milling of Inconel 718 using neural networks, Advances in engineering softw.42, 76-84, 2011.

2. Rao C. J, Nageswara Rao D, Srihari P, Influence of cutting parameters on cutting force and surface finish in turning operation, Procedia Engineering 64, 1405-1415, 2013.
3. Deep D. Vadalia, Alpesh H. Makwana, Investigation of cutting and tool wear in dry turning of Inconel 625, International Journal of advance engineering and research development, V.2, Issue 5, May 2015.
4. M. Nalbant, H Gökkaya, G. Sur, Application of taguchi method in the optimization of cutting parameters for surface roughness in turning, Materials and Design 28,1379-1385, 2008.
5. Hemant J, Jaya T, Ravindra B, Sanjay J, Avinash K, Optimization and evaluation of machining parameters for turning operation of Inconel 625, Materials Today: Proceedings 2, 2306-2313, 2015.
6. FujiaXu, YaohuiLv, Yuxin Liu, BinishiXu, Penge He, Effect of heat treatment on microstructure and mechanical properties of Inconel 625 alloy fabricated by pulsed plasma arc deposition, PhysicsProcedia 50, 48-54, 2013.
7. Venkatesan K, Ramanujam R, Saxena V, Chawdhury N, Choudhary V, Influence of cutting parameters on dry machining of Inconel 625 alloy with coated carbide insert- a statical approach, Journal of engineering and applied sciences, vol.9, no 3 , march 2014.

Experimental results of a hydrodynamic friction behaviour of a linear contact at low sliding velocity

A. Bouzana [1], *A. Guermat* [2], *F. Belarifi* [3*]
[1,3]*Institut de Maintenance et sécurité Industrielle- Université d'Oran2. B.P 170 El M'Naouer Oran, Algeria*
[2,3]*Laboratoire des Sciences de la Matière Condensé (Equipe de Tribologie et nano tribologie)- Université d'Oran1.B.P 1524 El M'Naouer Oran, Algeria*
bouzana_555@outlook.fr[1], aek_guermat@yahoo.fr[2]
**Corresponding author:* belarifi_farid@yahoo.fr[3]

Abstract: We propose in this work the experimental results of the lubricated friction behavior of linear contact (finite length) in isoviscous hydrodynamic regime. This study was made on a tribometer Plint - Cameron TE77, using a pure mineral oil lubricant (N175). without additives for three loads 20, 40 and 80 Newton. and a velocity, range varying from 0.05 to 0.4 ms-1, trials are held in pure sliding mode for a total distance of displacement L = 15mm. The studied contact is a cylinder/cylinder. The geometry of test pieces is part of a piston ring and a liner of a real engine. The first cylinder represents the male part with material of MKJet nuance having undergoes a surface coating by thermal projection (HVOF). the second cylinder represents the female part whose material is cast iron of nuance FGL, without surface treatment, and whose dimensions were adapted to minimize the computational error on the speed of sliding and the force of friction which is lower than 5%. Processing the results recorded for ten cycles with four hundred points per cycle to the extraction of average curves, enables us to plot the curves of friction according to velocity and thereafter the curve of Stribeck. The results show that we can get a total isoviscous regime for loads 20 and 40N, however for load 80N, this regime is partial, as it comes off the final curve from a speed value 0.1 m / s. the values of the friction coefficient varies for the three loads used between 0.004 and 0.017. These results show the possibility of obtaining a hydrodynamic regime with high load and low speed, with treatments suitable surfaces and are made to reduce wear and increase the lifetime of the mechanism.

Keywords: linear contact, hydrodynamic lubrication, friction, tribometer Plint-Cameron Up to

1. Introduction

The linear contacts such as gears, roller bearings, cams, are generally classified in the category of hertzian contacts[1], considering that they work under strong loads and, in this fact, the birth of strong deformations. In lubricated mode, that elements work in mixed lubrication or elastohydrodynamic mode[2]. On the other hand when it is a question of working under weak load (case of micro or nano mechanisms), the phenomenon of hertzian elasticity is null, we find the hydrodynamic regime [2,3]. We propose in this work the experimental results of the lubricated friction behavior of linear contact (finite length) in isoviscous hydrodynamic regime. This study was made on a tribometer Plint - Cameron TE77[4]. The geometry of test pieces is part of a piston ring and a liner of a real engine. Processing the results recorded for ten cycles with four hundred points per cycle to the extraction of average curves, enables us to plot the curves of friction according to velocity and thereafter the curve of Stribeck [5,6] This curve can be divided into three areas that are the lubrication boundary lubrication area (BL), hydrodynamic lubrication (HL) or elastohydrodynamic (EHD) and mixed lubrication (ML) which is the transition between the limit and hydrodynamic lubrication [6]. Figure 1 indicates the position of these various regimes on the Stribeck curve. This curve gives the total friction coefficient (*f*) as a function of the parameter Somerfield (*S*) [1,5]. who originally written.

$$S = \eta_0 \frac{V}{P} \tag{1}$$

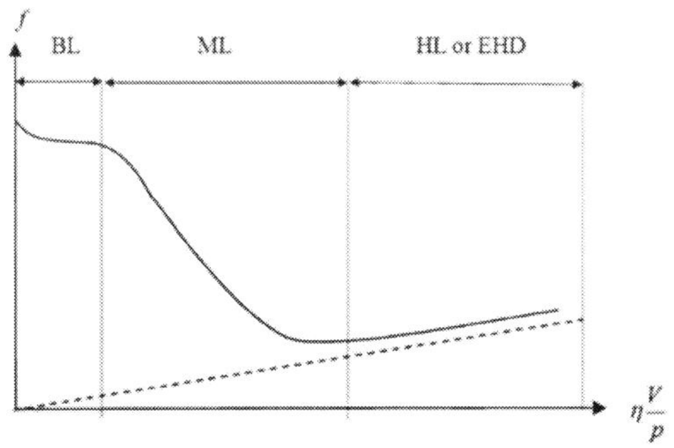

Figure 1. Stribeck curve

2. .Experimental study

This study is to make lubricated friction tests of a linear contact on a tribometer Plint - Cameron TE77(Fig.2a), using a pure mineral oil lubricant (N175). without additives for three loads 20, 40 and 80 Newton. and a velocity, range varying from 0.05 to 0.4 ms^{-1} trials are held in pure sliding mode for a total distance of displacement L = 15mm. This tribometer allows using samples of real parts, and its kinematic can be adaptable to be close to the operating conditions of the majority of contact types [5]. The figure2a shows the principle of the tribometer and the image of figure 2b the arrangement of test specimens and their characteristics. The possible test configurations in this version of the tribometer are planes contacts , cylinder on plane surface contact and Contact cylinder to cylinder.

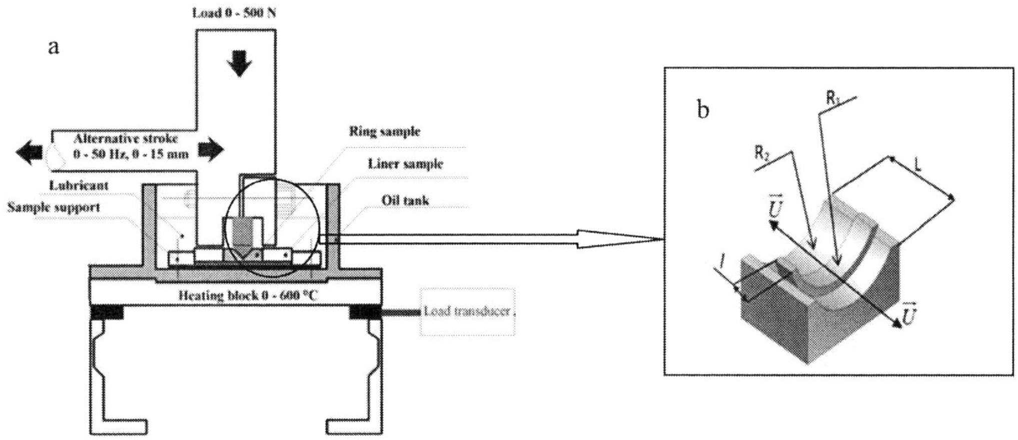

Figure 2.
a: Tribometer Cameron – Plint TE77 [4]
b: Characteristics of the test ($\vec{U}$: sliding velocity)

IOP Conf. Series: Materials Science and Engineering **295** (2018) 012012 doi:10.1088/1757-899X/295/1/012012

2.1. Test pieces

The studied contact is a cylinder-cylinder in pure sliding. The geometry of test pieces is part of a piston ring and a liner of a real engine (fig2b). The first represents the male part with material of MKJET nuance having undergoes a surface coating by thermal projection (HVOF), the second represents the female part whose material is cast iron of nuance FGL, without surface treatment and whose dimensions were adapted to minimize the computational error on the speed of sliding and the force of friction which is lower than 5%, whose characteristics are those of table I.

Table 1. Test characteristics

Geometry				Young Modulus		Lubricant (N175)
l (m)	R_1(m)	R_2(m)	L(m)	E_1 (GPa)	E_2 (GPa)	η (Pas)
4.5 10^{-3}	42.5 10^{-3}	42.9 10^{-3}	15 10^{-3}	590	160	0.085

2.2. Experimental procedure

The tests were prepared, carried out and exploited by developing a procedure that consists in the following stages:

- After a preliminary scanning of a certain number of loads and speeds combinations, we opted for loads of 20, 40 and 80N.
- The total distance of displacements of contact is 15mm with an instantaneous displacement of 4.5 mm and a hold time of ten cycles for frequencies of seven and nine Hertz (7HZ and 9HZ).

3. Results and discussion

The Figure 3, indicate the variation of friction for loads of 20N, 40N and 80N according to velocity variation. The Figure4, indicate the variation of friction according to the parameter of Somerfield (*S*).

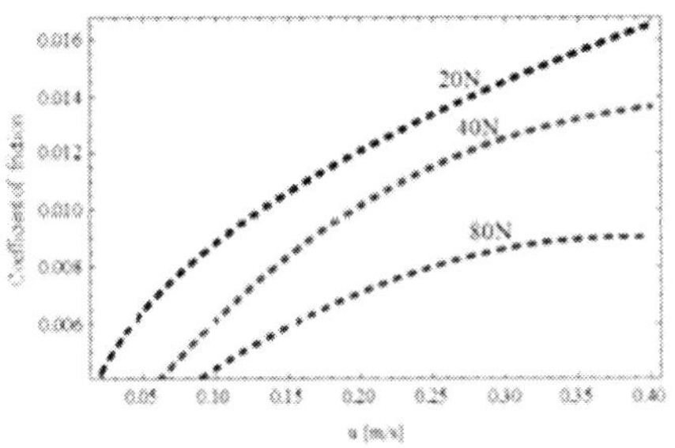

Figure 3. friction according to the sliding velocity

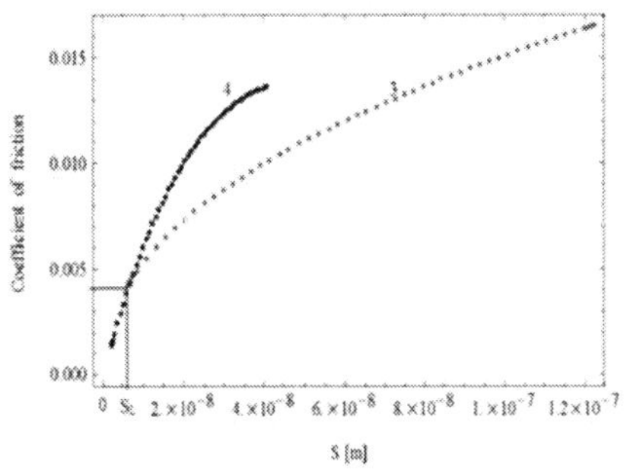

Figure 4. friction according to the Somerfield Parameter (S)
3: Curve for 20 and 40N
4: Curve for 80N

Concerning figure 3, shows a growth of friction whatever the load used. This growth, is inversely proportional to the load. The value limits accounts for 0.017. What shows that we summons in hydrodynamic mode. What enables us to plot the curve of Stribeck for these various loads indicated by the figure 4, that shows two distinct curves: the curve (3) represents the combination of loads 20 and 40N and the curve (4) represents the values for load of 80N. It is noticed that the curve (4) is

detached from the curve (3) at the value of SL ~ 510^{-9} ,for a friction coefficient of 0.0042. This curve shows that we have two modes. The First is rigid isoviscous whose SL represents the limit and the second is viscoelastic or rigid piezoviscous.
we think this is a viscoelastic field, because the experimental conditions used, the viscosity piezo has not had time to form, among other our model [6] confirms this situation. The experiment shows that, for the load of 80N this mode starts from a sliding speed of about $0.1 ms^{-1}$.
Concerning the curve (3) the isoviscous mode is observed and we recovers continuity of the curve for loads of 20N and 40N. We deduce the final curve of Stribeck given by figure 5, which represents the final curve of variation of the friction coefficient of according to the parameter S. For the rigid hydrodynamic mode, in the case of lubricated linear contact, we notices the increase of friction (f), with the increase of S which varies between the values $10^{-8}<S<1.2\ 10^{-7}$, for values of friction $0.004<f<0.017$.

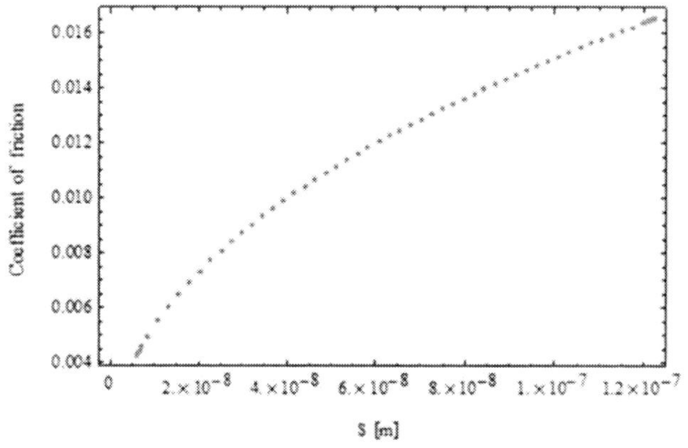

Figure 5. Final experimental curve of friction according to the Somerfield Parameter (S)

4. Confrontation

Confrontation of the Stribeck curves, regarding the results obtained by models[6] and experimental, is indicated by the figure 6, in which we notices convergence of the two curves up to the S value of S=4x 10^{-8} for a friction coefficient $f= 0.01$, subsequently the two curves are detached from one another, but remain converging on average for an error of about 5%.

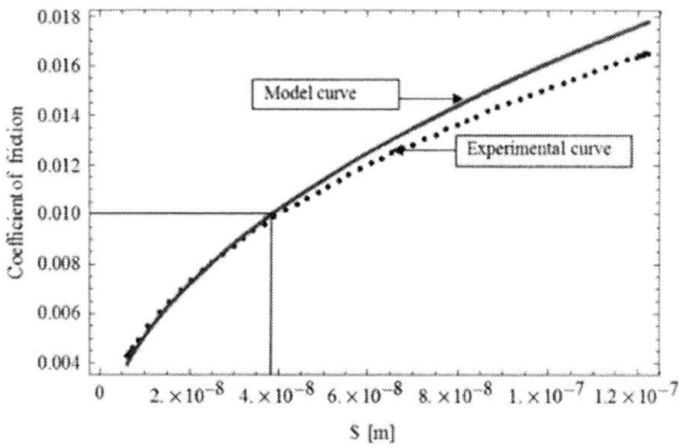

Figure 6. Confrontation Model[6]/Experimental results

5. Conclusion

The results show that we can get a total isoviscous regime for loads 20 and 40N, however for load 80N, this regime is partial, as it comes off the final curve from a speed value 0.1 ms^{-1}. the values of the friction coefficient varies for the three loads used between 0.004 and 0.017. The confrontation of the experimental Stribeck curve with model curve, shows same shape as the model. Comparison shows a total convergence for very low speeds below 0.15 ms^{-1}, after that we have a separation of the two curves but within an average convergence for an error of 5%. It is noted that the combination of material with an adequate surface treatment or surface coating, enables us to increase the values of loads and to stay in an isoviscous hydrodynamic mode. These results implies the possibility of obtaining a hydrodynamic regime with high load and low speed, with treatments suitable surfaces and are made to reduce wear and increase the lifetime of the mechanism.

References

1. K. F. Martin, A review of friction prediction in gear teeth. Wear, Vol. 49 201 - 238,1978
2. F. Belarifi, J. Blouet, G. Inglebert, A. Benamar, Confrontation between a mixed lubrication model and An experimental survey on the behaviour to the friction of spur gear teeth, Mechanic & Industry Vol. 7 pp. 527-536, 2006.
3. Bernard J. Hamrock, Film thickness for different regimes of fluid film lubrication. Bulletin of the University "Lulea"- Sweden (Lecture 4) , NASA Technical Memorandum TM-81550, 1980.
4. A. Guermat, G. Monteil, M. Bouchetara, Characterization of various coatings in terms of friction and wear for internal combustion engine piston rings; Mechanika. Vol.17(2),208-215, 2011.
5. E.R.M. Gelinck 1, D.J. Schipper, Calculation of Stribeck curves for line contacts; Tribology International, Vol.33, 175–181,2000
6. F. Belarifi, G. Inglebert, Numerical modelling of the hydrodynamic behaviour of linear contacts, Journal of the Balkan Tribological Association, vol. 21, no. 3, 630-639, 2015

A Performance Comparison Study of Uncoated and TiAlN Coated Carbide End Mill on Machining of the Al-35Zn Alloy

Bayraktar S[1*], Hekimoglu A P[1], Turgut Y[2], Haciosmanoglu M[1]
[1]Recep Tayyip Erdogan University, Department of Mechanical Engineering, 53100, Rize, Turkey
[2]Gazi University, Department of Manufacturing Engineering, 06500, Ankara, Turkey
senol.bayraktar@erdogan.edu.tr[1], ali.hekimoglu@erdogan.edu.tr[1], yturgut@gazi.edu.tr[2], murat_haciosmaoglu16@erdogan.edu.tr[1]

Abstract: In this study, Al-35Zn alloy was produced by permanent mold casting. To investigate the cutting performance of uncoated and TiAlN coated carbide end mills on this alloy, a series of tests were carried out in the CNC vertical machining center at a constant cutting speed, feed rate and depth of cut. The results obtained from the tests showed that uncoated carbide end mill have lower cutting force and surface roughness than TiAlN coated carbide end mill. These observations are discussed in terms of the alloys properties, cutting tool surfaces, and friction and wear behavior between the cutting tool and the material.

Keywords: Al-Zn alloy, Machinability, Adhesion wear, Milling

1. Introduction

Due to their some advantages over traditional industrial alloys such as cast iron, bronze and brass, aluminum-based alloys are heavily preferred, especially in the automotive, aerospace and aerospace industries [1-2]. The most important advantages of these alloys are easy and economical production, high specific strength, superior wear and corrosion resistance, and ideal tribological behavior even under overload and inadequate lubrication conditions [3-5]. Aluminum based alloys are used in the above mentioned industries especially for the manufacturing of spare parts such as sliding bearing, carburetor, cylinder lid, gearbox cover, engine feet, connecting rod and rim [6-7]. The surface finishing, service life and friction behavior of these parts depend on their surface roughness that reaches its final value after machining [8-9]. The surface finishing of these parts also significantly affects their fatigue strength, corrosion resistance, and tribological properties [10]. Therefore, the measurement and characterization of surface roughness in the aluminum based alloys is important for the optimization of machining operations. Many factors such as cutting speed, feed rate and cutting tool geometry effect the machining of these alloys [11]. These factors also have a great effect on the formation of cutting forces, surface roughness and build up edge (BUE) formation during the machining [12-13]. Formation of BUE and build up layer (BUL) are important determinants on the average surface roughness (Ra) values of the work pieces. As the hardness of the aluminum alloys increases, the surface roughness and the formation of BUE decrease [14]. Formation of BUE is accelerated by deactivating the coating, which causes the cutting tool geometry to change. On the other hand, the surface roughness decreased with the increasing of the BUE in short-time cutting operations [15]. In the study of turning the AA7075 alloy, it was observed that as the cutting speed increased, the formation of stacked chips on the cutting tool decreased, which led to a decrease in surface roughness [16]. It is known that in the milling of Al-7075 alloy, it is necessary to use low feed, high cutting speeds for optimum surface quality, and surfaces treated with carbide cutters are better than HSS and TiN coated cutters [17]. It has been suggested that Al-5083 alloy is milled with two-flutes carbide end mill with high cutting speed and low feed rate in terms of good surface quality [1]. In the case of Al-5083 alloy milling with uncoated end mills, it was found that cutting force and surface roughness decreased with increasing cutting speed, but burr height increased [2]. In this study, to determine of the machinability of the Al-35Zn alloy, it was tested with uncoated and TiAlN coated carbide end mills under constant cutting speed, feed rate and depth of cut conditions.

2. Material and method

The Al-35Zn alloy was prepared using commercially pure aluminium (99.8 wt.%) and high purity zinc (99.9 wt.%). These alloying elements were melted in an induction-melting furnace and poured at a temperature of approximately 680 °C into a steel mold having a conical shape. Samples taken for the structural analysis were prepared by standard metallographic methods and etched dispersed in a 2% NaOH solution and examined in light and scanning electron microscope (SEM). The cutting tests were performed with alloy samples having the dimensions of 80x50x10 mm and repeated at least three times. A technical drawing of cutting test sample is given in Fig. 1.The cutting parameters used in the tests are given in Table 1. The cutting tool performance in the milling of test samples with both uncoated and TiAlN coated carbide end mills coated with PVD method at 900 °C was investigated experimentally in terms of cutting force and surface roughness. The coated end mills have a hardness of 3300 HV, a coating thickness of 3 μm and a friction coefficient of 0.3-0.35.

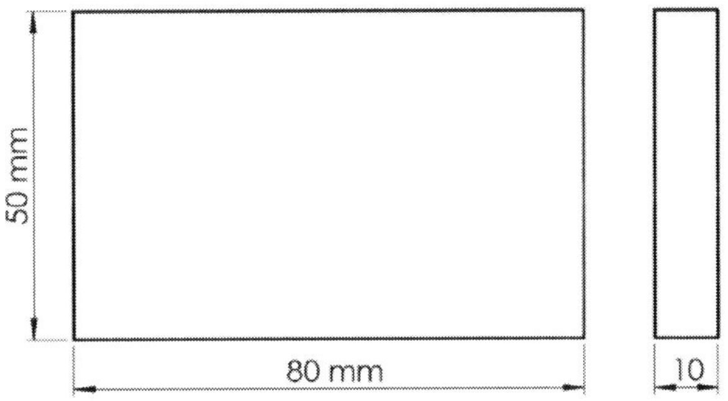

Figure 1. Technical drawing of cutting test sample

Table 1. Cutting parameters

Cutting speed (rev/min)	Feed rate (mm/tooth)	Depth of cut (mm)
1800	0.05	1.5

The cutting tests were carried out under dry conditions at a CNC vertical machining center of the Johnford-VMC 850 brand at a power of 7.5 kW. The schematic representation of the testing apparatus is shown in Fig. 2. Cutting tools were end mills with 6 mm diameter, 30° helix angle and h10 quality. The cutting forces generated during the milling of the work piece were measured using a Kistler 9273 model dynamometer and an amplifier. The cutting forces are calculated as the resultant of the forces generated in the X, Y and Z-axes. Mahr Perthometer M1 (MarSurf PS1) tracer surface roughness device (Fig. 3) was used for surface roughness measurements. Sampling length was determined as 0.8 mm and measurement length as 4 mm with the help of standard tables. Average surface roughness (Ra) values, one of the criteria for machinability, are based on the ISO 4287 standard [18] and the surface roughness values are calculated by the arithmetic mean of the deviations in the roughness profile on the machined surfaces. The values of shear force and surface roughness were determined by taking the average of the data obtained from three tests. Machined surfaces and chips examinations were performed with a SEM and optical microscope.

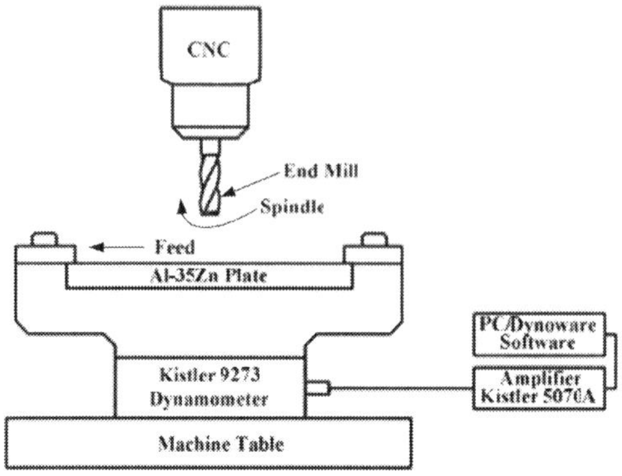

Figure 2. The schematic representation of the testing apparatus [2].

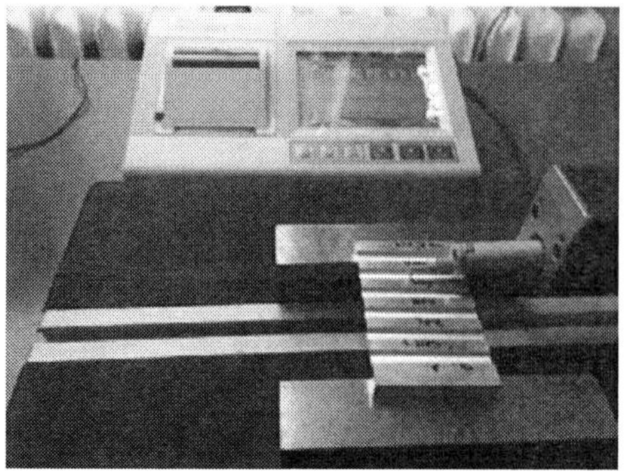

Figure 3. Surface roughness testing apparatus.

3. Results and Discussion

SEM micrograph showing the microstructure of the tested alloy is given in Fig. 4. It was observed that microstructure of this alloy consists of aluminum-rich α dendrites and zinc-rich η phase surrounding these dendrites, Fig. 4. Formation of these microstructures of the Al-Zn alloys was related to the phase transformations that take place during solidification of these alloys [19-21].

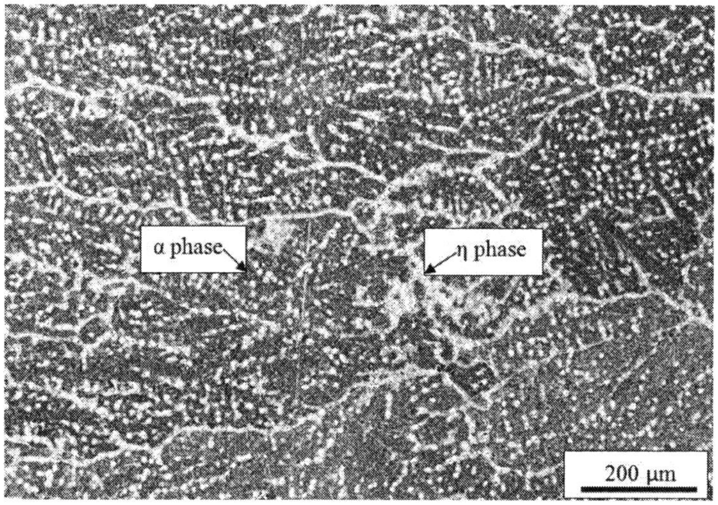

Figure 4. SEM micrograph showing the microstructure of the Al-35Zn alloy

The values of cutting force and surface roughness obtained from the cutting tests are given in Table 2. The cutting force and surface roughness values were found to be higher in the tests used coated cutting tools than in the tests used uncoated tools, Table 2. Coating can increase cutting performance of tools by increasing wear resistance of it. However, the coated cutting tools in the machining of aluminum-based alloys reduce the contact surface area between the work piece and the cutting tool by reducing the sharpness of the cutting edge. This leads to an increase in cutting forces and surface roughness, as seen in Table 2, in the machining of aluminum-based materials [22-23]. Cutting forces and surface roughness can be controlled with coated cutting tools with a lower grain size [24]. Thus, the cutting area is reduced by reducing the contact area between the cutting tool cutting edge and the work piece surface.

Table 2. Experimental results

Cutting tool	Cutting force (N)	Surface roughness (μm)
Uncoated carbide end mill	130.791	0.510
TiAlN Coated carbide end mill	141.218	1.161

Optical microscope photographs of chips obtained from the tests carried out with both cutting tools are given in Fig. 5a and b. These photographs show that dimensions of the chips are occurred in the test performed with the uncoated end mills bigger than the test performed with the coated end mills. This may be caused by low thermal conductivity of TiAlN coated end mill. Due to the low thermal conductivity of coated end mill, most of the heat occurred in the cutting zone transfer to chip. Therefore, chips bend more easily and their section becomes thinner.

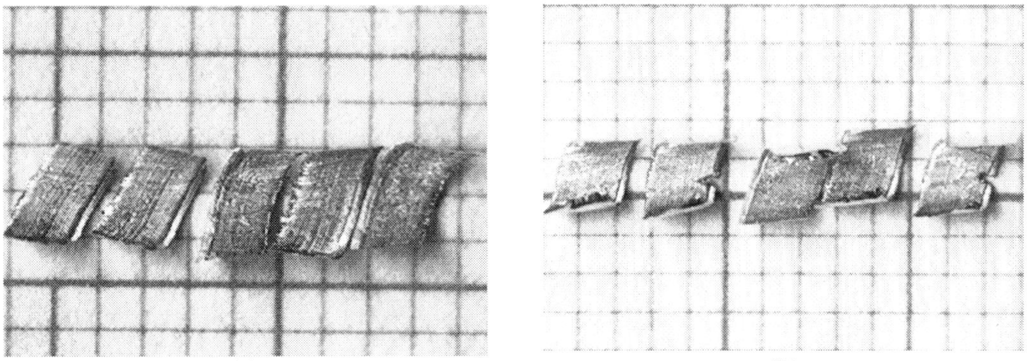

(a) (b)

Figure 5. Form of chips obtained from cutting test carried out with a) uncoated end mills b) coated end mills

It was observed that more wear materials are adhered to the cutting edge of coated carbide end mill compare to uncoated tools, Fig. 6. This is because transfer of heat generated in the cutting zone due to pressure and friction during the machining with the coated end mill is more difficult [25]. The difficulty of heat transfer increases the temperature in the cutting zone and leads to increment in ductility of the α and/or η phases, Fig. 4. This increase in ductility causes the chips to adhere to the cutting tools [26]. BUE formation in the cutting tool results in increased cutting force and adversely affects the quality of the surface quality (Fig. 3b and 4). It has been observed in many studies that cutting tools with a low coefficient of friction as well as high cutting speeds and low feed rates are recommended to prevent BUE [27].

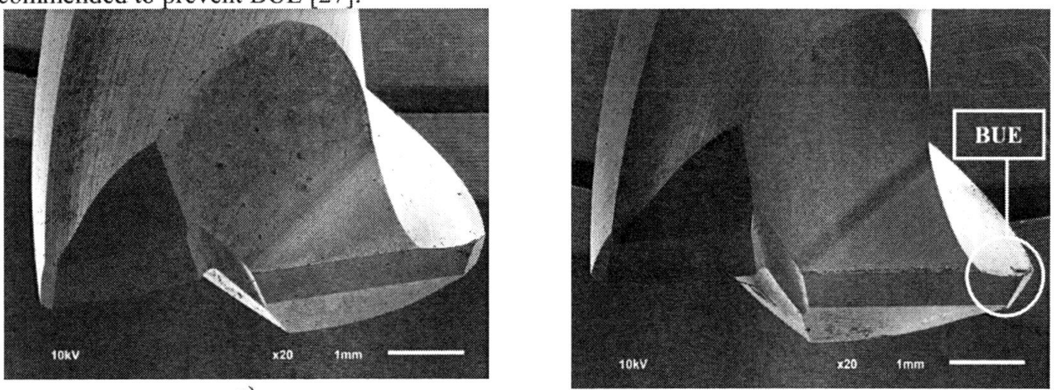

a) b)

Figure 6. a) Uncoated carbide end mill, b) Coated carbide end mill

Smearing and scratches were observed to be the main features of the machined surface of the alloy samples, Fig. 7. Smearing takes place by back transferring of the chips, scratches occur due to the removal of chips from the surface of the samples and contact of the abrasive particles adhered to the end mills to the sample surface during machining [28-31]. Smearing material due to effects of shear stress and frictional heat [32-33]. More smearing were observed in the tests carried out with coated end mills. This may be due to more heat build-up in the tests carried out with coated end mills than with the tests carried out with uncoated end mills.

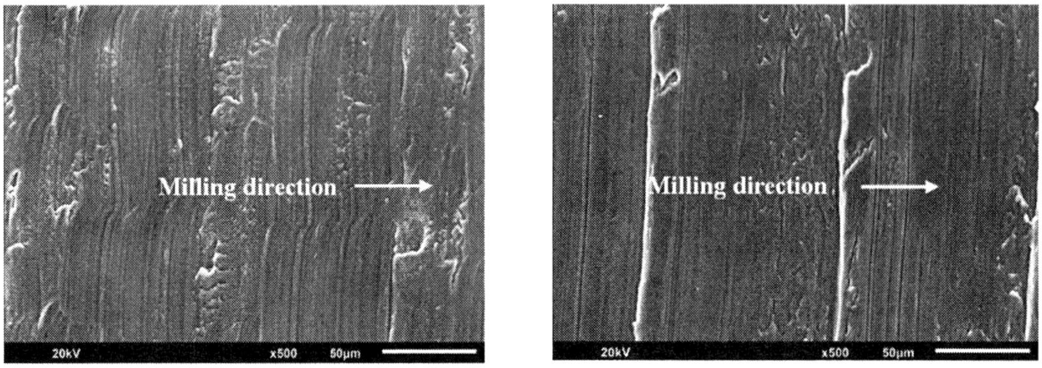

a) b)

Figure 7. SEM micrographs showing the cutted surfaces of the samples of the alloy tested with (a) uncoated end mills b) coated end mills

4. Conclusions

(1) Microstructure of the Al-35Zn alloy consists of aluminum-rich α dendrites and inter dendritic zinc-rich η phase.

(2) Machining of the Al-35Zn alloys with TiAlN coated cutting tools is require more cutting force than machining with uncoated tools.

(3) TiAlN coated cutting tools causes more surface roughness than uncoated tools in machining of the Al-35Zn alloy.

(4) Smearing and scratches are the main features of the machined surface of the Al-35Zn alloy.

(5) Build up edge (BUE) forms on cutting edge of the TiAlN coated cutting tools during the machining of the Al-35Zn alloy.

Acknowledgements
This work was supported by Research Fund of the Recep Tayyip Erdoğan University. Project Numbers: FLO-2016-612 and FHD-2017-737.

References

1. Karagiannis S., Stavropoulos P., Ziogas C., Kechagias J., Prediction of surface roughness magnitude in computer numerical controlled end milling processes using neural networks, by considering a set of influence parameters: an aluminium alloy 5083 case study, Proceedings of the Institution of Mechanical Engineers, Part B: Journal of Engineering Manufacture, Vol. 228, No 2, 233-244, 2014.
2. Bayraktar Ş., Turgut Y., Optimization of Cutting Force, Surface Roughness and Burr Height in Milling of Al-5083 Alloy, 7th International Symposium On Machining, Marmara University, İstanbul, 24-34, 2016.

3. Calayag T., Zinc Alloys Replace Bronze in Mining Equipment Bushings and Bearings, Mining Engineering, 727-728, 1983.
4. Geng H. Ma J., Friction and Wear of Al-Zn-Pb Bearings Alloy, Wear, Vol. 169, No 32, 201-207, 1993.
5. Mihaichuk W., Bess M.L., The ZA Die Casting Alloys, Society of Automotive Engineers, China, Paper No 860556, 560-568, 1986.
6. Savaşkan T., Murphy S., Mechanical Properties and Lubricated wear of Zn-25A1-Based Alloys, Wear, Vol. 116, No 23, 211-224, 1987.
7. Gross D.K., Zinc Alloys: Specifications and Processing, Society of Automotive Engineers, New York, Paper No 871952, 1069-1075, 1988.
8. Zhu Y., Yan B., Huan W., Bearing Wear Resistance of Monotectoid Zn-Al Based Alloy (ZA-35), Materials Science and Technology, Vol.11, No 1, 109-113, 1995
9. Savaşkan T., Çuvalcı H., Investigation of Tribological Properties of Zinc-Aluminum Based Bearing Alloys, 4th National Machine Design and Manufacturing Congress, Ankara, 328-336, 1990.
10. Erdoğan M., Demir Dışı Alaşımlar Mühendislik Alaşımlarının Yapı ve Özellikleri Cilt 1, Nobel Yayın Dağıtım, Ankara, 339-372, 2001.
11. Karayel D., Prediction and control of surface roughness in CNC lathe using artificial neural network, Journal of Materials Processing Technology, Vol. 209, 3125–3137, 2008.
12. Kim D.E., Hwang D.H., Experimental Investigation of the Influence of Machining Condition on the Contact Sliding Behavior of Metals, Journal of Manufacturing Science and Engineering, Korea, Vol.120, 395-400, 1998.
13. Jeelani S., Musial M., Dependence of Fatigue Life on the Surface Integrity in the machining of 2024-T351 Aluminum Alloy-Unlubricated Conditions, Journal of Materials Science, USA, 155-160, 1986.
14. Mizutani H., Yamada S., Takayanagi A., Okamura K., Mechanical Properties and Machinability of Various Aluminum Alloys, Keikinzoku/Journal of Japan Institute of Light Metals, Japan, Vol. 44, No 1, 16-21, 2000.
15. Gomez P.A., Alvarez A.M., Salguero J., Batista M., Marcos M., Analysis of the evolution of the Built-Up Edge and Built-Up Layer, Wear, Vol. 302, 1209-1218, 2013.
16. Sekmen M., Günay M., Şeker U., Effect on Formations of Built-up Edge and Built-up Layer, Surface Roughness of Cutting Speed and Rake Angle in the Machining of Aluminum Alloys, Journal of Polytechnic, Vol. 18, No 3, 141-148, 2015.
17. Hüseyinoğlu M., Tosun N., Effect of Minimum Quantity Coolant on Surface Roughness in Milling of 7075 Al Alloy, 5th International Advanced Technologies Symposium (IATS'09), Karabük, Turkey, 1321-1327, 2009.
18. ISO 4287, Geometrical Product Specifications (GPS)-Surface texture: profile method-terms, definitions and surface texture parameters, International Organisation for Standardisation", Geneva, 1997.
19. Hekimoğlu A.P., Savaskan T., "Structure and mechanical properties of Zn-(5-25) Al alloys", International Journal of Materials Research, Vol.105, 1084-1089, 2014,
20. Savaşkan T., Hekimoğlu A.P., "Microstructure and mechanical properties of Zn-15Al-based ternary and quaternary alloys", Materials Science and Engineering A-Structural Materials Properties Microstructure and Processing, vol.603, 52-57, 2014.
21. Savaşkan T., Hekimoğlu A.P., "Effect of quench-ageing treatment on the microstructure and properties of Zn-15Al-3Cu alloy", International Journal of Materials Research, vol.106, 481-487, 2015.
22. Quadrini F., Tagliaferri V., Polini R., Effect of Coating Roughness on Performance of Small CVD Diamond Coated Tools, Machining Science and Technology, Vol. 12, No 3, 390-404, 2008.
23. Qin F., Chou Y.K., Nolen D., Thompson R.G., Coating thickness effects on diamond coated cutting tools, Surface and Coatings Technology, Vol. 204, No 6, 1056-1060, 2009.

24. Heaney P.J., Sumant A.V., Torres C.D., Carpick R.W., Pfefferkorn F.E., Diamond coatings for micro end mills: enabling the dry machining of aluminum at the micro-scale, Diamond and Related Materials, Vol. 17, No 3, 223-233, 2008.
25. Nouari M., List G., Girot F., Gehin D., Effect of machining parameters and coating on wear mechanisms in dry drilling of aluminium alloys, International Journal of Machine Tools and Manufacture, Vol 45, No 12, 1436-1442, 2005.
26. Übeyli M., Karakaş M.S., Acır A., Ögel B., Study on performance of uncoated and coated tools in milling of Al–4%Cu/B4C metal matrix composites, Materials Science and Technology, Vol.23, No 8, 945-950, 2007.
27. Manna A., Bhattacharayya B., Influence of machining parameters on the machinability of particulate reinforced Al/SiC–MMC, The International Journal of Advanced Manufacturing Technology, Vol. 25, No 9, 850-856, 2005.
28. Savaşkan,T., Hekimoğlu, AP., Relationships between mechanical and tribological properties of Zn-15Al-based ternary and quaternary alloys, International Journal of Materials Research, Vol. 107, No 7, 646-652, 2016.
29. Oyman E., Aslantaş K., Pazarkaya İ., Investigation of parameters affecting chip morphology and shear band in orthogonal cutting process, Electronic Journal of Machine Technologies, Vol.10, No 2, 59-71, 2013.
30. Savaşkan T., Bican O., Dry Sliding Friction and Wear Properties of Al-25Zn-3Cu-(0-5)Si Alloys in the As-Cast and Heat-Treated Conditions, Tribology Letters Vol. 40, 327-336, 2010.
31. Savaşkan T., Hekimoğlu, A.P., Pürçek G., Effect of copper content on the mechanical and sliding wear properties of monotectoid-based zinc-aluminium-copper alloys, Tribology International Vol. 37, 45-50, 2004.
32. Savaşkan T., Maleki R.A., Friction and Wear Properties of Zn-25Al-Based Bearing Alloys
33. Tribology Transactions Vol. 57, 435-444, 2014.
34. Savaşkan T., Bican O., Dry sliding friction and wear properties of Al-25Zn-3Cu-3Si alloy
35. Tribology International Vol. 43, 1346-1352, 2010.

IOP Conf. Series: Materials Science and Engineering **295** (2018) 012014 doi:10.1088/1757-899X/295/1/012014

Graphical assessment of the linear contact steel on composite material at high temperature and pressure

Dorin Rus[1], Virgil Florescu[1], Florin Bausic[2], Robert Ursache[2], Anca Sasu[2]
1 Dorin Rus, Mechanical Department, University of Civil Engineering, Bucharest, Postcode 050153, Romania, rusdorin@gmail.com
1 Virgil Florescu, Mechanical Department, University of Civil Engineering, Bucharest, Postcode 050153, Romania, florescuvirgil@yahoo.com
2 Florin Bausic, Mechanical Department, University of Civil Engineering, Bucharest, Postcode 050153, Romania, florin.bausic@utcb.ro
2 Robert Ursache, Mechanical Department, University of Civil Engineering, Bucharest, Postcode 050153, Romania, robionut@gmail.com
2 Anca Sasu, Mechanical Department, University of Civil Engineering, Bucharest, Postcode 050153, Romania, anca.sasu@terranovaromtech.ro

Abstract: In this article we have tried to present a graphical assessment of the dry linear contact for composite materials reinforced with glass fibers as well as of the influence of the sliding speed, load and friction coefficient. Perpendicular loads, the contact temperature and the wear of the metal surface were recorded .The wear volume was calculated using the Archard relationship. Using the Archard relationship, the width of trace can be calculated in 3 locations. Numerous experimental trials were performed in connection to the wear of the metal surface, the contact temperature and the value of the friction coefficient. A connection between the evolution of the wear process and the dependency on the contact temperature and friction coefficient can be observed.

Keywords: hardness of steel, steel surface wear, plastics with glass fibers, dry friction linear contact, contact temperature

1. Introduction

One of the most sophisticated tribosystems is the one between a polymer polymer with short fibers (SGF) on steel, where the contact is dry. In this case, the input parameters are permanently modified. Based on comprehensive experimental tests, a study for the evolution of wear for a certain duration, as well as for certain loads and contact temperatures. The materials used were composite thermoplastic materials reinforced with glass fiber. The glass fiber content varied from a material to another. The evolution of wear was highlighted depending on the duration, the load, contact pressure and sliding speed. Starting from an extended study, we have tried to come up with a graphical representation of the wear process for a dry contact between polymers reinforced with glass fibers and C120 steel, respectively Rp3 steel. For experimental tests, Archard's relationship for adhesion wear was used.

Derjaguin-Muller-Toporov [1] (DMT) model. This model is correct for nanometrically sized bodies, with steel characteristics. The main characteristic of short fiber reinforced polymers (SFRP) is their high level of resistance to rapid loads. This occurs at a relatively low price compared to other materials.
Wang, et all, [2] have studied Nylon 1010 composite with MoS_2 filler and short carbon fibers, the study being carried out in a ring-block wear tester, using a dry contact. The following two aspects were observed: by adding the MoS_2 filler, it has led to an increase in wear, while adding the carbon fiber filler has led to a decrease in wear.
Chang and Friedrich [3] have noticed that the particles, respectively the nanoparticles do not entirely contribute to the film transfer, thus reducing the adhesion and, as a consequence, also reducing the friction coefficient.
Cho and Bahadur [4] have carried out a study on nanosized CuO-filled polyphenylene and on polyphenylene sulfide composites (PPS) reinforced with short carbon fibers (CF) and aramid fibers (Kevlar).
Vos, et all [5][6] have studied polyetheretherketone composites reinforced with short glass fibers and carbon fibers and have shown that the wear rate is influenced by the morphological structure of the matrix of the composite polymer.
Guo, et all [7] have used in their studies composites based on epoxy pitches filled with hybrid particles nano-SiO_2 and short pitches based on carbon fiber. A reduction of the friction coefficient was noticed in the case of hybrid polymers as compared to polymers with added nano-SiO_2 particles
U.S. Tewari, J. Bijwe [8] have highlighted the low manufacturing costs achieved through a composite polymer injection. [9] L. Chang, Z. Zhang show that the injection machines suffer from considerable wear due to the glass fibers.
Schwartz and Bahadur [10] have studied the transfer of the material film using infrared technology. They have observed the following phenomenon, an increase in the density of the polymer leads to an increase in the cohesion energy.
Li, et all [11] has studied analytically and experimentally the epoxide nano-composites, reinforced with short carbon fibers (SCF), of the nano-TiO_2 particles, of the powder of polytetrafluorethylen (PTFE) and graphite flakes, in order to understand the mechanism for adding filler to modify the wear parameters of the two epoxide nano-composites on metal counterpieces.
Chang,et all[12] studies the properties of composite materials, respectively the proprites of polyetheretherketone (PEEK) and polyetherimide (PEI), reinforced with short carbon fibers (CSA). They have determined that by adding submicron particles (TiO_2 and ZnS), for the high contact temperature of the pin-on-disk type tribometer, the wear rate has decreased.
L. Capitanu et all [13][14] have highlighted the behaviour of polyamide and polycarbonate reinforced with glass fibers (SGF) in friction on steel surfaces .
Kukureka, et all [15] have studied the wear of PA66 in rolling-sliding contact. For the polymer they added glass, carbon or aramid, and both for the glass fibers and for the carbon fibers a drop in the friction coefficient was noticed.
Stachowiak, et all [16] have studied the abrasive effect for the three body abrasion of metal samples, the tests were carried out on two ball-on-flat installations and modified pin-on-disk tribometer. The ball-on-flat teste yielded the most significant results.
Dwyer-Joyce [17] has noticed that during contact the wear due to contamination with solid compounds of lubricants occurs, the phenomenon closely resembling the abrasion with three bodies
No studies regarding the correlation between friction and use concerning the complex friction-wear phenomenon were shown. In our paper we have studied on Timken type couples (with linear contact), in conditions of dry sliding friction, the behavior of glass fiber reinforced composite materials under controlled loads and speeds. We have studied the influence of the percentage of glass fiber, as well as the influence of load and speed on the wear process.

2. Materials and methods

Wear and friction are analysed from several points of view, such as speed relative to load and stress. The two samples are cylindrical liner and flat sample.

2.1 Experimental Method

A Timken type couple with linear friction contact was used as experimental equipment. Thus the normal load and the contact temperature can be controlled. The friction couple is built out of a plastic cylinder which revolves at different speeds. The plastic piece rests on the polished surface of a steel plan disk. The cylinder has a diameter of 22.5 mm and a thickness of 10 mm.

The friction couple is built out of a cylindrical liner (1) and a flat disk (2). The liner is fixed by means of a nut (3) on the driving shaft (4). The disk sample is placed in a hole made in the elastic blade (5) (Figure 1)

An electric motor (7), the shaft (4) with rotation movement using trapezoidal belt.

Figure 1 shows the functional scheme (a) friction couple (b) and its installation within the experimental equipment (c). The way in which the liner moves against the plane sample is illustrated in Figure 1c.

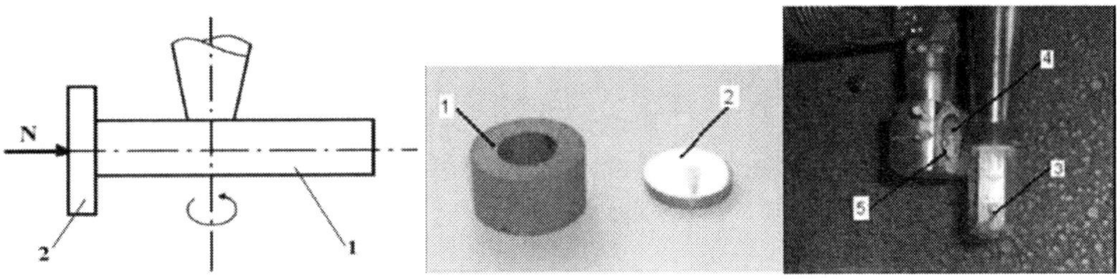

Figure 1. Functional scheme (a) (a) the way how the liner moves against the disk, (b) friction couple, and (c) its installation in the experimental equipment, where 1 - cylindrical liner; 2 – steel flat disk sample; 3 – nut; 4 – hole; 5 - knife-edge.

An electric motor (7), the shaft (4) with rotation movement using trapezoidal belt. The friction couple is built of a cylindrical liner (1) a flat disk (2). The liner is fixed by means of a nut (3) on the driving shaft (4). The disk sample is placed in a hole made in the elastic blade (5). An electric motor (7), the shaft (4) with rotation movement using trapezoidal belt transmission (6). The normal and tangential (friction) stresses are measured by means of resistive tensiometers, mounted on the elastic blade (5). The normal load to the elastic blade (5) is applied, through a calibrated spring system (8). The installation can register the friction force on an *X-Y* recorder. The duration is controlled with a clock and the contact temperature is measured with a miniature thermocouple (9), connected to a millivoltmeter. The installation can also study the wear using other radiometers techniques. For this purpose, the installation has a tank (10) assembled on a base (11) and a tube for collecting the radioactive wear particles (12). (Figure 2)

The unidirectional testing is investigations of metal surface wear. The tests were based on Hooke's law, at normal loads of 10; 20; 30; 40 and 50 N, loads which are adequate to some contact pressures all calculated considering the elastic contact hypothesis, meaning: 16.3; 23.5; 28.2; 32.6 and 36.4 MPa

Experimental tests were carried out on composite thermoplastic materials such as polyamides and polycarbonates reinforced with short glass fibers, whose content varies between 20 and 30%. Metal samples were manufactured from two steel types: C 120 steel hardened 59 HRC and Rp3 steel hardened 62 HRC

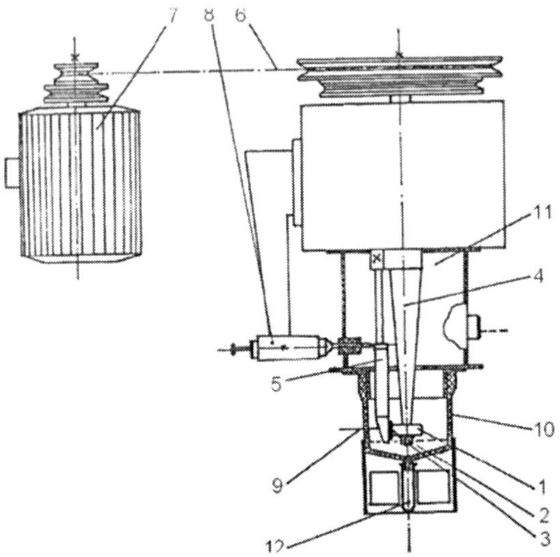

Figure 2. The scheme of experimental equipment with linear contact friction couple, Timken type

3. Results and discussion

All the tests were limited to an hour. The volume and wear depth for the wear process were determined. The curves for the volume of the worn metal material (V_u) (Tab.1) and for the depth of the worn metal material (h_u) (Tab.2), as well asa the normal load for each couple were charted.
The regression functions and the regression factor were calculated for each couple.

Tab. 1. The regression function between the volume of worn metal material (V_u) and the normal load (N)

Friction couple	v (cm/s)	Regression function	Correlation factor
Polyamide + 30% SGF/C120 steel	18.56	$V_u = 0.0005\ N^2 + 0.012\ N$	$R^2 = 0.9991$
Polyamide + 30% SGF/C120 steel	27.85	$V_u = 0.0004\ N^2 + 0.0188\ N$	$R^2 = 0.9996$
Polyamide + 30% SGF/C120 steel	37.13	$V_u = 0.0005\ N^2 + 0.0104\ N + 0.1423$	$R^2 = 1$
Polyamide + 30% SGF/C120 steel	55.70	$V_u = 0.0034\ N^2 - 0.0922\ N + 0.877$	$R^2 = 1$
Polyamide + 30% SGF/C120 steel	111.4	$V_u = 0.0006\ N^2 - 0.0205\ N + 0.272$	$R^2 = 1$
Polyamide + 30% SGF/C120 steel	153.57	$V_u = 0.0007\ N^2 + 0.1668\ N + 0.0675$	$R^2 = 1$
Polyamide + 30% SGF/ Rp3 steel	18.56	$V_u = 0.0003\ N^2 + 0.018\ N$	$R^2 = 0.9998$
Polyamide + 30% SGF/Rp3 steel	37.13	$V_u = 0.0004\ N^2 + 0.0077\ N + 0.2291$	$R^2 = 1$
Polyamide + 30% SGF/Rp3 steel	46.41	$V_u = 0.0003\ N^2 + 0.0236\ N$	$R^2 = 0.9997$
Polycarbonate + 20% SGF/C120 steel	27.85	$V_u = 0.0002\ N^2 + 0.0212\ N + 0.0138$	$R^2 = 1$

We have charted the characteristic curves for each couple.

Tab. 2. The regression function between the depth of the worn metal material (h_u) and the normal load (N)

Friction couple	v (cm/s)	Regression function depth	Correlation factor
Polyamide + 30% SGF/C120 steel	18.56	$h_u = 0.0007\,N^2 + 0.1099\,N$	$R^2 = 0.9977$
Polyamide + 30% SGF/C120 steel	27.85	$h_u = 0.0017\,N^2 + 0.0223\,N + 2.5401$	$R^2 = 1$
Polyamide + 30% SGF/C120 steel	37.13	$h_u = 0.0027\,N^2 - 0.0345\,N + 3.7722$	$R^2 = 0.9996$
Polyamide + 30% SGF/C120 steel	55.70	$h_u = 0.0038\,N^2 + 0.0017\,N + 4.0728$	$R^2 = 1$
Polyamide + 30% SGF/C120 steel	111.4	$h_u = 0.0005\,N^2 - 0.0607\,N + 5.0222$	$R^2 = 1$
Polyamide + 30% SGF/C120 steel	153.57	$h_u = 0.0062\,N^2 - 0.0569\,N + 5.1224$	$R^2 = 1$
Polyamide + 30% SGF/ Rp3 steel	18.56	$h_u = 0.0008\,N^2 + 0.0106\,N + 2.2026$	$R^2 = 0.9984$
Polyamide + 30% SGF/Rp3 steel	37.13	$h_u = 0.0032\,N^2 - 0.1452\,N + 5.7442$	$R^2 = 1$
Polyamide + 30% SGF/Rp3 steel	46.41	$h_u = 0.0022\,N^2 - 0.061\,N + 4.8439$	$R^2 = 1$
Polycarbonate + 20% SGF/C120 steel	27.85	$h_u = 0.002\,N^2 + 0.1141\,N + 2.6028$	$R^2 = 0.9400$

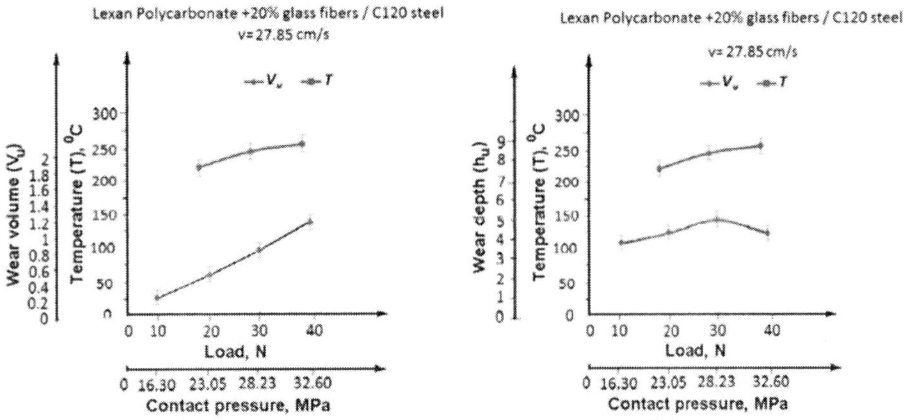

(a) (b)

Figure 3 Wear evolution as scar wear volume (a) and depth (b) function of the normal load and contact temperature and variation of contact temperature at the speed of 27.85 cm/s, for PC Lexan 3412 + 20% SGF / C120

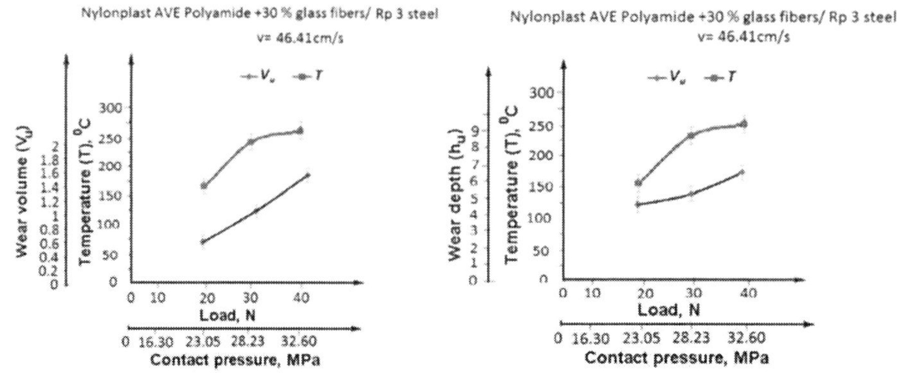

(a) (b)

Figure 4 Wear evolution as scar wear volume (a) and depth (b) function of the normal load and contact temperature and variation of contact temperature of speed of 46.41 cm/s for Nylonplast AVE Polyamide + 30% SGF / Rp3 steel

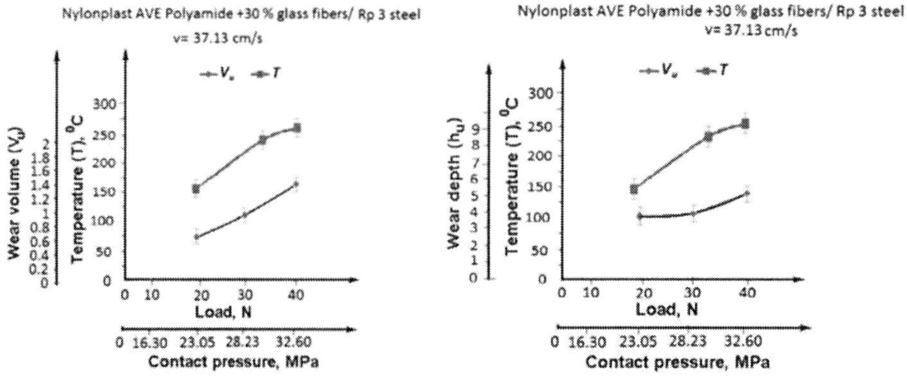

(a) (b)

Figure 5 Wear evolution as scar wear volume (a) and depth (b) function of the normal load and friction coefficient at the sliding speed of 37.13 cm/s for Nylonplast AVE Polyamide + 30% SGF / Rp3 steel

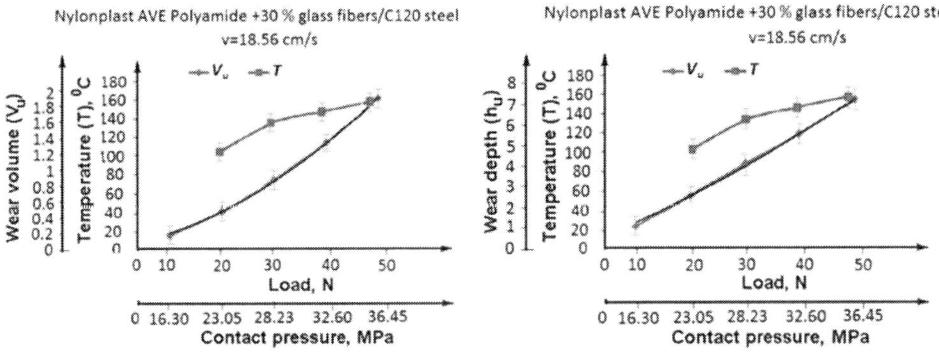

Figure 6 Wear evolution as scar wear volume (a) and depth (b) function of the normal load and contact temperature and variation of wear mode, at speed sliding of 18.56 cm/s for Nylonplast AVE Polyamide + 30% SGF / C120 steel

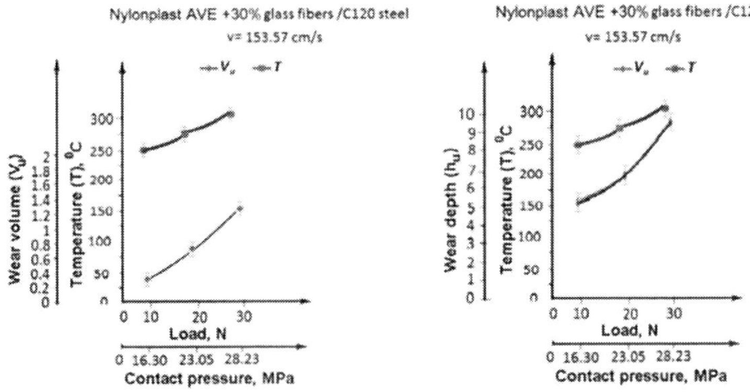

Figure 7 Wear evolution as scar wear volume (a) and depth (b) function of the normal load and contact temperature and variation of contact temperature at the sliding speed of 153.57 cm/s for Nylonplast AVE Polyamide + 30% SGF / C1

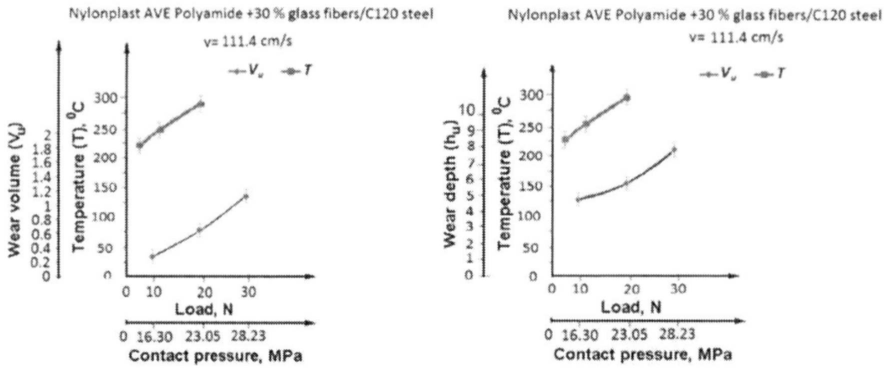

Figure 8 Wear evolution as scar wear volume (a) and depth (b) function of the normal load and contact temperature and variation of contact temperature at the sliding speed of 111.4 cm/s for Nylonplast AVE Polyamide + 30% SGF / C120 steel

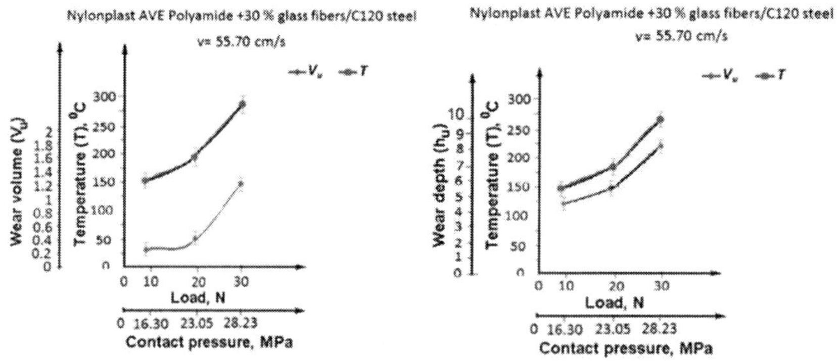

Figure 9 Wear evolution as scar wear volume (a) and depth (b) function of the normal load and contact temperature and variation of contact temperature at the sliding speed of 55.70 cm/s for Nylonplast AVE Polyamide + 30% SGF / C120 steel

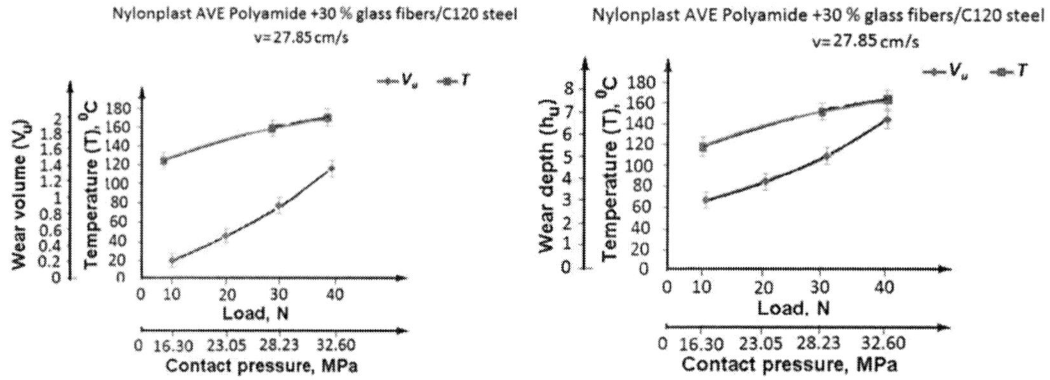

Figure 10 Wear evolution as scar wear volume (a) and depth (b) function of the normal load and contact temperature and variation of contact temperature at the sliding speed of 37.13 cm/s for Nylonplast AVE Polyamide + 30% SGF / C120 steel

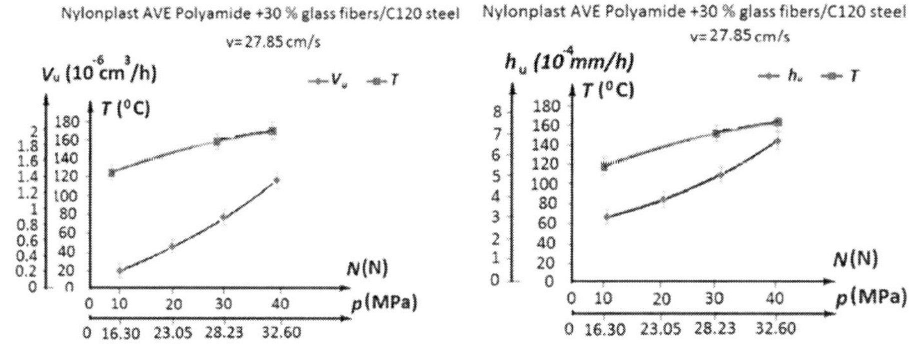

Figure 11 Wear evolution as scar wear volume (a) and depth (b) function of the normal load and contact temperature and variation of contact temperature at the sliding speed of 27.85 cm/s for Nylonplast AVE Polyamide + 30% SGF / C120 steel

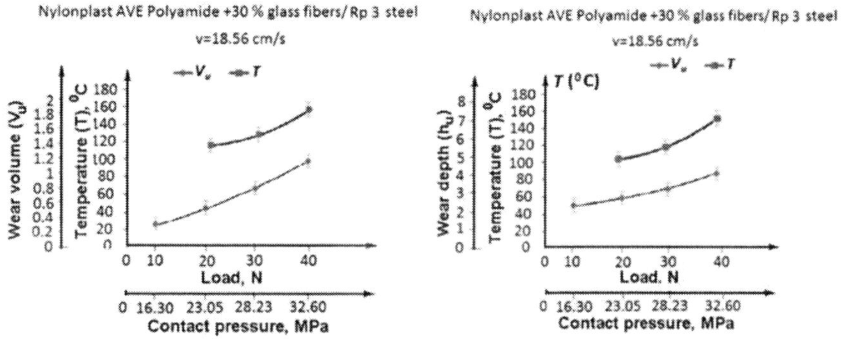

Figure 12 Wear evolution as scar wear volume (a) and depth (b) function of the normal load and contact temperature and variation of contact temperature at the sliding speed of 18.56 cm/s for Nylonplast AVE Polyamide + 30% SGF / Rp3 steel

4. Conclusion

From the information presented above we can draw some conclusions:

- the wear process of metal surfaces in dry friction contact against plastic materials reinforced with short glass fibers evolves over time, depending on loading, moving from the initial abrasive wear caused by glass fibers, at adhesion wear characterized especially by the transfer of plastic material on the metal surface, but also by corrosion;

It is difficult to establish a mathematical relationship between the input and output parameters. The linear dry contact between the composite material and steel is highly complex. By changing a single input parameter, all output parameters of the systems will change.

For this reason, a graphical representation of the phenomenon may give provide a more suggestive image of these phenomena when they occur, during the dry contact between the composite polymer and the metal surface.

The friction coefficient for the composite material and for the C 120 steel samples is higher than the one obtained at the surface of the steel samples. This is due to the differences in hardness between the two types of steel.

This phenomenon is so complex that the system evolves while the action of the loads leads from abrasion to adhesion wear and to corrosion. This phenomenon occurs simultaneously with the transfer of thermoplastic material unto the metal surface.

Acknowledgements
The authors would like to thank the University of Civil Engineering Bucharest, for its material and technical support offered in order to achieve these researches.

References

1. B.V. Deryagin, V.M. Muller, Yu.P. Toporov, Adhe-sive contact deformation of a single microelastic sphere. J. Colloid Interface Sci. 53 (1975) 314
2. J. Wang, M. G. Bai Songhao, G. Shirong, Investigation of the influence of MoS_2 filler on the tribological
3. L. Chang, K. Friedrich. Enhancement effect of nanoparticles on the sliding wear of the sliding wear of short fiber-reinforced polymer composites: A critical discussion of wear mechanisms, *Tribology International* 43 (2010)2355-2364.
4. S. Bahadur, The development of transfer layers and their role in polymer tribology. Wear 245 (2000) 92-99.
5. A.M. Hager, M. Davies, Advances in Composites Tribology: Short-fiber reinforced, high-temperature resistant polymers for a wide field of tribological applicants, Elsevier Science Publishers, BV, 1993, pp. 104-157.
6. H. Voss, K. Friedrich, On the wear behavior of short-fiber-reinforced PEEK composites, Wear 116 (1987) 1-18.
7. Q.B. Guo, M.Z. Rong, G.L. Jia, K.T. Lau, M.Q. Zang, Sliding wear performance of nano SiO_2 / short carbon fibre / epoxy hybrid composites. Wear 266 (2009) 658-665.
8. U.S. Tewari, J. Bijwe, Advances in Composites Tribology, Recent development in tribology of fiber reinforced
9. L. Chang, Z. Zhang, Tribological properties of epoxy-nanocomposites: 2. A combinative effect of short carbonfiber and nano-TiO_2, Wear 206 (2006) 869-878.
10. C.J. Schwartz, S. Bahadur, The role of deformability, filler-polymer bonding, and counterface material on the tribological behaviour of polyphenyl sulfide (PPS). Wear 251 (2003) 1532-1540.
11. C. Li, Z. Zhong, Y. Lin, K. Frederich, Tribological properties of epoxy nanocomposites: III. Characteristics of transfer films. Wear 262 (2007) 799-706.
12. L. Chang, Z. Zhang, H. Zhang, A.K. Schlarb, On the Sliding wear of nanoparticles filled polyamide 66, Composites Science and Technology 66 (2006) 3188-3198.
13. Dorin Rus and Lucian Capitanu "Wear and Contact Temperature on Steel Surface in
14. Linear Dry Friction Contact with Polimers with SGF" Journal of Mechanics Engineering and Automation 5 (2015) 554-566 doi: 10.17265/2159-5275/2015.10.004 David Publishing
15. Rus Dorin "The contact temperature between steel surface in linear dry friction contact with polimers" Sinteze de Mecanica Teoretica si Aplicata; Bucharest 7.2 (2016): 157-166© Matrix Rom
16. S.N. Kukureka, C.J. Hooke, M. Rao, P. Liao, Y.K. Chen, The effect of fibre reinforcement on the friction and wear of polyamide 66 under dry rolling-sliding contact. Tribology International 32 (1999) 107-116. Properties of carbon fiber reinforced nylon 1010 composites. Wear 255 (2003) 774-779.
17. G.B. Stachowiak, G.W. Stachowiak. The effects of particle characteristics on three-body abrasive wear. Wear 249 (2001) 201-207.
18. G.B. Stachowiak, G.W. Stachowiak. The effects of particle characteristics on three-body abrasive wear. Wear 249 (2001) 201-207.

An analytical interpretation of the high temperature linear contact between composite materials reinforced with glass fibers and steel

Dorin Rus[1], Virgil Florescu[1], Florin Bausic[2], Robert Ursache[2], Anca Sasu[2]
1 Dorin Rus, Mechanical Department, University of Civil Engineering, Bucharest, Postcode 050153, Romania, rusdorin@gmail.com
1 Virgil Florescu, Mechanical Department, University of Civil Engineering, Bucharest, Postcode 050153, Romania, florescuvirgil@yahoo.com
2 Florin Bausic, Mechanical Department, University of Civil Engineering, Bucharest, Postcode 050153, Romania, florin.bausic@utcb.ro
2 Robert Ursache, Mechanical Department, University of Civil Engineering, Bucharest, Postcode 050153, Romania, robionut@gmail.com
2 Anca Sasu, Mechanical Department, University of Civil Engineering, Bucharest, Postcode 050153, Romania, anca.sasu@terranovaromtech.ro

Abstract: In this paper we have tried to present the influence of the metal surface wear and of the contact temperature on the evolution of the sliding speed, of the normal load and of the friction coefficient. We have performed numerous experimental trials that have highlighted the dependency between load and wear in relation to the friction coefficient. A dry linear friction couple was used with a large range of loads and speeds, simulating real-life working conditions: temperature, sliding speed, contact pressure. We have made a connection between the theoretical case and the experimental results arising from the use of the "wear imprint method" for the volume and depth of wear.

Keywords: plastic material transfer, steel surface wear, plastics with glass fibers, linear contact, dry friction hardness of steel.

1. Introduction

Thermoplastic materials with glass fiber are characterized by high plasticity under high pressure and temperature, and they can be machined so that they return to their initial shape after being cast. These thermoplastic materials are biphasic, consisting of a polymer mass and reinforcement materials such as glass fibers, in order to obtain good mechanical characteristics as a result of the fact the stresses are transmitted through the glass fibers. At the same time the polymer does not allow a big deformation, limiting the significant deformation of the composite material. Also, it is good to use alkali-free glass fibers for a slow degradation in time and oxides can be added to maintain the high values of the elasticity module.

In specialist papers, values are given for the friction coefficient for different couples, plastic material/ plastic material, plastic material/ metal. The values are given for couples with lubricant or with dry

contact. The couples are formed from reinforced composite materials or composite materials without reinforcement.
Shen and Dumbleton [1] have studied comparatively the tribological behaviour of the polyoxymethylene and of the high density polyethylene (UHMWPE). They suggest a relationship for the calculation of the wear as follows:

$$h = kpx$$

Where: h – linear wear; k - wear factor; p - nominal pressure; x - sliding distance.
Bartenev, Lavrentiev [2] establish that for the metal/plastic contact, the increase of the friction force is proportional to the sliding speed. This dependency is shown through the increase of the friction force together with the normal load. This was also demonstrated by Vinogradov
Jacobi [3] presents friction coefficient values between 0.04 and 0.5 for glass fiber reinforced polyamides. Barlow [4] shows that, for lubricated surfaces, an increase of the friction coefficient occurs with the increase of the relative sliding speed at the level of the friction surface.

Bilik [5] has shown that, for the couple polyamides/ steel, the friction coefficient is not constant. The coefficient depends on several factors: temperature, the ruggedness of the surface, the sliding speed and the contact pressure. Lancaster and Evans [6] have studied the tribological behaviour of glass fiber reinforced composite polymers, hydrodynamically lubricated, and notice the decrease of $\eta v d^3 / N$ for beak type couplings.

Clerico [7] has discovered that for the polyamide/ metal couple the value of the friction coefficient is higher for short periods of operation than for longer periods of operation, due to the viscoelastic behaviour of the composite material.
Lancaster [8] has studied different natural glass fiber reinforced polymers. He has shown that the friction coefficient decreases with the decrease in the ruggedness of the metal surface. .Jost [9] shows that for polyamide/ metal couples the most encountered type of wear is adhesion wear.
Hrusciov and Babicev [10,11] have shown that, if we increase the polymer content for the plastic material reinforced with glass fiber/ steel couplings an increase of the microcutting component of the friction force occurs.

The Johnson-Kendall-Roberts (JKR) model [12] shows that for composite materials a strong abrasive component occurs upon contact, and several models of the contact have been drawn up together with the description of the adhesion contact.
Bely [13], Bartenev and Laventiev [14] have discovered that if the glass fibers inside the polymer matrix do not have the same orientation, the friction coefficient increases. Bowden and Tabor [15] have shown the importance of distributing forces on the Hertzian contact surface with an elliptical pressure distribution. In this situation, the central area will be the most affected area.
Watanabe et al [16] shows the influence of the friction coefficient depending on the temperature and the transfer of the plastic material unto steel.
Myshkin [17] mentions that, at the surface of the composite material, the links are viscous-elastic and that the abrasive friction component is higher that the adhesive friction component.

Therefore we can infer that the friction process for composite materials is complex, because numerous parameters influence the friction process, with or without lubricants. In specialist literature there is a heterogeneous approach, because different test installations, different experimental conditions and different materials are used.

2. Materials and methods

2.1 Analytical Method

The functioning under load of the Timken type friction couple (with linear contact) subjected to a load and in the presence of a relative sliding movement of the bushing made of glass fiber reinforced plastic on the surface of the steel sample highlights the occurrence of a wear imprint on the plane metal surface, the imprint being graphically represented in (Figure 1)

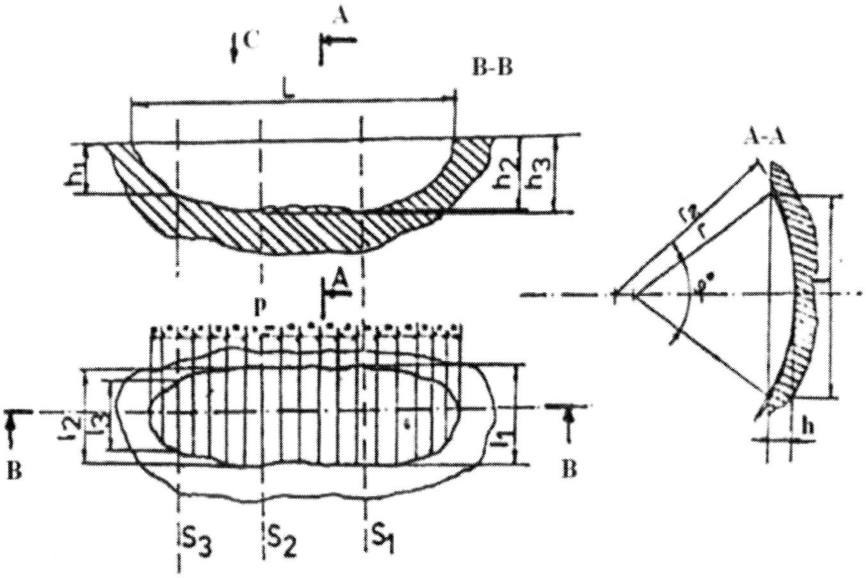

Figure 1. The imprint scheme for Timken friction couples

Knowing Coulomb's law that the friction force F_f is direct proportional to the normal force N:

$$F_f = \mu N \tag{1}$$

More studies have shown that μ, the friction coefficient is not only dependent on the normal load applied. Relations for variations of the friction force, depending on the load applied:

$$F_f = aN + bN^n \tag{2}$$

$$F_f = a + bN \tag{3}$$

$$F_f = a + bN^n \tag{4}$$

$$F_f = kN^n \tag{5}$$

Where n is subunitary.
We express the friction coefficient for the plastic materials:

$$\mu = \tau_f / p_c \tag{6}$$

Where τ_f represents the shear strength of the softer material

p_c represents the flow pressure of the same material.

Because $p_c = HB/3$, we have:

$$\mu = 3\tau_f / HB \tag{7}$$

The bush is rigid and accounting for the generally low non-uniformity of the imprint, considered as being formed by a series of cylindrical sectors having the length q. The area of the lateral surface of the cylindrical sector is a circle segment:

$$S_i = 0.5r^2\left(\pi\varphi_i^0/180° - \sin\varphi_i\right) \tag{8}$$

Where S_i – lateral surface of the crystal sector;
φ_i - the angle;
r - the circle radius.
The radius r could not be identified with the cylindrical bush radius for the plastic / metal couples.
It is possible due to the elastic deformation of the bush under loading conditions and we illustrated this by the sketch in (Figure 2)

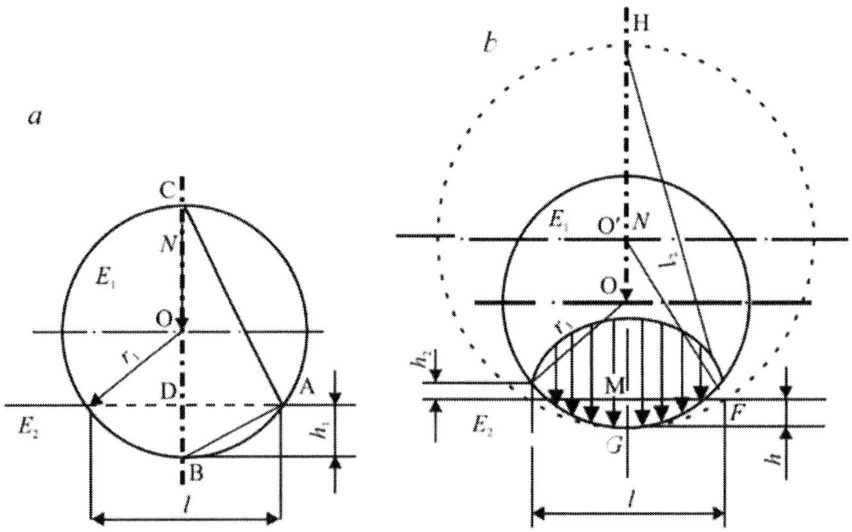

Figure 2. The elastic deformation of the cylindrical liner on the contact area for Timken friction couples (a – theoretical b - practical)

Using r_1 for the non-deformable liner radius and r_2 for the radius – in the contact area – of the deformed liner, we see in (Figure 2b):

$$r_2 > r_1 \tag{9}$$

Increasing the bush radius in the contact area leads to the decrease of the depth of the wear imprint from h_1 - which would appear if the elastic deformation of the bush would be neglected, to the value h with the quantity h_2:

$$h_2 = h_1 - h \tag{10}$$

Using l for the width of the wear imprint, from ΔABC it is:

$$(2r_1 - h_1)h_1 = l^2/4 \tag{11}$$

Because the value of the depth h_1 is very small, the term is negligible, we write:

$$h_1 = l^2/8r_1 \tag{12}$$

Similarly, in ΔFGH , we have:

$$(2r_2 - h_2)h_2 = l^2/4$$

Using the same assumption, for the term h_2, we obtain:

$$h_2 = l^2/8r_2 \tag{13}$$

Introducing (12) and (13) in (11) we have the following results:

$$h = l^2(1/r_1 - 1/r_2)/8 = l^2(r_2 - r_1)/8r_1r_2 = l^2/8r \tag{14}$$

Where r is the equivalent curvature radius:

$$1/r = 1/r_1 - 1/r_2 = (r_2 - r_1)/r_1r_2 \tag{15}$$

From (14) we have:

$$(r_2 - r_1)/r_1r_2 = l^2/8h_2 \tag{16}$$

Considering the frictional couple is loaded in the elastic domain with an elliptic distribution of stresses, the Hertz formula is:

$$l^2/4 = 8Nr(1-v^2)/\pi EL \tag{17}$$

Where:
v - Poisson ratio
L - the length of the wear imprint
E - equivalent Young modulus
Using index 1 for quantities related to the cylindrical bush, and index 2 for those related to plane half-couple, the equivalent elasticity modulus is:

$$1/E = 0.5\left[(1-v_1^2)/E_1 + (1-v_2^2)/E_2\right] \tag{18}$$

$$E = 2E_1E_2/0.91(E_1 + E_2) \tag{19}$$

From (17) we can express the width of the wear imprint:

$$l = 4\left[2Nr(1-v^2)/\pi EL\right]^{1/2} \tag{20}$$

IOP Conf. Series: Materials Science and Engineering **295** (2018) 012015 doi:10.1088/1757-899X/295/1/012015

Introducing in (20) the equivalent elasticity modulus and the equivalent radius expressions, the numerical value of Poisson ratio is:

$$h_2 = 0.527N(E_1 + E_2)/LE_1E_2 \tag{21}$$

Considering (12) and (21), we have for the depth of the wear imprint:

$$h = (l^2/8r_1) - 0.527N(E_1 + E_2)/LE_1E_2 \tag{22}$$

The wear imprint is the sum of cylindrical sectors, expanding in series the relation (21), neglecting the high-order terms and reducing the similar terms, the area of the lateral surface of a sector is:

$$S_i = r^2\phi^3/12 \tag{23}$$

Replacing in the relation above the angle ϕ with the ratio l/r and accounting for (15) and (16):

$$S_i = l^3(r_2 - r_1)/12r_1r_2 = 2lh_2/3 \tag{24}$$

Replacing the value of h_2 obtained (22) in (24), we obtain the expression for the area of lateral transversal surface of a cylindrical sector:

$$S_i = 0.35l(E_1 + E_2)Nl/E_1E_2L \tag{25}$$

The volume of worn metal material will be:

$$V_u = \sum_{i=1}^{n}(S_iq_i) = 0.351(E_1 + E_2)Nl_m/E_1E_2 \tag{26}$$

Where l_m is the mean width of the wear imprint.
With this value we can obtain the volume of worn metal material V_u and the mean value of the depth of removed layer h_{mu}.

We have some correlation functions between wear scar volume (V_u) .The wear scar volume and the wear scar depth are according to the normal load (N) in polynomial forms :

$$T = \sum_{i=0}^{2} a_iN^i;$$

$$V_u = \sum_{i=0}^{2} b_iN^i; \tag{27}$$

$$h_u = \sum_{i=0}^{2} c_iN^i,$$

And functions between friction coefficient μ and normal load N are

$$\mu = A\ln N + B \tag{28}$$

Where a_i, b_i, c_i and A, B are determined from regression functions.

From relations (13) we obtain

$$V_u = \frac{b_2}{a_2}T + \left(b_1 - \frac{b_2}{a_2}a_1\right)N + c_0 - \frac{b_2}{a_2}a_0, \tag{29}$$

or, after substituting N from (28) and replacing in (29) we have

$$V_u = \frac{b_2}{a_2}T + \left(b_1 - \frac{b_2}{a_2}a_1\right)\exp[(\mu - B)/A] + c_0 - \frac{b_2}{a_2}a_0. \tag{30}$$

Similarly, the wear scar depth is

$$h_u = \frac{c_2}{a_2}T + \left(c_1 - \frac{c_2}{a_2}a_1\right)N + c_0 - \frac{c_2}{a_2}a_0, \tag{31}$$

$$h_u = \frac{c_2}{a_2}T + \left(c_1 - \frac{c_2}{a_2}a_1\right)\exp[(\mu - B)/A] + c_0 - \frac{b_2}{a_2}a_0. \tag{32}$$

Also, contact temperature as a function of μ such:

$$T = \sum_{i=0}^{2} a_i \exp[(\mu - B)i/A] \tag{33}$$

Considering that wear is adhesive, in case of linear polymer/steel contact, we consider the volume of wear material according to J.F. Archard's relationship (18) [18]. Taking into account (8) and (9) :

$$V_u = kNvt = \frac{b_2}{a_2}T + \left(b_1 - \frac{b_2}{a_2}a_1\right)N + c_0 - \frac{b_2}{a_2}a_0 \tag{34}$$

and from (18) and (26) :

$$V_u = kNvt = \frac{b_2}{a_2}T + \left(b_1 - \frac{b_2}{a_2}a_1\right)\exp[(\mu - B)/A] + c_0 - \frac{b_2}{a_2}a_0. \tag{35}$$

Writing the Archard relationship for the wear depth (20), from (25) and (27):

$$h_u = k^* pvt = \frac{c_2}{a_2}T + \left(c_1 - \frac{c_2}{a_2}a_1\right)N + c_0 - \frac{c_2}{a_2}a_0 \tag{36}$$

From (20) and (21):

$$h_u = k^* pvt = \frac{c_2}{a_2}T + \left(c_1 - \frac{c_2}{a_2}a_1\right)\exp[(\mu - B)/A] + c_0 - \frac{b_2}{a_2}a_0 \tag{37}$$

3. Results and Discussion

The relationships (34), (35) and (36), (37) show the connection between the output parameters V_u, T, μ and the input parameters of the wear process N, v, t, respectively between h_u, T, μ and p, v, t. These relationships show the complexity of the wear process for metal surfaces during the linear dry friction

contact with short glass fiber reinforced polymers [19].
By changing an input parameter, all the other parameters will be modified. This modification changes all output parameters of the system due to the numerous factors which contribute to the phenomenon themselves.
In conclusion, the phenomena that occur have a high level of complexity and we can illustrate this by means of a drawing (Fig .3) which helps us see the complexity of the phenomenon.

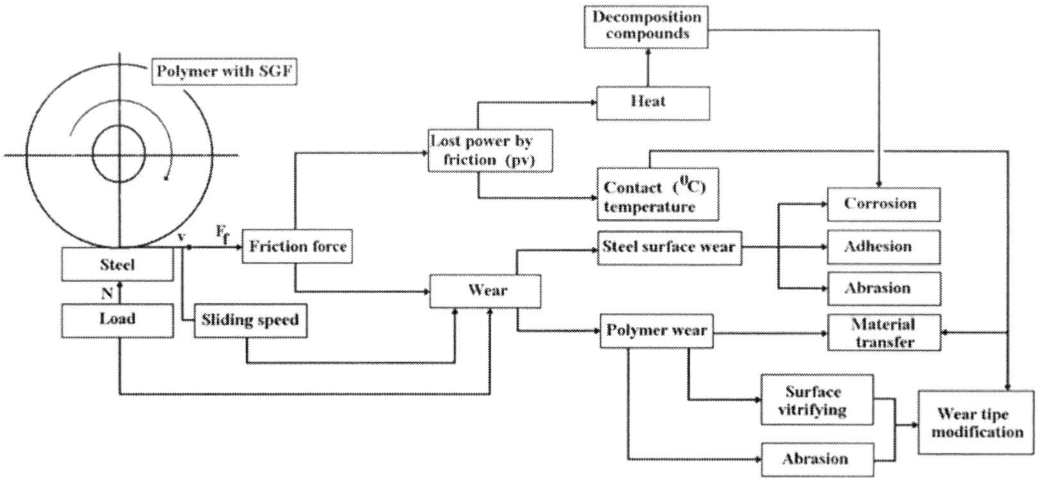

Figure 3. The complexity of the evolution of the friction – wear process for the linear dry contact polymer + SGF/steel.

4. Conclusions

The studies carried out concerning the friction and wear of metal surfaces belonging to friction couples, in linear contact with short glass fiber reinforced thermoplastic material / steel, were aimed at highlighting the influence of the input parameters of the tribosystem (normal load – contact pressure, relative sliding speed, type of friction, the characteristics of the materials coming into contact), on the output parameters of the tribosystem (the volume of the worn metal material, the depth of the worn material, the contact temperature and the friction coefficient).
According to tribological knowledge published at international level up to this point, intuitively, there is a clear influence of the friction coefficient on the wear of the metal surfaces, however no theoretical connection between the two values has been achieved.
By presenting and comparing the variation functions of the friction coefficient, of the contact temperature, of the volume of worn metal material and of the depth of wear depending on normal load, we have sought and we considered that we have succeeded to show that between the input parameters of the tribosystem with linear contact (normal load, relative sliding speed, physical and mechanical characteristics of the materials coming into contact, the state of the surface) and the output parameters (friction coefficient, contact temperature and wear) there is an interdependency which changes constantly during the friction process.
Presenting the evolution of the tribosystem with dry linear contact, SGF-reinforced polymer / steel is very useful for tribological research. This actually shows the closed loop between the input and output parameters, taking into consideration the modification of the initial parameters (and especially the modification of the state of the metal surface after wear).

Acknowledgements

The authors would like to thank the University of Civil Engineering Bucharest, for its material and technical support offered in order to achieve these researches.

References

1. Shen C., Dumbleton J.H. "The Friction and Wear Behavior of Polyoximethylene in Convection with Joint Replacement", *Wear 38(2)*: 291-303 (1978).
2. Bartenev G.M., Lavrentiev V.V., Konstantinova N.A. The actual Area and friction Properties of Elastomers under Frictional Contact with Solid Surfaces, *Wear 18(6)*: pp.439-448 (1971).
3. Jacobi H.R. "Neue Erkenntnisse über Gleitenschaften von Polyamiden", *K*unststoffe, 47 (5): 234-239 (1957).
4. Barlow D.A. "Some Observations of the Friction Properties of Vinyl", *Wear 20(2)*: pp.151-157 (1972).\
5. Bilik M. 1965 Parî trenia metal-plastmassa v maşinah i mechanizma*h*, Izd. Maşinostroienie, Moskva.
6. Lancaster J.K., Evans D.C. "Wear of Polymers in *Treatise on Mat. Sci. And Technology*". Academic Press, 1979, pp. 13-86.
7. Clerico M. "A Study of the Friction Wear of Nylon against Metal", *Wear, 13 (3)* pp. 183-197 (1969).
8. Lancaster J.K. "Lubrication of Carbon Fibre reinforced Polymers", *Wear 28(3)*. pp. 315-352 (1972).
9. Jost H. Zum Verschleissverhalten von Polyamid, *Plaste und Kautschuk 17(4)*. pp. 266-270 (1970).
10. Hrusciov M.M. "Resistance of Metals to Wear by Abrasion, as related to Hardness", *Proc. Conference on Lubrication and Wear, Inst. Mech. Eng.*, London, 1957: 979-986.
11. Hrusciov M.M, Babicev M.A., 1982 "Resistance to Abrasive Wear of structurally Inhomogeneous Materials". Friction and Wear in Machinery, part 2, *ASME, New York, 1958*, pp.5-23.
12. Briscoe B.J., "Friction and Wear of Polymer Composites", Elsevier, Amsterdam, 1986, p. 25.
13. Bely V.A., Sviridenok A.I. et.al. "*Friction and Wear in Polymer based Materials*", Pergamon Press
14. Bartenev G.M., Lavrentiev V.V. "*Trenie i iznos polimerov*", , 1972, Izd. Himia, Leningrad.
15. Bowden F.P., Tabor D. "The Friction and Lubrication of Solids, part I-II", Clarendon Press, Oxford, 1964
16. Watanabe M., Korusawa M., Marsubara K. "The Frictional Properties of Nylon", *Wear 32 (3)*. pp 185-191 (1968).
17. Myshkin N.K., Petrokovets, M.I. Kovale A.V., "Tribology of polymers: Adhesion, friction, wear, and mass-transfer", *Tribology*

18. Dorin RUS, Lucian CAPITANU, Liliana–Laura BADITA "A qualitative correlation between friction coefficient and steel surface wear in linear dry sliding contact to polymers with SGF" *International 38* (2005) 910-921 *Friction* 2(1): 47–57 (2014)
19. Rus Dorin "The contact temperature between steel surface in linear dry friction contact with polimers" Sinteze de Mecanica Teoretica si Aplicata; Bucharest 7.2 (2016): 157-166. © Matrix Rom

IOP Conf. Series: Materials Science and Engineering **295** (2018) 012016 doi:10.1088/1757-899X/295/1/012016

Research concerning the balancing of a plane mechanism

Bădoiu D[1], Petrescu M G[2], Antonescu N N[3], Toma G[4]
[1]Petroleum-Gas University of Ploiesti, Department of Mechanical Engineering, 100680, Ploiesti Romania
[2]Petroleum-Gas University of Ploiesti, Department of Mechanical Engineering, 100680, Ploiesti Romania
[3]Petroleum-Gas University of Ploiesti, Department of Mechanical Engineering, 100680, Ploiesti Romania
[4]Petroleum-Gas University of Ploiesti, Department of Mechanical Engineering, 100680, Ploiesti Romania
dorin.badoiu@gmail.com[1], pmarius@upg-ploiesti.ro[2], nnantonescu@upg-ploiesti.ro[3], georgeta_tm@yahoo.com[4]

Abstract: By statically balancing of the plane mechanisms and especially those functioning at high speeds is being pursued the decrease of the value of the resultant force of all inertia forces that work on the component elements, thus obtaining a significant decrease in vibrations and shocks during the functioning. On the other hand, the existence of balancing masses which ensure the balancing of the mechanism leads to increased gauge and its mass. In this paper are presented some possibilities of statically balancing a plane mechanism which is composed of three independent contours. First is analyzed the case when the mechanism is totally balanced. Then a solution is proposed for a partial balancing of the mechanism based on the balancing of the first harmonic of the inertia force developed in a piston of the mechanism. Finally, are presented some simulation results concerning the variation of the value of the resultant inertia force during a cinematic cycle when the mechanism is unbalanced and when it is partially balanced. Also, it is analyzed the variation of the motor moment when the mechanism is unbalanced and when is totally and partially balanced.

Keywords: plane mechanism, inertia forces, balancing

1. Introduction

During the operation of the mechanisms, the forces and moments of inertia that work on the component elements vary over time, which leads to the occurrence of dynamic forces in joints whose value increases as the mechanism is faster. When the values of these dynamic forces exceed certain limits, a premature wear of the joints comes out that generate vibrations, sometimes even shocks during operation. By statically balancing of the mechanisms is being pursued the decrease of the value of the resultant force of all inertia forces that work on the component elements [1, 2]. This is achieved by introducing some balancing masses mounted in the prolongation of the component elements which leads to the increase in the gauge of the mechanism and its mass. So, the problem of balancing the mechanisms has to be studied very carefully. Finding solutions of partial balancing that lead to the diminution of the values of the resultant inertia force without a significant increase in the gauge of the mechanism is highly indicated.

In this paper are presented some possibilities of statically balancing a plane mechanism. First, is analyzed the mechanism kinematics. Then a statically total balancing solution of the mechanism, when the resultant inertia force is cancelled, is presented. Also, it is proposed a solution for a partial balancing of the mechanism based on the balancing of the first harmonic of the inertia force developed in a piston of the mechanism. Finally, are presented some simulation results concerning the variation of the value of the resultant inertia force during a cinematic cycle when the mechanism is unbalanced and when it is partially balanced. Also, it is analyzed the variation of the motor moment when the mechanism is unbalanced and when is totally and partially balanced.

2. Mechanism kinematics

In figure 1 the analyzed plane mechanism is presented. With $C_i, i=\overline{1,7}$, were noted the mass centres of the component elements. The dimensions of the component elements are: $OA = 0.14$ m; $OC_1 = 0.07$ m; $AD = 0.9$ m; $AB = 0.45$m; $BF = 0.8$m; $BE = 0.4$m; $FG = 0.5$m; $FC_6 = 0.25$m; $O_2E = 0.6$m; $EC_5 = 0.3$m. The coordinates of the G point as against the Oxy coordinate system are: $x_G = 0.75$ m, $y_G = 0.7$ m. The nominal angular speed of the leader element of the mechanism is $\omega_1 = 20$ rad/s. The positional and cinematic analysis of the mechanism was realized by applying the method of the projection of the closed and independent vectorial contours [3, 4].

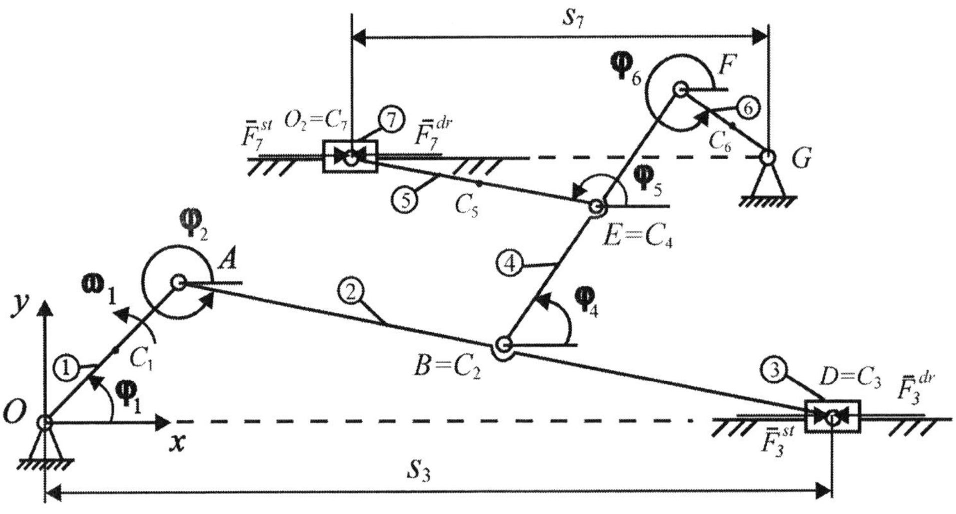

Figure 1. Plane mechanism.

The mechanism has three closed and independent contours: *O-A-D-O*, *O-A-B-F-G-O* and *O-A-B-E-O$_2$-G-O*. By projecting on the x and y axes the vectorial closing equations corresponding to the three contours the following systems of equations were obtained:

$$\begin{cases} l_1 \cdot \cos\varphi_1 + l_2 \cdot \cos\varphi_2 - s_3 = 0 \\ l_1 \cdot \sin\varphi_1 + l_2 \cdot \sin\varphi_2 = 0 \end{cases} \tag{1}$$

$$\begin{cases} l_1 \cdot \cos\varphi_1 + AB \cdot \cos\varphi_2 + l_4 \cdot \cos\varphi_4 + l_6 \cdot \cos\varphi_6 - x_G = 0 \\ l_1 \cdot \sin\varphi_1 + AB \cdot \sin\varphi_2 + l_4 \cdot \sin\varphi_4 + l_6 \cdot \sin\varphi_6 - y_G = 0 \end{cases} \tag{2}$$

$$\begin{cases} l_1 \cdot \cos\varphi_1 + AB \cdot \cos\varphi_2 + BE \cdot \cos\varphi_4 + l_5 \cdot \cos\varphi_5 + s_7 - x_G = 0 \\ l_1 \cdot \sin\varphi_1 + AB \cdot \sin\varphi_2 + BE \cdot \sin\varphi_4 + l_5 \cdot \sin\varphi_5 - y_G = 0 \end{cases} \tag{3}$$

where: $l_1 = OA$; $l_2 = AD$; $l_4 = BF$; $l_5 = EO_2$; $l_6 = FG$.

By solving the equations system (1) results the angle φ_2 and the displacement s_3, then by solving the equations system (2) results the angles φ_4 and φ_6, and finally by solving the equations system (3) results the angle φ_5 and the displacement s_7. The kinematics study of the mechanism was transposed into a computer program using the Maple programming environment [5] that holds o lot of packages of powerful symbolically computation functions, so that solving the three equations systems became extremely easy. Once established the angles values mentioned above, it were determined the coordinates of the various points of the mechanism, including those of the mass centres of the component elements. The projections on the x and y axes of the speeds and of the accelerations of any point on the mechanism may be calculated by deriving in relation to time the expressions of its coordinates, for example in the case of the mass centres $C_j, j=\overline{1,7}$, by using the following relations:

$$\begin{cases} (v_{C_j})_x = \dot{x}_{C_j} = \dfrac{dx_{C_j}}{d\varphi_1}\dfrac{d\varphi_1}{dt} = \omega_1 \cdot \dfrac{dx_{C_j}}{d\varphi_1} \\ (v_{C_j})_y = \dot{y}_{C_j} = \dfrac{dy_{C_j}}{d\varphi_1}\dfrac{d\varphi_1}{dt} = \omega_1 \cdot \dfrac{dy_{C_j}}{d\varphi_1} \end{cases} \tag{4}$$

$$\begin{cases} (a_{C_j})_x = (\dot{v}_{C_j})_x = \varepsilon_1 \cdot \dfrac{dx_{C_j}}{d\varphi_1} + \omega_1^2 \cdot \dfrac{d^2x_{C_j}}{d\varphi_1^2} \\ (a_{C_j})_y = (\dot{v}_{C_j})_y = \varepsilon_1 \cdot \dfrac{dy_{C_j}}{d\varphi_1} + \omega_1^2 \cdot \dfrac{d^2y_{C_j}}{d\varphi_1^2} \end{cases} \tag{5}$$

The angular speeds and accelerations $\omega_j, \varepsilon_j, j=2,4,5,6$, can be determined in the same way by deriving in relation to time the expressions of the angles $\varphi_j(\varphi_1), j=2,4,5,6$:

$$\begin{cases} \omega_j = \dot{\varphi}_j = \dfrac{d\varphi_j}{d\varphi_1}\dfrac{d\varphi_1}{dt} = \omega_1 \cdot \dfrac{d\varphi_j}{d\varphi_1} \\ \varepsilon_j = \ddot{\varphi}_j = \varepsilon_1 \cdot \dfrac{d\varphi_j}{d\varphi_1} + \omega_1^2 \cdot \dfrac{d^2\varphi_j}{d\varphi_1^2} \end{cases} \tag{6}$$

3. Static balancing of the mechanism

It is presented a solution of statically total balancing of the mechanism by using the method of static concentration of the mass of the component elements [1]. In figure 2 is presented the model of the mechanism with concentrated masses. Because it was considered the mass centres $C_j, j=1,2,4,5,6$, at the middle of the elements, it results that the concentrated masses m_{1O} and m_{1A} are equal to half of the mass of element *1*, m_{2A} and m_{2D} are equal to half of the mass of element *2*, m_{4B} and m_{4F} are equal to half of the mass of element *4*, m_{5E} and m_{5O_2} are equal to half of the mass of element *5* and m_{6F} and m_{6G} are equal to half of the mass of element *6*.

First it is added the balancing mass m_{E1} in the prolongation of the element *2* so that:

$$m_{E1} \cdot r_{E1} = m_D \cdot l_2 + m_B \cdot AB \tag{7}$$

where: $m_D = m_3 + m_{2D}$ and $m_B = m_{4B}$, m_3 being the mass of the piston *3*.

From equation (7) it is calculated the balancing mass m_{E1} and the length of its arm r_{E1}, by imposing the value of one of the two unknowns. Neglecting the mass of the arm of length r_{E1}, it results that in point A will be now concentrated a mass whose value is $M_A = m_{E1} + m_A + m_D$, where $m_A = m_{1A} + m_{2A}$.

It is added the balancing mass m_{E2} in the prolongation of the crank *1* so that:

$$m_{E2} \cdot r_{E2} = M_A \cdot l_1 \tag{8}$$

The balancing mass m_{E2} and the length of its arm r_{E2} are calculated in the same way, by imposing the value of one of them. Neglecting the mass of the arm of length r_{E2}, it results that in point O is now concentrated a mass whose value is $M_O = m_{E2} + M_A + m_O$, where $m_O = m_{1O}$. Then, it is added the balancing mass m_{E3} in the prolongation of the element *5* so that:

$$m_{E3} \cdot r_{E3} = m_{O_2} \cdot l_5 \tag{9}$$

where $m_{O_2} = m_{5O_2} + m_7$, m_7 being the mass of the piston *7*. After setting the values for m_{E3} and r_{E3} by imposing the value of one of them in equation (9), it results (neglecting the mass of the arm of length r_{E3}) that in point E is concentrated a mass whose value is $M_E = m_{E3} + m_{O_2} + m_E$, where $m_E = m_{5E}$.

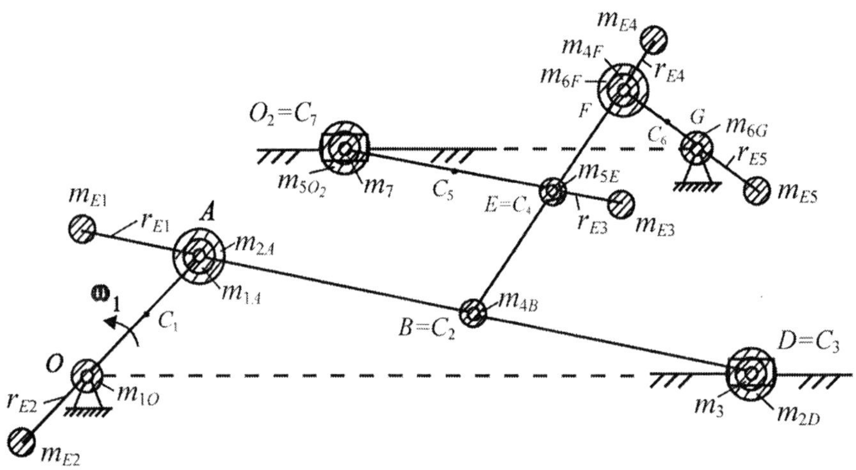

Figure 2. Solution of statically total balancing of the mechanism.

It is added the balancing mass m_{E4} in the prolongation of the element *4* so that:

$$m_{E4} \cdot r_{E4} = M_E \cdot EF \tag{10}$$

equation from which results the value for m_{E4} and r_{E4}, by imposing the value of one of them. So, in point F is now concentrated a mass whose value is $M_F = M_E + m_F + m_{E4}$, where $m_F = m_{4F} + m_{6F}$. Finally, it is added the balancing mass m_{E5} in the prolongation of the rocker *6* so that:

$$m_{E5} \cdot r_{E5} = M_F \cdot l_6 \tag{11}$$

wherefrom results the values of the balancing mass m_{E5} and the length of its arm r_{E5}, by imposing the value of one of them in equation (11). Neglecting the mass of the arm of length r_{E5}, it results that in point G is now concentrated a mass whose value is $M_G = m_{E5} + M_F + m_G$, where $m_G = m_{6G}$.
It can be observed that after adding the balancing masses the entire mass of the mechanism has been concentrated in the points O and G. So, the mass centre of the mechanism occupies a fix position between O and G during operation, and in this way the resultant force of all inertia forces that work on the component elements is cancelled. On the other hand it is obvious that introducing the balancing masses leads to the increase in the gauge of the mechanism and its mass. That is why it is necessary

finding solutions of partial balancing that lead to the diminution of the values of the resultant inertia force without a significant increase in the gauge of the mechanism.

Further, it is presented in figure 3 such a partial balancing solution in the case of the analyzed mechanism. In this case, first it is added the balancing mass m_E in the prolongation of the crank *1* so that:

$$m_E \cdot r_E = m_A \cdot l_1 \tag{12}$$

The balancing mass m_E and the length of its arm r_E are calculated by imposing the value of one of them in equation (12).

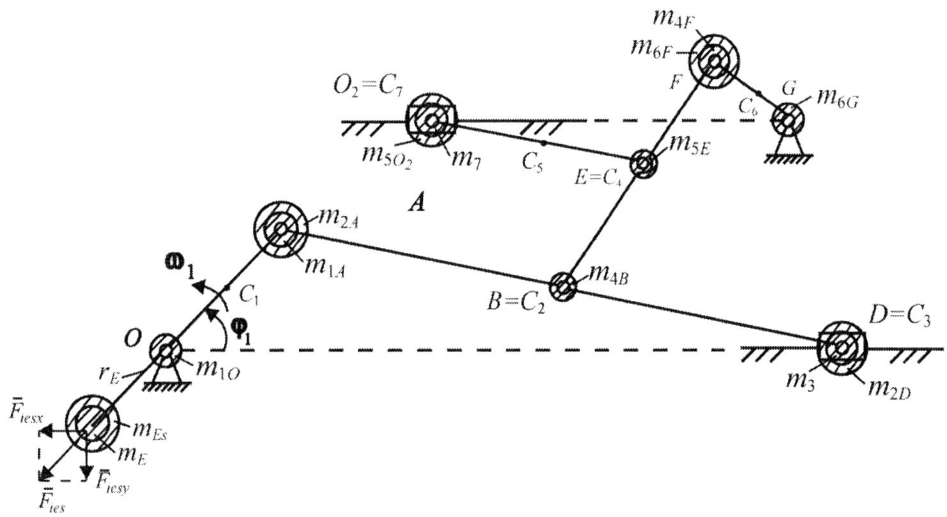

Figure 3. Solution of partially balancing of the mechanism.

Then the balancing mass m_E is supplemented with the mass m_{Es} so that the horizontal component of the inertia force developed by it ($\overline{F}_{iesx}$), whose value is $m_{Es} \cdot r_E \cdot \omega_1^2 \cdot \cos\varphi_1$, balances the first harmonic of the inertia force developed by the mass m_D: $\overline{F}_{iD} = -m_D \cdot \overline{a}_D$, whose expression is [1, 6]: $m_D \cdot l_1 \cdot \omega_1^2 \cdot \cos\varphi_1$, $\overline{a}_D$ being the acceleration of point *D*. From this equality it results: $m_{Es} = m_D \cdot l_1 / r_E$. It can be observed that this partial balancing solution leads to an insignificant increase in the gauge of the mechanism. However, the inertia forces developed by the other concentrated masses of the model of the analyzed mechanism remain unbalanced, as well as the higher harmonics of the inertia force $\overline{F}_{iD}$ and the projection on the *y* axis of the inertia force $\overline{F}_{ies}$.

4. Simulation results

It was considered that the elements of the mechanism are made of aluminium, the elements *1,2, 4, 5* and *6* having circular cross section with a radius of 0.015 m. The mass of each of the pistons 3 and 7 is 1.5 kg.

In the case of statically total balancing of the mechanism we have imposed the lengths of the arms of the balancing masses in this way: $r_{E1} = 0.8 \cdot l_2$; $r_{E2} = l_1$; $r_{E3} = 0.8 \cdot l_5$; $r_{E4} = 0.8 \cdot l_4$ and $r_{E5} = 1.2 \cdot l_6$. So, the balancing masses have resulted with the following values: $m_{E1} = 2.948$ kg; $m_{E2} = 6.3$ kg; $m_{E3} = 2.59$ kg; $m_{E4} = 4.226$ kg and $m_{E5} = 9.555$ kg.

In the case of the partial balancing solution by imposing $r_E = l_1$, it has been obtained $m_E + m_{Es} = 3.351$ kg.

In figure 4 it is presented the variation on a cinematic cycle of the total inertia force F_{itot} in the case when the mechanism is unbalanced (curve *1*) and when it is partially balanced (curve *2*). The total inertia force F_{itot} has been calculated by considering its projections on the x and y axes with the relation: $F_{itot} = (F_{itotx}^2 + F_{itoty}^2)^{1/2}$, where:

$$\begin{cases} F_{itotx} = -\sum_{j=1}^{7} m_j \cdot (a_{C_j})_x \\ F_{itoty} = -\sum_{j=1}^{7} m_j \cdot (a_{C_j})_y \end{cases} \tag{13}$$

when the mechanism is unbalanced and:

$$\begin{cases} F_{itotx} = -\sum_{j=1}^{7} m_j \cdot (a_{C_j})_x - (m_E + m_{Es}) \cdot (a_E)_x \\ F_{itoty} = -\sum_{j=1}^{7} m_j \cdot (a_{C_j})_y - (m_E + m_{Es}) \cdot (a_E)_y \end{cases} \tag{14}$$

when the mechanism is partially balanced. In relations (13) and (14) with m_j was noted the mass of the element j and with $(a_E)_x$ and $(a_E)_y$ were noted the projections on the x and y axes of the acceleration of the point where is concentrated the balancing mass $m_E + m_{Es}$.

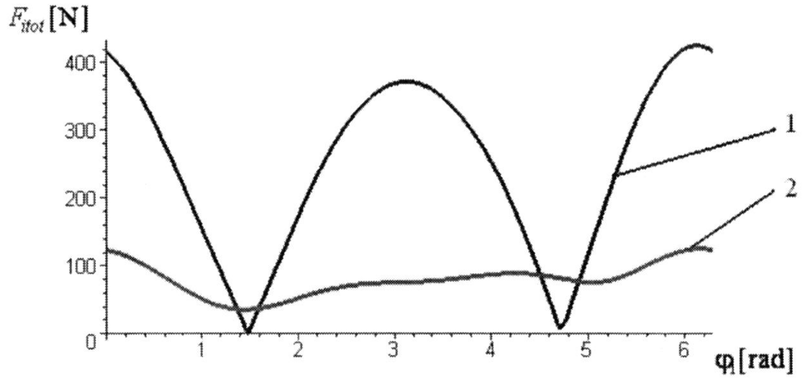

Figure 4. The variation on a cinematic cycle of the total inertia force F_{itot} in the case when the mechanism is unbalanced (curve *1*) and when it is partially balanced (curve *2*).

Figure 4 shows a significant decrease of the total inertia force in the case of partial balancing of the mechanism under conditions of very small increase of its gauge.
Also, it was analyzed the variation of the motor moment when the mechanism is unbalanced and when is totally and partially balanced.
The variation on a cinematic cycle of the motor moment M_m may be obtained by expressing the dynamic equilibrium in instantaneous powers of all the forces and moments that work on the component elements of the mechanism [1, 7]. So, when the mechanism is unbalanced the motor moment may be calculated with the following relation:

$$\overline{M}_m \cdot \overline{\omega}_1 + \sum_{j=1}^{7} \overline{G}_j \cdot \overline{v}_{C_j} + \sum_{j=1}^{7} \left(\overline{F}_{ij} \cdot \overline{v}_{C_j} + \overline{M}_{ij} \cdot \overline{\omega}_j \right) + \overline{F}_3 \cdot \overline{v}_D + \overline{F}_7 \cdot \overline{v}_{O_2} = 0 \tag{15}$$

where: $G_j = m_j \cdot g, j = \overline{1,7}$, are the weight forces corresponding to the component elements ($g = 9.81\,\text{m/s}^2$ is the gravitational acceleration); $\overline{F}_{ij} = -m_j \cdot \overline{a}_{C_j}, j = \overline{1,7}$, are the inertia forces; $\overline{M}_{ij} = -I_{C_j} \cdot \overline{\varepsilon}_j, j \in \{1,2,4,5,6\}$, are the inertia moments, where $I_{C_j}, j \in \{1,2,4,5,6\}$, are the inertia mass moments of the elements; $\overline{F}_j, j = 3,7$, are the technological forces equal to $\overline{F}_j^{dr}, j = 3,7$, when the j piston moves to the right and equal to $\overline{F}_j^{st}, j = 3,7$, when the movement is to the left (figure 1); $\overline{v}_D$ and $\overline{v}_{O_2}$ are the speeds of the points D and O_2, respectively.

The simulations were performed considering $F_3^{dr} = 1500$ N; $F_3^{st} = 800$ N; $F_7^{dr} = 800$ N and $F_7^{st} = 1500$ N.

When the mechanism is statically total balancing the relation (15) becomes:

$$\overline{M}_m \cdot \overline{\omega}_1 + \sum_{j=1}^{7} \overline{G}_j \cdot \overline{v}_{C_j} + \sum_{j=1}^{5} \overline{G}_{Ej} \cdot \overline{v}_{Ej} + \sum_{j=1}^{7} \left(\overline{F}_{ij} \cdot \overline{v}_{C_j} + \overline{M}_{ij} \cdot \overline{\omega}_j \right) + \sum_{j=1}^{5} \overline{F}_{iEj} \cdot \overline{v}_{Ej} + \overline{F}_3 \cdot \overline{v}_D + \overline{F}_7 \cdot \overline{v}_{O_2} = 0 \quad (16)$$

where: $G_{Ej} = m_{Ej} \cdot g, j = \overline{1,5}$, are the weight forces corresponding to the balancing masses; $\overline{F}_{iEj} = -m_{Ej} \cdot \overline{a}_{Ej}, j = \overline{1,5}$, are the inertia forces corresponding to the balancing masses; $\overline{v}_{Ej}, \overline{a}_{Ej}, j = \overline{1,5}$, are the speeds and the accelerations, respectively of the points where the balancing masses are concentrated.

When the mechanism is partially balancing the calculus relation of the motor moment becomes:

$$\overline{M}_m \cdot \overline{\omega}_1 + \sum_{j=1}^{7} \overline{G}_j \cdot \overline{v}_{C_j} + \overline{G}_E \cdot \overline{v}_E + \sum_{j=1}^{7} \left(\overline{F}_{ij} \cdot \overline{v}_{C_j} + \overline{M}_{ij} \cdot \overline{\omega}_j \right) + \overline{F}_{iE} \cdot \overline{v}_E + \overline{F}_3 \cdot \overline{v}_D + \overline{F}_7 \cdot \overline{v}_{O_2} = 0 \quad (17)$$

where: $G_E = (m_E + m_{Es}) \cdot g$; $\overline{F}_{iE} = -(m_E + m_{Es}) \cdot \overline{a}_E$; $\overline{v}_E, \overline{a}_E$, are the speed and the acceleration, respectively of the point where the balancing mass $m_E + m_{Es}$ is concentrated.

Figure 5 shows the variation on a cinematic cycle of the motor moment M_m when the mechanism is unbalanced (curve *1*) and when is totally balanced (curve *2*) and partially balanced (curve *3*).

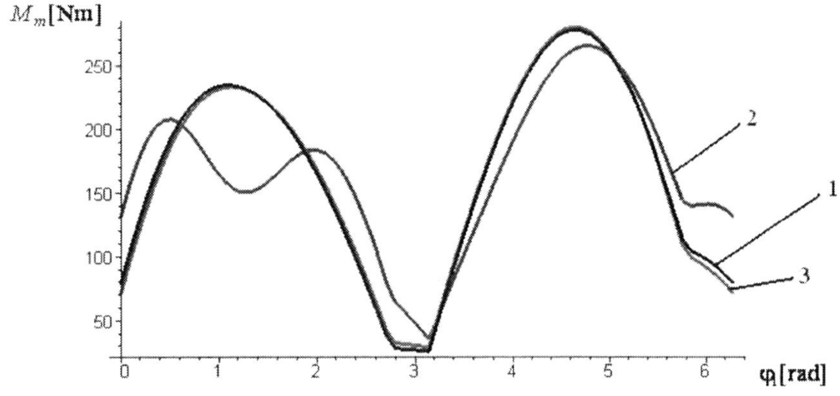

Figure 5. The variation on a cinematic cycle of the motor moment: mechanism unbalanced (curve *1*); mechanism totally balanced (curve *2*); mechanism partially balanced (curve *3*).

Figure 5 shows that by partially balancing the mechanism the variation of the motor moment practically does not change, while in the case of totally balancing the extreme values of the motor moment have slightly decreases.

5. Conclusions

In this paper were presented some possibilities of statically balancing a plane mechanism. After presenting a statically total balancing solution of the mechanism it was proposed a solution for a partial balancing based on the balancing of the first harmonic of the inertia force developed in a piston of the mechanism. The simulations results showed a significant decrease of the total inertia force in the case of partial balancing of the mechanism under conditions of very small increase of its gauge. Also, it was analyzed the variation of the motor moment when the mechanism is unbalanced and when is totally and partially balanced. The simulations results showed in the case of totally balancing that the extreme values of the motor moment have slightly decreases compared to cases where the mechanism is unbalanced or partially balanced.

References

1. Bădoiu D., The Dynamic Analysis of Mechanisms and Machines (in Romanian), Didactical and Pedagogical Publishing House, Bucuresti, 2003.
2. Shigley J.E. and Uicker Jr.J.J., Theory of Machines and Mechanisms, McGraw-Hill, 1981.
3. Bădoiu D., Structural and Cinematic Analysis of the Mechanisms (in Romanian), Technical Publishing House, Bucuresti, 2001.
4. Bădoiu D., Petrescu M.G., Toma G., On the analysis and synthesis of the quadrilateral mechanism, Journal of the Balkan Tribological Association, vol. 20, 161-168, 2014.
5. Monagan M.B., Geddes K.O., Heal K.M., Labahn G., Vorkoetter S.M., McCarron J. and DeMarco P., Maple Introductory Programming Guide, Maplesoft, a division of Waterloo Maple Inc., 2005.
6. Manolescu N., Kovacs Fr. and Oranescu A., Mechanisms and Machines Theory (in Romanian), Didactical and Pedagogical Publishing House, Bucuresti, 1972.
7. Bădoiu D., Toma G., On a dynamic optimization problem of the quadrilateral mechanism, Journal of the Balkan Tribological Association, vol. 22, 250-260, 2016.

Operational limit conditions of the spur gears in lubricated modes

S. Benilha [1], ***F. Belarifi*** [2*]
[1,2]Institut de Maintenance et sécurité Industrielle- Université d'Oran2. B.P 170 El M'Naouer Oran, Algeria
[2]Laboratoire des Sciences de la Matière Condensé (Equipe de Tribologie et nano tribologie) - Université d'Oran1,B.P 1524 El M'Naouer Oran, Algeria
benilha.samir@gmail.com[1],
[*]Corresponding author: belarifi_farid@yahoo.fr[2]

Abstract: The calculation of the gear teeth resistance, shows the using of a certain number of coefficients determined experimentally and which are accepted by the various international standards. However, this kind of calculation determines the gears by excess material and does not support the tribological parameters of operation. We propose in this work the support of these parameters, to determine the limit operation conditions of the spur gears, using the equivalent geometry. This is represented by two cylinders, which geometrically models of the contact between two teeth of a gear and whose lubrication is generally in mixed lubrication mode. The concept of Mc cool is used to determine the distribution of the load and the friction force, which are distributed in liquid (elastohydrodynamic) and solid domains and interact with each other. The phenomenon of interaction between the two domains is used, to predict the tribological limit conditions of operation. The proposed model is based on the resolution of elastohydrodynamic equations for the determination of load and friction as well as the deduction of mixed friction by tracing the Stribeck curve. This is calculated by the model of the decomposition of the patterns profile of rough surfaces in contacts. The results of non-dimensional calculations allow us to deduce the boundary conditions and can be adapted for any type of gear pair defined according to pre-established operating conditions.

Keywords: —: lubrication, gears, mixed lubrication, Stribeck curve, limit condition

1. Introduction

The design of gears includes calculations of resistance of materials, the geometrical aspect, manufacture and the checking. Among all these parameters, it is essential to know with precision the stresses being in the tooth of gears to prevent certain risks of rupture. Consequently, several theoretical and experimental methods were developed, starting from the end of the 19th century. O.Corina[1] indexed in its thesis the various methods of calculating of resistance. This reveals that the various standards AGMA and ISO that those take into account the parameters of fatigue and flexural strength. AGMA 218.01 names is frequently used in North America. and normalizes it ISO 6336-1: 2006 are used in Europe; In the two standards of calculation we notice the existence of several experimental parameters for the assumption of responsibility of the tribological parameters[2][3]. into this work we introduce the tribological parameters, for the determination of the limiting conditions of operation of the spur gears, by using the equivalent geometry and the theory of mixed lubrication. our approach is based on the basic equations of the material resistance with which we combine the various equations of elastohydrodynamic lubrication to have adequate limiting conditions of operation.

2. Static resistance of a gear teeth

We consider the resistance of the tooth of a spur gear height H, width L and thickness E as indicated by figure 1. It is supposed that the effort F is applied at the end of the tooth which represents the most unfavourable case and which only the component tangential T produces a bending of the tooth, that we regard as a fixed beam.

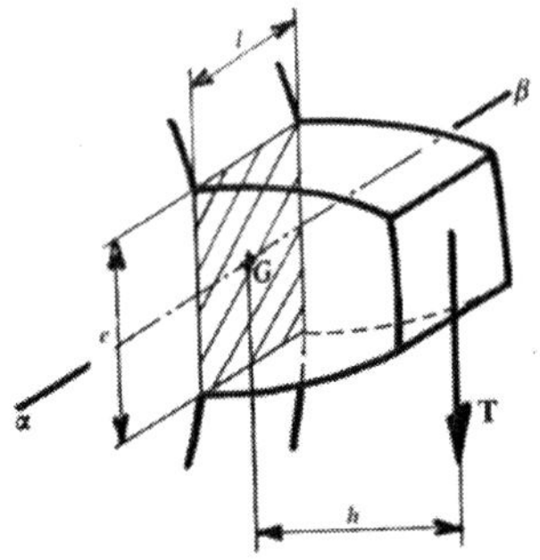

Figure1: gear teeth resistance

The equation of the maximum bending moment is written:

$$Mf_{max} = T.h \tag{1a}$$

The tangential stress on diameter according to the load is written:

$$T = W\cos\alpha \tag{1b}$$

The equation of the bending moment becomes:

$$Mf_{max} = W.h\cos\alpha \tag{2}$$

The flexural strength condition of the tooth is written:

$$\frac{I}{v} \geq \frac{Mf_{max}}{\sigma_{max}} \tag{3a}$$

who is written:

$$Mf_{max} \leq \left(\frac{I}{v}\right)\sigma_{max} \tag{3b}$$

the quadratic moment of a square surface is written:

$$\frac{I}{v} = \frac{l.e^2}{6} \tag{4}$$

The condition of Hertzian pressure resistance is written:

$$\sigma_{max} = 0.3\,p_0 \tag{5a}$$

With:

$$p_0 \leq 0.6HB \tag{5b}$$

The equation (5a) becomes:

$$\sigma_{max} \leq 0.18HB \tag{5c}$$

Thus the equation of the moment (3b), becomes

$$Mf_{max} \leq 0.03(HB)le^2 \tag{6}$$

By replacing the equation (2) in the equation (6), the equation of the load is written:

$$W \leq \frac{0.03(HB)le^2}{h\cos\alpha} \tag{7}$$

3. Variation of the load on teeth in lubricated contacts

We modelize geometrically the contact between two teeth as being the contact between two cylinders or two half rolls as indicated by figure 2 and their lubrications is generally in mode of mixed lubrication, and the concept of Mc cool is often used (figure 3) to determine the burden-sharing in the contact [3].

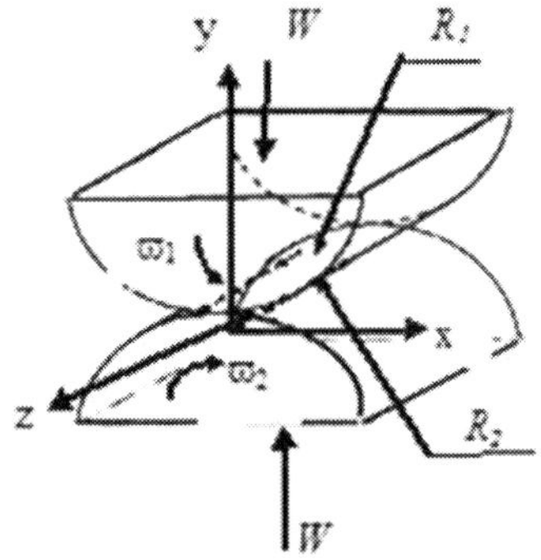

Figure 2. Geometrical model

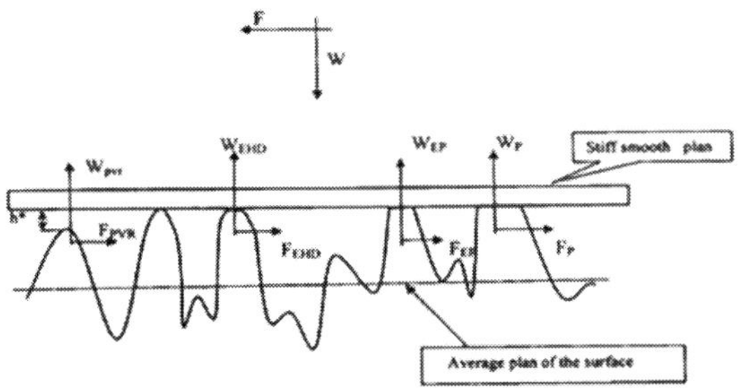

Figure 3. Mc Cool Model

We note in this model the influence of the asperities in the movement of a body compared to the other and causes the load (*W*) to be distributed in a liquid lift (W_L) and the other solid (W_S)
The asperities the least far away from the smooth plan, become deformed geometrically in an elastic range to form the elastohydrodynamic field (EHD), whose resultant is W_{EHD}. The sum of these efforts contributes to the lift of viscous origin, dependant amongst other things the speed of slip and on the lubricant. This field generates a liquid friction (f_L), which in the case of tightens in its majority the gears towards an elastohydrodynamic friction (f_{EHD}), whose limiting value does not exceed 0.02[4], and is written:

$$f_{EHD} = \frac{F_{EHD}}{W_{EHD}} \tag{8}$$

The asperities intercepted by the plan smoothes are subjected to a crushing, which can be weak of elastic type (W_E), moderately crushed of elastoplastic type (W_{EP}) and lastly of plastic type (W_P) for the asperities the most requested. The sum of these efforts contributes to a lift of the solid type independent speed.
The solid lift is written:

$$W_S = W_E + W_{EP} + W_P \tag{9}$$

It results a solid friction force which is written:

$$F_S = F_E + F_{EP} + F_P \tag{10}$$

The coefficient of total solid friction is written:

$$f_S = \frac{F_S}{W_S} \tag{11}$$

According to the authors [3] the coefficient f_S vary between 0.12 to 0.09
The total load, results from the sum of the elastohydrodynamic and solid load, and is written:

$$W = W_{EHD} + W_S \tag{12}$$

The total friction force, results from the sum of the elastohydrodynamic and solid force, and is written:

$$F = F_{EHD} + F_S \tag{13}$$

It results a global coefficient of friction which is written:

$$f = \frac{F}{W} = \frac{F_{EHD}+F_S}{W_{EHD}+W_S} \tag{14a}$$

Who is written

$$f = \frac{f_{EHD}W_{EHD}+f_S W_S}{W_{EHD}+W_S} \tag{14b}$$

By replacing the equation (14b) in the equation (12), the solid load (Ws) is written:

$$W_S = W_{EHD}\frac{(f - f_{EHD})}{(f_S - f)} \tag{15}$$

The equation of thetotal load (12) becomes:

$$W = W_{EHD}\left(1+\frac{(f - f_{EHD})}{(f_S - f)}\right) \tag{16a}$$

Thus:

$$\frac{W}{W_{EHD}} = \left(1+\frac{(f - f_{EHD})}{(f_S - f)}\right) \tag{16b}$$

Let us write: $\overline{W} = \frac{W}{W_{EHD}}$ Thus the equation (16b) is written:

$$\overline{W} = \left(1+\frac{(f - f_{EHD})}{(f_S - f)}\right) \tag{17}$$

Same manner let us divide the equation (7) by W_{EHD}, it becomes

$$\frac{W}{W_{EHD}} \leq \frac{0.03(HB)le^2}{W_{EHD}h\cos\alpha} \tag{18a}$$

Who is written too

$$\overline{W} \leq \frac{0.03(HB)le^2}{W_{EHD}h\cos\alpha} \tag{18b}$$

Let us introduce the dimensional parameter K ,who is written:

$$\mathrm{K} = \frac{0.03(HB)le^2}{h\cos\alpha}$$

The equation (18b) becomes:

$$\overline{W} \leq \mathrm{K}\left(\frac{1}{W_{EHD}}\right) \tag{19}$$

For a module of cutting, a material and a geometry of the tooth which is fixed, the parameter K is constant. It is noticed that the inequation (19) depends on the load in the elastohydrodynamic field. The latter is determined by the resolution of the equation given by Belarifi et al[3].

$$H_e\widehat{W}_{EHD}^{\,0.1} - 0.63\widehat{W}_{EHD} - H_0 = 0 \tag{20a}$$

Dowson equation film thickness [3] are used at the center for linear contact

$$\widehat{H}c = 3.06\widehat{U}^{0.69}\widehat{G}^{0.56}\widehat{W}_{EHD}^{-0.1} \tag{20b}$$

We introduce:

$$\widehat{H}c = H_e\widehat{W}_{EHD}^{-0.1} \tag{20b}$$

With:

$$\widehat{H}c = \left(\frac{h_c}{R_e}\right) ; \widehat{U} = \left(\frac{\eta_0 U}{E^* R_e}\right); \widehat{G} = \left(\alpha E^*\right) ; \widehat{W}_{EHD} = \left(\frac{W_{EHD}}{E^* l R_e}\right); R_e = \frac{R_1 R_2}{R_1 + R_2} ; E^* = \frac{E_1 E_2}{E_1 + E_2}$$

These equations are valid, as long as load W_{EHD} max; remain lower than the elastoplastic mode ($W_{EHDMax} < W_{Ep}$), and is written:

$$\widehat{W}_{EHD}^{\lim} = 2.18\left(\frac{R_{pe}}{E^*}\right)^2 \tag{21}$$

Knowing that:

$$HB \le 2.77 Rpe \tag{22a}$$

Finally the elastohydrodynamic limiting load is written:

$$\widehat{W}_{EHD}^{\lim} = 16.73\left(\frac{HB}{E^*}\right)^2 \tag{22b}$$

4. Results and discussions

The various computational results indicated by figures 4 and 5, were established for the various characteristics indicated to table 1, for a thickness of film in the center $h_0 = 10^{-6}$ m, solid friction ($fs = 0.12$) [3], elastohydrodynamic friction ($f_{EHD} \le 0.02$), slip velocity (Vg 0.5 ms^{-1}) and hertzian pressures variable.

Table 1. Characteristics of material and lubricant

geometry of the solid		Young Modulus	Lubricant		
Re	*l*		η	α	τ_0
4.7 10^{-3}m	6 10^{-3}m	231 GPa	0.324Pa s	19 10^{-9} Pa^{-1}	2.5 MPa

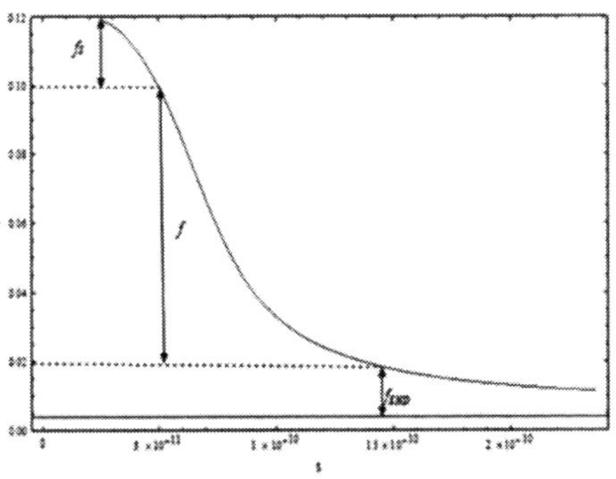

Figure 4. Stribeck curve

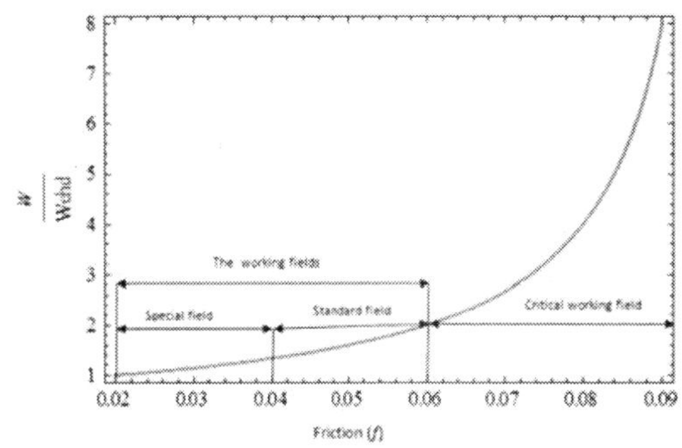

Figure 5. Operational Limit condition

The curve of Stribeck (figure 4), is plotted according to the model of Robbe-Valloir & al [5][6], indicates the variation of the coefficient of friction (*f*) according to the parameter of Somerfield ($S = \frac{\eta V}{p}$),the values of frictions are introduced into the equation (17), to determine the adimensional limits loads operation, indicated by figure 5. We notice on this curve two distinct area.
The first area is known as working area, it is included for varying friction coefficients from 0.02 to 0.06 for adimensional loads varying between values 1 to 2. We distinguish in this area two parts.
- A special field : requires specific conditions to guarantee friction lower than 0.04, which requires very good surface qualities (shaving), we notice that we work in minimal values of the limiting adimensional load what requires an abundant lubrication and a high speed without reaching the phenomenon of resonance.
- A Standard field: we can work under reasonable conditions where frictions varies between 0.04 to the 0.06. The adimensional load can reach the maximum value of 2. Our various calculations showed that this part can be used for all the types of gears.

The second area represents the critical field of operation, where frictions vary from 0.06 to 0.09 in this area the load believes in an exponential way. It is noticed that tends some mixed lubrication towards a boundary lubrication.

5. Conclusion

The model proposed, defined the various ranges operation of gears by giving the limiting adimensional loads according to lubricated frictions. Our various work on tribometer pin- disc confirms that we are in mixed lubrication but great difficulties persist in to guarantee good conditions of operation, those is noticed by the Stribeck curve, which shows low numbers of Somerfield which due to strong the hertzian pressures applied. These low values will tend to bring closer the couple to gears towards boundary lubrication what is to be avoided. Thus the knowledge of the limiting loads defined in this work will make it possible to avoid the phenomena of rupture or seizing, because we will manage to define the optimal operating ranges beforehand.

References

1. C. Oancea, Analyse des dents d'engrenages droits par la méthode des potentiels complexes," Thèse de Doctorat. univérsité de Laval (Québec-Canada), 1997.
2. G. Henriot, Engrenages : détermination des charges sur les dentures et calculs de résistance," Technique de l'ingénieur Transm. puissance mécanique engrenages liens souples, vol. base docum, no. ref. article : bm5623, 2002.
3. F. Belarifi, J. Blouet, G. Inglebert, and A. Benamar, Confrontation between a mixed lubrication model and an experimental survey on the behaviour to the friction of spur gear teeth, Mechanics &. Industry., vol.7, 527–536, 2006
4. F. Belarifi and G. Inglebert, Numerical modelling of the hydrodynamic behaviour of linear contacts, Journal of the Balkan Tribological Association., vol. 21, no. 3, 630-639, 2015.
5. F. Robbe-Valloire, B. Paffoni, and R. Progri, Load transmission by elastic, elasto-plastic or fully plastic deformation of rough interface asperities, Mechanics of. Materials., vol. 33, no. 11, 617–633, 2001.
6. F. Robbe-Valloire, R. Progri, B. Paffoni, and R. Gras, Theoretical prediction and experimental results for mixed lubrication between parallel surfaces, Boundary and Mixed Lubrication Science and Applications Proceedings of the 28th Leeds-Lyon Symposium on Tribology, Vol. 40, 129–137, 2002.

Ex-Vivo Cow Skin Viscoelastic Effect for Tribological Aspects in Endoprosthesis

Subhi K A[1, 2], Tudor A[1], Hussein E K[2], Wahad H[1], Chisiu G[1]
[1]University POLITEHNICA of Bucharest, Dept. of Machine Elements and Tribology, 060042, Bucharest, Romania
[2]AL-FURAT AL-AWSAT Technical University, Iraq
qusay_abunoor@yahoo.com[1,2], andreitudor17@gmail.com[1],

Abstract: Abstract: The viscoelastic behavior of ex-vivo cow skin was experimentally studied by applied load from different indenter types (circle, square and triangle, all types have the same area) for different times (10 sec, 30 sec, and 60 sec). The viscoelastic tests were carried out using a UMT series (UMT-II, CETR Corporation). The experimental results collected at different operating conditions showed that the cow skin has a higher reaction against the triangle indenter compared to the other shapes. Whereas the hysteresis of cow skin was lower at low applied load time and it's increased when the time increased.

Keywords: Ex-vivo cow skin, viscoelastic, friction.

1. Introduction

According to the latest reports of much international organization, there is a huge number of amputee persons all over the world due to many factors like diabetes, landmines, etc. besides to the overpopulation, all these reasons makes a large demand on prosthetic limbs and consequently too many associated medical problems are rises up for example ulcer formation, trauma, and keratinization in the stump human skin as a result of direct contact between the residual skin and the internal layer of the socket, so that encourage researchers to deeply start investigation in such effective and humanitarian topic, here and below a detailed literature review with miscellaneous related aspects.
Li W. et al [1] used a piece of rabbit skin as a testing material with the aid of the UMT- II tribometer at a sampling rate of 20 KHz, so in order to study healing process, two phenomena have been examined, they are skin self-rehabilitation and self-adaptation to friction trauma where a simple simulation process considered to simulate rubbing conditions between rabbit skin and prosthetic socket. Final results indicate the formation of the so-called keratinization on the direct contact region between the two mating surfaces after several alternations of the trauma which leads to the coefficient of friction decrement, this is, of course, happening because of a viscoelastic feature of such skin.
[2] Żak M. et al started studying sectional layers of the domestic pig skin, then examining the most effective mechanical properties for such particular skin taking into account the random direction of the collagen fibres. Many uniaxial tensile tests were done to find out the expected essential features with respect to its sectional structure. Concluding facts indicated that, there were no considerable differences in the acquired values of the maximum tensile strength and the associated modulus of elasticity (Young's modulus).
Khatyr F. et al [3] suggested a comprehensive single axis tensile test for a viscoelastic material by studying creep and relaxation behavior of in-vivo human skin. The nominated area is forearm skin for 63 volunteers with different ages where the skin has been considered as an orthotropic material. The final conclusion was that the gained results based on the viscoelastic model are independent, on the type of the applied load or it is true to say that principal direction of experiments varies from one

volunteer to another one, means that Young modulus could be considered as a function for the dermatologist and skin care centres.
Kussay et al [4] suggested a mathematical model used for measuring a hysteresis component of friction between human skin which obeys Voight mechanical model and a rigid sphere taking into account different friction parameters such as the applied load, probe velocity, the strength of the interface between skin and the artificial material. The deformation (hysteresis) component of the skin friction model is developed by using the Mathcad software. The gained numerical results showed that the adhesive component of the skin friction is greater than the hysteresis component at all operating conditions.
[5] Li W. et al studied improvement of friction conditions between the prosthesis and the stump in case of below knee amputation, where maximum direct contact happened due to load transfer depending on the total weight of the amputee person which in turn causes many dermatological problems, so two phenomena have been studied in this paper, they are skin self-rehabilitation and self-adaptation to friction trauma based on specific friction testing and viscoelastic properties of human skin in assistance with UMT-II tribometer machine. Final investigations showed the formation of keratinization in the two mating surfaces region and consequently ascending variation in coefficient of friction happening.
Soldatenkov I. [6] calculated the dominated component (deformation) component of the friction force for a viscoelastic substrate based on energy approach. A modified punch has been used to measure friction force with different surface finishing and geometry conditions, final results proved that friction conditions is a function of punch smoothness, or friction forces increased as the punch surface is less smooth and vice-versa taking into account tip geometry. Chisiu G. et al [7] a technique was used by a device (UMT 2 tribometer (CETR, Campbell, CA, USA)) to study the behavior of UHMWPE and to verify the analytical results of the Zener model.
Cheng L. et al [8] has selected a three-element viscoelastic material (polymer) which is named Kelvin-Voight mechanical model in his theoretical research to study indentation effect by employing an axisymmetric flat ended indenter and two types of tests were conducted including flat punch creep test and load relaxation test. The attained facts proved that a reasonable quantitative method for evaluating viscoelastic and instantaneous elastic features from experimental expectations, in addition, this proposed solution could be applied to indentation process on the two cases bulk polymers and polymer coating based on the acceptable ratio of the employed punch radius to the coating thickness.
Lyubicheva A. et al [9] employed a pin on disk wear rig to investigate tribological properties of viscoelastic materials, where the analytical model of direct contact between indenter with a spherical tip in touch with rotating disk has been used. The main objective of this work is to examine mechanical (viscoelastic) properties of cow skin in order to create new trends in using such available skin in the prosthetic industries to ensure comfortable direct contact conditions between human skin and prosthetic socket for both upper and lower limbs.

2. Materials and Methods

In order to conduct the experimental part of this paper a well-known rig has been used named UMT-II Tribometer, which has many excellent facilities in this topic, so mechanical properties and viscoelastic characteristics of cow skin are measured according to the following procedure:
A fresh specimen of bull ox aged about one-year cow skin is prepared with the following dimensions (50 x 45 x 3.5) mm where a sterile water is used to wash this tissue without any chemical reactions and shaved well in one direction as shown in figure (1).
The other requirement for this particular test is preparing a three different in shape geometrical tips of indenter (circular, rectangular, triangular) cross-section, manufactured from commercial brass with the equally cross-sectional area of about 16 mm^2, as indicated in the figure (2) below.
The UMT-II Tribometer device used to calculate the mechanical properties of the skin (see figure 3).The tribometer includes a part zoom from contact the tip with the cow skin accordingly, clarifies in detail how to fix this piece of cow skin on the platform of the employed rig, the first step was fixing the skin without any tension or extension, after that starting with the circular stylus fitted on the upper

jaw followed by the other two styluses. Starting with the first moment of touching between the stylus and the skin to consider this first touching point as a reference i. e. zero load point, then starting with a gradual load to reach 1 mm depth in the skin within three different time duration initiating with (10, 30, and 60) seconds. The desired goal is to find out the induced mechanical reaction with all above-mentioned cases taking into account the three geometric tips.

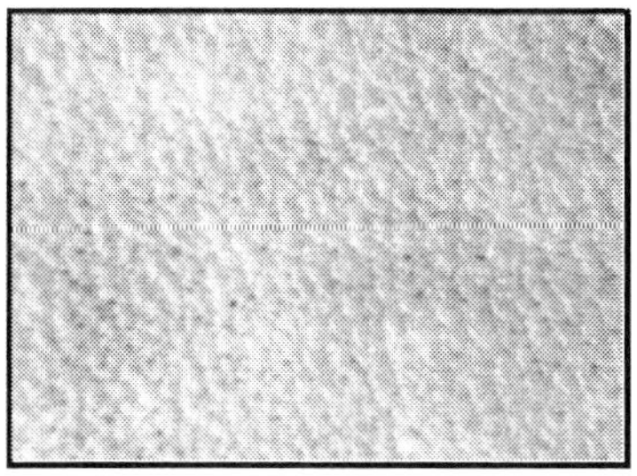

Figure 1. Image of the utilized skin.

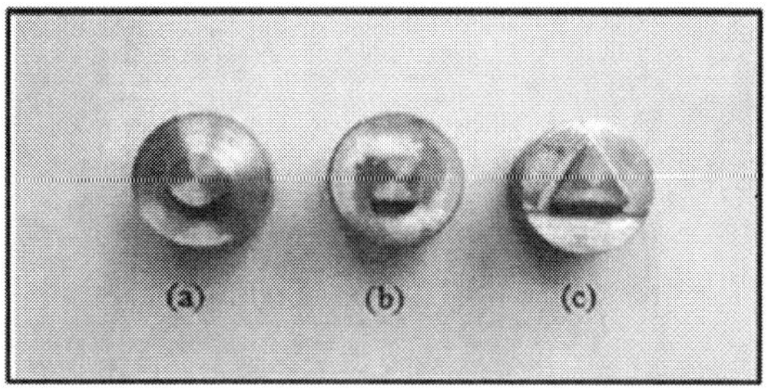

Figure 2. The utilized tips of indenter: (a) Circle; (b) Square; (c) Triangle.

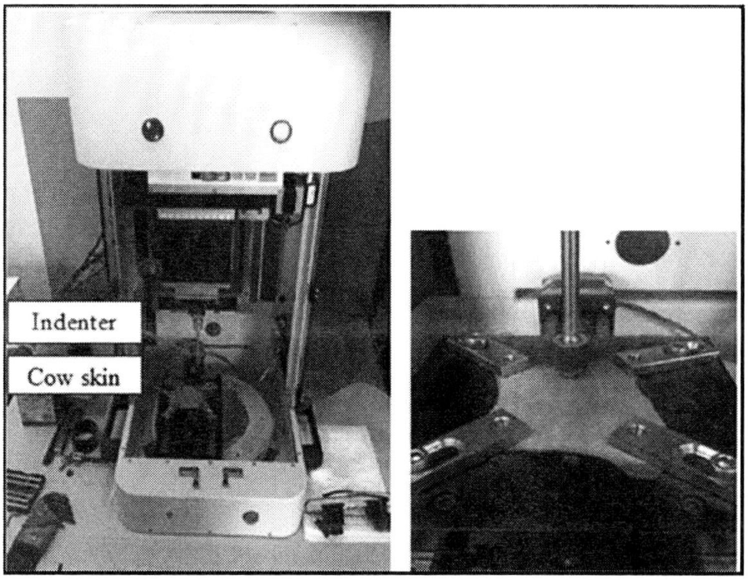

Figure 3. Image of the test device during the experiment explaining the contact with the skin

3. Results and Discussion

Based on the steps described in the previous section, the results obtained from different geometric shapes over periods of time (10, 30, 60) seconds respectively. The variation of normal force with respect to indenter deep given in figures 4a, 4b, and 4c. The results were obtained from different indenter shapes (circle, square and triangle), the load was applied on the cow skin for time (10, 30, and 60) seconds, separated. As shown in these figures, the cow skin reaction reduced when the applied load time increased. The explanation of this behavior is due to the viscoelastic phenomena of the skin. The maximum cow skin reaction against applied load was when used the triangle shape, whereas the lowest reaction register when used square shape at all operating conditions. The normal force reaction of cow skin against the triangle shape was equal to 5.25 N, 3.1N and 1.4 N, for circle shape equal to 2.4N, 1.2N and 0.9N, for square equal to 1.5N, 1.1N and 0.6N at applied load time 10, 30, and 60 seconds respectively.

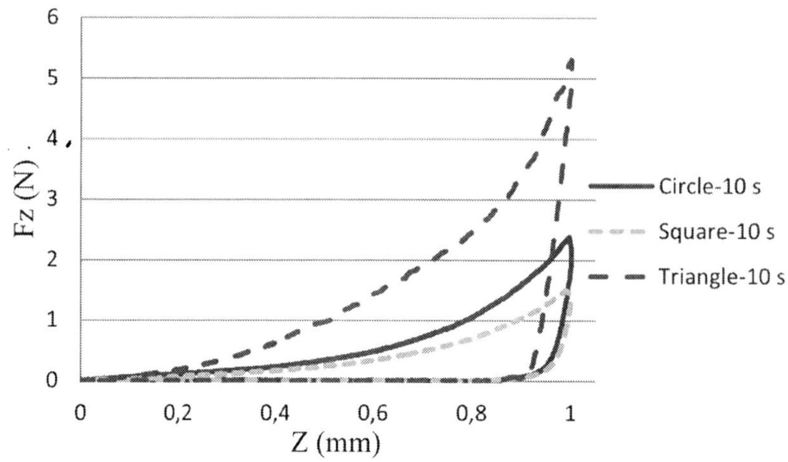

(a)

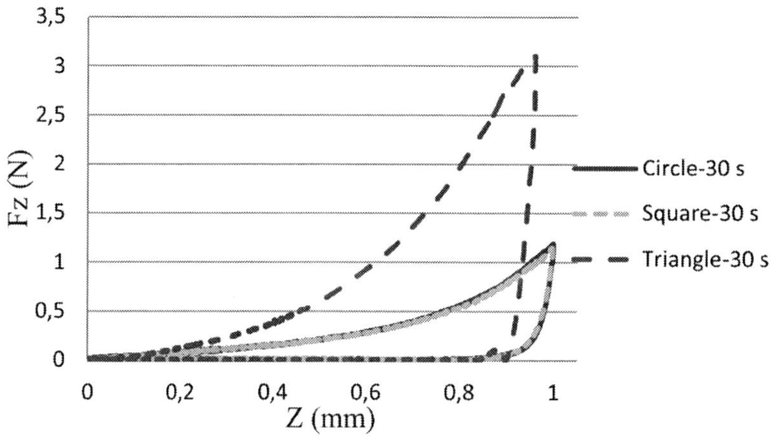

(b)

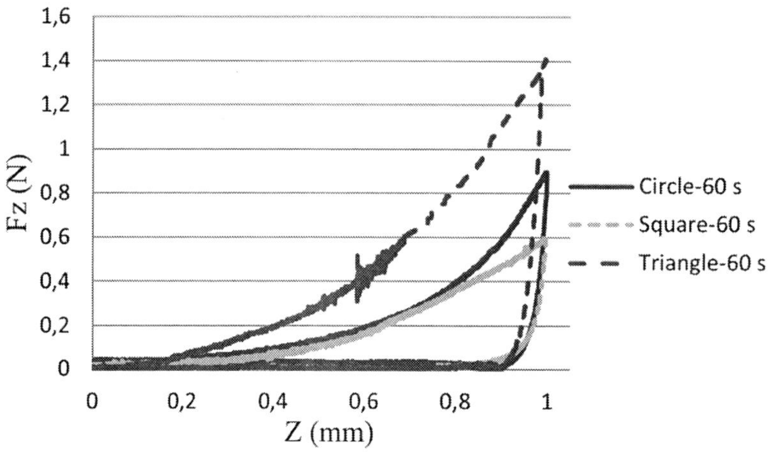

(c)

Figure 4. Variations of the cow skin reaction respect with depth

The hysteresis of cow skin was determined by using below equations:

$$\alpha_h = \frac{A}{Area\max} \tag{1}$$

$$A = (Z_{i+1} - Z_i).(Fz_{i+1} - Fz_i) \tag{2}$$

$$Area\max = \frac{1}{2}(Z_{\max}.F_{\max}) \tag{3}$$

where: α_h: hysteresis parameter [%], A: area of the curve [mm^2], Fz normal force reaction (N), Z: depth (mm)

The hysteresis of cow skin when applied the load from three different shapes with respect to applied time shown in figure 5. The hysteresis of cow skin was lower at low applied load time and it's increased when the time increased. The lowest hysteresis of cow skin was with the triangle shape, the explanation of this behavior due to the pressure generated from the edges or sides sharp angles resulting in an increase the skin reaction.

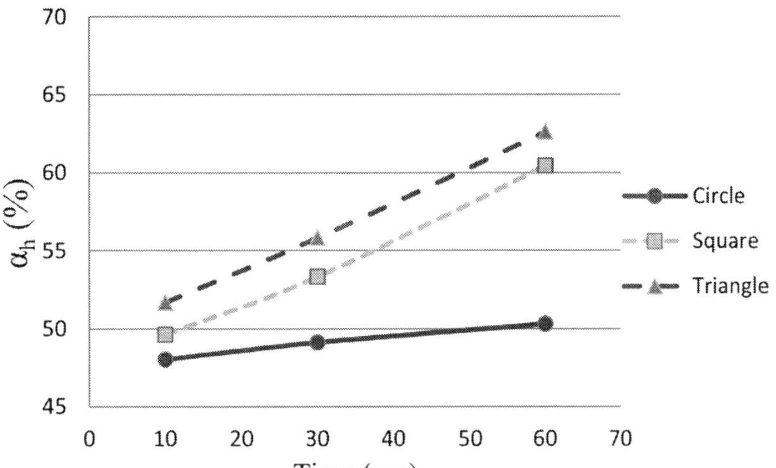

Figure 5. Variations of hysteresis parameter respect with depth

The same procedure above was repeated at the same operating conditions except for the time, the load was applied for 60 seconds, and then left the load for 60 seconds on the cow skin and then removed the load slowly for 60 seconds. The results of this test shown in figure 6. As shown in this figure that the cow skin reaction was significantly lower against the load applied to all tip shapes of the indenter. Again, the maximum cow skin reaction against applied load when used the triangle shape and equal to 6.2N, whereas the lowest reaction was equal to 1.6 N register when used square shape at all operating conditions. The variations of cow skin reaction with respect to time when applied the load from different tip shapes of indenter shown in figure 7. It's clearly shown from this figure that the cow skin reaction becomes approximately zero value after 120 second from the load applied; this may be due to the skin creep.

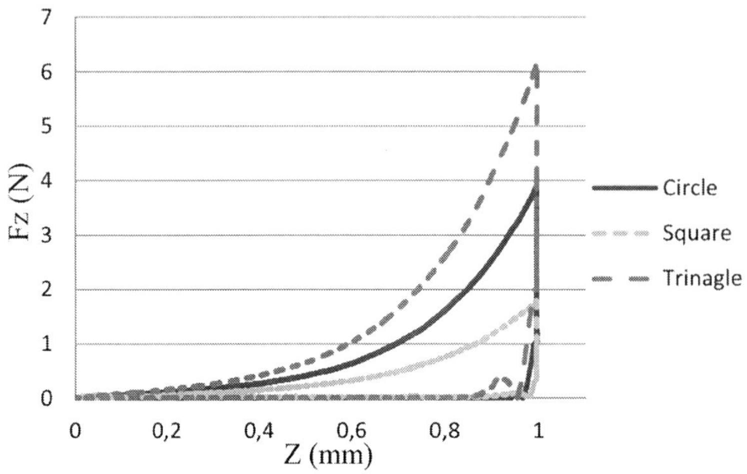

Figure 6. Variations of the cow skin reaction respect with depth in the case stay 60 second

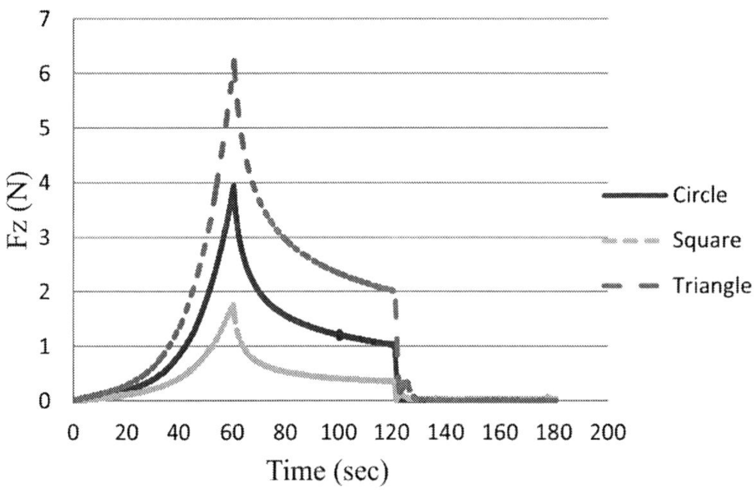

Figure 7. Variations of the cow skin reaction respect with time in the case stay 60 second

4. Conclusion

In the current study, the viscoelastic behavior of ex-vivo cow skin was experimentally studied by applied load from different tips of indenter (circle, square and triangle, all types have the same area) for different times (10 sec, 30 sec, and 60 sec).

The important findings are summarized:

• The cow skin reaction reduced when the applied load time increased. The maximum cow skin reaction against applied load was when used the triangle shape, whereas the lowest reaction register when used square shape at all operating conditions.

• The hysteresis of cow skin was lower at low applied load time and it's increased when the time increased, the lowest hysteresis of cow skin was registered when used triangle shape compared to other shapes.

• The cow skin reaction was significantly lower against the load applied to all tip shapes when the load still 60 seconds on the skin. Again, the maximum reaction was 6.2N against a load applied when used the triangle shape, whereas the lowest reaction was equal to 1.6 N register when used circle shape. Equations and mathematics

Acknowledgements

The first author of this paper acknowledges the Iraqi government for its financial support and Department of Machine Elements and Tribology, University POLITEHNICA of Bucharest to provide them with all requirements for testing.

References

1. Li, W., Qu, S. X., Zheng, Y. J., Pang, Q., Zheng, J., & Zhou, Z. RSkin self- adaptation to friction trauma under reciprocal sliding conditions. Tribology International, vol. 44(12), 1782-1789, 2011.
2. Żak, M., Kuropka, P., Kobielarz, M., Dudek, A., Kaleta-Kuratewicz, K., & Szotek, S. Determination of the mechanical properties of the skin of pig foetuses with respect to its structure. Acta of Bioengineering and Biomechanics, vol. 13(2), 37-43, 2011.
3. Khatyr, F., Imberdis, C., Vescovo, P., Varchon, D., & Lagarde, J. M. Model of the viscoelastic behaviour of skin in vivo and study of anisotropy. Skin research and technology, vol.10 (2), 96-103, 2004.
4. Kussay A. S, Tudor A., Hussein E. K., and Wahad H. S. The adhesion and hysteresis effect in friction skin with artificial materials, 13th International Conference on Tribology, ROTRIB'16, IOP Conference Series: Materials Science and Engineering. vol. 174, 012018, 2017.
5. Li, W., Pang, Q., Lu, M., Liu, Y., & Zhou, Z. R. Rehabilitation and adaptation of lower limb skin to friction trauma during friction contact. Wear, vol. 332, 725-733, 2015.
6. Soldatenkov, I. A. Calculation of the deformation component of the force of friction for a standard elastoviscous base. Journal of Friction and Wear, vol. 29(1), 7-14, 2008.
7. Chisiu G, Laurian T., Tudor A., Indentation tests to study the mechanical tribological properties of UHMWPE, U.P.B. Sci. Bull., Series D, vol. 73, Iss. 4, 2011.
8. Cheng, L., Xia, X., Yu, W., Scriven, L. E., & Gerberich, W. W. Flat-punch indentation of viscoelastic material. Journal of Polymer Science Part B: Polymer Physics, vol.38 (1), 10-22, 2000.
9. Lyubicheva, A., & Soldatenkov, I. On the model for identification of wear law parameters of viscoelastic materials based on ball-on-disk wear tests. Tribology International, vol. 55, 77-80, 2012.

Dissipative Properties of EHD Lubricant Film

Fedorov S V

Kalinigrad State Technical University , Department of Theory of Mechanisms and Machines and Machine Parts, 236022, Kaliningrad, Russia fedorov@klgtu.ru

Abstract: For the case of the failure of the lubricant film at hydrodynamic lubrication a common thermodynamic theory of strength is considered. According to this theory the failure occurs when the internal energy density (potential and thermal components) in the volume of material reaches a constant for a given material. A special case of this theory is considered when only the density of heat (kinetic) component of internal energy is taken into account. Temperature condition determines the limit state for liquid lubricants - mineral oils. When analyzing the regularities of friction at EHD lubrication the state and properties of the oil film at the condition of irregular and hydrostatic compression. The original structural model of oil film at EHD lubrication in the form of the rotary oscillating cells with elastic interactions to each other is proposed. It is similar to the Rayleigh-Benard cells and corresponds to the cellular hypothesis of J. Gibbs for the case of equilibrium and reversible process. It is quite possible that the size of the cells have an order of about nano level. The oil film dissipates energy in the direction of relative motion of bodies. This oil film has the highest dissipative properties.

Keywords: friction, energy, dissipation, nanostructure, lubricant film

1. Introduction

We all know the friction is a global nature phenomenon of the energy transformation. Friction is subjected to energy balance equation and with thermodynamic point of view [1, 2] is the process of two interrelated, oppositely directed and concurrent trends operating in a strained contact. According to the energy balance scheme (figure 1) for plastic deformation and fracture [3-5] presented below, equations for friction work W_f , frictional force F and friction coefficient μ (without lubrication) has view:

$$W_f = \Delta U_e + Q = \Delta U_{e_1} + \Delta U_{e_2} + \Delta U_{T_1} + \Delta U_{T_2} + \vec{Q}_1 + \vec{Q}_2, \quad (1)$$

$$\dot{W}_f = \dot{U}_e + \dot{Q} = \dot{U}_{e_1} + \dot{U}_{e_2} + \dot{U}_{T_1} + \dot{U}_{T_2} + \dot{\vec{Q}}_1 + \dot{\vec{Q}}_2, \quad (2)$$

$$F_l = \frac{\Delta U_e}{l} + \frac{Q}{l} = \frac{\Delta U_{e_1} + \Delta U_{e_2}}{l} + \frac{Q_1 + Q_2}{l}, \quad (3)$$

$$F_v = \frac{\dot{U}_{e_1} + \dot{U}_{e_2}}{v} + \frac{\dot{Q}_1 + \dot{Q}_2}{v} = F_{mechanical} + F_{molecular}, \quad (4)$$

$$\mu_l = \frac{\Delta U_{e_1} + \Delta U_{e_2}}{Nl} + \frac{Q_1 + Q_2}{Nl} = \mu_{adapt} + \mu_{dis} = \mu_{adapt} + \mu_{T(dis)} + \mu_{\dot{Q}(dis)}, \quad (5)$$

IOP Conf. Series: Materials Science and Engineering **295** (2018) 012019 doi:10.1088/1757-899X/295/1/012019

$$\mu_v = \frac{\dot{U}_{e_1} + \dot{U}_{e_2}}{Nv} + \frac{\dot{Q}_1 + \dot{Q}_2}{Nv} = \mu_{deformation} + \mu_{adhesion}, \tag{6}$$

where $\Delta U_e = V_f \Delta u_e$; $Q = V_f q$; $\dot{U}_e = V_f \dot{u}_e$; $\dot{u}_e = \mathrm{d}u_e/\mathrm{d}t$; V_f - is the deformable (friction) volume; μ - friction coefficient; μ_{adapt} - adaptive friction coefficient; $\mu_{T(dis)}$ and $\mu_{\bar{Q}(dis)}$ - static and dynamical components of dissipative friction coefficient; ΔU_T - thermal component of internal energy; N - normal load; l - distance of friction. The latent energy density Δu_e is an integral parameter of tribostate and damageability (failure Δu_e^*)).

Thus, viewed thermodynamically, the work done by friction forces W_f (the friction power $\dot{W}_f$), the friction force F and the friction coefficient μ may be classified conventionally into two specific components with different kinetic behaviour [3-5].

The first component is associated with microscopic mechanisms of adaptive type and relates to the change of latent (potential) energy ($\Delta u_{e_1}, \Delta u_{e_2}$) of various elementary defects and damages that are generated and accumulate in the deformable volumes of materials friction pair (figure 2).

This energy is a unique and integral characteristic of the submicro–and microstructural transformations that occur in plastically strained materials [3-5]. This energy is a measure of strain hardening and damageability of materials.

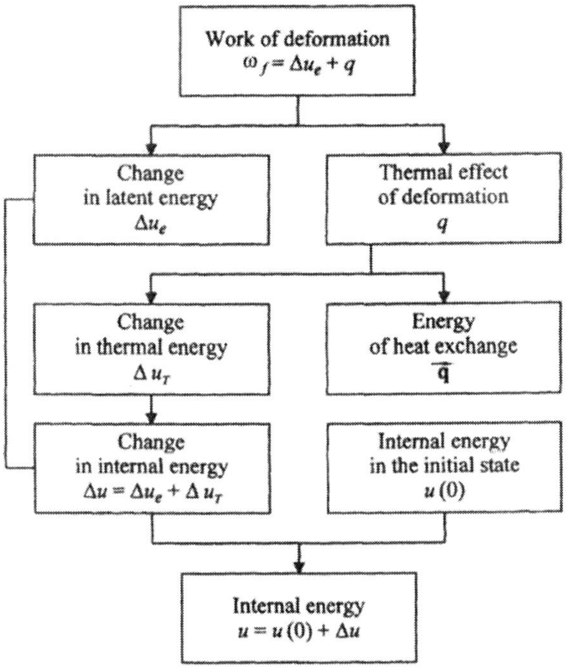

Figure 1. Scheme of the energy balance for the plastic deformation of a solid body [3-5].

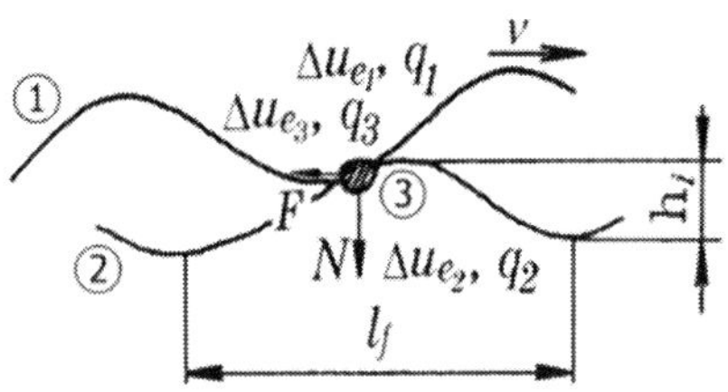

Figure 2. Schematic view of elementary friction's contact [1, 2].

The second component is associated with microscopic mechanisms of dissipative type and relates to dynamic recovery processes in which latent energy and frictional heat are released (q_1, q_2). This energy originates in the motion and destruction of various elementary defects of opposite signs, the egress of these defects to the surface, the healing of reversible submicroscopic discontinuities, etc.

The ratios of the components Δu_{e_1} and Δu_{e_2} as well as q_1, q_2 of the balance vary over a wide range, depending on the physical, chemical, and structural properties of the materials that comprise the friction couple and the friction conditions.
During this transformation the friction surfaces are heated and the heat of friction is transfered to oil film. In the course of time an oil film may fail.

2. Energy criterion of oil film failure

2.1. Thermodynamic theory of strength

Thermodynamic condition [3, 4] for local failure has view

$$u(\bar{r}_*, t_*) = u(\bar{r}_*, 0) + \int_0^{t_*} \dot{u}(\bar{r}_*, t)\,\mathrm{d}\,t = u_* = \text{const}. \tag{7}$$

Here $u(\bar{r}_*, 0)$ - represents the density of the internal energy within the local macrovolume of the material in the initial (prior to deformation, t=0) state; $\dot{u}(\bar{r}_*, t)$ - represents the specific power of the internal energy sources in the local volume responsible for the failure; $\bar{r}_*$ - is the parameter which characterizes the coordinates (x_*, y_*, z_*) of the local volume responsible for failure.
In accordance with the modern thermodynamic concepts relating to the strength and fracture of materials [3,4], having a fundamental nature, failure of a volume of a material (critical "defectness") occurs when the internal energy density u (potential u_e and thermal u_T components) in this volume reaches the critical value u_*, constant for a given material. The criterion u_* is a single-valued and integral characteristic of limiting damage ("defectness") of a material. Thermal failure of the material is the particular case of this theory when the change of the potential component of the internal energy is negligibly small and can be neglected.
The temperature, however, is a quantitative measure of macroscopic manifestation of change of the density of the thermal (kinetic) u_T component of the internal energy of the material. We shall use these concepts to describe the case of oil film failure during friction with account for the kinetic characteristics of internal energy accumulation by liquid materials (oils).

2.2. Thermodynamic failure condition for solids

This condition [3,4] has the form

$$u = u_0 + \Delta u = u_*. \tag{8}$$

Here

$$u_0 = u_{e0} + u_{T0}, \tag{9}$$

$$\Delta u = \Delta u_e + \Delta u_T. \tag{10}$$

u - density of internal latent energy in volume of material; $u_{(0)}, \Delta u$ - internal energy of material in initial state and its change during friction; $u_{e(0)}, u_{T(0)}$ - potential and thermal components of internal energy of material in the initial state ($t = 0$); $\Delta u_e, \Delta u_T$ - change of potential and thermal components of internal energy of material during friction.

2.3. Thermodynamic failure condition for liquid oil films

We can say the liquid materials cannot accumulate the energy of elementary defects as solids. Therefore, in the particular case for liquid oil films u_{e0} and Δu_e are equal to zero, then the failure condition (8) can be written in the form

$$u_T = u_{T0} + \Delta u_T = u_T^*. \tag{11}$$

Thus, equation (11) is the energetic condition of oil film failure in the contact [6]. In accordance with this equation the oil film fails if the density of the thermal (kinetic) energy u_{T} in its volume reaches the critical value u_{T}^*. Upon reaching this value the oil loses its lubricating properties and friction transitions to the regime of friction without lubrication.

As the condition of the absence of scuffing as a consequence of lubricant film failure we take the relation

$$u_{\text{T}} = u_{\text{T0}} + \Delta u_{\text{T}} < u_{\text{T}}^*. \tag{12}$$

Dividing both, sides of (12) by the oil density ρ and the average oil heat capacity $\bar{c}_{\text{p}}$, we obtain

$$T = T_0 + \Delta T < T_{\text{cr}} = T_*. \tag{13}$$

Here

$$u_{\text{T0}} = \rho\,\bar{c}_{\text{p}} T_0 = \int_0^{T_0} \rho\, c_{\text{p}} dT;$$

$$\Delta u_{\text{T}} = \rho\,\bar{c}_{\text{p}} \Delta T = \int_{T_0}^{T_*} \rho\, c_{\text{p}} dT;$$

$$u_{\text{T}}^* = \rho\,\bar{c}_{\text{p}} T_* = \int_0^{T_*} \rho\, c_{\text{p}} dT, \tag{14}$$

T_* - temperature of oil film failure; $T_0, \Delta T$ - surface temperature and. temperature flash.

Equation (13) is the well-known H.Block relation [6] for seizure, where T_{cr} is the critical temperature, upon reaching which the oil loses its lubricating properties and seizure of the surfaces takes place.

Table 1. Energetic Characteristics of Mineral Oils [7].

Oil	ρ, kg/m^3	T_{cr}, ^{0}C [9]	u_{T}^*, MJ/m^3
Instrument Vaseline, GOST*1805-51 (MVP)	878	100	617
Transporter, GOST 982-50	880	180	751,6
Spindle, GOST 1642-50 (AU)	891	165	756
Compressor, GOST 5546-54 (KhF-12)	-	160	727,4
Turbine, GOST 32-53 (ZOUT)	895	120	663,2
Medicinal Vaseline, GOST 3164-52	875	20	483,4
Turbine, GOST 32-53 (ZOUT)	-	140	693
Industrial, GOST 5289-51(50)	906	140	705,5
Diffusion, GOST 7904-56 (DI)	-	20	491,7
Autotractor, GOST 1862-42 (AK-10)	920	150	733,6
Autotractor, GOST 1862-42 (AK-15)	930	140	724,2
Naphthene-paraffin fraction of MS-20 oil (NPS MS-20)	-	155	718,2
Aviation, GOST 1013-49 (MZS)	890	140	693
Aviation from Groznenskll crude, GOST 1012-29	905	210	824,2
Cylinder, TUM NP 233-47 (bright stock)	-	140	693
Ethyiene glycol	1110	125	833

Table 1 presents the values of u_{T}^* for known values of T_{cr} for mineral oils [8].

3. The calculation correlation with scuffing terms under lubricated friction

To establish the connection between the energetic characteristics of oil film failure in the contact of rubbing bodies and the friction process parameters we can use the relations presented in [9] for calculating the average temperature rise in the sliding contact friction zone. Since in the case of friction with lubrication the temperature of the thin lubricant film reflects the thermal state of the friction surfaces, then with accuracy adequate for practical purposes the values calculated for the friction surface temperature can be referred to the thermal state (temperature) of the oil film as well.
Thus, to calculate the average temperature rise in the friction zone for sliding contact on an area in the form of a square with side $2l$ we take a relation [9], which at high sliding speeds has the form

$$\Delta T = \frac{1{,}064\omega}{\lambda_1}\left(\frac{2a_1 l}{v}\right)^{1/2}. \tag{15}$$

In this case the relations (15) transform to the form

$$\Delta u_\mathrm{T} = \frac{1.064\omega_\mathrm{lub}^* \rho \bar{c}_\mathrm{p}}{\lambda_1}\left(\frac{2a_1 l}{v}\right)^{1/2}; \tag{16}$$

$$u_\mathrm{T}^* = u_\mathrm{T0} + \frac{1.064\omega_\mathrm{lub}^* \rho \bar{c}_\mathrm{p}}{\lambda_1}\left(\frac{2a_1 l}{v}\right)^{1/2}; \tag{17}$$

$$\omega_\mathrm{lub}^* = \frac{\left(u_\mathrm{T}^* - u_\mathrm{T0}\right)\lambda_1}{1{,}064\rho \bar{c}_\mathrm{p}}\left(\frac{v}{2a_1 l}\right)^{1/2}. \tag{18}$$

Here ω_lub - specific friction power in friction with lubrication; ρ - density of oil; $\bar{c}_p$ - average heat capacity of oil; l - length of contact spot; a_1, λ_1 - thermal and heat conductivity of bearing material; λ_2 - heat conductivity of counter-body material.
The condition (18) is the quantitative criteria characterizing the moment of transition from friction with lubrication to the friction regime without lubrication. Because of breakdown of the lubricant film, i.e., upon satisfaction of the condition $\omega_\mathrm{lub} = \omega_\mathrm{lub}^*$ the oil loses its lubricity - the ability to intensely reduce the surface energy (passivate the surface) - and the friction regime without lubrication takes place.
Analysis of equation (18) shows that the scuffing criterion ω_lub^* depends significantly on the critical value of the oil internal (thermal) energy density u_T^* and on the sliding speed (other conditions being the same). The larger the value of u_T^*, the larger the specific friction power ω_lub^* at which oil film failure occurs, which creates conditions for increase of the area of real contact of dry unlubricated surfaces. The Influence of increase of the sliding speed v on increase of ω_lub^* shows up as a consequence of increase of the volume (flow rate) of the lubricant pumped through the friction zone, which leads to improvement of heat transfer and as a result of this to reduction of the temperature in the friction zone and the parameter u_T0. The thermo physical properties of the shaft and ring materials $a_1, \lambda_1, \lambda_2$ have an influence on the parameter ω_lub^*.

4. HD experiment and results

The experimental evaluation of the energetic condition of oil film failure during friction was performed on a full-scale stand with a pulsing load created by the hydraulic technique, which makes it possible to simulate the conditions of operation of the diesel engine connecting rod sliding bearings [7, 10]. The tests were made with the actual bimetallic sliding bearings (bushings) with a layer of antifriction material of the alloys AlSn20-1 and CuPb30 (D=80 mm, H=34 mm, h_B=2,5 mm). Here H - bearing width; ε - diametral clearance; h_B - bearing thickness. The sliding bearings were tested in a pair with a steel

40NiCrMo6 (DIN) (induction hardened, HRC 50-55) shaft. Mark SAE40 diesel oil was used to lubricate the bearings. The experiments were made with shaft sliding velocity 4 m/sec. Special design of thermocouple has determined the real temperature into oil jet (stream) flowing out of connecting rod bearing (figure 3).
The limiting load capacity of the friction pair was determined on the basis of marked increase of the friction power W_f and the temperature T_{oil} of the oil leaving the connecting rod bearing (figure 3) in the given loading stage p_{cr}, which was taken as the limiting (critical) value. The results of the experiments are shown in Table 2.

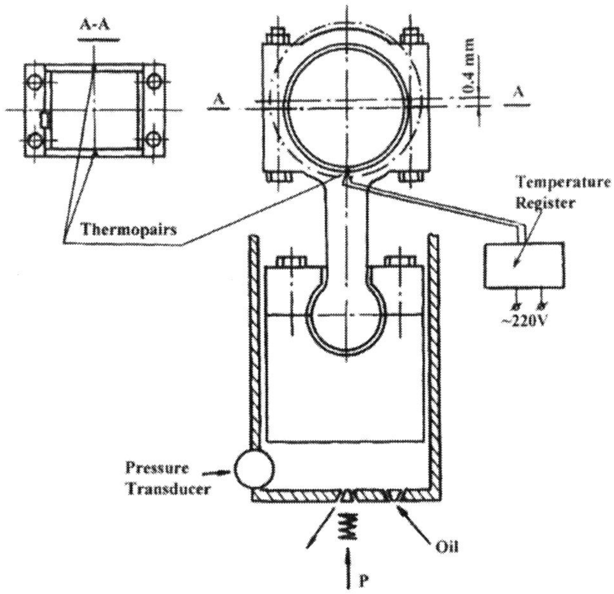

Figure 3. The disposition of thermopairs for measuring of temperature T_{oil} of the oil leaving the connecting rod bearing.

Table 2. Results of Experimental Studies of Load Capacity of AlSn20-1and CuPb30 Alloys on Full-Scale Test Stand [10].

Material	AlSn20-1	CuPb30
p_{cr}, MPa	48,0	35,0
W ,W	5157	9375
μ	0,011	0,027
T_{oil}, ^{0}C	120	115
$A^{exp} \cdot 10^{-4}$, m^2	14,4	12,5
$A^{calc} \cdot 10^{-4}$, m^2	13,3	11,9
ω^*_{lub}, MW/m^2	3,6	7,5

From the obtained experimental data (ω^*_{lub}, l, v) we determined the oil film internal energy density change Δu_T in the shaft-bushing contact at the moment of scuffing from formula (16); the friction surface temperature flash $\Delta T_f \cong \Delta T_{oil}$:

$$\Delta T_{\text{oil}} = \frac{\Delta u_{\text{T}}}{\rho \bar{c}_{\text{p}}}; \tag{19}$$

the running value of the oil film internal (thermal) energy density

$$u_{\text{Tf}} = T_{\text{oil}} \rho \bar{c}_{\text{p}}; \tag{20}$$

the critical value of the oil film internal (thermal) energy density and the friction surface temperature at the moment of scuffing

$$u_{\text{T}}^{*} = u_{\text{T f}} + \Delta u_{\text{T}}; \tag{21}$$

$$T_{\text{cr}} = \frac{u_{\text{T}}^{*}}{\rho \bar{c}_{\text{p}}}. \tag{22}$$

Here $T_{\text{oil}}, \Delta T_{\text{oil}}$ - running value of oil temperature and critical value of oil temperature change in volume of oil film.
In the calculations it was assumed that the temperature T_{oil} of the oil leaving the connecting rod bearing reflects the average bearing friction surface temperature.
The results of calculation of the oil film energetic characteristics and the friction surface temperature are shown in table 3. For the given diesel oil grad (table 3) the critical values u_{T}^{*} and T_{cr} are, respectively, 733 MJ/m^3 and 150°C.

Table 3. Results of Calculation of Oil Film Energetic Characteristics Corresponding to Moment of Scuffing [10].

Material	AlSn20-1	CuPb30
u_{T}	682	673
Δu_{T}^{*}	43	54
u_{T}^{*}	725	727
T_{oil}	120	115
ΔT_{oil}	25,5	31
T_{cr}	145,5	146

Thus the experiment results confirm reliably the assumed energetic scuffing condition (11). The critical magnitude of the oil film internal (thermal) energy density for a given inactive mineral oil grade is independent of the bearing material grade and the friction process conditions and is a physical constant of the tribosystem scuffing process.
The obtained results agree well with the fundamental concepts of the ergodynamic theory of strength [3-5], in accordance with which the integral measure of material susceptible to damage is the critical (constant for a given material) value of the internal energy density, and also with the H.Block hypothesis [6] on constancy of the overall seizure temperature for a given mineral oil grade.

5. EHD lubrication hypothesis

It is known that the EHD lubricant type has significantly higher performance (capacity for work), than HD lubricant type. Modern tribology [11] high performances lubricant when EHD explains with the effect of increasing the viscosity of the lubricant when it is compacted. It is known that for most mineral oils use high pressures can lead to a huge increase in viscosity. The oil between the surfaces at high pressures in the contact region behaves virtually like a solid separating layer. This process explains why many mechanisms in practice operate under much severer conditions than the classical theory would allow.

Naturally the question arises about the viability of EHD lubrication in its temperature aspect. Is it possible to apply temperature failure criterion of an oil film that successfully explains the loss of efficiency at HD lubricant?

We can examine the Hertzian scheme of contact under friction (figure 4).

In proximity of Hertzian contact with size of a material point A due to elasticity (Poisson effect) two equilibrium (elastic) hemisphere (roughness) - on roller and on the surface are formed. If there is a thin film of lubricant between the surfaces this two roughnesses operate as elements of a gear pump. Roughness of roller captures lubricant, which is getting into the zone of Hertzian contact of material point A and is hydrostatically compressed and then pressed out into disclose contact on surface roughness. Here, we may consider these three stages of formation of the equilibrium contact roughnesses at EHD lubrication on an example of mechanical model of EHD lubricant (figure 5, it is turned by 180° to figure 4).

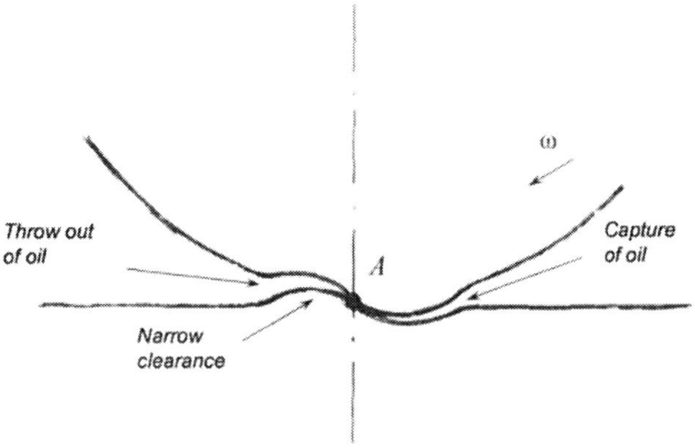

Figure 4. Model of elastic perfectly smooth roller on elastic and smooth surface.

In the first stage (a) an oil film is captured with elastic roughness into the cavity under the roughness. In the second stage (б) an oil film is hydrostatically compressed in this cavity. At the third stage (c) a narrow gap begins to form between these two roughnesses and compressed lubricant flows at high speed from the field of hydrostatic compression, creating a peak of pressure and then dropping the stresses in oil film up to zero.

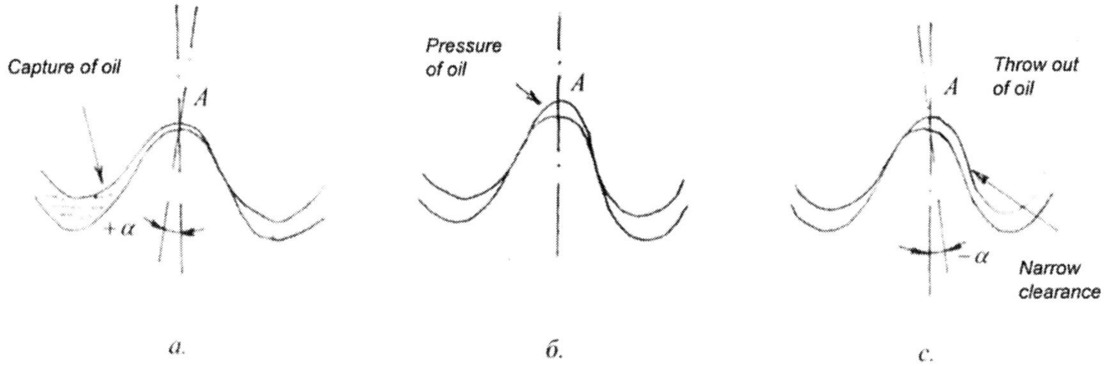

Figure 5. Three stages of forming contact with EHD lubrication.

Proposed model for Hertzian lubricant contact is fully corresponds with the known curve of pressure distribution in contact EHD (figure 6).

Left and central parts of pressure distribution in elasto-hydrodynamical film of lubricant correspond to the areas of capture and compression. To the right of pressure peak part corresponds to condition of high-speed flow of compressed lubricant through a narrow gap on the third stage of forming contact of EHD lubrication.

The question about the nature and state of the oil film in EHD lubrication is directly linked to the question of temperature fracture criteria for lubricant. The logic is here quite simple. Why does under EHD lubrication where the loads exceed many times the load of HD lubricant, lubricating film remains healthy?

When analyzing the friction regularities at EHD lubrication it is necessary to pay attention to the two independent and possible effects. The first one is regularity of plastic deformation (states and properties) at Hertzian contact of solids. The second one is the state and properties of the oil film in the condition of irregular and hydrostatic compression.

Let us consider the first. It is deformation of the Hertzian contact. Hertzian contact of friction should be considered as a contact in the most fully developed evolution of plastic deformation. The degree of structural adaptations of such contact can be maximum [12]. In essence, we can have a perfect elasto-visco-plastic deformation. Such contact has the maximum capacity for work and temperature effect on contact is minimal. Almost totally equilibrium (rotary elastically, interactive) nano-crystal structures of contact (mechanical quantum (figure 7) [13-16]) have thermal effect in the form of temperature growth on the temperature level is close to absolute zero. Such microflash is unable to heat and destroy the oil film. Irrelevance of lubricants is very often supposed for such contacts [11]. But on the other hand the problem of contact temperature cycling and the possibility of gradual heat accumulation require a lubricant which will remove the heat.

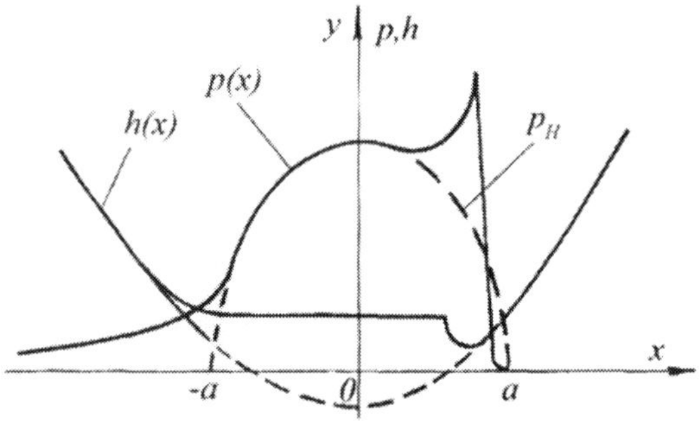

Figure 6. Model of Ertel-Grubin EHD film: p_H -Hertzian pressure; $p(x)$- current pressure layer; $h(x)$ -the current thickness of the lubricating layer; a - half width of contact area.

Figure 7. Model of elementary dissipative nanostructure of friction (8103 atomic cubical cells) [13].

Thus, despite of the fact that Hertzian contact works at stresses at the threshold of the strength limit, within the contact between structural elements (mechanical quanta, as subtribosystems) interaction is realized at the level of the perfect elasticity. Elastic mutual rotations of nanostructures define the essence for models as abnormally low friction and wearlessness [14]. Such state of tribocontact we may examine as state of selforganization, nanoquantum solid lubricant [15].

Now, as to the second - it is about state and properties of lubricants at EHD lubrication. Quasi solid state of the oil film in full irregular hydrostatic compression can give unique lubrication properties.

First, let's take into account the classic force-separation curve of interatomic interaction (figure 8). Hydrostatic compression of lubrication film makes the destruction process more difficult (lubricant state under point P). It is so, because stretching (attractive force, see point Q) is necessary for destruction (breaking of atomic bond).

Secondly, quasisolid state of oil film may analyze from the perspective of solids friction.

Therefore, the regularities of shear deformation of lubricant (oil) may be similar with regularities that take place under friction of solids, namely, one may remember the structural-energy diagram (figure 9) for evolution of the rubbing surfaces [1, 2, 12, 16].

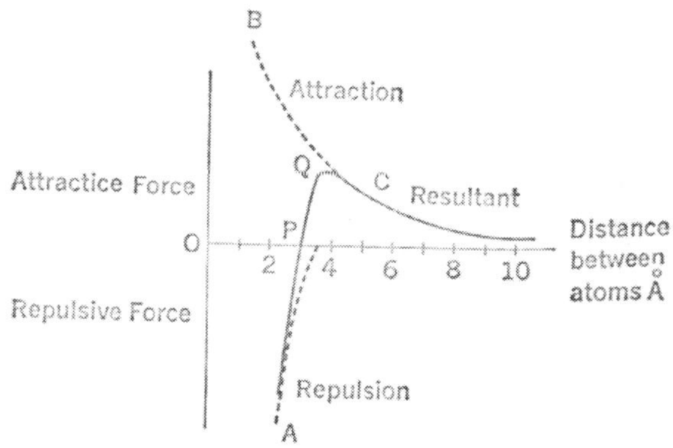

Figure 8. Sketch, showing the way in which the force between two atoms varies with the distance between them.

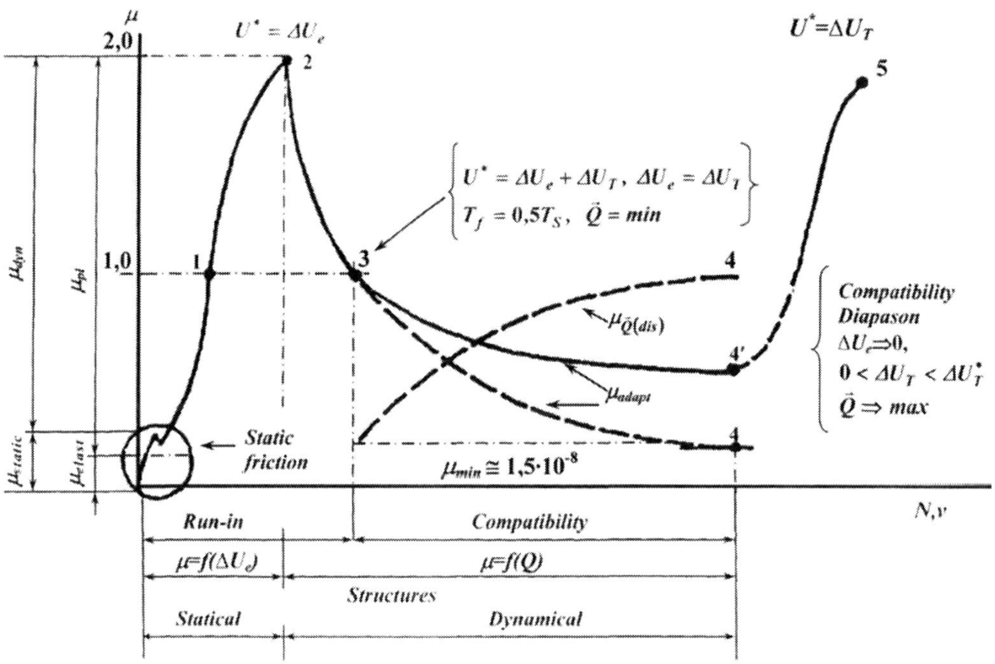

Figure 9. Structural-energy diagram for evolution of rubbing surfaces [1, 2, 12-16].

Here may well work mechanism for dynamic dissipation of energy with energy balance component $\vec{q}^*$ (figure 1), i.e. EHD lubricant can have properties of dissipative structures of friction such by friction of solids. These dissipative structures can be formed at the gap mouth of the expiration of oil where the lubrication hydrostatic compression has irregularity at the expense of stretching (figure 10). Between the second and third stages of the process occurs a momentary restructuring lubricant from the most compact to a new nanostructured one. A phase transition takes place. A volume front of thermal (temperature) blow is momentarily formed at that. Actually this blow is a cause the formation of maximum internal pressure peak that precedes the development and process of the third phase –

IOP Conf. Series: Materials Science and Engineering **295** (2018) 012019 doi:10.1088/1757-899X/295/1/012019

laminar flow of nanostructured lubricant in the narrow clearance. At that a resistance of lubricant flow drops to zero. Internal stresses drop to zero that the chart of Ertel-Grubin is demonstrated. This condition is a perfect, elastic, inner balance between structural (nano) elements of lubricant.

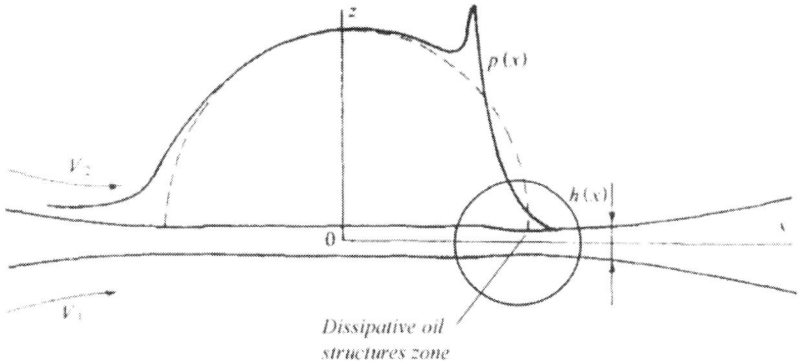

Figure 10. Dissipative oil structures zone under EHD lubrication.

It momentary and volume front of temperature blow can't warm up the lubricant film. It is only mechanism of energy transformation for transition the lubricant film state from state with latent elastic energy Δu_e^* of hydrostatic compression to state with latent elastic energy $\vec{q}^*$ of nanostructured flow. Practically all accumulated energy Δu_e^* of the first stage is converted to energy $\vec{q}^*$ of the third stage owing to the structural transformation of the oil film.

Correspondingly, high density of energy flux (friction power) at EHD lubrication stimulates the formation of similar structures of quick heat dissipation.

Here it is quite possible to propose structural model of oil film (figure 11) with EHD lubrication in the form of rotary oscillating cells with elastic interactions with each other.

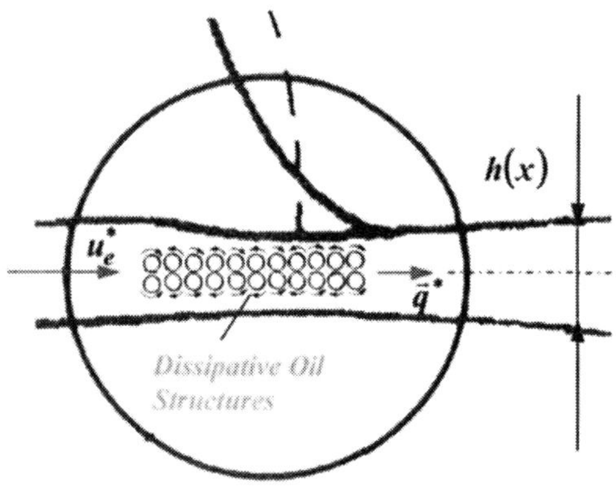

Figure 11. Cellular (nano-structured) model of EHD lubricant into a narrow gap between two surfaces.

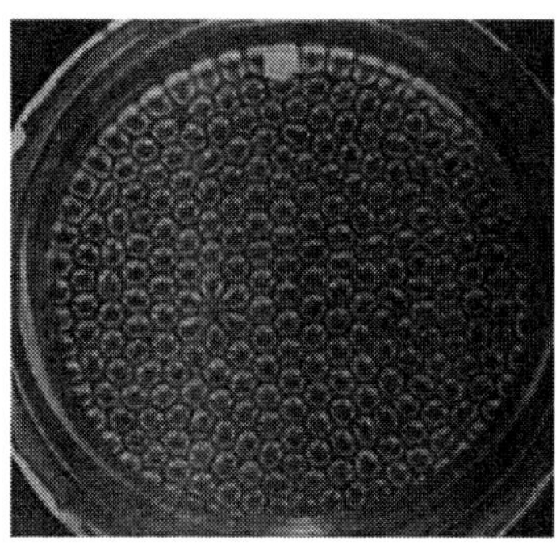

Figure 12. Rayleigh-Benard cells.

This is similar to the Rayleigh-Benard cells (figure 12) in the case of a powerful stationary heating of liquids. For static heating condition the Rayleigh-Benard cells pass the heat in perpendicular direction to friction surfaces. It is possible that the size of these cells has an order is about nano-scale level such as fullerene' form or mechanical quantum (figure 7) [12-16]. It corresponds to the cellular hypothesis

of J. Gibbs for case of equilibrium and reversible process. Accordingly, this oil film has maximum dissipative properties.
At EHD lubrication this high structured oil film dissipates energy in the direction of relative motion of bodies (figure 11).

Here all accumulated energy Δu_e^* in compressed zone (the work of external forces) is almost entirely scattered on the dissipative structures (subtribosystems) of oil film.

6. Conclusions

Thus, neither the first nor the second effects Hertzian contact do not give the temperature effect of friction. When EHD lubrication there is we have no temperature effect at destruction of lubricant.

The criterion of the ultimate state of oil film at EHD is a critical density $\vec{q}^*$ of dissipation energy, which is equal to the critical density of the internal thermal energy u_T^* of lubricant. The concept of temperature for such film has almost no meaning. Here you should write ratio

$$\Delta u_T = u_T^* - u_{T0} = 0. \qquad (23)$$

This is associated with a particular (unique) structural state of oil film. The presence of dissipative component $\vec{q}^*$ presupposes the equality condition of zero temperature flashes $\Delta T = 0$ and consequently $\Delta u_T = 0$ too.

Therefore, the critical state of the film cannot be achieved until the dissipative structures lubricating film operate. Loadability of EHD lubrication film is very high because of the elastic and reversible energy dissipation.

As a result, we can write the following - the equations for critical power of friction for HD and EHD lubricants:

$$\dot{W}_{HD} = \dot{V}\Delta u_T^* = \dot{V}\rho c_p (T^* - T_0). \qquad (24)$$

$$\dot{W}_{EHD} = \dot{V}\cdot\vec{q}^*. \qquad (25)$$

Here $\dot{V}$ is the amount of lubricant in contact at a time.

This result is fully in line with the practice. EHD lubrication capacity for work should be significantly higher than HD lubrication

$$\dot{W}_{EHD} >> \dot{W}_{HD}, \qquad (26)$$

Because

$$\vec{q}^* >> \Delta u_T^*. \qquad (27)$$

If you write down the numerical values Δu_T^* and $\vec{q}^*$ for engine oil, for example, M14V (Table 3) then we get the following experimental results for AlSn-20 and CuPb30 (in brackets):

$$\Delta u_T{}^* = 43(54)\,\mathrm{MJ}/m^3,$$

and

$$\vec{q}^* = u_T^* = 725(727)\,\mathrm{MJ}/m^3.$$

In the most general case, the thermal model of the ultimate state of oil film with HD and EHD lubricants have the following forms:

HD lubricant - $\dot{W}_{HD} = \dot{V}q^*$, here, $q^* = \Delta u_T^*$, and $\vec{q}^* = 0$; (28)

EHD lubricant - $\dot{W}_{EHD} = \dot{V}q^*$, here, $q^* = \vec{q}^* = u_T^*$, and $\Delta u_T^* = 0$. (29)

Excess of specific friction power above critical level would lead to turbulence of the flow, laminar flow destruction of dissipative structures and consequently to loss of load capacity of EHD. It will be a critical state.

IOP Conf. Series: Materials Science and Engineering 295 (2018) 012019 doi:10.1088/1757-899X/295/1/012019

References

1. Fedorov S.V., General model of friction. Journal of Friction and Wear, vol. 14, no. 3, 20-28, 1993.
2. Fedorov S.V., The Foundations of Triboergodynamics and Physico-Chemical Prerequisits of Compatibility Theory, Kaliningrad State Technical University Press [in Russian], Kaliningrad, 416, 2003.
3. Fedorov, V.V., Ergodynamic concept of failure. 1, Basic statements of the ergodynamics of deformed bodies, Criteria of failure ductility, Strength Of Materials (Translated from Russia), vol. 23, no. 8, 883-889, 1991.
4. Fedorov V. V., Thermodynamic Aspects of Strength and Fracture of Solid Bodies, Science [in Russian], Tashkent, 168, 1979.
5. Fedorov V. V., Kinetics of Damage and Fracture of Solid Bodies, Science [in Russian], Tashkent, 186, 1985.
6. Blok H., Les Temperatures de Surface dans des Conditions de Graffsage Sour Extreme Pression, in: Proceeding of Materials on Second Cong. Mondial du Petrole, Paris, France, 111-114, 1937.
7. Fedorov S.V., Energy criterion of oil film failure at friction, Tribology in Industry, vol. 36, no. 4, 384-393, 2014.
8. Matveevskli R.M., Temperature Method for Evaluating Limiting Lubricity of Machine Oils [in Russian], USSR Science Academia, Moscow, 143, 1956.
9. Jaeger J.C., Applied Mechanics and Machine Construction (Russian translation), USSR Science Academy, Moscow, no. 6, 17-22, 1952.
10. Fedorov S.V., Thermodynamic aspects of the metal scuffing process in boundary lubrication, Journal of Friction and Wear, vol. 9, no. 3, 12-20, 1986.
11. Bowden F.P., Tabor D., Friction. An introduction to tribology, Anchor Press, New York, 178, 1973.
12. Fedorov S.V., Energy model of sliding friction coefficient and generalized regularities of tribosystems evolution, Transactions of the 19th International Colloquium Tribology «Tribology-Industrial and Automotive Lubrication», TAE, Esslingen, Germany, 2014, www.tae.de info@tae.de.
13. Fedorov S.V., The mechanical quantum of dissipative friction structures is the elementary tribonanostructure, IV WTC2009, Kyoto, Japan, 926, 2009.
14. Fedorov S.V., Nanotribology: Nano-quantum model of dissipative structures of friction. Book of Abstracts, Viennano'15, 6th Vienna International Conference «Nano-Technology. Focus: Tribological Aspects», Wiener Neustadt, Austria, 19-22, 2015, ISBN 978-3-901657-52-8.
15. Fedorov S.V., Self-organized nano-quantum solid lubricant, OilDoc Conference & Exhibition, Lubricants, Maintenance, Tribology, Book of Synopses, Bavaria, Germany, 114, 2015.
16. Fedorov S.V., Generalized energy model of sliding friction coefficient and regularities of tribosystem evolution, V WTC2013, Torino, Italy, 2013, ISBN 9788890818509.

Research concerning the evaluation of the connection forces in the joints of the sucker rod pumping units mechanism

***Bădoiu D*[1]*, Petrescu M G*[2]*, Antonescu N N*[3]*, Toma G*[4]

[1]*Petroleum-Gas University of Ploiesti, Department of Mechanical Engineering, 100680, Ploiesti Romania*
[2]*Petroleum-Gas University of Ploiesti, Department of Mechanical Engineering, 100680, Ploiesti Romania*
[3]*Petroleum-Gas University of Ploiesti, Department of Mechanical Engineering, 100680, Ploiesti Romania*
[4]*Petroleum-Gas University of Ploiesti, Department of Mechanical Engineering, 100680, Ploiesti Romania*

dorin.badoiu@gmail.com[1], pmarius@upg-ploiesti.ro[2], nnantonescu@upg-ploiesti.ro[3], georgeta_tm@yahoo.com[4]

Abstract: At present, the sucker rod pumping installations are the most used in the case of the wells in production, when an eruptive exploitation is not possible. The practice has demonstrated that an important role in increasing safety in the operation of the pumping units has the design of the various component bearings because of the extremely high values of the connection forces to which they are loaded. That is why it is necessary to establish as accurately as possible the values of these connecting forces. In the paper is analyzed the dynamics of a conventional pumping unit mechanism. The dynamic study which allows establishing the values of the connecting forces in the joints is performed within the Assur structural groups. The dynamic analysis was implemented into a computer program using Maple programming environment and finally it has been presented some simulation results in the case of a C-320D-256-100 pumping unit.

Keywords: pumping unit, connection forces, dynamics

1. Introduction

At present, since most wells can not provide an eruptive exploitation, this is accomplished by pumping using the sucker rod pumping installations [1]. The functioning analysis of these installations must take into account many factors of which we can mention: the configuration of the pumping unit, the dimensions of the component elements, the pumping speed and the size of the sucker rod and of the tubing which contributes to the greatest extent to the load of the pumping unit. An important role in increasing safety in the operation of the pumping units has the design of the various component bearings because of the extremely high values of the connection forces to which they are loaded. In this regard the values of these connecting forces have to be established as accurately as possible. In this paper it is analyzed the dynamics of the mechanism of the conventional pumping units, thus establishing the variation on the cinematic cycle of the values of the connecting forces. The dynamic study is performed within the *Assur* structural groups [2-4]. A computer program using Maple programming environment [5] has been developed and some simulation results regarding the variation on a cinematic cycle of the values of the connection forces in the joints of a C-320D-256-100 pumping unit are presented.

2. Pumping unit mechanism dynamics

A conventional pumping unit which is part of a sucker rod pumping installation is shown in figure 1 and in figure 2 is presented the cinematic scheme of its mechanism.

Figure 1. Conventional pumping unit.

With C_1, C_2 and C_3 were noted the mass centres of the cranks, connection rods and of the rocker, respectively; m_{CGR} is the total mass of the counterweights; m_{L1} is the mass of the two crank pin bearings; m_{L2} is the mass of the equalizer bearing; m_{tr} is the mass of the equalizer considered to be concentrated in the point *B*; m_{CB} is the mass of the horsehead considered to be concentrated in the point *D'* on the rocker.

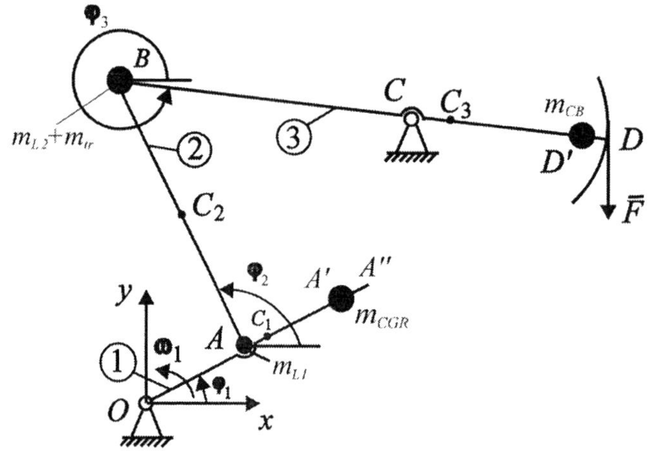

Figure 2. Cinematic scheme of the mechanism of a conventional pumping unit.

For establishing the position of the connection rods *2* and of the rocker *3*, the angles φ_2 and φ_3 are calculated by solving the following equations system obtained by projecting the vectorial equation corresponding to the closed contour *O-A-B-C-O* on the *x* and *y* axes:

$$\begin{cases} l_1 \cos\varphi_1 + l_2 \cos\varphi_2 + l_3 \cos\varphi_3 - x_C = 0 \\ l_1 \sin\varphi_1 + l_2 \sin\varphi_2 + l_3 \sin\varphi_3 - y_C = 0 \end{cases} \tag{1}$$

where $l_1 = OA$; $l_2 = AB$, $l_3 = BC$ and x_C and y_C are the coordinates of the point C where is located the centre bearing.
The angular velocities and the angular accelerations of the connection rods and of the rocker may be calculated by deriving in relation to time the expressions of the angles $\varphi_2(\varphi_1)$ and $\varphi_3(\varphi_1)$, with the following relations:

$$\omega_j = \dot{\varphi}_j = \frac{d\varphi_j}{d\varphi_1}\frac{d\varphi_1}{dt} = \omega_1 \frac{d\varphi_j}{d\varphi_1};\ j = 2,3 \tag{2}$$

$$\varepsilon_j = \ddot{\varphi}_j = \varepsilon_1 \frac{d\varphi_j}{d\varphi_1} + \omega_1^2 \frac{d^2\varphi_j}{d\varphi_1^2};\ j = 2,3 \tag{3}$$

where ω_1 and ε_1 are the angular velocity and the angular acceleration, respectively, of the cranks.
Having established the values of the angles φ_2 and φ_3 it were determined the coordinates of the various points on the mechanism, including those of the mass centres of the component elements and of the point D where acting the force F (figure 2) at the polished rod. The projections on the x and y axes of the speeds and of the accelerations of any point on the mechanism may be calculated by deriving in relation to time the expressions of its coordinates, for example in the case of the point D, by using the following relations:

$$\begin{cases} (v_D)_x = \dot{x}_D = \frac{dx_D}{d\varphi_1}\frac{d\varphi_1}{dt} = \omega_1 \cdot \frac{dx_D}{d\varphi_1} \\ (v_D)_y = \dot{y}_D = \frac{dy_D}{d\varphi_1}\frac{d\varphi_1}{dt} = \omega_1 \cdot \frac{dy_D}{d\varphi_1} \end{cases} \tag{4}$$

$$\begin{cases} (a_D)_x = (\dot{v}_D)_x = \varepsilon_1 \cdot \frac{dx_D}{d\varphi_1} + \omega_1^2 \cdot \frac{d^2x_D}{d\varphi_1^2} \\ (a_D)_y = (\dot{v}_D)_y = \varepsilon_1 \cdot \frac{dy_D}{d\varphi_1} + \omega_1^2 \cdot \frac{d^2y_D}{d\varphi_1^2} \end{cases} \tag{5}$$

The mechanism comprises a single *Assur* structural group, namely a dyad composed of the connection rods and the rocker. Figure 3 shows the loads corresponding to the elements of the dyad, where $G_j = m_j \cdot g; j = 2,3$, are the weight forces of the connecting rods and of the rocker and m_2 and m_3 are their masses; $g = 9.81\,\text{m/s}^2$ is the gravitational acceleration that acts in the opposite direction of the y axis; $G_{L1} = m_{L1} \cdot g$ is the weight force of the two crank pin bearings (it has been considered that only half of the weight of these bearings is concentrated on the connecting rods, the rest remaining concentrated on the cranks); $G_{L2} = m_{L2} \cdot g$ is the weight force of the equalizer bearing; $G_{tr} = m_{tr} \cdot g$ is the weight force of the equalizer; $G_{CB} = m_{CB} \cdot g$ is the weight force of the horsehead; $\overline{F}_{ij} = -m_j \cdot \overline{a}_{C_j}; j = 2,3$, are the inertia forces corresponding to the connecting rods and to the rocker and $\overline{a}_{C_j}, j = 2,3$ are the accelerations of their mass centres; $\overline{M}_{ij} = -J_{C_j} \cdot \overline{\varepsilon}_j; j = 2,3$, are the inertia moments, where $J_{C_j}, j = 2,3$, are the mass moments of inertia corresponding to the connecting rods and to the rocker; $\overline{F}_{iL1} = -m_{L1} \cdot \overline{a}_A$, $\overline{F}_{iL2} = -m_{L2} \cdot \overline{a}_B$, $\overline{F}_{itr} = -m_{tr} \cdot \overline{a}_B$ and $\overline{F}_{iCB} = -m_{CB} \cdot \overline{a}_{D'}$ are the inertia forces corresponding to the two crank pin bearings, equalizer bearing, equalizer and to the horsehead, respectively; $\overline{F}$ is the force acting at the end of the polished rod. Figure 3 shows also the projections on x and y axes of the connection forces from the two crank pin bearings ($2\overline{F}_{12x}$ and $2\overline{F}_{12y}$) and of the connection force from the centre bearing ($\overline{F}_{03x}$ and $\overline{F}_{03y}$).

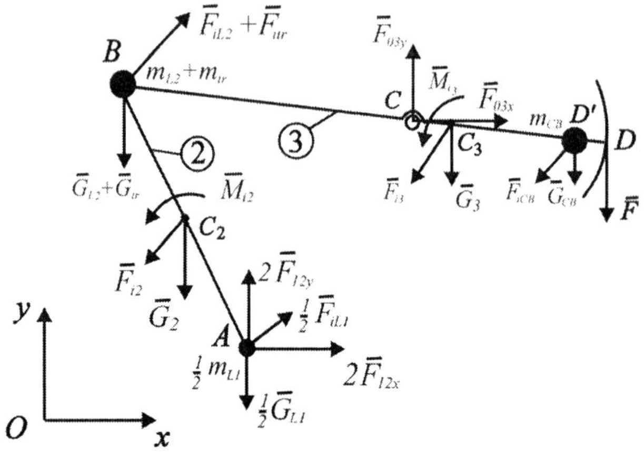

Figure 3. The load scheme of the dyad made up of the connecting rods and the rocker.

The connection forces F_{12x} and F_{12y} may be calculated by solving the equations system obtained from the following dynamic equilibrium equations in moments:

$$\left(\sum M_B\right)_2 = 0 \ \Rightarrow \ 2\overline{BA}\times\left(\overline{F}_{12x}+\overline{F}_{12y}\right)+\frac{1}{2}\overline{BA}\times\left(\overline{G}_{L1}+\overline{F}_{iL1}\right)+\overline{BC_2}\times\left(\overline{G}_2+\overline{F}_{i2}\right)+\overline{M}_{i2}=0 \quad (6)$$

$$\begin{aligned}\left(\sum M_C\right)_{2-3} = 0 \ \Rightarrow \ & 2\overline{CA}\times\left(\overline{F}_{12x}+\overline{F}_{12y}\right)+\frac{1}{2}\overline{CA}\times\left(\overline{G}_{L1}+\overline{F}_{iL1}\right)+\overline{CC_2}\times\left(\overline{G}_2+\overline{F}_{i2}\right)\\ & +\overline{CB}\times\left(\overline{G}_{L2}+\overline{F}_{iL2}+\overline{G}_{tr}+\overline{F}_{itr}\right)+\overline{CC_3}\times\left(\overline{G}_3+\overline{F}_{i3}\right)\\ & +\overline{CD'}\times\left(\overline{G}_{CB}+\overline{F}_{iCB}\right)+\overline{CD}\times\overline{F}+\overline{M}_{i2}+\overline{M}_{i3}=0\end{aligned} \quad (7)$$

The projections on x and y axes of the connection force $\overline{F}_{32}$ (F_{32x} and F_{32y}) from the equalizer bearing acting on the equalizer may be established from the following dynamic equations in forces:

$$\left(\sum F_x\right)_2 = 0 \ \Rightarrow \ F_{32x}+2F_{12x}+\frac{1}{2}F_{iL1x}+F_{i2x}+\frac{1}{2}F_{iL2x}+F_{itrx}=0 \quad (8)$$

$$\left(\sum F_y\right)_2 = 0 \ \Rightarrow \ F_{32y}+2F_{12y}+\frac{1}{2}F_{iL1y}+F_{i2y}+\frac{1}{2}F_{iL2y}+F_{itry}-\frac{1}{2}G_{L1}-G_2-\frac{1}{2}G_{L2}-G_{tr}=0 \quad (9)$$

In the previous equations it has been considered that the mass of the equalizer bearing is distributed evenly on the rocker and the equalizer.
The projections on x and y axes of the connection force $\overline{F}_{03}$ from the centre bearing may be calculated from the following dynamic equilibrium equations in forces:

$$\left(\sum F_x\right)_3 = 0 \ \Rightarrow \ F_{03x}+F_{23x}+\frac{1}{2}F_{iL2x}+F_{i3x}+F_{iCBx}=0 \quad (10)$$

$$\left(\sum F_y\right)_3 = 0 \ \Rightarrow \ F_{03y}+F_{23y}+\frac{1}{2}F_{iL2y}+F_{i3y}+F_{iCBy}-\frac{1}{2}G_{L2}-G_3-G_{CB}-F=0 \quad (11)$$

where $\overline{F}_{23x}=-\overline{F}_{32x}$ and $\overline{F}_{23y}=-\overline{F}_{32y}$.

Figure 4 shows the loads corresponding to the cranks, where $G_1 = m_1 \cdot g$ is the weight force of the cranks (m_1 is their mass); $\overline{F}_{i1} = -m_1 \cdot \overline{a}_{C_1}$ is the inertia force corresponding to the cranks and $\overline{a}_{C_1}$ is the acceleration of their mass centres; $G_{CG} = m_{CG} \cdot g$ and $\overline{F}_{iCG} = -m_{CG} \cdot \overline{a}_{A'}$ are the weight force and the inertia force, respectively, corresponding to the counterweights; $\overline{a}_{A'}$ is the acceleration of point A' where the total mass of the counterweights is considered to be concentrated. Also, figure 4 shows the projections on x and y axes of the connection forces from the two joints connecting the two cranks and the output shaft of the reducer ($2\overline{F}_{01x}$ and $2\overline{F}_{01y}$), the connection forces from the two crank pin bearings ($2\overline{F}_{21x} = -2\overline{F}_{12x}$ and $2\overline{F}_{21y} = -2\overline{F}_{12y}$) and the motor moment M_m at the cranks shaft.

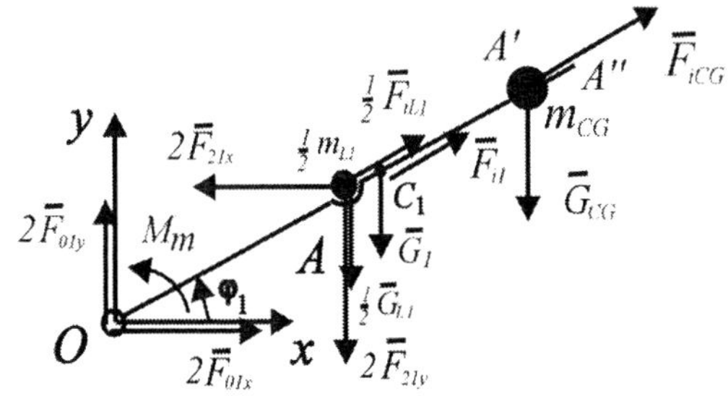

Figure 4. The load scheme of the cranks.

The projections on x and y axes of the connection force $\overline{F}_{01}$ may be established from the following dynamic equilibrium equations in forces:

$$\left(\sum F_x\right)_1 = 0 \;\Rightarrow\; 2F_{01x} + 2F_{21x} + \frac{1}{2}F_{iL1x} + F_{i1x} + F_{iCGx} = 0 \tag{12}$$

$$\left(\sum F_y\right)_1 = 0 \;\Rightarrow\; 2F_{01y} + 2F_{21y} + \frac{1}{2}F_{iL1y} + F_{i1y} + F_{iCGy} - \frac{1}{2}G_{L1} - G_1 - G_{CG} = 0 \tag{13}$$

Next, the motor moment at the cranks shaft can be established from an equation in moments calculated relative to the point O:

$$\left(\sum M_O\right)_1 = 0 \;\Rightarrow\; \overline{M}_m + 2\overline{OA} \times \left(\overline{F}_{21x} + \overline{F}_{21y}\right) + \frac{1}{2}\overline{OA} \times \left(\overline{G}_{L1} + \overline{F}_{iL1}\right) + \overline{OC_1} \times \left(\overline{G}_1 + \overline{F}_{i1}\right) + \overline{OA'} \times \left(\overline{G}_{CG} + \overline{F}_{iCG}\right) = 0 \tag{14}$$

3. Simulation results

The cinematic and dynamic analysis of the conventional pumping units mechanism has been transposed into a computer program using Maple programming environment. Further, are presented some simulation results in the case of a C-320D-256-100 pumping unit. The dimensions of the component elements of a C-320D-256-100 pumping unit (figure 2) produced by *Lufkin* [6] are: OA = 42 in. (1.0668 m); AB = 132 in. (3.3528 m); BC = 111.07 in. (2.8211 m); CD = 129 in. (3.2766 m). The coordinates of the point C are [6]: $x_C = 111$ in. (2.8194 m) and $y_C = 136$ in. (3.4544 m). The total length of the cranks is: OA'' = 95 in. (2.413 m). The functioning angular speed of the cranks expressed in rotations per minute is equal to 5 rot/min. The simulations have been accomplished by considering the following values of the other parameters involved in the calculation of the connection forces and of the motor moment: OA' = 65 in. (1.651 m); CD' = 110 in. (2.794 m); m_{L1} = 55 kg; m_{L2} = 125 kg; m_{tr} = 400 kg; m_{CB} = 443 kg; q_1 = 566 kg/m; q_2 = 25.7 kg/m; q_3 = 217.5 kg/m (q_1, q_2 and q_3 are the linear masses of the cranks, connecting rods and of the rocker, respectively). In figure 5 is presented the variation on a cinematic cycle of the force F acting at the end of the polished rod.

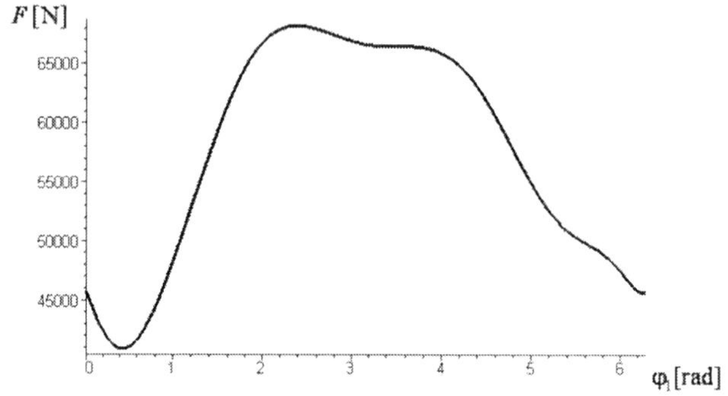

Figure 5. The variation on a cinematic cycle of the force F acting at the end of the polished rod.

In figure 6 it is presented the variation on a cinematic cycle of the motor moment M_m and in figures 7÷10 are presented the variations on a cinematic cycle of the connection forces F_{01}, F_{12}, F_{32} and F_{03}, respectively. The curves *1* correspond to the case when the angular speed of the cranks is equal to 5 rot/min and the curves *2* and *3* correspond to the cases when the angular speed of the cranks is equal to 7 rot/min and 10 rot/min, respectively. In these two last cases we have considered an increase of the force F values compared to the initial situation when the cranks speed is 5 rot/min, with 8% and 19.5%, respectively.

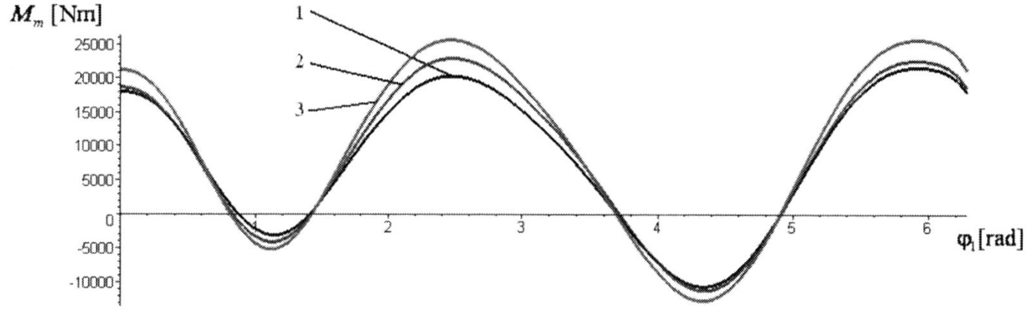

Figure 6. The variation on a cinematic cycle of the motor moment M_m when the angular speed of the cranks is equal to 5 rot/min (curve 1), 7 rot/min (curve 2) and 10 rot/min (curve 3).

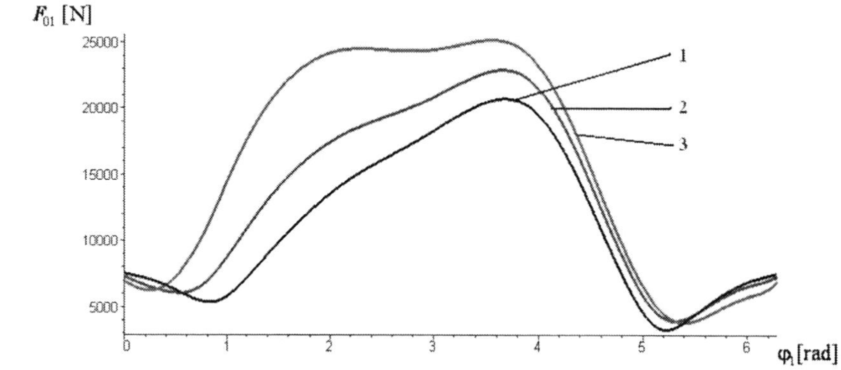

Figure 7. The variation on a cinematic cycle of the connection force F_{01} when the angular speed of the cranks is equal to 5 rot/min (curve 1), 7 rot/min (curve 2) and 10 rot/min (curve 3).

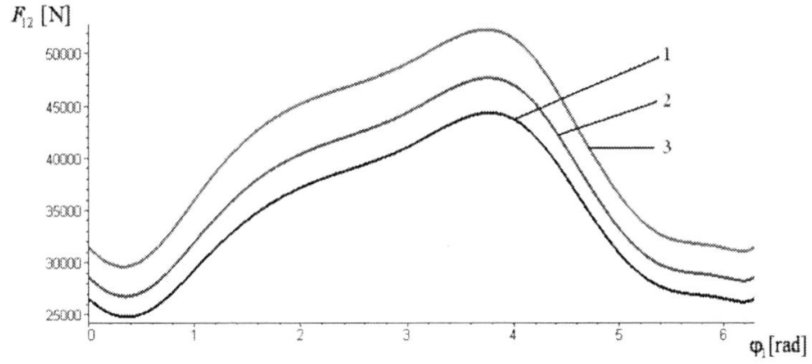

Figure 8. The variation on a cinematic cycle of the connection force F_{12} when the angular speed of the cranks is equal to 5 rot/min (curve 1), 7 rot/min (curve 2) and 10 rot/min (curve 3).

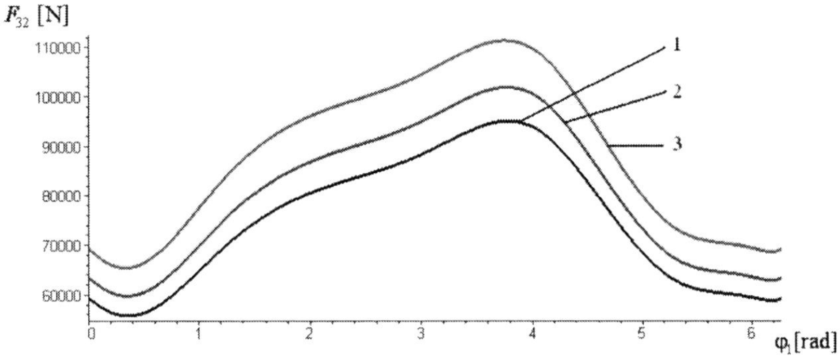

Figure 9. The variation on a cinematic cycle of the connection force F_{32} when the angular speed of the cranks is equal to 5 rot/min (curve 1), 7 rot/min (curve 2) and 10 rot/min (curve 3).

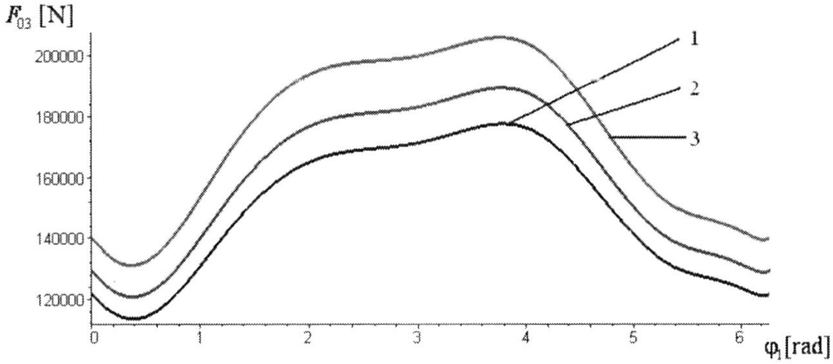

Figure 10. The variation on a cinematic cycle of the connection force F_{03} when the angular speed of the cranks is equal to 5 rot/min (curve 1), 7 rot/min (curve 2) and 10 rot/min (curve 3).

Figure 6 shows that the motor moment variation curve retains the same allure in all three analyzed cases and its extreme values are relatively close. Figures 7÷10 show that the connecting forces in the bearings of the analyzed pumping unit mechanism have extremely high values, in this regard highlighting the F_{03} connecting force in the central bearing. On the other hand it can be noticed that with the increase of the values of the force F there is an important increase of the maximum values of all the connecting forces in the bearings of the pumping unit mechanism.

4. Conclusions

In this paper it was analyzed the dynamics of the mechanism of the conventional sucker rod pumping units. The dynamic study was first performed within the component dyad composed of the connecting rods and the rocker and then it was continued on its leader cranks. A computer program using Maple programming environment has been developed and so it was possible to establish the variation of the values of the connecting forces in the mechanism bearings and of the motor moment on the cinematic cycle, simulation results being presented in the case of a C-320D-256-100 pumping unit. So, it have been highlighted the extremely high values of these connecting forces, in particular the values of the connection force acting in the centre bearing. Also it have been highlighted the increases in the maximum values of these connection forces with the increase of the values of the force at the polished rod.

References

1. Popovici A., Equipment for the exploitation of oil wells (in Roumanian), Technical Publishing House; Bucuresti, 1989.
2. Bădoiu D., The Dynamic Analysis of Mechanisms and Machines (in Romanian), Didactical and Pedagogical Publishing House, Bucuresti, 2003.
3. Bădoiu D., Petrescu M.G., Toma G., On the analysis and synthesis of the quadrilateral mechanism, Journal of the Balkan Tribological Association, vol. 20, 161-168, 2014.
4. Bădoiu D., Toma G., On a dynamic optimization problem of the quadrilateral mechanism, Journal of the Balkan Tribological Association, vol. 22, 250-260, 2016.
5. Monagan M.B., Geddes K.O., Heal K.M., Labahn G., Vorkoetter S.M., McCarron J. and DeMarco P., Maple Introductory Programming Guide, Maplesoft, a division of Waterloo Maple Inc., 2005.
6. 2008/2009 General Catalog Lufkin Oilfield Products Group website, 2017, Available: http://infohost.nmt.edu/~petro/faculty/Kelly/Lufkin.pdf

The Influence of The Temperature on Dry Friction of AISI 3315 Steel Sliding Against AISI 3150 Steel

Odabas D.
Erciyes University, Department of Mechanical Engineering, 38039, Kayseri, Turkey
odabasd@erciyes.edu.tr

Abstract: In this paper, the effects the influence of frictional heating on the wear of AISI 3315 Steel were investigated experimentally using a pin-on-ring geometry. All the tests were carried out in air without any lubricant. In order to understand the variation in frictional coefficient and temperature with load and speed, the friction tests were carried out at a speed of 1 m/s and loads in the range 115-250 N, and at a speed range 1-4 m/s, a load of 115 N. The sliding distance was 1500 m. The bulk temperature of the specimen was measured from the interface surface at a distance of 1 mm from the contact surface by using type K thermocouples (Ni-Cr-Ni). The coefficient of friction was determined as a function of test load and speed. The steady state coefficient of friction of the test material decreases with increasing load and speed due to the oxide formation. But the unsteady state coefficient of friction increases with an increase in load and speed.

Keywords: Sliding wear, bulk temperature, coefficient of friction.

1.Introduction

Friction occurs whenever two solid bodies slide against each other. It takes place by a variety of mechanisms in and around the real area of contact between the sliding bodies. It is through frictional processes that velocity differences between the bodies are accommodated. It is also through these processes that mechanical energy is transformed into internal energy or heat, which causes the temperature of the sliding bodies to increase. Although the word tribology was first coined almost 450 years after the death of Leonardo da Vinci, it is clear that Leonardo was fully familiar with the basic tribological concepts of friction, lubrication and wear [1]. Amontons and Coulomb who has established the earliest mechanical friction theory have found the basic friction laws as follows [2-3];

1.The friction is independent of the apparent area of contact between the bodies.

2.The friction force F_s is proportional to the normal load F_N between the bodies and there is a constant between F_s and F_N as known the friction coefficient $\mu = F_s/F_N$

3.The frictional coefficient is independent of the sliding velocity.

It is well known that the most important factor in friction is the interaction of surface asperities and the sliding involves the riding of rigid asperities on one surface over the other. Tabor [3] assumed that adhesion was the main source of friction and that surface asperities were deformed themselves elastically, plastically, viscoelastically or brittle. The friction process is an energy process, and studies have shown, a very complex phenomenon. In the common phenomenological view of the friction process, it has appeared that friction depends on the properties of mated materials, sliding speed, load and environmental conditions [5]. Some studies showed that increasing normal load increases the coefficient of friction [6]. Other studies showed that increasing this test parameter decreases this friction characteristic (7-9). In addition, some authors studied the effect of sliding speed on the tribological behaviour of a number of steels; they found that increasing sliding speed decreases the coefficient of friction. It has been studied the wear of medium carbon steel with different microstructures and found that oxidative wear was prevalent under low normal loads. Once the normal load reached a critical value, a mild to severe wear transition occurred, and subsequently an extrusive wear prevailed and abruptly increased the wear rate but reduced the coefficient of friction [10]. Some authors found that the coefficient of friction decreased with increasing surface hardness [11]. Moreover, it has been reported that lower friction is usually associated with harder surfaces [6]. A great deal of research has been carried out concerning the sliding friction characteristics of metals at the low loads and speeds [12-17]. Also the experimental conditions were chosen so that the temperature rise was minimal. Hence thermal aspects could be neglected. However, it is well known that it is at temperatures above one-half of the absolute melting point of metals [12]. The effect of temperature on the friction behaviour of metals cannot be neglected because oxidation plays an increasingly important role in high load and speed sliding friction [17]. This work aims to investigate the effects of normal load, sliding speed and frictional heating on the coefficient of friction of AISI 3315 steel dry sliding against AISI 3150 steel.

2.Experimental details

2.1. Material

The friction characteristics of the AISI 3315 steel used in manufacturing many machine elements such as gears, bearings and piston liners were investigated. The test samples were cut from the same batch of hot rolled materials, and they were machined with 16 mm long and 15 mm in diameter (i.e. pin). The AISI 3150 steel which is a hot working tool steel has been chosen as the counter test material (i.e ring). The counter test samples were rings 16 mm thick with 50 mm outer and 20 mm inner diameter. The nominal contact area was about 1.77×10^{-4} m^2. The sample configuration is illustrated in Fig.1 and Fig.2 AISI 3315 contains % C 0.15, % Mn 0.7, % Si 0.21, % P0.03, % S 0.029, % Cr 0.8, % Ni 3.5, and AISI 3150 contains % C 0.48, % Mn 0.85, % Si 0.2, % P 0.02, % S 0.21, % Cr 0.85, % Ni 1.25. The pin specimens is homogenised and its hardness is 260 HV_{30}. The ring homogenised and hardened in oil, so its hardness is 750 HV_{30}. As shown in Fig.1, the surfaces of the pins were machined to conform to the surface of the ring in order to provide a more realistic contact model. Thus, the investigation of the frictional coefficient for both the unsteady state and steady state sliding regimes could be possible. The pin-on-ring geometry used here has many advantages over other sliding configurations. First, conformity of the sliding surfaces enables a calculation of the nominal normal pressure, which remains constant during the entire test. Second, the sliding velocity (which determines the rate of heat generation) at the contact is the same over the entire contact are. Finally, it is possible to calculate the surface temperature with simple assumptions, because the sliding geometry does not change during the wear test and the heat flow into the specimen is at the most two dimensional. Both the specimen and the ring were grinded before testing. After grinding, the average roughness values R_a of the pins and rings were measured as 0.5 μm and 0.3 μm respectively. The specimen and the ring were cleaned with trichloroethylene before testing.

2.2. Test methods

All the tests were carried out at room temperature (30 °C) in dry sliding friction. The friction tests were carried out at a speed of 1 m/s and loads in the range 115-250 N, and at a load of 115 N and speeds in the range 1-4 m/s. The sliding distance was 1500 m. The wear tests were carried out on a wear machine working by pin-on-ring sliding principle. It is possible to measure the frictional force by means of a deflection beam, one end of which is fixed on the machine plate and the other is free (Fig.1). The maximum deflection of the free end of the beam is calculated as $Y_{max}=F_s L_k^3/3EI$ [m]. Where F_s is the frictional force, L_k is the effective length of the beam (0.14 m), E is the elastic module

IOP Conf. Series: Materials Science and Engineering **295** (2018) 012021 doi:10.1088/1757-899X/295/1/012021

of the beam material ($21x10^{10}$ N/m^2), and I is the inertia moment of the beam section ($I=bh^3/12 =3.5x10^9$ m^4). Thus, the experimental frictional force is obtained as $F_s=80x10^4.y_{max}$ [N]. The deflection of the beam was measured by a displacement transducer and all the deflection values were multiplied by a gain factor of 1.20. These values, then, used to calculate the frictional force. The experimental frictional coefficient μ was obtained from $\mu=F_s/F_N$ formula. Several theories have been proposed in the past to account for the temperature rise at sliding contacts [17]. Unfortunately, the theories cannot be made use of in practical situations because the assumptions made in those theories are not fully valid. Measurement of the thermoelectric e.m.f. produced by the sliding pair, infrared techniques, physical and chemical changes associated with the temperature rise and the thermocouple technique are some of these. Because of its simplicity, the last technique is used in this work although there are some limitations. In this technique, temperature is measured away from the surface.

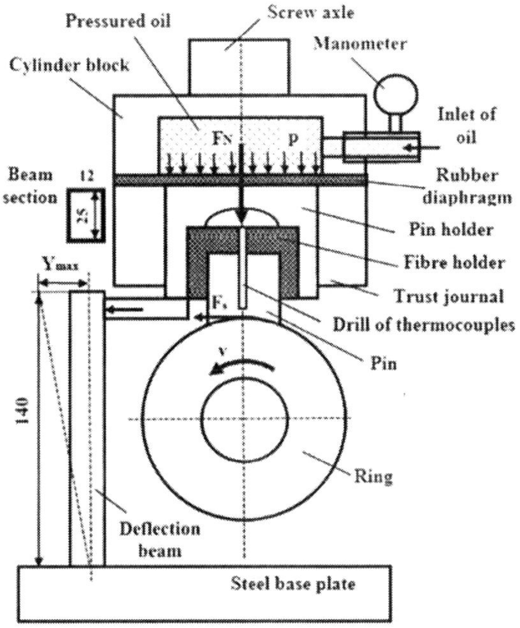

Figure 1. The hydrostatic loading and pin holder assembly.

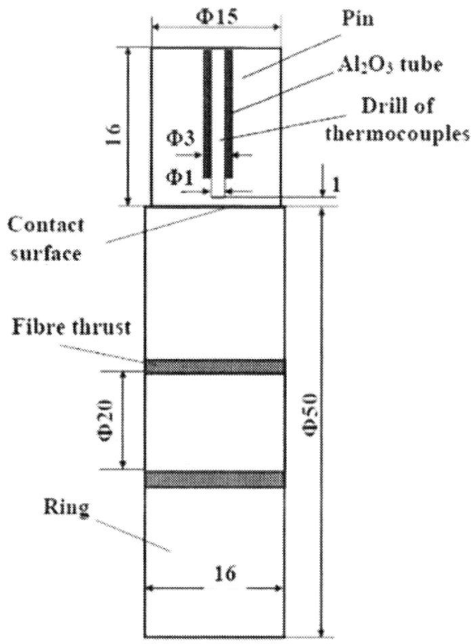

Figure 2. Test rig and the dimensions of samples assembly.

Figs 1 and 2 show the schematic arrangement of the ring, the pin and the thermocouples. As a result of frictional work, heat is generated at the interface and flows into the pin and the ring. The amount is dependent on the geometry and on the thermal, mechanical and tribological properties of both the pin and the ring. As shown in Figure 1, the amount of energy dissipated on sliding a distance L is given by $\mu.F_N.L$. The rate of energy dissipation is therefore given by $\mu.F_N.L/dt=\mu.F_N.v=F_s.v$. Where F_s is friction force, v is the sliding speed at the interface and μ is the coefficient of friction. The factor of pv is the power per unit contact area [2]. The heat release intensity in the friction zone depends on friction power and is determined by relationship $Q=\mu.F_N.v$. Where μ is the friction coefficient, F_N is the load per unit of length, v is the slip velocity [18]. The frictional heat per unit area generated in the sliding contact is given by $q=\mu.p.v$. The combination of contact pressure (p) and sliding velocity (v), referred to as the pv factor is an important parameter in the field of mechanical wear. The energy dissipated (frictional heat) in the mechanical wear process is proportional to pv, which has a significant effect on the mechanical wear. The bulk temperature of the pin specimen was measured from the interface surface at a distance of 1 mm from the contact surface by using type K thermocouples (Ni-Cr-Ni). The type K is the most common type of thermocouple. It is inexpensive, accurate, reliable, and has a wide temperature range. The type K is commonly found in nuclear applications because of its relative radiation hardness. Maximum continuous temperature is around 1100 °C. Thermocouples were placed in one hole drilled in the pin holder at a distance of 1 mm from the contact surface [15]. Each test was repeated with three sets of the pin and ring samples.

3.Results

3.1. The effects of sliding speed and load on the coefficient of friction and the temperature

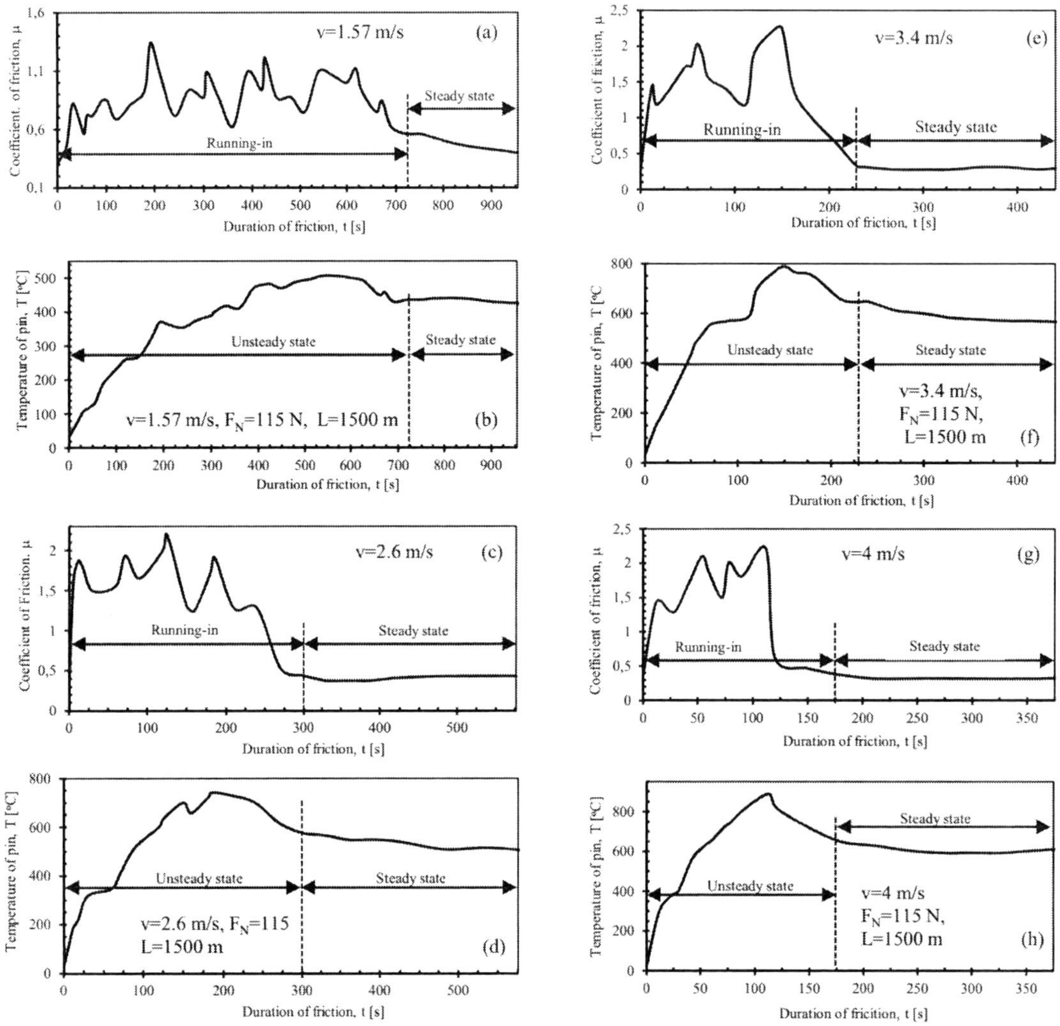

Figure 3. Comparison of the frictional coefficient and temperature of the pin.

In order to understand the variation in frictional coefficient and bulk temperature of pin with sliding speed, the tests were performed at a speed range 1-4 m/s, a load of 115 N and a sliding distance of 1500 m under dry friction conditions. During the tests, the friction coefficient and pin bulk temperature were continuously recorded throughout the test by a multichannel recorder as shown in Figure 3. The experimental results for general friction state were given in Table 1. Figures 3a, 3c, 3e, and 3g show the variation in coefficient of friction with duration of friction. It is possible to calculate three types of frictional coefficient for a sliding distance of 1500 m by using the recorded diagrams in these figures as follows : the minimum frictional coefficient (μ_{min}), the average frictional coefficient (μ_m) and the maximum frictional coefficient (μ_{max}). Figures 3b, 3d, 3f and 3h show the pin temperature recorded by the thermocouples placed at 1 mm distance from the sliding contact. It is possible to define three types of bulk temperature of pin for a sliding distance of 1500 m in these figures as follows: the minimum temperature (T_{min}), the average temperature (T_m) and the maximum temperature (T_{max}). The general pattern of the graphs of frictional coefficient are similar to that for temperature of pin.

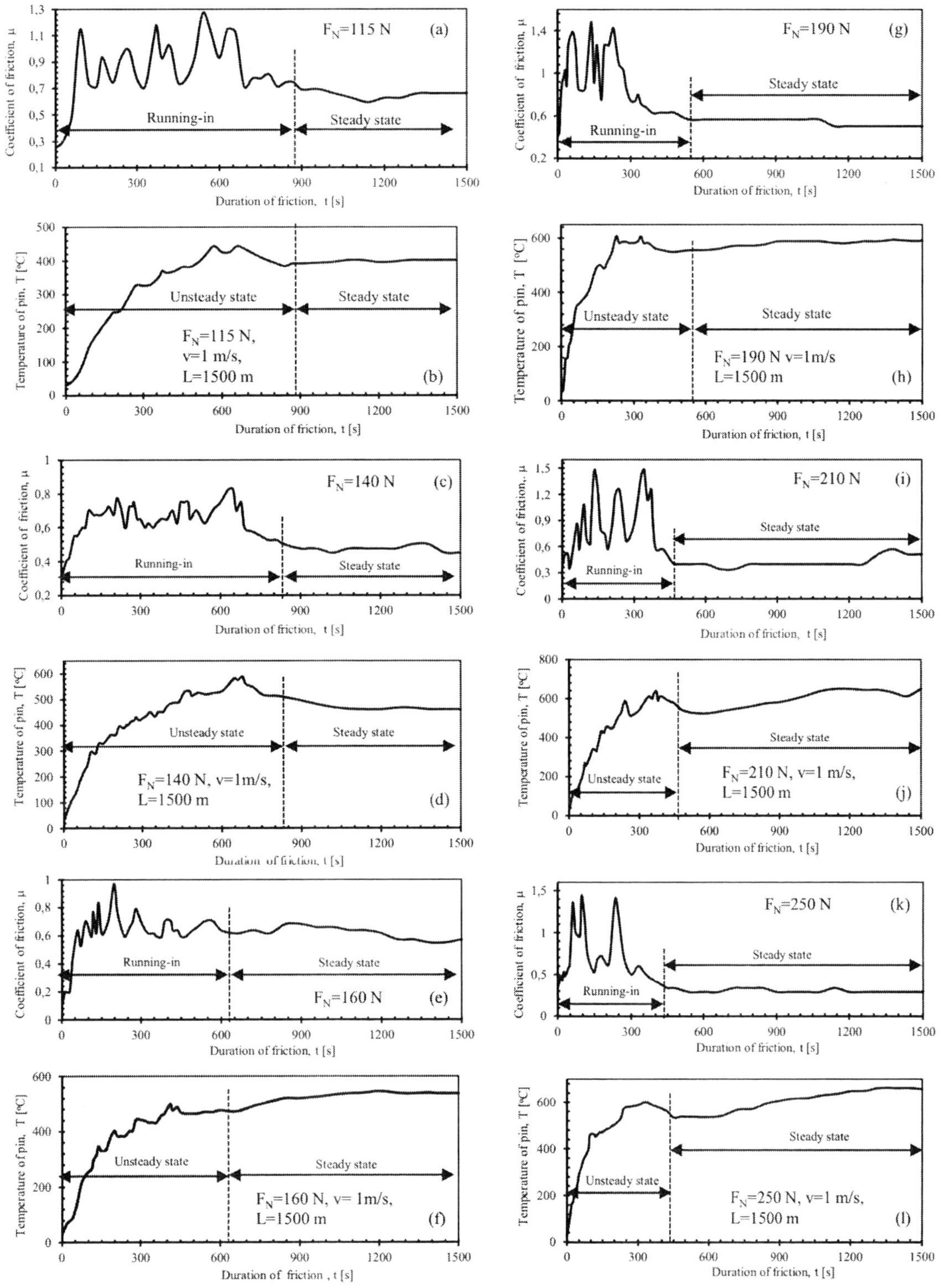

Figure 4. Comparison of the frictional coefficient and temperature of the pin.

Table 1. The influence of sliding speed on the coefficient of friction and the temperature of pin for cumulative friction.

Speed v (ms^{-1})	p (Nm^{-2}) x10^{-4}	pv factor ($Nm.m^{-2}.s^{-1}$) x10^{-4}	Time t (s)	Friction coefficient μ_{min}	μ_{max}	μ_m	Pin temperature T_{min} (°C)	T_{max} (°C)	T_m (°C)	Friction heat q ($Nm.m^{-2}.s^{-1}$) x10^{-4}
1		64.97	1500	0.248	1.271	0.795	30	444	337	51.65
1.57		102	955	0.312	1.341	0.802	30	506	358	81.80
2.6	64.97	168.9	577	0.25	2.184	1.159	30	742	493	195.75
3.4		220.89	441	0.28	2.246	1.072	30	788	541	236.79
4		259.88	375	0.32	2.20	1.039	30	888	587	270.01

In order to compare the variation in the coefficient of friction and the bulk temperature with load, tests were carried out at a sliding distance of 1500 m, a speed of 1 m/s, and loads in the range 115-250 N under dry friction conditions. Figures 4a, 4c, 4e, 4g, 4i and 4k show the variation in coefficient of friction with duration of friction. The experimental results for general friction state were given in Table 2. Figures 4b, 4d, 4f, 4h, 4j, and 4k show the pin temperature recorded by the thermocouples placed at 1 mm distance from the sliding contact.

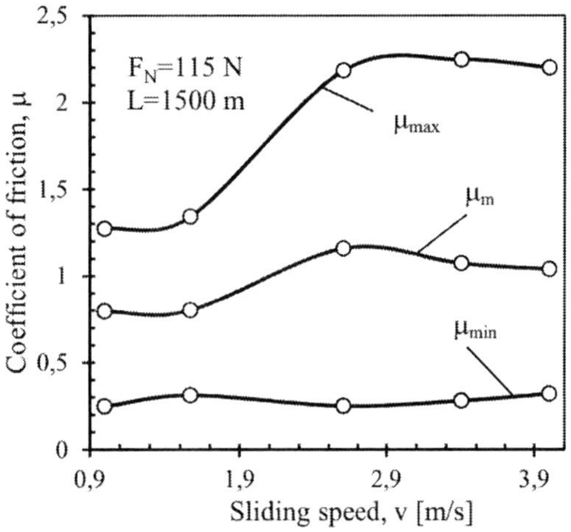

Figure 5. The variation in coefficient of friction with speed

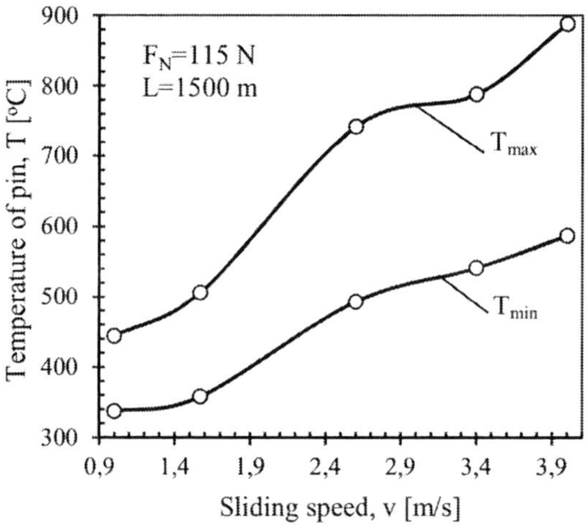

Figure 6. The variation in temperature of pin with speed.

Figure 5 shows the dependence of frictional coefficient on the sliding speed for general friction. In general, the average frictional coefficients increase as the speed increases in range 1-4 m/s. According to the third law of friction, the frictional coefficient would be independent of the sliding velocity. The reason for this is that changes arising from strain rate and friction heating are expected with increasing speeds. However, the maximum frictional coefficient is directly proportional to the speed. The slope of the curves of average frictional and minimum frictional coefficient decreases with increasing speed. The average and minimum frictional coefficients increase slightly. The average maximum and minimum frictional coefficients belong to unsteady state sliding. Minimum frictional coefficient is defined as static friction coefficient at the beginning of sliding. So, the contact between pin and ring continues as metal-metal and the surfaces are roughed severely. The flakes of metallic debris produced between surfaces adhere to contacting surface as shown in Figure 7. Since plastic deformation and wear normally do occur, the general contact situation changes with time. Figure 8 shows the optical micrograph of subsurface for the worn surface of AISI 3315 steel. It can be seen that the subsurface is affected by frictional forces and heat generated during the friction. The thickness of the mechanically affected layer increases with an increase in load because the plastic deformation of the subsurface is increased with the temperature [15]. The plastic deformation occurred at the surface during the friction can be seen clearly from the Figure 8. The grains of the subsurface is deformed and elongated to the rolling direction.

Figure 7. The flakes of metallic debris on surfaces of pins for 115 N and 3.4 m/s after 10 s. (x400)

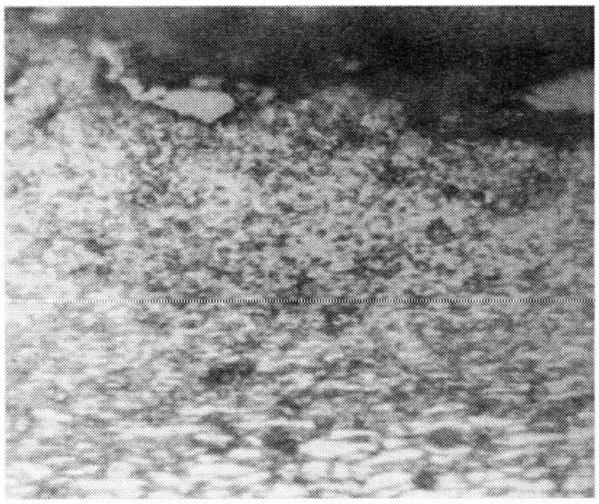

Figure 8. Optical micrograph of subsurface for the worn surface of AISI3315 steel (x200, Sliding direction is from left to right)

Table 2. The influence of load on the coefficient of friction and the temperature of pin for general friction.

Load F_N (N)	p (Nm^{-2}) x10^{-4}	pv factor ($Nm.m^{-2}.s^{-1}$) x10^{-4}	Friction coefficient μ_{min}	μ_{max}	μ_m	Pin bulk temperature T_{min} (°C)	T_{max} (°C)	T_m (°C)	Frictional heat q_s ($Nm.m^{-2}.s^{-1}$) x10^{-4}
115	64.97	64.97	0.248	1.271	0.797	30	444	342	51.78
140	79.09	79.09	0.30	0.825	0.613	30	426	388	48.48
160	90.39	90.39	0.10	0.969	0.614	30	546	402	55.49
190	107.34	107.34	0.399	1.482	0.795	30	606	481	85.33
210	118,64	118.64	0.342	1.482	0.695	30	650	484	82.45
250	141.24	141.24	0.564	1.440	0.288	30	662	489	40.67

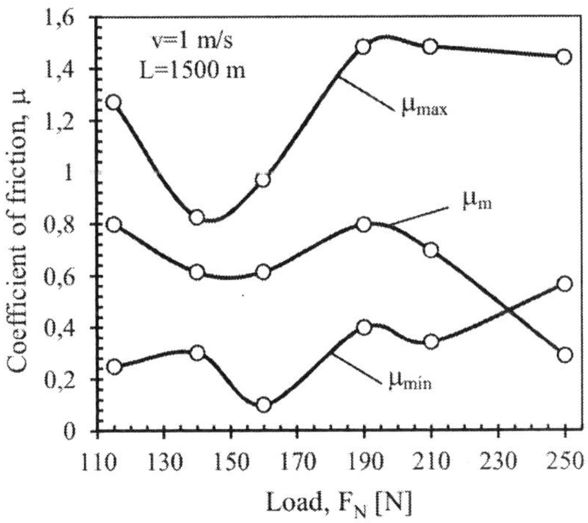

Figure 9. The variation in coefficient of friction with load for general friction.

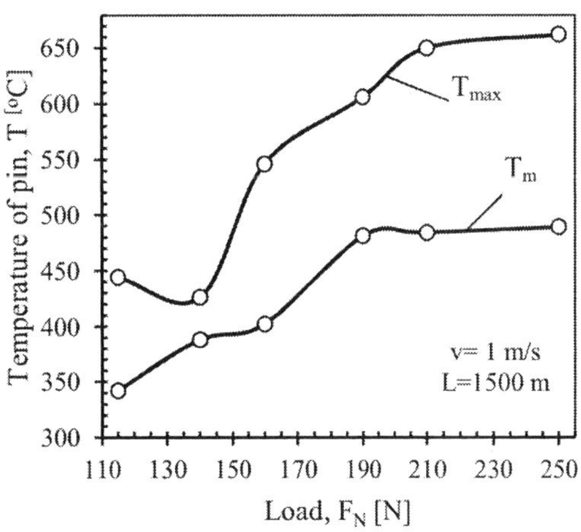

Figure 10. The variation in temperature of pin with load for general friction

Figure 9 shows the relationship between the frictional coefficients and the load for general friction. In general, the average frictional coefficient decreases slightly as the load increases in range 115-250 N because of the temperature rise the surface oxidizes. But, the average maximum and minimum frictional coefficients increase slightly as the load increases in range 115-250 N. Figure 10 shows the relationship between the temperature of pin and the load for general friction. The temperature increases with an increase in the load as shown in this figure. The reason for this is that changes arising from strain rate and friction heating are expected with increasing loads. It is shown clearly in Figures 3 and 4 that the friction is occurred two stages between steel-steel pairs for pin-on ring test system as running-in (unsteady state) and steady state friction.

3.2. The effects of sliding speed on the unsteady (running-in) friction

As shown in Figures 3, the recorded temperature and frictional coefficient increase with time. At the beginning of sliding, till the first few minutes of test, the frictional coefficient usually fluctuates very much. This friction behaviour is defined as unsteady state sliding. However, steady state conditions are soon reached and remained constant thereafter. It is possible to define this friction behaviour as steady state friction. The unsteady state and steady state is limited by the vertical lines. The vertical lines represent the limit between the unsteady state and steady state as shown in these Figures. The experimental results of unsteady friction are given in Table 3.

Table 3. The influence of sliding speed on the coefficient of friction and the temperature of pin for unsteady friction.

Speed v (ms^{-1})	pv factor ($Nm.m^{-2}.s^{-1}$) x10^{-4}	Time t (s)	Friction coefficient			Pin temperature			Friction heat q ($Nm.m^{-2}.s^{-1}$) x10^{-4}
			μ_{min}	μ_{max}	μ_m	T_{min} (°C)	T_{max} (°C)	T_m (°C)	
1	64.97	900	0.248	1.271	0.857	30	444	304	55.67
1.57	102	700	0.312	1.341	0.861	30	506	340	87.82
2.6	168.9	270	0.53	2.184	1.55	30	742	474	261.79
3.4	220.89	200	0.312	2.246	1.409	30	788	527	311.23
4	259.88	125	0.468	2.20	1.49	30	888	572	387.22

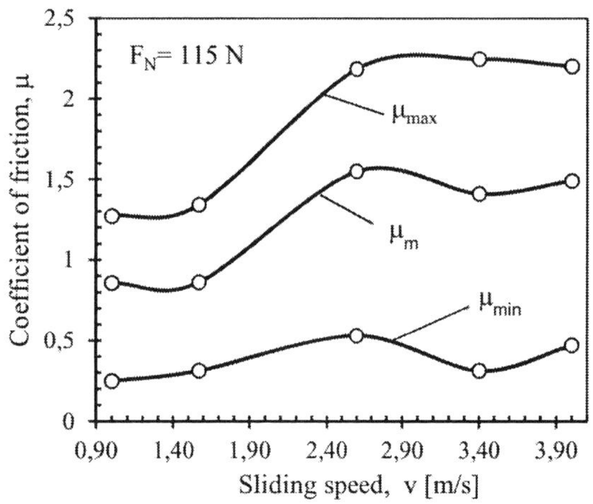

Figure 11. The variation in unsteady friction coefficients with speed.

Figure 12. The flakes of metallic debris on surfaces of pins for 115 N and 3.4 m/s after 120 s. (x200)

Figures 11 shows the variation in unsteady friction coefficients with speed. Generally, in the unsteady state regime, the friction coefficient increases slightly with an increase in speed. In this regime, the contact between pin and ring continues as metal-metal and the surfaces are roughed severely. The flakes of metallic debris produced between surfaces adhere to contacting surface during unsteady state sliding as shown in Figure 13. Since plastic deformation and wear normally do occur, the general contact situation changes with time.

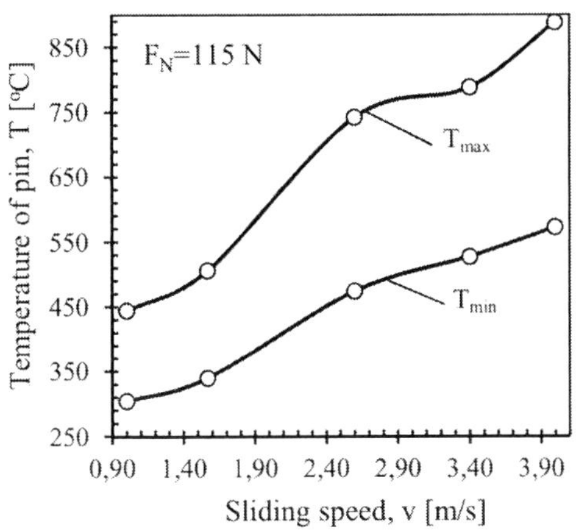

Figure 13. The variation in temperature of pin for unsteady friction with speed.

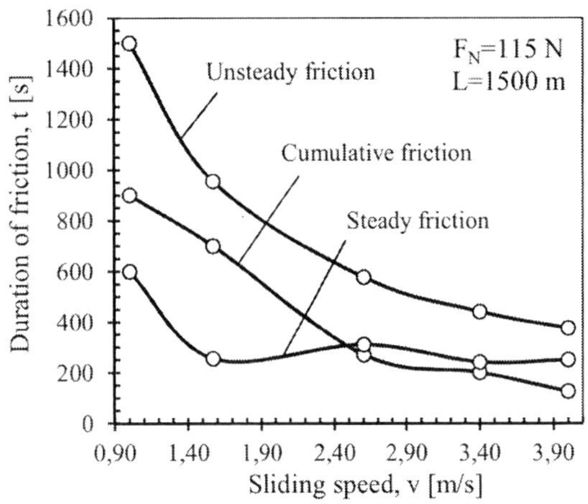

Figure 14. The variation in duration of friction with speed

Figures 13 show the variation in the temperature of pin with speed for unsteady friction. Generally, in the unsteady state regime, the temperature increase linearly with an increase in speed. The reason for this is that changes arising from strain rate and friction heating are expected with increasing speeds. Figure 14 shows the variation in the duration of unsteady friction steady friction and with speed for a distance of 1500 m. The duration of friction decreases with an increase in speed for three friction state. The reason of this can be explained as duration of test decreases with an increase in speed and the steady state sliding is reached earlier at because of formation of protective oxide film on surfaces.

3.3. The effects of sliding speed on the steady friction

As shown in Figures 18, after the first few minutes of test, steady state conditions are reached and frictional coefficient and temperature remain constant. It is possible to define this friction behaviour as steady state friction. The vertical lines represent the limit between the unsteady state and steady state. The experimental results of steady friction were given in Table 4.

Table 4. The influence of sliding speed on the coefficient of friction and the temperature of pin for steady friction.

Speed v (ms^{-1})	pv factor ($Nm.m^{-2}.s^{-1}$) x10^{-4}	Time t (s)	Friction coefficient μ_{min}	μ_{max}	μ_m	Pin temperature T_{min} (°C)	T_{max} (°C)	T_m (°C)	Friction heat q ($Nm.m^{-2}.s^{-1}$) x10^{-4}
1	64.97	600	0.594	0.806	0.690	384	422	399	44.82
1.57	102	255	0.330	0.842	0.456	425	458	436	46.51
2.6	168.9	310	0.374	0.495	0.425	495	566	512	71.78
3.4	220.89	241	0.280	0.363	0.315	542	612	561	69.58
4	259.88	250	0.320	0.495	0.401	594	648	617	104.21

Figure 15 shows the relationship between the steady frictional coefficient and the sliding speed. In general, the average and minimum frictional coefficients decrease slightly as the speed increases in range 1-3.4 m/s because of formation of protective oxide film on surfaces as shown in Figure 13. But, the maximum frictional coefficient decrease linearly to a minimum value as the speed increases in range 1-3.4 m/s. And then, the frictional coefficients increase maximum as at a speed of 4 m/s because of the oxide film with critical thickness is broken up and the contact of surfaces occurs metal-metal as shown in Figure 14. As shown in Figure 16, temperature of pin increase linearly as the speed increases. Figure 17 shows the variation of the average friction coefficients with speed for three friction states.

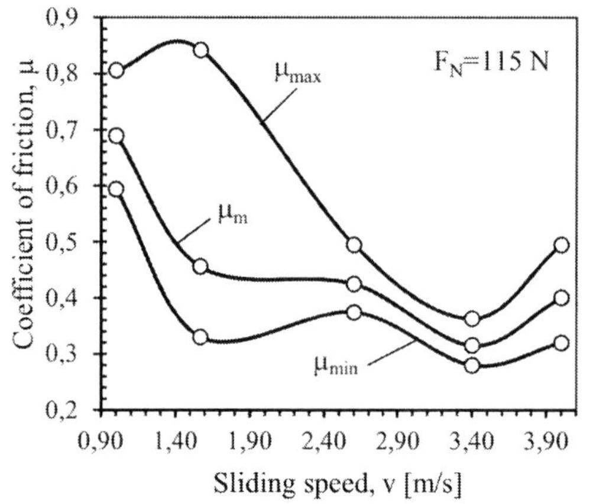

Figure 15. The variation in steady friction coefficients with speed.

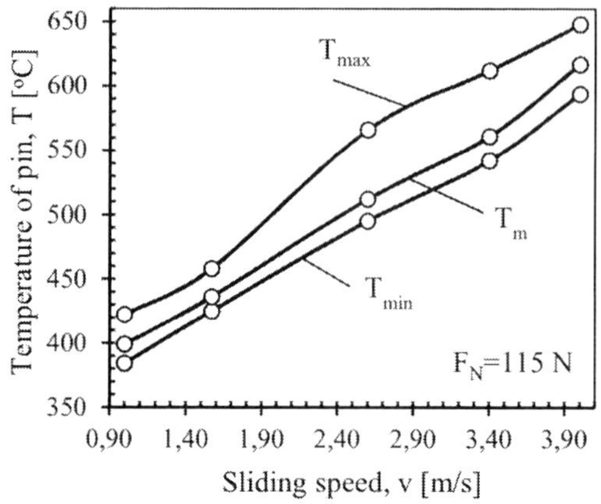

Figure 16. The variation in temperature of pin for steady friction with speed

As shown in Figure 17, the friction coefficient for unsteady state is higher than that for the other two friction state. It can be seen that the unsteady and cumulative friction coefficients increase linearly as the speed increases in range 1-2.6 m/s and then they decrease slightly in range 2.6-4 m/s. However, the average steady friction coefficient decreases slightly with an increase in speed to 4 m/s. Figure 18 shows the variation of the average temperature of pin with speed for three friction states. As shown in figure, the average temperature for unsteady state is lower than that for the other two friction state. It can be seen that the temperatures linearly increase with an increase in speed.

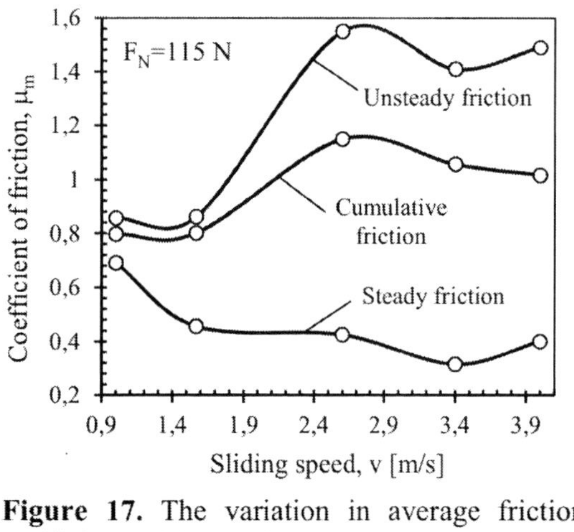

Figure 17. The variation in average friction coefficients with speed.

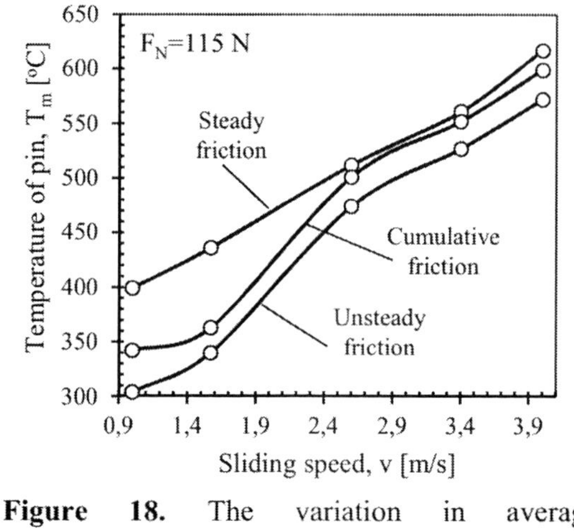

Figure 18. The variation in average temperature of pin with speed

Figure 19 shows the variation in frictional heat with speed in general friction state. As shown in this figure, frictional heat of unsteady and cumulative friction increases with increasing speed. However, the frictional heat for steady increases slightly with an increase in speed. Figure 20 shows the variation in frictional heat with load in general friction state. Friction is the process of conversion of energy. The conversion of external mechanical energy into the energy of internal processes occurs in the process of friction. As shown in Figure 20, frictional heat of unsteady and cumulative friction increases with increasing load and reaches a maximum at 190 N because of the average frictional coefficient increases with load and reaches a maximum at 190 N. After the load of 190 N, the cumulative frictional heat decreases with increasing load to a load of 250 N because the average frictional coefficient decreases with load to a load of 250 N.

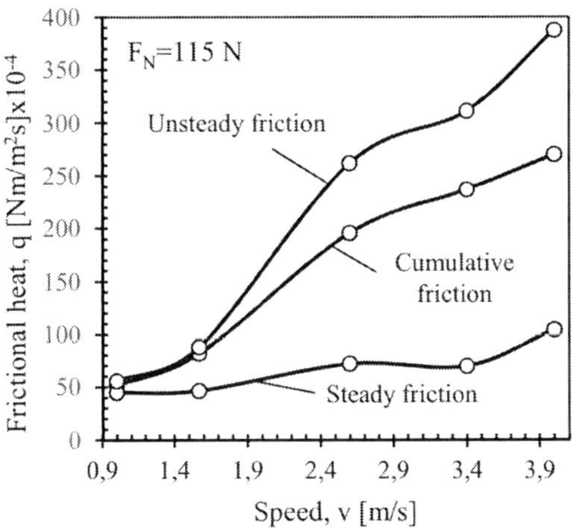

Figure 19. The variation in frictional heat with speed.

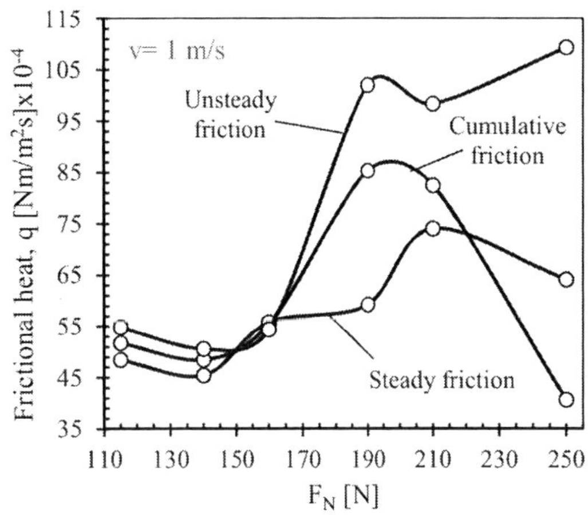

Figure 20. The variation in frictional heat with load for three friction state

4. Conclusions

The conclusion derived from the present study may be summarised as follows

1.In general, the frictional coefficient decreases slightly as the load and speed increases because of the temperature rise the surface oxidizes.

2.The thickness of the mechanically affected layer increases with an increase in load because the plastic deformation of the subsurface is increased with the temperature.

3.The friction is occurred two stages between steel-steel pairs for pin-on ring test system as running-in (unsteady state) and steady state friction. The duration of unsteady state friction decreases with an increase in load and speed.

4.The steady state coefficient of friction decreases with an increase in load and speed because of the formation of oxide film on the surfaces. In general, the average frictional coefficients increase as the speed increases.

5.The temperature increases with an increase in the speed. Generally, in the unsteady state regime, the friction coefficient increases slightly with an increase in speed.

6.Generally, in the unsteady state regime, the temperature increase linearly with an increase in speed. The duration of friction decreases with an increase in speed.

7.In general, the steady frictional coefficients decrease slightly as the speed increases because of formation of protective oxide film on surfaces.

8.The temperature of pin increase linearly as the speed increases for steady friction. The frictional heat increases with increasing speed.

References

1. Hutchings I.M., Leonardo da Vinci's studies of friction, Wear, 360-361, 51-66, 2016.
2. Halling J., Principles of Tribology, Macmillan, London,72-92, 1983.
3. Tabor D., Friction -the present state of our understanding, Trans. ASME,. Journal of lubrication technology, vol. 103, 169-179, 1981.
4. Hisakado T., On the mechanism of contact between solid surfaces, 4th report surface roughness on dry friction Bulletin of the JSME, vol.7, number 4, 223-226, 1984.
5. Do Z. and You-Dai X., Two-body micro cutting wear model Part IV: Theoretical analysis of friction, Wear, 129, 67-79, 1989.
6. Autay R., Kchaou M. and and Dammak F., Friction and wear behavior of steels under different reciprocating sliding conditions, Tribology Transactions, 55, 590-598, 2012.
7. Abouei, V., Saghafian, H., and Kheirandish, S., Effect of microstructure on the oxidative wear behavior of plain carbon steel, Wear, 262, 1225–1231, 2007.
8. Bahrami, A., Mousavi Anijdan, S. H., Golozar, M. A., Shamanian, M., and Varahram, N. , Effects of conventional heat treatment on wear resistance of AISI H13 tool steel, Wear, 258, 846–851, 2005.
9. Hardell, J., Kassfeldt, E., and Prakash, B., Friction and wear behaviour of high strength boron steel at elevated temperatures of up to 800 ◦C, Wear, 264,788–799, 2008.
10. Wei, M. X. Wang, S. Q., Cui, X. H., and Chen, K. M., Characteristics of extrusive wear and transition of wear mechanisms in elevated-temperature, Tribology Transactions, 53,888–896, 2010.
11. Kim, M. H., Rhee, K. Y., Paik, Y. N., Hong, J. S., and Ham, Y. S., Experimental investigation on the mechanical behavior of high-frequency induction-hardened mild carbon, SPS5 steel, Materials Science and Engineering A, 485, 31-38, 2008.
12. Z. Rymuza, Energy concept of the coefficient of friction, Wear 199, 87-96, 1996.
13. Quinn T. F. J., Review of oxidational wear, part I; the origins of oxidational wear, Tribology International, vol. l6, 257-271, 1983.
14. Chen Z. and Thomson P.F., Friction against superplastic aluminium alloys, Wear, 201, 227-232, 1997.
15. Odabaş D., Theoretical and experimental investigation of the effects of thermomechanical factors on the wear behaviour of AISI 3315 steel under dry sliding conditions, Ph. D. Thesis, Erciyes University, Engineering Faculty, Kayseri, Turkey (In Turkish), 1991.
16. Ling F. F. and Simkins T.E., Measurement of pointwise juncture condition of temperature at the interface of two bodies sliding contact, J. Basic Eng., 85, 481-487, 1963
17. Saka N., A, Eleiche M. and Suh N. P., Wear of metals at high sliding speeds, Wear, 23, 109-125, 1977.
18. Tareq M. A. A., Oksana O.M., Rudolph G. M., Temperature resistance of the boundary lubrication layers under rolling with slippage condition, International Journal of Mechanical Engineering and Applications, 5(2), 78-86, 2017.

Roughness Influence on Initiation of Fretting Fatigue Scar of Ti-6Al-4V Alloy

Capitanu L.[1]***, Badita L.L.***[2]***, Florescu V.***[3]***, Tiganesteanu C.***[1]
[1]*Institute of Solid Mechanics of the Romanian Academy 010141, Bucharest, Romania*
[2]*National Institute of Research and Development for Mechatronics & Measurement Technique, 021631, Bucharest, Romania*
[3]*University of Civil Engineering, 020396, Bucharest, Romania*
lucian.capitanu@yahoo.com[1], badita_l@yahoo.com[2], florescuvirgil@yahoo.com[3], tctin@yahoo.com[1]

Abstract: This paper reports on the experimental studies undertaken to detect the early stage when appears the fretting wear of the Ti-6Al-4V alloy used for the hip prostheses. Wear is a critical aspect for estimating the fretting fatigue. Studies were performed on samples of special shape, in order to be able to study the influence of in contact surfaces roughness on the durability to fretting. Fretting buffers, with roughnesses R_a of the contact surface of 0.015 and 0.045 μm, and Ti-6Al-4V samples with roughnesses R_a = 0.045 μm, R_a = 0.075 μm and R_a = 0.19 μm, were used. Testing periods of 3 seconds, 1 minute and 5 minutes were selected to capture the moment of the fretting scar appearance, long before these initiate the eventual fretting cracking. Simultaneously with fretting wear of the surface, the friction coefficient was also measured. From the in time evolution determinations of the fretting wear, it resulted that, under the experimental conditions used, the minimum wear occurs at a certain value of the roughness and not at the minimum roughness. Surprisingly, the minimum friction coefficient does not coincide with the minimum fretting wear.

Keywords: Ti-6Al-4V alloy, roughness, fretting damage, wear, friction

1. Introduction

Fretting is a phenomenon that occurs at the interface between two bodies in contact, which are pressed against each other in the presence of cyclic load, which gives rise to a small relative displacement. At the interface there is a complex interaction between wear, corrosion and fatigue. Thus, this phenomenon involves many aspects of contact mechanics, multiaxial fatigue, tribology, and materials science. Directly or indirectly, the damage due to the fretting fatigue is caused by several factors and interdependent variables, such as the amplitude of the relative displacement, normal load, tangential and axial loads, material's nature, friction coefficient, cyclic frequency, temperature and the environment.

There are several application domains where this phenomenon occurs, such as aerospace industry, human body implants, automobiles, etc. Designing at fatigue without considering the effect of fretting will certainly lead to premature and unexpected failure in many applications.

Fretting can be classified into two categories: fretting fatigue and fretting wear. Also, corrosion, mentioned in several studies as the third manifestation of the fretting [1], was accepted in the whole

area of fretting damage and incorporates the two manifestations of fretting fatigue and fretting wear [2].
In recent years, several studies have used multiaxial fatigue models to establish correlations with the experimental data of fretting fatigue. The most common multiaxial fatigue models are: Dang Van (DV), Smith-Watson-Topper (SWT) and Fatemi and Socie (FS) [2-7]. Lykins et al. [6] investigated the behavior of fatigue crack initiation by fretting of the titanium alloy, Ti-6Al-4V, using the SWT parameter.
Tribo-corrosion continued to be an area of many investigations. In particular, the concepts and influence of the third body during the tribo-corrosion processes, as well as the significance of the experimental techniques for studying problems caused by tribo-corrosion, have been investigated. A typical tribo-corrosion action is the materials degradation due to mechanical, electrochemical and synergistic effects. These effects are distinct areas of scientific study, but they occur together to cause the wear of a material. A typical example of mechanical experiment includes measurements of tangential load. However, some electrochemical investigations have been performed to help understand the synergistic effects. These synergies have been analyzed to help understand corrosion, both as a mechanical phenomenon and as a mechanical-electrochemical coupled phenomenon [8].
To predict the crack's propagation in a fretting wear case, Fidrici et al. [9] also used the SWT parameter. Moreover, to match the predicted life with the experimental life, a new factor has been added that takes into account the effect of the residual compression stresses induced by the stricking blow. Araujo and Nowell [10] have demonstrated that the use of the SWT and FS approaches to estimate the life of fretting fatigue initiation of Ti-6Al-4V is not appropriate if the calculations are based exclusively on the stress-strain behavior of the surface at extremely tensioned points. This suggested that these methods would not be adequate for assessing the life of fretting fatigue initiation in tests with different geometries, and that the reason for the poor performance of the prediction methodology is the presence of the effects of the high stress gradient. They proposed two methods of averaging, which, combined with a multiaxial fatigue criterion, would give a less conservative prediction of the life of fretting fatigue initiation. The first method is an average of the SWT and FS parameters over a characteristic depth on the critical plane. The second method is an average of the stresses in a characteristic elementary volume. Both methods of calculating the average produced sensible similarly results.
Giannakopoulos et al. [11] reported that the fretting problem can be approached as a problem of simple fatigue subjected to a localized stress concentration. They developed an analytical model for fretting fatigue at a corner of a rounded mandrel in contact with a substrate, and an analogy with the initiation of the fatigue crack at the notch of a tip was made. They analyzed the similarities and differences between the stress concentration factors at the edge of the mandrel-on-flat unexpectedly rounded contact and the tip of a blunt crack. Notch-like methodology provided a direct connection between the life of the fretting fatigue crack initiation of a flat mandrel with rounded corners and the life of the simple fatigue crack initiation of a smooth sample made of the same material. Model predictions were compared to the fretting fatigue experiments on Ti-6Al-4V showing good concordance.
Madge et al. [12] have shown that the wear is a critical aspect for estimating the fretting fatigue and can have a major effect on the predicted evolution of the stresses and deterioration parameters (e.g. SWT), and, thereafter, on the predicted fatigue life and the position's failure. They performed a finite elements simulation (removed element simulation), which simultaneously combines a fretting wear model and the SWT parameter of fretting fatigue, which includes the effect of the stress peak.
From the various approaches mentioned above, one can conclude that is still missing a more comprehensive analysis of the fretting fatigue phenomena. The data currently available on different materials and test conditions are insufficient to allow taking a decision on a suitable model.
The models chosen for examination are two models of fatigue based on deformation. The first approach is the SWT model, and the second approach is Morrow's model. The first model is already used in fretting fatigue, while the latter was first tested for fretting fatigue situations by Buciumeanu et al., [1]. In both models, the effects of average stress are included. Morrow's model was modified to introduce the local stresses involved in the fretting fatigue process (normal, tangential and axial).

Moreover, a new parameter, which is a stress concentration factor, K_t, which includes the effect of the affected area, called here, the "effect of the fretting scar", has been introduced in both models.
In accordance with the direction of relative movement under a ball-on-flat (BOF) contact configuration, four basic fretting modes can be distinguished, namely: tangential, radial, rotational and torsional fretting. Rotational fretting can be defined as the relative movement induced by the micro-amplitude rotation that takes place in the pair of contacts with tightly fixing in the alternative loading medium. Rotational fretting phenomena resulting in wear of the contact surface exist in many mechanical equipment and instruments such as cars, rail vehicles, planes and biomedical implants, and wear greatly induces the reduction of component life, and huge economic losses. Researches on rotating fretting wear under the BOF contact configuration were performed by Zheng et al., [13]. However, the contra-pair contact configurations, in which the fretting damage occurs in practice, are often not BOF contacts, but also other contact configurations. According to the modern tribological theory, the behaviors of contact stress, of contact stiffness and of contact residues are quite different under different contact configurations. The contact configuration plays a very important role in the fretting wear. In order to understand the rotating fretting mechanisms in the different contact configurations, rotating fretting tests with ball-on-concave (BOC) and ball-on-flat (BOF) contacts were performed at the same maximum contact stress and, respectively, normal loads.
Many research works have been done on the wear mechanisms by fretting and fatigue. It has been achieved the understanding of some basic mechanisms of fretting, allowing the development of various possible palliatives. However, to date, there has been no general agreement on the mechanisms and modeling of fretting. For example, the mechanical, chemical, tribological and electrical processes between the two contact surfaces during the fretting were not fully understood, because the most models and wear theories are based on the observation of deterioration by the fretting wear. The direct or in situ observation of the fretting wear process implies the procurance of some time-dependent changes and casts more light on the actual fretting mechanisms [14]. Fretting, in a total hip prosthesis, results from the destruction of the passive oxide layer from the metal, thus leading to increased corrosion and to the residues generation, such as polymer and/ or metal oxides particles resulting in serious dysfunctions of the hip joint [15]. The cyclic loading due to the daily human walking, and the differences between the mechanical properties of the femoral stem and bone cement give rise to the separation of the two materials. This unsoldering leads to secondary effects, such as degradation and cracking of cement. Residues, including metal oxides and ions, invade bone tissues through cracks and finally induce inflammation in the bone tissues
This study presents the experimental and theoretical research on the fretting behavior of the Ti-6Al-4V alloy samples used for the manufacture of total hip prostheses.

2. Experimental evaluation

2.1. Experimental assembly for testing

Fretting tests were performed on a universal servo-hydraulic testing machine, MTS Bionix, figure 1.

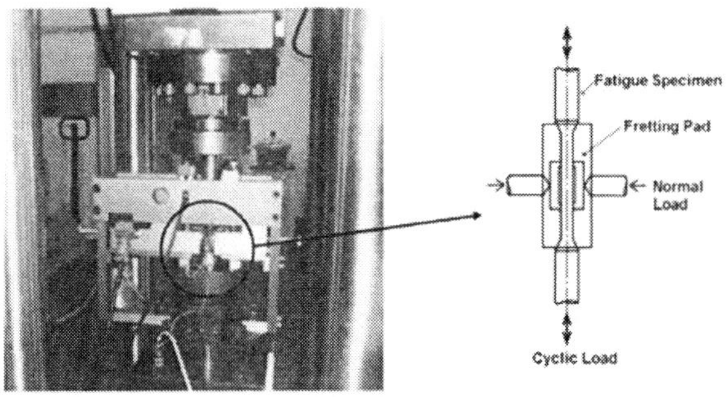

Figure 1. MTS Bionix servo-hydraulic dynamic testing machine

The testing machine allowed application of a 600 kN load under static conditions and 500 kN for fatigue loading. The maximum cyclic loading frequency of the machine is 50 Hz. The equipment operates at ambient temperature in the laboratory environment. To perform fretting fatigue tests, on the machine was a special fretting device that has two load cells that allow (prior to testing) application of normal and tangential contact loads (during the test) and their recording/ measurement. A 1000 N loading cell is used to set and measure the pre-tangential/ tangential load and a 2500 N loading cell to set/ measure the normal load. The load is applied by two buffers, which are pressed perpendicularly on the flat faces of the sample by a pair of compression springs, which are loaded with adjustable screws. Also, two vertical compression springs, which action on the buffer, are used to apply the pre-tangential load (to avoid the buffer's movement).
"Bridge" type fretting buffers were used, which have as main advantage the simplicity with which a normal fatigue specimen can be used, both at bending and at cyclic traction. The bridges are simply fixed to the sides of the specimen by a ring or a similar arrangement and the cyclic deformation of the specimen causes a relative movement between the bridge legs and the specimen.
The buffers were made of RUL 1 bearing steel, hardened and rectified to a roughness R_a of 0.15 μm. The tangential load varies depending on the axial load of the machine. Due to the system's symmetry, loads are measured only in one part. The axial cyclic load applied to the sample was monitored with the machine's loading cell.

2.2. The geometry of the sample and buffer

Fretting fatigue experiments were performed using the geometries of sample and buffer, which are shown in figure 2, and figure 3 shows the scheme of the sample-buffer contact test configuration.

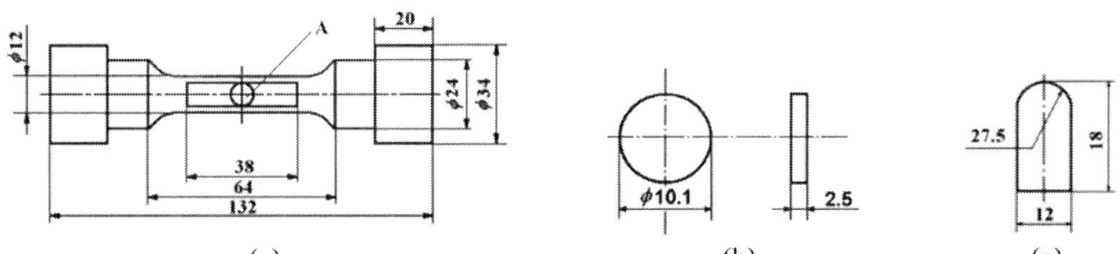

Figure 2. Geometry of the specimen, sample and fretting test buffer.

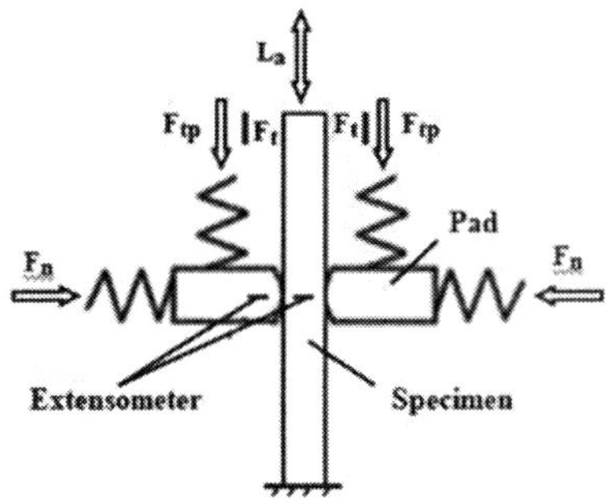

Figure 3. Scheme of the test configuration of the sample-buffer contact: F_n– normal loading of the buffer; F_{tp}– tangential pre-loading of the buffer; F_t– tangential loading of the buffer and L_a– axial load of the machine.

The specimen (figure 2) has a special shape, with two flat opposite faces (A), milled, rectified and polished in the middle of the sample. In this situation, the two contact areas on the flat sides of the sample were generated. In the middle of these areas, holes with diameter of 10.1 mm and depth of 2.5 mm were drilled. Fretting samples were introduced into these places. This was done in order to study the influence of in contact surfaces roughness to the durability at fretting. Several buffers and fretting samples were produced, which had been finished with roughness R_a= 0.015 µm. The spherical head of the buffers had a roughness of 0.05 µm. Prior to testing, the samples and buffers surfaces were ultrasonically cleaned in alcohol. Figure 3 schematically shows the configuration of the sample-buffer contact testing in the fretting fatigue tests.

2.3. Experimental procedure

Under static conditions, the compressive stresses due to point contact, p_{max} and p_{med} (maximum and average contact pressure) are given by:

$$p_{max}^3 = 1.5\, PE^2 \,/\, \pi r^2 (1 - \mu^2)^2 \quad (1)$$

$$p_{med} = \frac{P}{\pi a^2} \quad (2)$$

while the radius a of the circular contact surface is:

$$a^3 = 1.5(1 - \mu^2)P\frac{r}{E} \quad (3)$$

where P is the load; a the radius of the contact surface and r the radius of the spherical head of the buffer.

For coupling components made of steel, the values of the previous equations become:

$$p_{max} \approx 5800\sqrt[3]{P} \quad (4)$$

$$p_{med} \approx 1700\sqrt[3]{P} \quad (5)$$

$$a \approx 0.09\sqrt[3]{P} \quad (6)$$

Particular attention was given to the finishing of the specimens friction surfaces. The state of the surface, being defined by topography, microstructure of the superficial layer and the oxidation state, has a major contribution to the wear process. Due to the complexity of surface creation through an abrasion process, the safest way to ensure a consequent reproducible surface is strictly by adhering to all stages of the finishing process.

The following stages have been established: shape transformation, finishing transformation, thermal treatment, rectification, fine adjustment and ultimately super-finishing of the work surface. While in all the stages, until the fine final adjustment, traditional manufacturing techniques are used, it is worth mentioning that the overall intensity of the process is deliberately maintained at a low level to protect the superficial structure of the material. The super-finishing stage uses the metallographic polishing techniques and includes: wet polishing with abrasive paper of 32 µm and 17 µm granulation, polishing with diamond paste of 6 µm and 1 µm granulation, and, finally, wet polishing using alumina suspension that has a granulation of 2000 Å. The surfaces are then cleaned with alcohol and distilled water and then are dried. The storage of the finished samples is done in sealed containers on a layer of silica gel.

After the super-finishing stage, four different roughness values were obtained for Ti-6Al-4V samples surfaces: R_a = 0.015 µm, R_a= 0.045, R_a = 0.075 µm and R_a = 0.190 µm. Surfaces roughness was determined using a profilometer with parametric transducer and graphical recording, Perth-O-Meter. The tool allows not only the recording of surface profiles, but also the determination of the R_a and *r.m.s* values, defined as:

$$R_a = \frac{1}{l}\int_0^l |y|dx \tag{7}$$

$$r.m.s = \left(\frac{1}{l}\int_0^l y^2 dx\right)^{1/2} \tag{8}$$

The tests mainly followed the evolution of fretting scar according to the normal load applied under different roughness conditions of the surface and normal loading, but at constant tangential loading. Experimental tests were performed under lubrication conditions with SBF (simulated body fluid) from Hyclone Inc., USA, which has a composition close to that of the human serum.

3. Experimental results

Fretting fatigue tests were performed at a stress ratio R = 0.1 where $R = F_n / F_t$. All tests were performed at ambient temperature, in the laboratory environment, as well as at a cyclic frequency of 4 Hz. The profiles and microphotographs of the initial surfaces topography with initial roughness R_a = 0.015 µm (A1-A2), R_a = 0.045 µm (A3-A4), R_a = 0.075 µm (A5-A6) and R_a = 0.19 µm (A7-A8) are illustrated in figure 4.

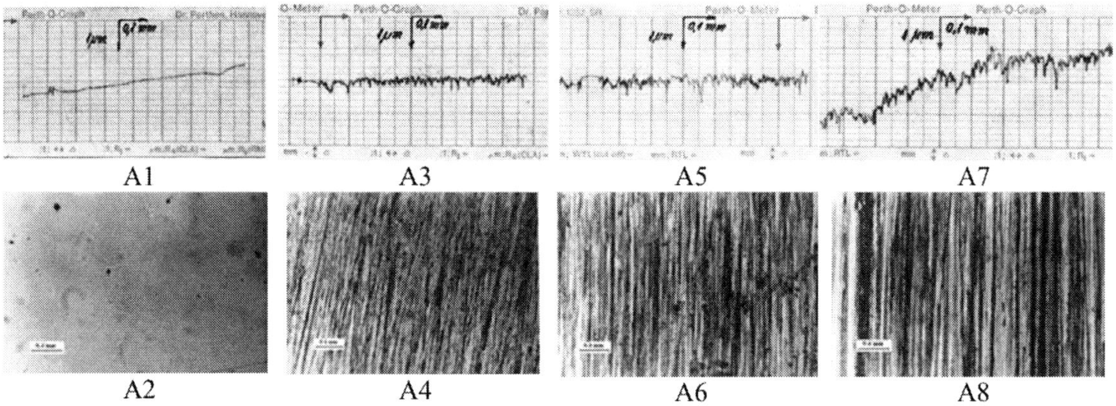

Figure 4. Profiles and microphotographs of the initial surfaces topography with roughness R_a = 0.015 µm (A1-A2), R_a = 0.045 µm (A3-A4), R_a = 0.075 µm (A5-A6) and R_a = 0.19 µm (A7-A8).

In order to properly view the wear of the fixed surface, depending on the roughness of the coupling, the following solution was used: the coupling asperities were concentrated on one of the surfaces, especially on the mobile one. Fixed surface has always had the minimum achievable roughness, i.e. approximately $R_a \approx 0.015$ µm. As is well known, the composite roughness of the coupling, expressed by standard deviations, σ, is:

$$\sigma^2 = {\sigma_1}^2 + {\sigma_2}^2 \tag{9}$$

where σ_1, σ_2 represent the standard deviations of the two surfaces.
If one of the surfaces has a much smaller roughness, i.e. $\sigma_1 << \sigma_2$, then $\sigma \approx \sigma_1$. Thus it is possible to study the influence of roughness on wear, even by changing the roughness of a single surface. In these conditions and under a load P = 30 N, the relative velocity u = 1.74 m/s and at the temperature θ = 50°C of the lubricant volume, the evolution of the surface wear was developed in function of time for the following roughnesses of the mobile surface: R_a = 0.015 µm; R_a = 0.045 µm; R_a= 0.075 µm and R_a = 0.19 µm.

3.1. Evolution of the contact surface state in fretting conditions

The consequence of fretting is the fretting scar that forms in the area of the buffer-sample contact and which can evolve by initiating the fretting crack until the sample breaks over time. The initiation phase takes almost 80% of the coupling's runtime. In figure 5, some fretting wear scars obtained under the above mentioned conditions (at time $t = 5$ min) are shown selectively to estimate the reproducibility of the results.

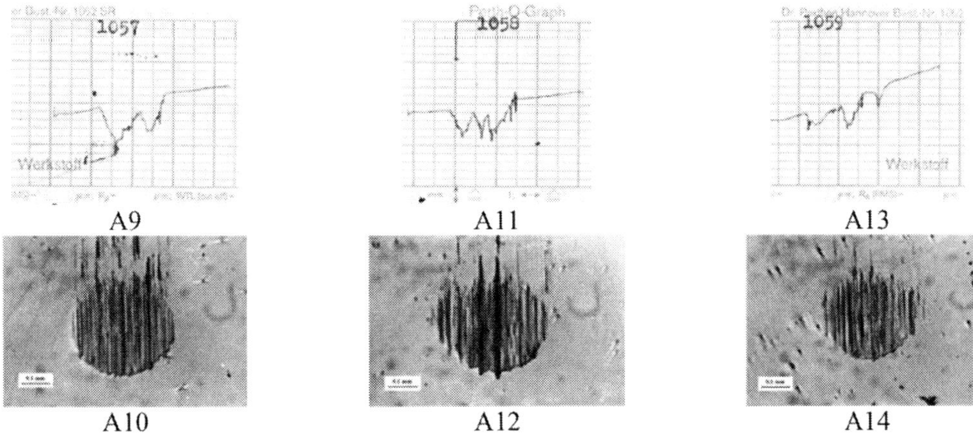

Figure 5. Central transversal profile and image of fretting wear scar. $R_a = 0.015$ µm, $t = 5$ min.

The scar volume for the three experimental determinations is: 1057 specimen: $V = 8.50 \times 10^{-5}$ mm^3; 1058 specimen: $V = 6.11 \times 10^{-5}$ mm^3; 1059 specimen: $V = 4.00 \times 10^{-5}$ mm^3, resulting an average volume $\overline{V} = 6.2 \times 10^{-5}$ mm^3. For wear measurements, a deviation of 25% is completely satisfactory. Figure 6 shows the evolution of the wear depending on time, for the roughness of the buffer $R_a = 0.045$ µm. Trial times of 3 seconds, 30 seconds, 1 minute and 5 minutes were selected, each performed with another coupling. This solution was chosen to also monitor the change of the buffer surface state. Figure 6 shows that, except for the first 5 minutes, the wear remains at the same level.

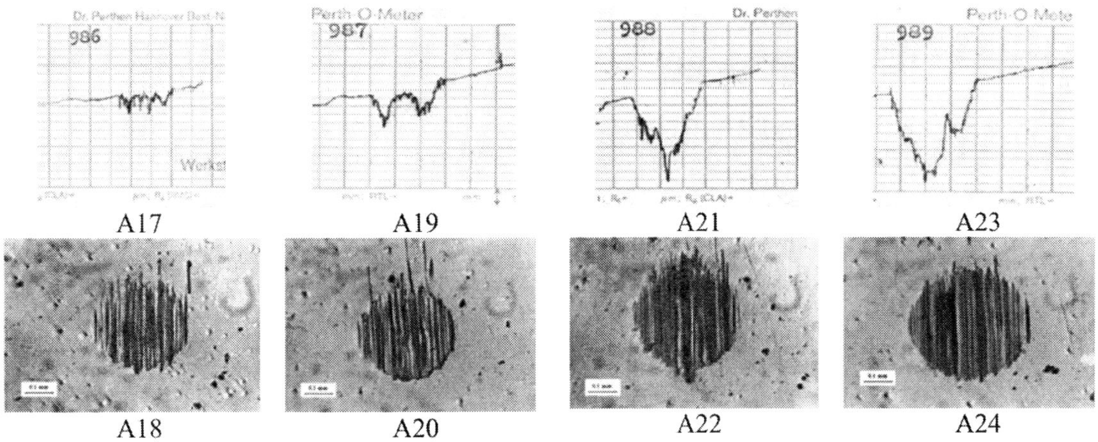

Figure 6. Central transversal profile and image of fretting wear scar $R_a = 0.045$ µm, $t = 5$ min.

In figure 7 is shown the transversal profile and the image of the fretting wear scar for the sample with $R_a = 0.045$ µm, $t = 5$ min (sample 703); $R_a = 0.045$ µm, $t = 5$ min, but with the buffer from the previous determination (total operating time $t = 10$ min, sample 704); $R_a = 0.045$ µm, $t = 5$ min, with the buffer from the previous determination (total operating time $t = 20$ min, sample 705).

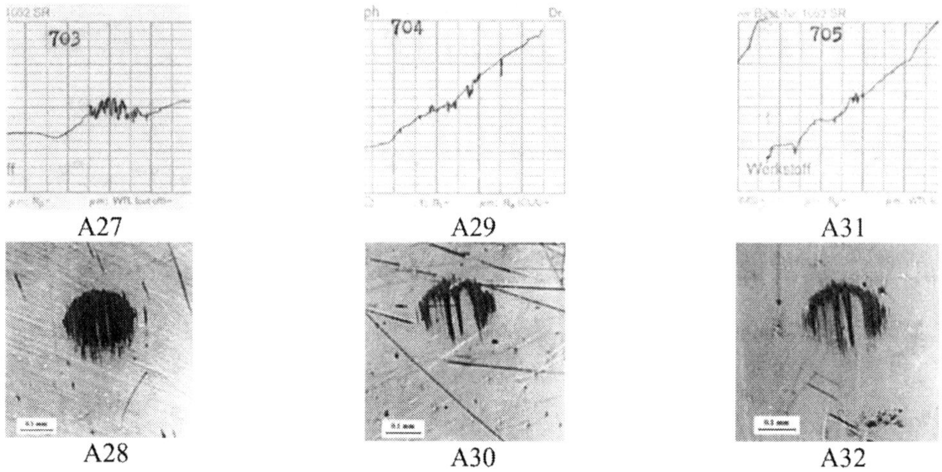

Figure 7. Profile and image of the fretting wear scar for samples with R_a = 0.045 μm, t = 5 min. (sample 703); R_a = 0.045 μm, t = 5 min., with the buffer from the previous determination (total operating time t = 10 min. (sample 704); R_a = 0.045 μm, t = 5 min., with the buffer from the previous determination–total operating time t = 20 min. (sample 705).

The evolution of fretting wear depending on time for different roughnesses is presented in the double-logarithmic diagram in figure 8.

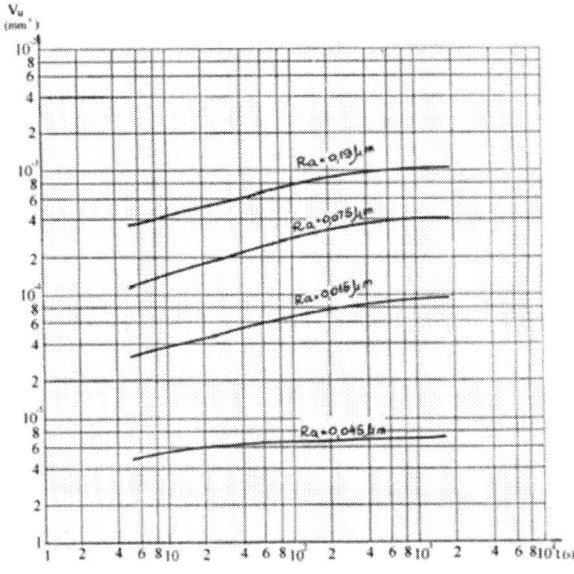

Figure 8. Evolution of wear versus time, for different roughness R_a.

By increasing the operating time from 5 minutes to 30 minutes, the wear increases by only about 10%. It is worth noting the rapid reduction of the wear velocity over time, excepting the surface with roughness R_a = 0.045 μm. In the first 3 seconds, between 25% and 50% of the wear occurs in 30 minutes. This evolution of wear is explained by the surfaces running, which leads to a change in the lubrication regime. Note that the amount of wear on the plate is caused by the initial wear (in the first few seconds of operation). This observation allows the use of the wear value at t = 5 min as a representative quantity for the existing operating conditions. During this operation period, the values scattering is small.

In the case of surfaces with R_a = 0.045 μm, the wear is so low that during the entire period of time used, the lubrication conditions remain roughly unchanged.
The evolution of the fretting wear function of time, for roughness R_a = 0.075 μm (samples 829, 830 and 832) is shown in figure 9.

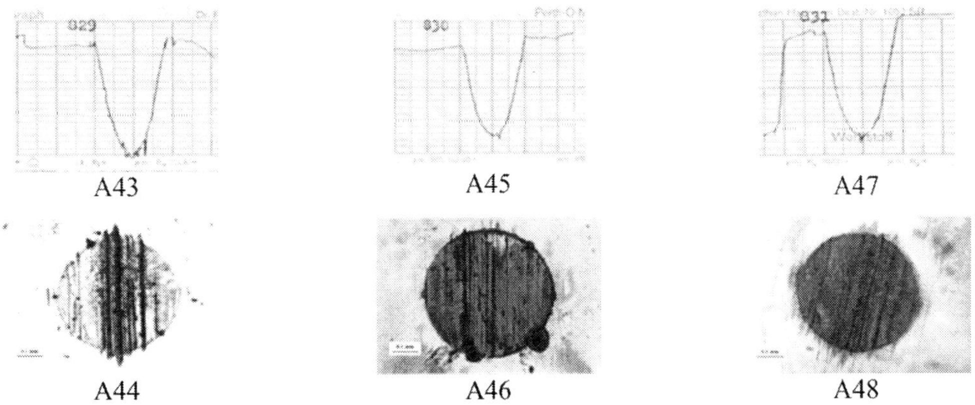

Figure 9. Evolution of the fretting wear function of time, for R_a = 0.075 μm (samples 829, 830, 832).

Different roughnesses used have caused not only a difference between the amount of wear volume of the scar, but also of the wear scar appearance (the type of wear). In the case of the surfaces with R_a = 0.015 μm, R_a = 0.075 μm (t = 3 s) and R_a = 0.19 μm (t = 3 s), the wear is of adhesive type (metallic shape with pronounced scratches). In the case of surfaces with R_a = 0.045 μm, the fretting wear type is predominantly oxidative. While the fretting wear velocity is reduced due to surface conformation, the surfaces with R_a = 0.075 μm and R_a = 0.19 μm also undergo in oxidative wear regime.
The evolution of the fretting wear function of time, for roughness R_a = 0.19 μm (samples 835, 849, and 850) is shown in figure 10.

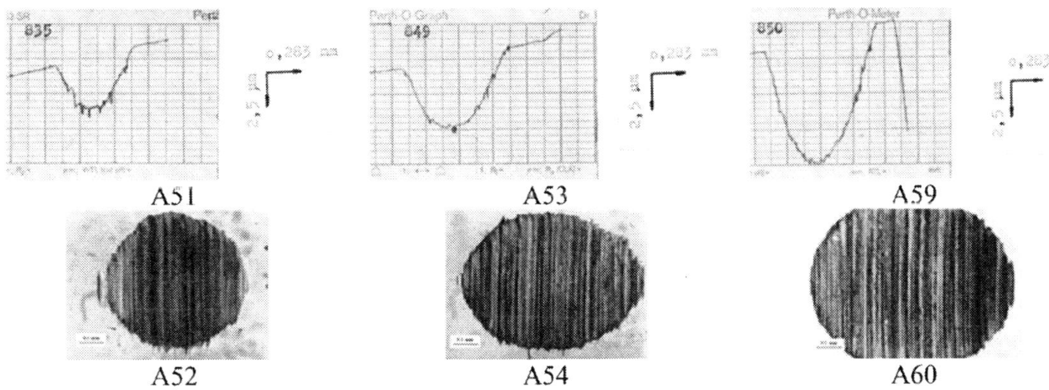

Figure 10.Evolution of the fretting wear function of time, for roughness R_a = 0.19 μm (samples 835, 849, and 850).

For the roughnesses R_a = 0.015 μm, R_a = 0.075 μm and R_a = 0.19 μm, the results obtained for the volume of the worn material are in accordance with the contact deformation, determined by the parameter of the lubricant film h_{min} / σ. For surfaces with roughness R_a = 0.045 μm, an influence of the running-in was observed. After the first 5 minutes of operation, the entire contact surface is covered with oxide. The scar obtained after another 5 minutes, with the same buffer, has a special shape, the oxidative wear area is limited to half of the loaded surface (samples A52 and A54).
Figure 11 shows the central profile and the image of the fretting wear scar R_a = 0.19 μm, t = 30 min, sample 997 (new buffer).

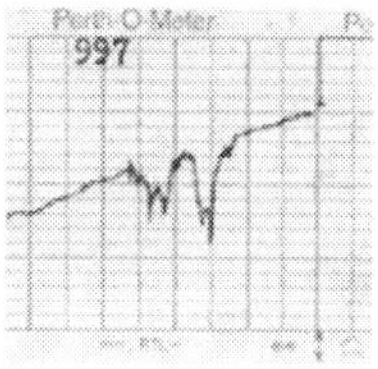

A61

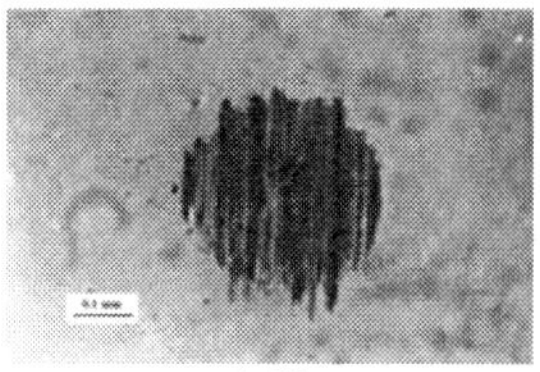

A62

Figure 11. Central transversal profile and image of wear scar. R_a = 0.19 μm, t = 30 min., sample 997, (new buffer).

Next, figure 12 shows the effect of the running-in on the fretting wear behaviour of the surface with R_a = 0.19 μm, with the same buffer as in the previous test, compared to figure 13, that shows the effect of the running-in on the wear behavior of the surface with roughness R_a = 0.045 μm, using a new buffer.

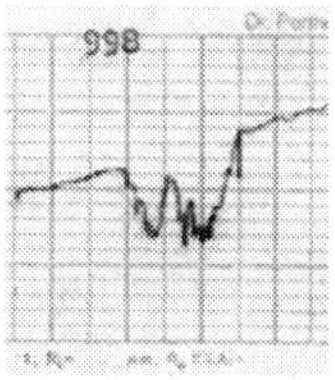

A63

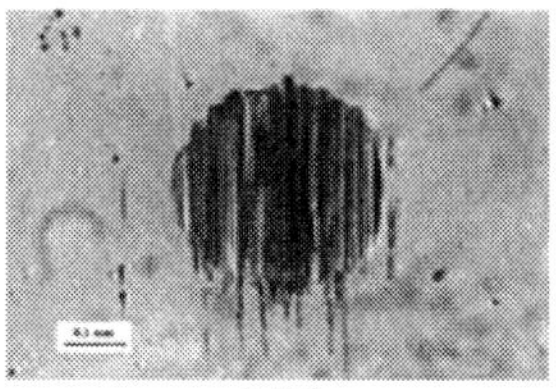

A64

Figure 12. The effect of running-in on the fretting wear behaviour of the surface with R_a= 0.19 μm, t = 5 min, sample 998. The buffer from previous determination was used.

In the experimental conditions of figures 11 and 12, it is clearly observed the start of the fretting scars elongation in the sliding direction (from top to bottom A61 / A62 and A63 / A64) and the typical W-shape of the transversal profile of the scar, noted in many of the papers published in the field [16-20].
Different roughnesses used have caused not only a difference between the values of the fretting worn volume, but also of the wear scar appearance (type of wear). In the case of surfaces with R_a = 0.015 μm, R_a = 0.075 μm (t = 3 s) and R_a = 0.19 μm (t = 3 s), the wear is of adhesive type (metallic shape with pronounced scars). In the case of surfaces with R_a = 0.045 μm, the predominant type of wear is the oxidative one. While the wear velocity is reduced as a result of the surface conformation, surfaces with R_a = 0.075 μm and R_a = 0.19 μm also go under oxidative wear regime.
In the case of super-finished surfaces with R_a = 0.015 μm no favorable influence of the running-in is observed. For surfaces with roughnesses of R_a = 0.045 μm, a favorable influence of the running-in was observed (figure 13).

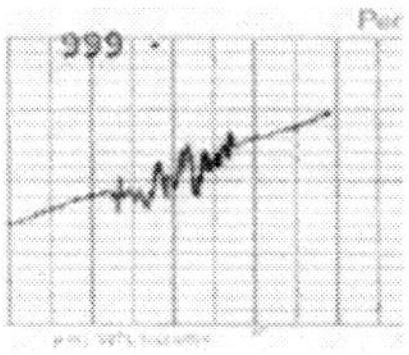

A65

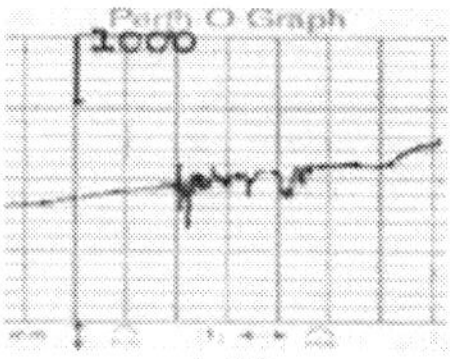

A67

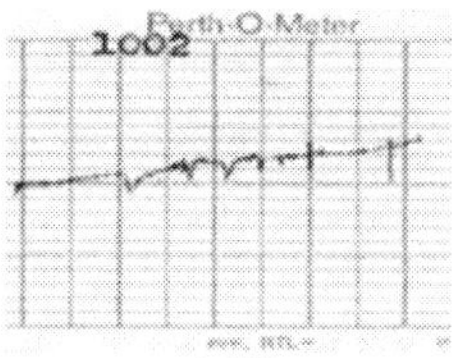

A69

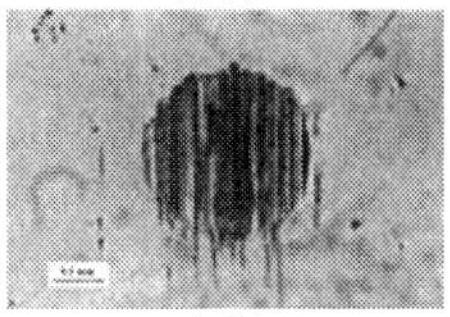

A64

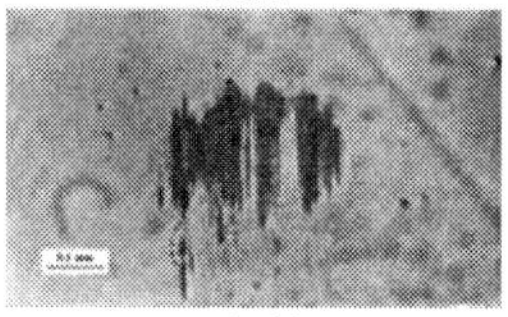

A68

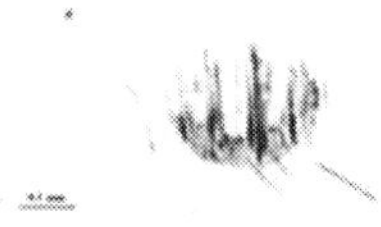

A70

Figure 13. The effect of the running-in on the fretting wear behaviour of the surface with R_a= 0.045 μm, t = 5 min, samples 999 (with new buffer), 1000 and 1002 (with the buffer from the previous determination).

After the first 5 minutes of operation, it was observed that the entire contact surface was coated with oxide. The scar obtained after another 5 minutes, with the same buffer, has a special shape, the oxidation wear area being limited to half of the loaded surface (sample 1002) – figure13.

3.2. Influence of the initial roughness on the fretting wear and of the friction coefficient

The minimum value of the surface wear must coincide with the minimum value of the surface roughness. A large number of wear measurements were performed with four roughnesses: $R_a = 0.015$ µm; $R_a = 0.045$ µm; $R_a = 0.075$ µm and $R_a = 0.19$ µm. The average values of the volume of fretting worn material for the four roughnesses are

$$R_a = 0.015\ \mu m \rightarrow V_u = 7.0 \times 10^{-5}\ mm^3 \rightarrow \mu = 0.038;$$
$$R_a = 0.045\ \mu m \rightarrow V_u = 7.0 \times 10^{-6}\ mm^3 \rightarrow \mu = 0.050;$$
$$R_a = 0.075\ \mu m \rightarrow V_u = 3.7 \times 10^{-5}\ mm^3 \rightarrow \mu = 0.038;$$
$$R_a = 0.190\ \mu m \rightarrow V_u = 1.0 \times 10^{-3}\ mm^3 \rightarrow \mu = 0.078.$$

Simultaneously with the production of fretting wear scars on the surface of the Ti-6Al-4V alloy samples, the coefficient of friction was also measured.
The effect of the running-in on the fretting wear behavior of the surface was also analyzed. This was done by using the same buffer for several successive tests.
In figure 14 are shown representative images of the running-in effect on the fretting wear behavior of the surface with the buffer from the previous determination, for three of the four roughnesses (samples 1077, 1065 and 1067).

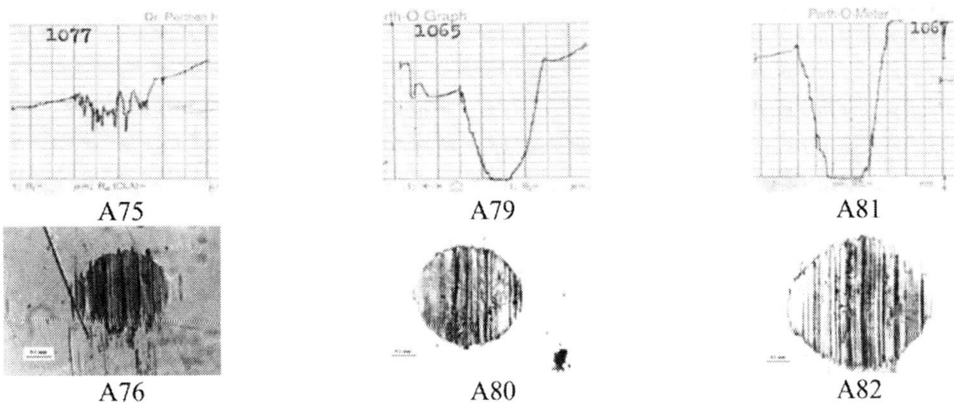

Figure 14. Effect of the running-in on the fretting wear behaviour of the surface with $R_a = 0.19$ µm, $t = 5$ min, samples 1077, 1065 and 1067 (with the buffer from previous determination).

The existence of an optimal roughness can be explained either by an effect on the lubricant film or by a change in the mechanical properties of the surface. In this case, at the optimum roughness, the reduction of the h/σ ratio is compensated by increasing the wear resistance of the surfaces.

4. Conclusions

In the experimental conditions used, the fretting contact between the buffer and the sample was a point contact, which produced a fretting wear scar that evolved rapidly even during the short time considered (maximum 30 minutes).
The influence of the sample and buffer roughnesses on the initiation and evolution of the fretting wear scar was studied. The geometry of the test configuration allowed the investigation of finished surfaces at roughnesses $R_a = 0.015$ µm, $R_a = 0.045$ µm, $R_a = 0.075$ µm and $R_a = 0.19$ µm.

From the determinations of fretting wear evolution in time under the experimental conditions used, it was found that in the case of the shortest experiment (5 minutes) the fretting wear scar is visible.
The recording of fretting scars allowed calculation of the volume of worn material by fretting and correlation with the value of the friction coefficient recorded during each test.
It has been found that there is a minimum of the wear curve resulted for roughness R_a = 0.045 μm.
A double-logarithmic evolution of fretting wear function of time was determined for different roughnesses analyzed. The used roughnesses caused not only a difference in the value of the scratch scar removal volume but also the appearance of wear scar (type of wear). Used roughnesses have caused not only a difference between the values of the worn volume of the fretting scar, but also of the wear scar appearance (type of wear).
After 3 seconds, in the case of surfaces with R_a = 0.015 μm, R_a = 0.075 μm and R_a = 0.19 μm (t = 3 s), the fretting wear is of adhesive type (metallic shape with pronounced scratches). In the case of surfaces with R_a = 0.045 μm, the type of fretting wear is predominantly oxidative. During the running-in, the fretting wear velocity is reduced as a result of the surface conformation, and surfaces with R_a = 0.075 μm and R_a = 0.19 μm are undergoing in oxidative wear regime.
It has been observed that by increasing the operating time from 5 minutes to 30 minutes, the wear increases only with about 10%. It is worth noting the rapid reduction of the wear velocity over time, excepting the surface with roughness R_a = 0.045 μm, where in the first 3 seconds, between 25% and 50% of the wear occurs in 30 minutes. At surfaces with roughness R_a = 0.045 μm, a significant influence of the running-in was observed. After the first 5 minutes of operation, the entire contact surface is covered with oxide. The scar obtained after another 5 minutes, with the same buffer, has a special shape, the oxidative wear area being limited to half of the loaded surface.
In the experimental conditions R_a= 0.015 μm, t = 5 min and R_a = 0.19 μm, t = 30 min it is clearly observed the start of the fretting scars elongation in the sliding direction, as well as the typical W-shape of the transversal profile of the scar, noted in many of the papers published in the field.
It was found the existence of a minimum of the resulting wear curve for roughness R_a = 0.045 μm, but which surprisingly does not coincide with the minimum friction coefficient.
A mathematical relationship between the friction coefficient and wear can not be established.

References

1. Buciumeanu M., Crudu I., Palaghian L., Miranda A.S., Silva F.S., Influence of wear damage on the fretting fatigue life prediction of an Al7175 alloy, International Journal of Fatigue, vol. 31, 1278–1285, 2009.
2. Araujo J.A, Nowell D, Vivacqua R.C., The use of multiaxial fatigue models to predict fretting fatigue life of components subjected to different contact stress fields, Fatigue& Fracture of Engineering Materials and Structures, vol. 27, 967–978, 2004.
3. Madge J.J, Leen S.B, Shipway P.H., The critical role of fretting wear in the analysis of fretting fatigue, Wear vol. 263, 542–551, 2007.
4. Salerno G., Magnabosco R., de Moura Neto C., Mean strain influence in low cycle fatigue behaviour of AA7175-T1 aluminum alloy, International Journal of Fatigue, vol. 29, 829–835, 2007.
5. Szolwinsky M.P, Farris T.N., Observations, analysis and prediction of fretting fatigue in 2024-T351 aluminium alloy, Wear, vol. 221, 24–36, 1998.
6. Lykins C.D, Mall S., Jain V.K., Combined experimental–numerical investigation of fretting crack initiation, International Journal of Fatigue, vol. 23, 703–711, 2001.

7. Proudhon H., Fouvry S., Buffiere J.Y., A fretting crack initiation prediction taking into account the surface roughness and the crack nucleation process volume, International Journal of Fatigue, vol. 27, 569–579, 2005.
8. Landolt, D., Mischler, S., Tribocorrosion of passive metals and coatings, Woodhead Publishing Limited, 2011.
9. Fidrici V., Fouvry S., Kapsa P., Perruchaut P., Prediction of cracking in Ti-6Al-4V alloy under fretting-wear: use of the SWT criterion, Wear, vol. 259, 300–308, 2005.
10. Araujo J.A, Nowell D., The effect of rapidly varying contact stress fields on fretting fatigue, International Journal of Fatigue, vol. 24, 763–775, 2002.
11. Giannakopoulos A.E, Lindley T.C, Suresh S. Chenut C., Similarities of stress concentrations in contact at round punches and fatigue at notches: implications to fretting fatigue crack initiation, Fatigue &Fracture of Engineering Materials & Structures, vol. 23, 561–571, 2000.
12. Madge J.J., Leen S.B., Shipway P.H., The critical role of fretting wear in the analysis of fretting fatigue, Wear, vol. 263: 542–551, 2007.
13. Zheng J.F., Yang S., Shen M.X., Mo J.L., Zhu M.H., Study on rotational fretting wear under a ball-on-concave contact configuration, Wear, vol. 271, 1552-1562, 2011.
14. Fu, Y.Q., Wei, J., Batchelor, A.W., Some considerations on the mitigation of fretting damage by the application of surface-modification technologies, Journal of Materials Processing Technology, vol. 99, 231-245, 2000.
15. Kim K., Geringer J., Pellier J., Macdonald D.D., Fretting corrosion damage of total hip prosthesis: Friction coefficient and damage rate constant approach, Tribology International, vol. 60, 10–18, 2013.
16. Batchelor A. W., Stachowiak G. W., Stachowiak G. B., Control of fretting friction and wear of roping wire by laser surface alloying and physical vapour deposition coatings, Wear, vol. 152, 127-150, 1992.
17. Geringer J., Forest B., Combrade P., Fretting-corrosion of materials used as orthopaedic implants, Wear, vol. 259, 943–951, 2005.
18. Geringer J., Macdonald D.D.,Modeling fretting-corrosion wear of 316L SS against poly(methyl methacrylate) with the Point Defect Model: Fundamental theory, assessment, and outlook,ElectrochimicaActa, vol. 79, 17– 30, 2012.
19. Garcia D.B., Grandt A.F. Jr.,Fractographic investigation of fretting fatigue cracks in Ti–6Al–4V, Engineering Failure Analysis, vol. 12, 537–548, 2005.
20. Magaziner R.S., Jain V.K., Mall S., Wear characterization of Ti-6Al-4V under fretting – reciprocating sliding conditions, Wear, vol. 264, 1002-1014, 2008.

Analysis of Failures of High Speed Shaft Bearing System in a Wind Turbine

Michał Wasilczuk [1], *Rafał Gawarkiewicz* [2], *Bartosz Bastian* [3]
[1] *Gdansk University of Technology, Faculty of Mechanical Engineering, mwasilcz@pg.gda.pl*
[2] *Gdansk University of Technology, Faculty of Mechanical Engineering, gawar@pg.gda.pl*
[3] *Gdansk University of Technology, Faculty of Mechanical Engineering, barbasti@pg.gda.pl*

Abstract: During the operation of wind turbines with gearbox of traditional configuration, consisting of one planetary stage and two helical stages high failure rate of high speed shaft bearings is observed. Such a high failures frequency is not reflected in the results of standard calculations of bearing durability. Most probably it can be attributed to atypical failure mechanism. The authors studied problems in 1.5 MW wind turbines of one of Polish wind farms. The analysis showed that the problems of high failure rate are commonly met all over the world and that the statistics for the analysed turbines were very similar. After the study of potential failure mechanism and its potential reasons, modification of the existing bearing system was proposed. Various options, with different bearing types were investigated. Different versions were examined for: expected durability increase, extent of necessary gearbox modifications and possibility to solve existing problems in operation.

Keywords: wind turbines; gearbox failures; bearing failures; high speed shaft bearing systems; design modification

1. Introduction

In recent years wind power has been the sector with the fastest growth in comparison to other sectors. Average year growth rate amounted to 20% in Europe and 30% in the USA in 2000-2010 decade. Also in Poland a fast growth of installed power has been observed and is expected (Figure 1) [1].

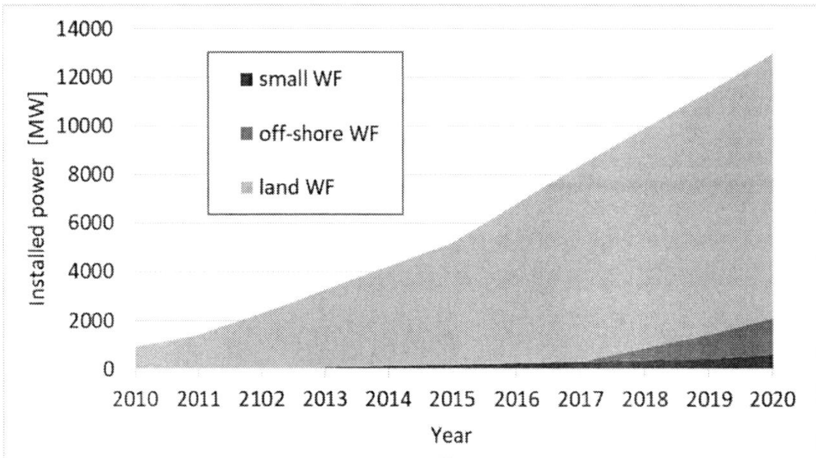

Figure 1. Increase in power installed in Polish wind farms [1]

Despite of modern character of wind farms equipment, wind energy sector faces durability problems with serious failures occurring much earlier than theoretical durability assumptions. Scale of the problem can be illustrated by an example of press release in which Siemens in 2014 informed that its profit has been decreased by 54% because of customers claims dealing with inadequate durability of wind turbine components [2]. Also in Poland the owners and operators of wind farms face serious durability problems so they undertake various steps to find sources of problems and improve the situation.

Most of the wind turbines are of gearbox type with the drivetrain configuration such as presented in Figure 2 [3]. Power from the rotor is transmitted to the generator (15) through the rotor shaft (4) – connected with the multiplicating gearbox (6) of the ratio of approximately 100, by the friction shaft-hub joint and further by the flexible coupling (12) and composite sleeve (not marked on the drawing). A disc brake (7) is situated at the gearbox output (and at the generator input).

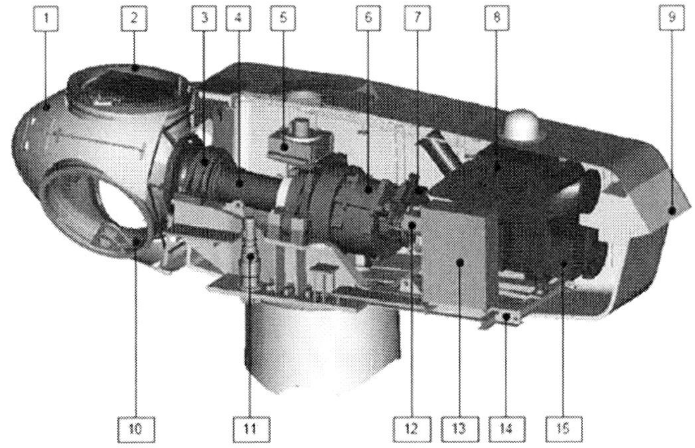

Figure 2. Typical configuration of a wind turbine drivetrain [3]

For the analysis of the durability issues it is important to give a brief introduction into the structure of the gearbox, as well. The most common configuration is shown in Figure 3 [4].

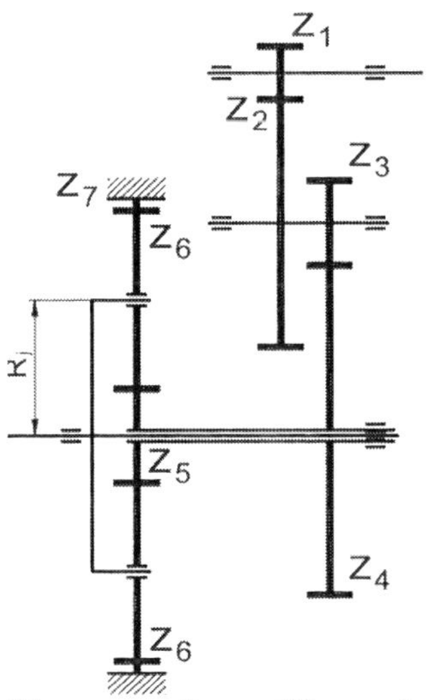

Figure 3. Kinematic diagram of the gearbox [4]

It consists of one planetary stage and two stages of helical gears. The slow speed input shaft is an integral part of the carrier of the planetary gears, the outer ring (z_7) is a stationary part connected to the housing. With the use of planetary gear (z_6) the torque is transmitted to the central wheel (z_5) connected with a slow speed shaft. On the same shaft helical gear z_4 is mounted. From this gear torque is transmitted to gear z_3 of an intermediate shaft (IMS). High speed shaft (HSS) is driven by gear z_2 of IMS shaft meshed with pinion z_1 which is integral part of HHS.

2. Failure statistics and mechanisms

The consequences of wind turbine failures are very expensive, and numerous research institutes are conducting wide range of studies aimed at solving existing issues. One of the findings was that, although the failures of electrical equipment failures are the most frequent, the gearbox failures are responsible for the largest downtime and cost (Figure 4) [5]

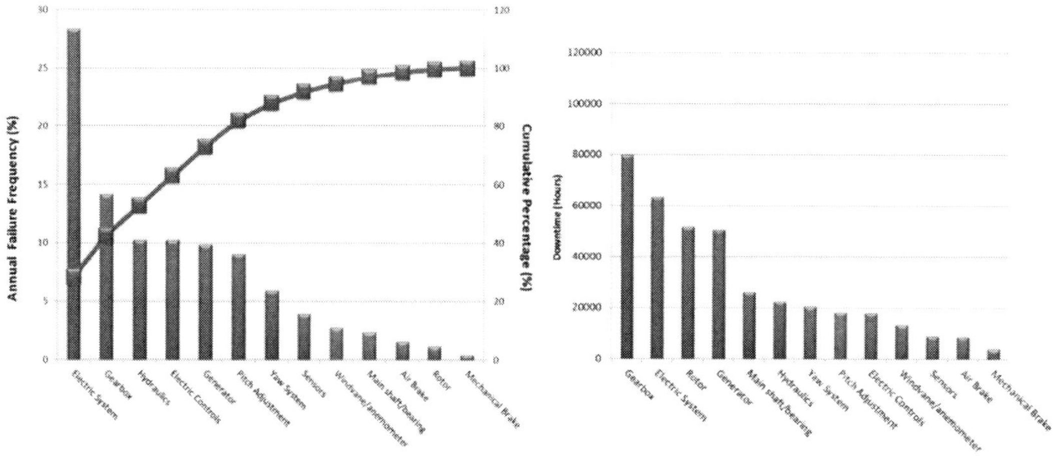

Figure 4. Failure frequency and downtime caused by particular wind turbine component [5]

Out of the gearbox failures these of the high speed shaft (HSS) rolling bearings are most common, which was confirmed by data collected by NREL in 2014 presented at Figure 5 [6].

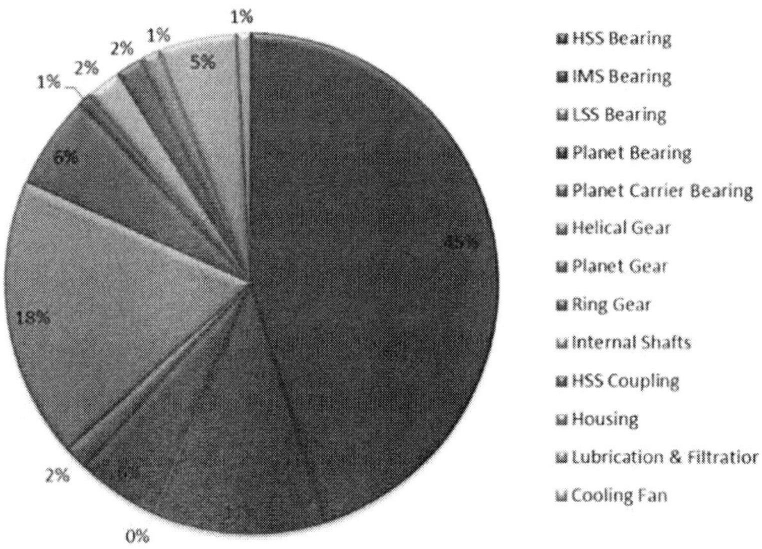

Figure 5. Failure rate of particular elements of wind turbine gearbox [6]

In the analysis of data of occurring failures made by other researchers, it was found that:

- Most of the turbine gearbox issues are generic in nature, and are not related to any specific wind turbine or gearbox manufacturer. Failures can be connected with converged construction solutions. Therefore, failures may be associated with design assumptions rather than with calculation methods used for evaluation of the durability of individual components [7].
- In most of the cases, gearbox failures do not originate as gear failure or gear-tooth design deficiencies, but as bearing failures. The observed failures appear to initiate at specific bearing locations, and later advance into the gear teeth as bearing debris and excess clearances cause surface wear and misalignments. Transmission failure due to gear related quality issues is assessed at up to 10% [7]
- The highest number of gearbox failures originate from bearings failure, even using the best bearing design practices and calculation of rating life, results are not reflected in field data of existing mechanisms [7]. Existing failures show atypical failure mode for the bearings, not taken into account in considered calculation methods [7].

Extremely unsteady operating conditions are characteristic for wind turbines and it is considered to be one of the reasons of frequent failures [8] – Figure 6 shows an example of changes of power generated by a wind turbine with four peaks of over 1000 kW and one drop to 400 kW in just 120 seconds.

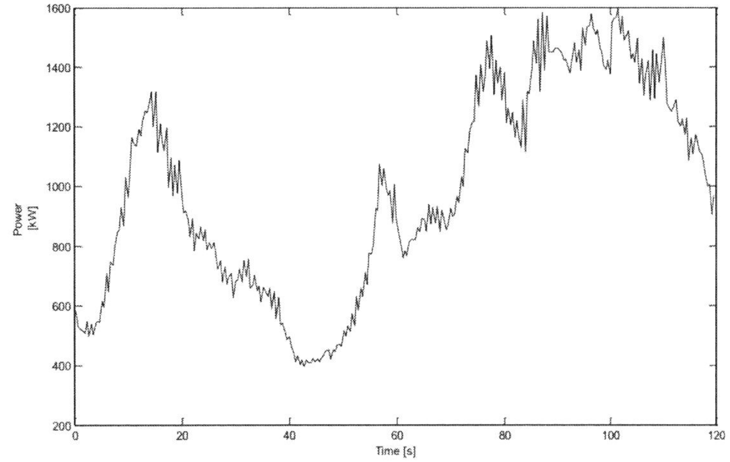

Figure 6. Example of power output changed in a wind turbine in 120 s [8]

Roller element bearings are vulnerable to the conditions of loading and unloading, which causes sliding of the rolling elements [9]. As explained in Figure 7 [9] the rolling elements leaving the loaded zone are disengaged from the inner raceway due to centrifugal forces and their rotational speed is lowered considerably. Then entering the loaded zone again the elements become accelerated, but their inertia opposes such acceleration, so the elements slide with respect to the raceways and excessive heat is generated causing overheating, structural changes and even adhesion in metallic contact areas. This failure mechanism is often referred to as smearing or adhesive wear. Structural changes of bearing material caused by repeated overheating is also one of the failure modes characteristic for wind turbine bearings, known as WEA (white etching areas) [10]. As pointed out by Erichello [10] there are several hypotheses for the root cause of these failures including not only sliding, but also impact loads, hydrogen embrittlement; electrostatic discharge, corrosion fatigue, and adiabatic shear. None of these hypotheses has been proven, and it is a field of current research.

The sliding of the roller elements affects larger bearings and bearing with high rotational speeds, which is the case of HSS bearings in the gearboxes of wind turbines. Avoiding too loose fits and too low loads are the methods of preventing this failure mode. In the bearing design procedures, specifying minimum bearing load is the only means of preventing sliding of the rolling elements [11]. In the machines with fluctuating loads fulfilling these specifications may be impossible.

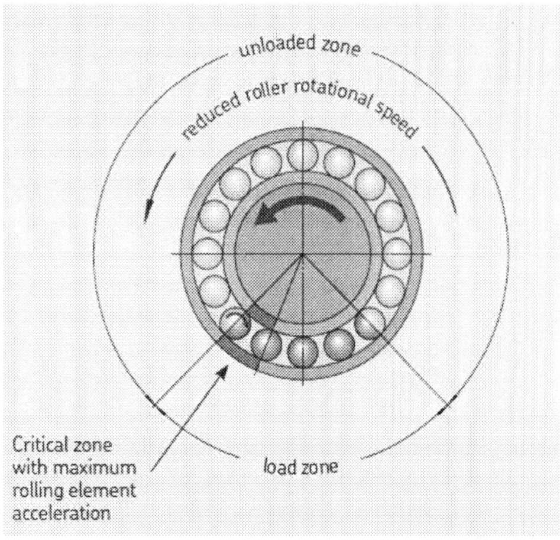

Figure 7. Mechanism of bearing failure caused by sliding of rolling elements [9]

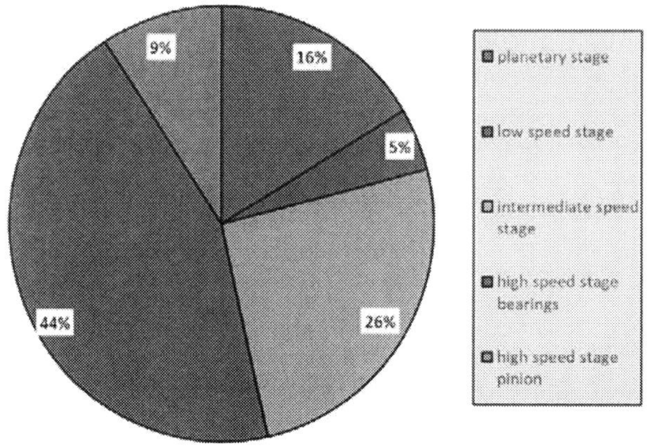

Figure 8. Percentage of failure of particular turbine elements in the analysed Polish wind farm [4]

The other way in which unsteady operating conditions affect failures of the machines is increased level of machine vibration. Changes in condition monitoring signals due to changes in load can be larger than changes resulting from developing failures, which makes it more difficult to detect them at early stages than in machines in steady operating conditions. The delay in detecting failures increases their extent, downtime and cost.

An analysis of the reasons of failures in 1.5 MW wind turbines installed in one of Polish wind farms presented in Figure 8 showed that the percentage of particular reasons was similar to that observed worldwide, with 44% of the failures occurring in the high speed shaft, mostly in the bearings [4].

3. Current design

Figure 9 shows the high speed shaft with its bearing arrangement of the studied wind turbine. The shaft is located on three bearings: first, on the left hand side there is a cylindrical roller bearing (designated as U – Upwind), providing radial support, and on the right hand side there is a system of two bearings: a second cylindrical roller bearing (D- Downwind) and four-point contact bearing, playing a role of locating bearing. In operation it was found that the downwind cylindrical roller bearing has the highest failure rate. Larger number of failures in comparison to upwind bearing is not consistent with higher theoretical rating life L_{10} shown in Table 1.

Table 1. Theoretical durability and recorded number of failures of the HSS bearings

	Durability L_{10} [years]	Number of failures
Upwind bearing	2,4	3
Downwind bearing	9,4	16

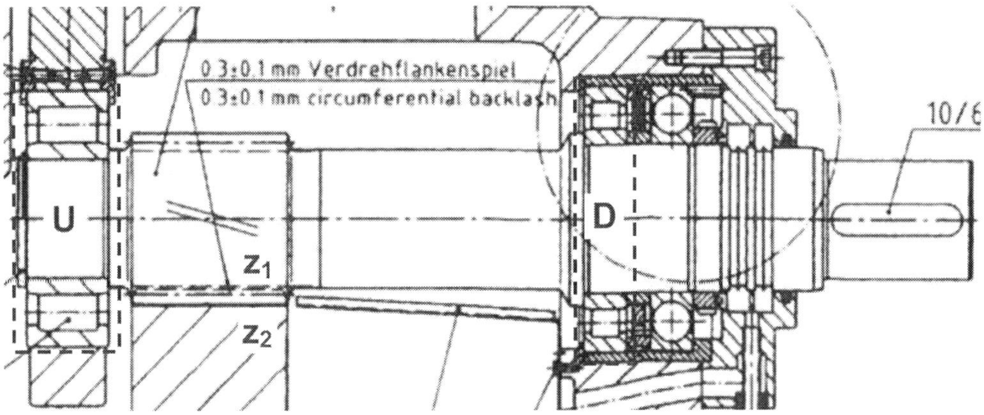

Figure 9. Existing bearing system of the high speed shaft in the analysed wind turbine [4]

Inconsistency may be caused by the failure mechanism not considered by the standard bearing durability calculations, i.e. by smearing (adhesive wear) caused by sliding of the rolling elements. Sliding of the rolling elements is the reason of overheating of the bearing rings and structural changes of the bearing material. Plastic deformation of the material (Figure 10), being possible only at elevated temperature was observed in the microscopic image of the destroyed bearing ring [4]. This may indicate smearing caused by sliding of the rolling elements as a possible failure mechanism in the analysed case.

In the literature several remedies are suggested [12], including DLC coated bearing rollers, cleaner steels and lower roughness of bearing rings. It is also pointed out that pre-loaded taper roller bearings exhibit less roller sliding than cylindrical roller bearings.

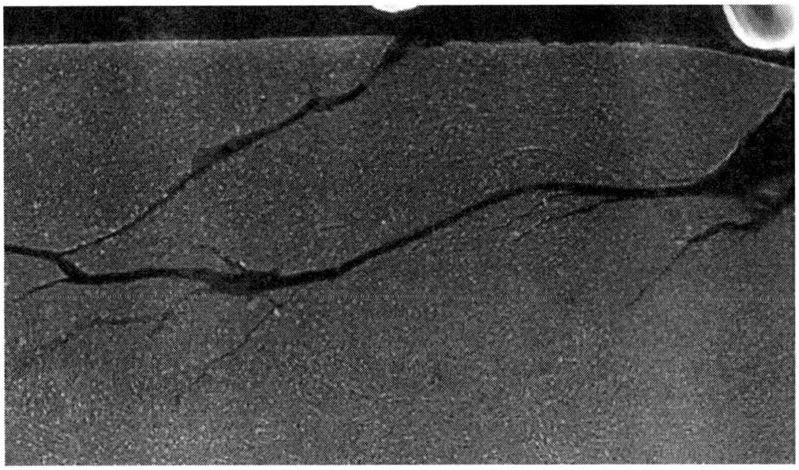

Figure 10. Plastic deformations under the surface of destroyed bearing ring [4]

4. Proposed modifications

Because of large number of failures and evidence that smearing is a probable reason of failures a redesign of the bearing system of the high speed shaft was considered. Quite an obvious assumption of the planned retrofit was minimizing necessary modifications of the gearbox design. Especially changes in the gearbox housing, shaft with pinion should be limited. Design changes should comprise small bearing mounting elements, such as bearing bushing, gaskets, housing cover. Various possible bearing configurations were considered utilizing bearings of the outer and inner rings diameters unchanged, i. e. shaft diameter of 150 mm and housing diameter of 270 mm. Three of the proposed versions with satisfactory life rating values are shown in Figure 11 [13]

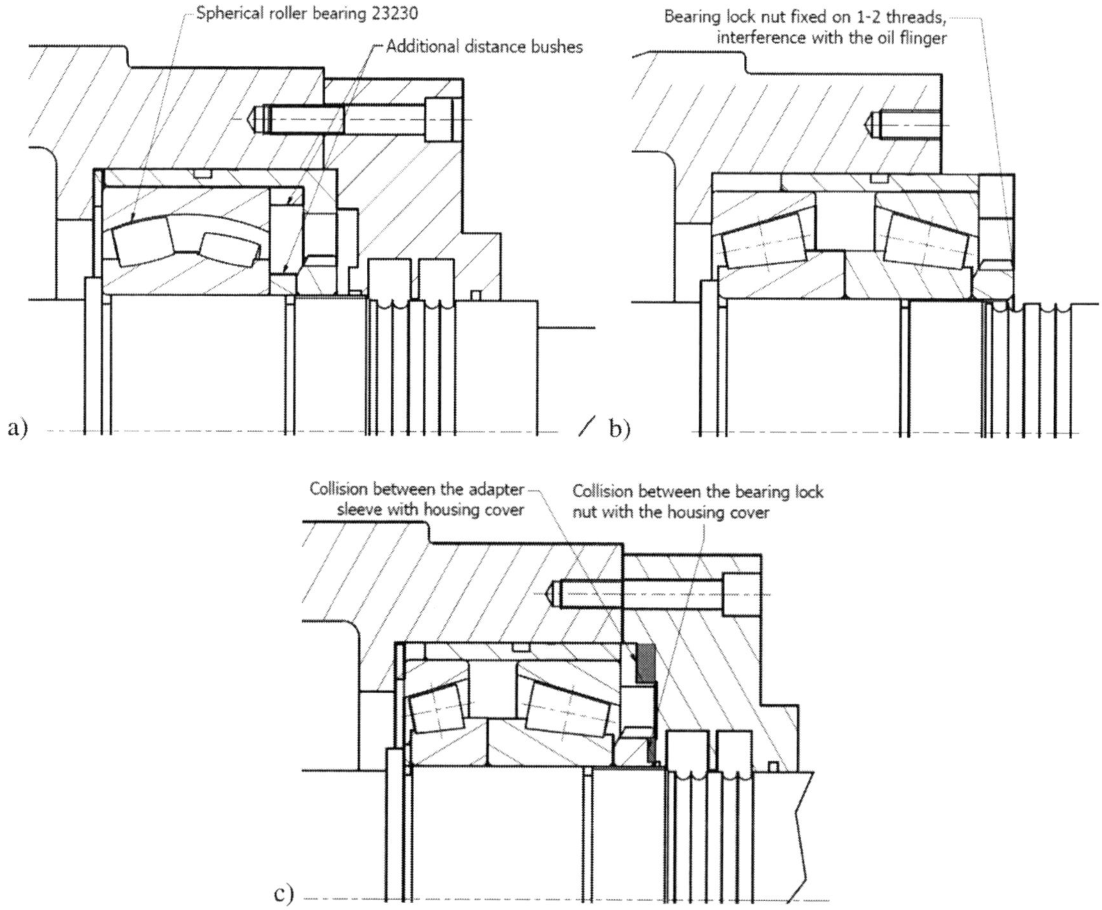

Figure 11. Proposed bearing arrangements: a) Spherical roller bearing 23230 – version A, b) Set of two identical taper roller bearings 32230-A in X arrangement – version B, c) Set of two different taper roller bearings 32230-A, 30230-A in X arrangement – version C [13]

As it can be seen in the figure, only one design (Case A) with spherical roller bearing 23230 (Figure 11 a) fits into the housing without any modification concerning the housing and the cover, this design only requires using an additional distance sleeve, but its use does not provide conditions for limited sliding of the rollers, because in two row spherical roller bearing loaded with substantial axial load and small radial load one row of rollers is unloaded, so despite good theoretical rating life this version was discarded.

Version B - a pair of 32230-A bearings shown in Figure 11 b can not be used because of excessive size in axial direction. Bearing lock nut is improperly mounted - most probably it would be fixed only on 1-2 threads and the nut would interfere with the oil flinger. Furthermore, bearing from the side of the lock nut, is partially seated on the threaded area itself, which will not provide proper fit. At the figure, the housing cover is not presented, due to other existing conflicts.

In case C incorporating a set of two tapered roller bearings of different size, shown in Figure 11 c, there is an expected collision between the bearing lock nut and the adapter sleeve with the housing cover. However, modification of the shape of the housing cover can be done, so the mounting will be possible. Other possibility to accomplish such mounting, is to design special bearing lock nut and an adapter sleeve, to avoid interference with the cover. Bearings in X arrangement are shown in the figure. O-type arrangement is not presented, because an increased axial length of such a bearing arrangement would make the use of a lock nut impossible, and would not ensure proper operation of the flinger seal arrangement.

5. Conclusions

The gearboxes of the analysed turbines demonstrated similar problems to those reported worldwide – high failure rate, especially of the roller element bearings of the high speed shaft, and inconsistency of the rated life predictions with the operational data. The results of the study on wear patterns, including microscopic analysis of the material structure of the broken bearings suggested that smearing and adhesion due to sliding of the rolling elements may be one of the important reasons of premature failures. In such case, according to the literature, the use preloaded sets of taper roller bearings is one of the solutions. Out of presented concepts, preferred version is the one of bearing pair consisting of two bearing of different widths. It can be characterized by satisfactory durability, high probability of elimination of the existing sliding problem. Necessary modifications of the gearbox are relatively small, and they do not require machining of the housing. The new parts can be prepared earlier and the turbine retrofit can be carried out during the nearest turbine maintenance stop without dismantling the whole gearbox from the turbine nacelle. Now the retrofit project is on its way and will be implemented in one of the gearboxes in which a major repair will be carried out.

References

1. Wiśniewski G. et al.: Vision of development of wind energy in Poland to 2020 (in Polish). Institute of Renewable Energy, Warsaw 2009 – report downloaded from http://www.ieo.pl/pl/raporty/doc_details/333-qwizja-rozwoju-energetyki-wiatrowej-w-polsce-do-2020-r.html
2. http://www.offshorewind.biz/2014/05/07/siemens-energy-division-profit-down-54-pct/
3. *http://www.energa-wytwarzanie.pl*
4. Wasilczuk M., Gawarkiewicz R., Libera M., Wasilczuk F., Kinal G.: Bearing systems of wind turbines - maintenance problems (in Polish). *Tribologia 4/2015(4-2015):187-198 · 10/ 2015*
5. Sheng S. (2013) Report on Wind Turbine Subsystem Reliability — A Survey of Various Databases, presentation from: http://www.smartgridinformation.info/pdf/5527_doc_1.pdf
6. Sheng S. (2014) Gearbox Reliability Database: Yesterday, Today, and Tomorrow, materiały z konferencji Wind Turbine Tribology Seminar 2014, dostępne pod adresem: http://www.nrel.gov/docs/fy15osti/63106.pdf
7. W. Musial, S. Butterfield: Improving Wind Turbine Gearbox Reliability, 2007 European Wind Energy Conference, Milan, Italy, May 7–10, 2007
8. T. Barszcz, R. Gawarkiewicz, A. Jabłoński, M. Sękal, M. Wasilczuk: Knocking sounds in the wind turbine gearbox during slowing down – case study. Paper presented at international conference ICTD/CMMNO Gliwice September 2016.
9. Bearing-Damage-and-Failure-Analysis, SKF brochure downloaded from: https://pl.scribd.com/doc/299743401/14219-en-Bearing-Damage-and-Failure-Analysis
10. Errichello R., Budny R., Eckert R.: Investigations of Bearing Failures Associated with White Etching Areas (WEAs) in Wind Turbine Gearboxes Power Transmission Engineering 8,2 (2014), pp. 38-44
11. SKF General Catalogue
12. Doll G. L., Tribological challenges in wind turbine technology, Wind Turbine Tribology Seminar, Broomfield, Colorado, USA, November 15-17, 2011
13. Bastian B., Gawarkiewicz R., Wasilczuk M.: Analysis of the possibilities of high speed shaft bearing system durability increase. Tribologia 3/2016 (3-2016), pp. 37-47

Stress Analysis and Fatigue Behaviour of PTFE-Bronze Layered Journal Bearing under Real-Time Dynamic Loading

***Duman M S*[1]*, Kaplan E*[2]*, Cuvalcı O*[3]**
[1,3]*KTU, Department of Mechanical Engineering, 61080, Trabzon, Turkey*
[2]*Gümüşhane University, Department of Mechanical Engineering, 29100, Gümüşhane, Turkey*
duman@ktu.edu.tr[1], *erdarkaplan@gumushane.edu.tr*[2], *ocuvalci@ktu.edu.tr*[3]

Abstract: The present paper is based on experimental studies and numerical simulations on the surface fatigue failure of the PTFE-bronze layered journal bearings under real-time loading. 'Permaglide Plain Bearings P10' type journal bearings were experimentally tested under different real time dynamic loadings by using real time journal bearing test system in our laboratory. The journal bearing consists of a PTFE-bronze layer approximately 0.32 mm thick on the steel support layer with 2.18 mm thick. Two different approaches have been considered with in experiments: (i) under real- time constant loading with varying bearing widths, (ii) under different real-time loadings at constant bearing widths. Fatigue regions, micro-crack dispersion and stress distributions occurred at the journal bearing were experimentally and theoretically investigated. The relation between fatigue region and pressure distributions were investigated by determining the circumferential pressure distribution under real-time dynamic loadings for the position of every 10° crank angles. In the theoretical part; stress and deformation distributions at the surface of the journal bearing analysed by using finite element methods to determine the relationship between stress and fatigue behaviour. As a result of this study, the maximum oil pressure and fatigue cracks were observed in the most heavily loaded regions of the bearing surface. Experimental results show that PTFE-Bronze layered journal bearings fatigue behaviour is better than the bearings include white metal alloy.

Keywords: Journal bearing, plastic bearing, surface fatigue, hydrodynamic load, stress distribution

1. Introduction

Typical journal bearings are widely used as the main bearings, connecting rod bearings of crank shaft and cam-shaft bearings undertake many critical tasks in internal combustion engines. Journal bearings are subjected to increased dynamic loads throughout their service life which caused fatigue failure. Battelle and at all studies in 1982 showed that 80% to 90% of all structural damages occur on the bearings because of fatigue mechanism. In this case, the design and use of journal bearings operating under dynamic loading condition, shows that the fatigue behavior is an important factor to be investigated. Fatigue performance of the journal bearing is adversely affected by complex factors, i.e. materials, structures, loading conditions and so on. Traditionally, multi-layered journal bearings with steel backing give the required combination of properties for the bearing [1, 2]. One of the representative sliding materials is Polytetrafluoroethylene (PTFE)-based composites has been widely used as self-lubricating liner of the journal bearings, possess the properties of PTFE; i.e. lower friction coefficient, excellent corrosion resistance, and significant reduction in maintenance. However, PTFE composites have the disadvantages of low mechanical strength, poor thermal conductivity, high wear rate and lower stiffness [3-4]. In order to overcome disadvantages, filler materials like fiberglass; bronze fillers, molybdenum disulphide, graphite and carbon are used [5]. A number of studies have been carried out on the tribological effects of PTFE-based composites for journal bearing applications. PTFE-bronze composite material was experimentally tested as journal bearing under dry conditions by Tevrüz [6]. These tests showed that the coefficient of friction decreased with increasing contact pressure while the friction coefficient increased with increasing temperature. Iwai and et al [7] show that PTFE based composites journal bearing friction is better than white metal (babbit) under lubricated conditions. Researchers have been done many studies to improve performance and properties of PTFE based composite bearings. However, there is limited number of studies to investigate fatigue behaviour of PTFE based composite liner. Lu et al [8] theoretically analysed different fatigue damage mechanisms such as surface fatigue wear, impact effect, and sweep deformation of the spherical plain bearing including PTFE liner. They also experimentally validated their theoretical results under static loading. Fatigue behavior and tribological test of a journal bearing under real-time dynamic loading conditions can be experimentally performed on journal bearing test rig in our university. The main aim of this study is that, evaluating stress analysis and fatigue behaviour of PTFE-bronze layered of journal bearing under real-time dynamic loading conditions to obtain more information for surface fatigue failure, oil film pressure and stress distribution on the journal bearings by compering experimental and simulation results.

2. Experimental Details

2.1. Test bearing and experimental conditions

Commercial 'Permaglide Plain Bearings P10' type journal bearing was used as the test bearing with the constant diameter of 47.775 mm and with varying widths of 10, 12 and 14 mm. The journal bearing consists of a PTFE-bronze layer approximately 0.32 mm thick on the steel support substrate with 2.18 mm thick. PTFE-Bronze layered journal bearing consist of four layers as seen figure 1 [9]: (i) the outer surface and the edges of steel substrate are overlaid with tin a thickness of 2 μm to protect from corrosion; (ii) steel substrate used for giving form and supporting bearing materials; (iii) porous bronze layer is sintered on the steel substrate; (iv) PTFE-lead compound layered on top of the sintered porous bronze layer with approximately thickness of 0.01-0.03 mm. During this process, PTFE–lead compound also filled the gaps inside the sintered porous bronze layer. After all above manufacturing process, the roughness of the PTFE surface measured approximatelyRa = 2μm. The bearing parameters and lubricating properties used during the experimental study are given in table 1. Rotating speed of the journal was chosen as 1240 rpm. Oil supply holes of the bearing were manufactured to become 45° to the natural horizontal axis. Commercial engine mineral oil 'Shell Tellus 27' was used as lubricant for journal bearing and the output temperature of the lubricant which coming from test bearing was constantly maintained at 50 °C by using cooler system.

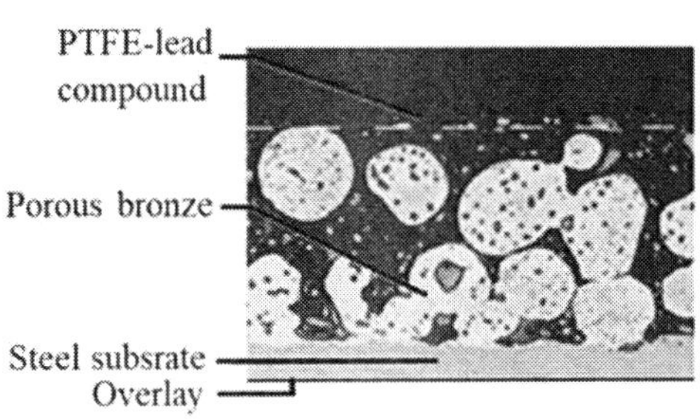

Figure 1. Micro-section of the test bearing.

Table 1. Bearing parameters and lubricant properties.

Bearing parameters	Value	Units
Bearing lengths (L)	10, 12, 14	mm
Inner diameter of bearing (D)	47.775	mm
Normal bearing clearance (c)	0.0375	mm
Shaft diameter (d)	47.700	mm
Rotational Speed	1240	rpm
Viscosity of motor oil (at 50 °C)	20.03	cp

2.2. Test rig and loading of the test bearing

The experimental studies were performed on real-time loading journal bearing test rig having the different pattern of dynamic loadings. The test rig and its hydraulic system for the loading were described fully in great detail elsewhere [10-11]. The current bearing test rig allows the bearing to be tested at different loads, rotating speeds and lubricating conditions. Furthermore, it is possible to determine wear life, fatigue behaviour, and journal centre orbit of the bearings under real-time dynamic loads. The dynamic loads can be applied by four hydraulic cylinders likewise in motor bearings. The journal bearing test rig is schematically shown in figure 2. During the bearing loading time, dynamic pressures inside the four loading cylinders were measured by using four separate P3M/100 type pressure sensors mounted on the loading cylinder an integrated with the full bridge measuring system. The nominal pressure measurement range of each sensor is 10 MPa. The signals taken from sensors from the two loading cylinders which are on the same axis but opposite direction were combined by an electrical connection to give a pressure difference between the two opposite loading cylinders. These signals for four sensors were transferred to the amplifier (KWS/T-5 type) for each axis. The pressure difference between the loading cylinders was obtained with each amplifier outputs by means of taking to separate socket of the oscilloscope. The variation of the polar load distribution was taken from the oscilloscope screen via combining two channels of oscilloscope. The hydraulic loading unıt generates a maximum 5 MPa working pressure at the loading cylinders. The sensors were calibrated at these pressure ranges with static pressures being gradually changed. The obtained calibration curve of Sensor has ability to continuously read in pressure shocks and sudden pressure increases inside the journal bearing. By using this ability of the test jig, experimental pressure distribution inside the journal bearing can able to obtain under different loading, rotating speed and lubricating conditions.

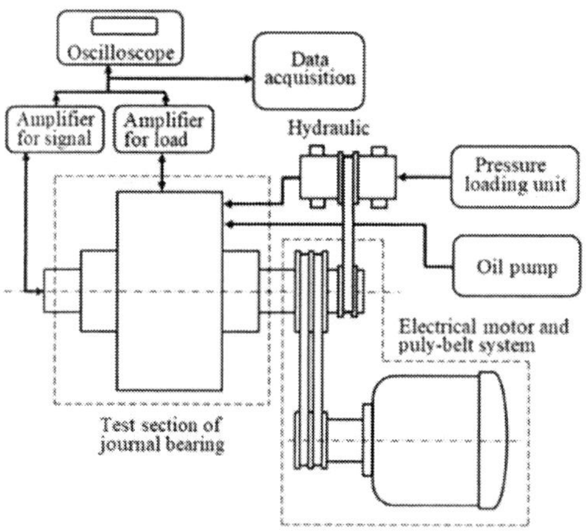

Figure 2. Journal bearing test rig.

The pressure sensor placed on the journal was read pressure distribution along the circumference of the bearing at any position during the test. The position change mechanism of the sensor was developed by Bıyıklıoğlu [10-11]. The oil film pressure values taken from sensor were recorded on the computer by every 2 degrees of the shaft during each load cycle via this mechanism. Thus, the pressure distributions around the bearing were obtained for each dynamic load. The sensor was calibrated by means of hand pump which provides constant pressure before the experimental study. The signals were taken from sensor transferred to the amplifier and after pressure distributions were determined via connecting between the amplifier and oscilloscope.

3. Finite element analysis

Finite element analyses were carried out to investigate the relationship between stresses and fatigue behaviour of the bearing. To this end, stresses and deformations of the bearing were calculated for different pressure distribution by using experimental pressures. 2D finite element model of the journal bearing [12] was designed in ANSYS software package and used for stress and deformation calculation. The half bearing is modelled by using total number of 7800 S4R quadratic elements. For calculation, it is assumed that pressures do not change in the axial direction of the journal bearing and there is no displacement along the bearing length, and bearing width. The sliding layer is considered as composite material because of the thickness of the PTFE running-in layer is very thin PTFE-lead and bronze mixed layer. The elastic modulus and poison ratio of the composite are recalculated. For the calculation, elastic modulus and poison ratio of the composite are taken as $E_c = 46$ GPa and $\nu_c = 0.373$ respectively. Also, $E_s = 210$ GPa and $\nu_s = 0.3$ are taken for steel substrate. The oil inlet holes are assumed being outside the loading zone of the half bearing.

4. Results and discussion

4.1. Experimental results

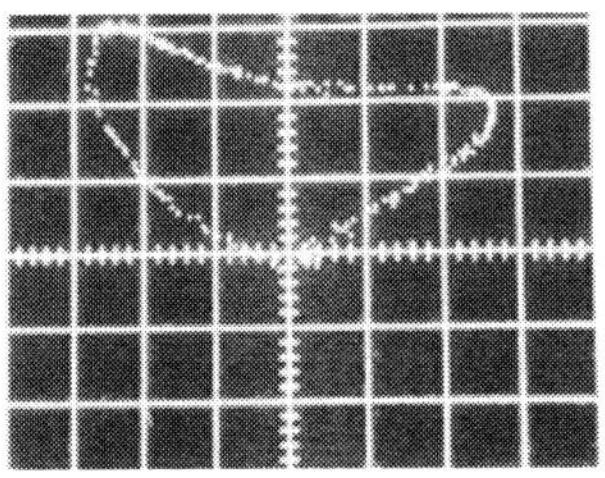

Figure 3. Polar load signal taken from oscilloscope screen during Test-I.

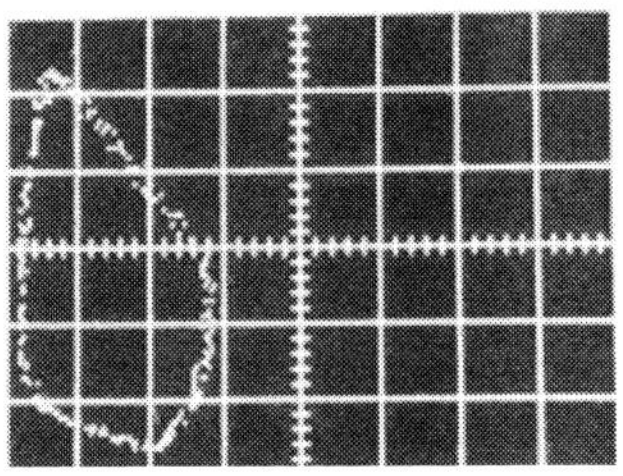

Figure 4. Polar load signal taken from oscilloscope screen during Test-II.

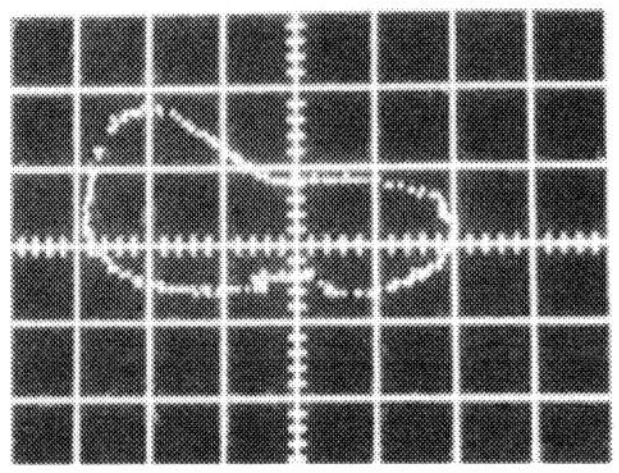

Figure 5. Polar load signal taken from oscilloscope screen during Test-III.

Journal bearing was tested under three different loads applications to the journal bearing during experimental studies. During the all load application are presented real time on polar coordinate system on the oscilloscope screen. The variation of polar loads applications are given in figure 3, figure 4 and figure 5. The oil film pressure changes at each crank angle in the shaft orbit were calculated using the solution method [13] for each experiment performed by considering the dynamic loads and bearing sizes applied to the journal bearing. The Reynolds boundary conditions were used for orbit of the shaft centre calculation for each experiment. Furthermore, the experimental results were also compared by the Ricardo calculation method [14].
Two different main variables have been considered with in experimental studies: (i) under constant loading with varying bearing widths, (ii) under different loadings at constant bearing widths. Fatigue regions, micro-crack dispersion, and stress distributions occurred at the journal bearing were investigated for each experiment. Experimental study is investigated under three different tests. Test-I: Test bearing with width of 10, 12, and 14 mm are investigated under constant loading as seen

figure 3. Test-II: Test bearing with the width of 10 mm were analysed under loading characteristics as shown in figure 4. Test-III: Test bearing with the width of 10 mm were again analysed under loading characteristics as shown in figure 5. Figure 3 illustrated that vertical component of the applied load to the journal bearing looks like in the form of fluctuating change but horizontal component of the applied load to the journal bearing looks like follows in the form of sinusoidal change for Test-I. As seen from figure 4, component of the applied load acts just like sinusoidal motion along the horizontal axis and as a general variable load along the horizontal axis during Test-II. During Test-III, it is seen that the load components of both horizontal and vertical axes are sinusoidal and cosine shaped but act as a cycled polar load with a larger width along one axis as seen from figure 5. Variations of the horizontal and vertical loads as a function of crank angle are plotted for 10 mm width bearing for loading of Test-I type. Variations of total, vertical and horizontal real-time dynamic load changes are illustrated with respect to crank angle at figure 6. Maximum dynamic load effecting into the bearing occurs between the crank angles of 238° and 267°. But the maximum load effects into the bearing with value of approximately 285558 N at the crank angle of 255° for type of load applied for Test-I. The variation of loads as a function of crank angle during Test-II and Test III are given in the reference [15]. In the case of Test-I type of dynamic loading applied the bearing having three different widths to investigate what is the relation between width of the bearing and fatigue life and fatigue regions of the bearing. The polar load diagram, the journal orbit for journal of the main bearing and the fatigue regions for 10 mm bearing are only plotted graphically in figure 7. The other experimental results such as Test-I with 12 and 14 mm width bearing and Test-II and Test-III with different dynamic loading for 10 mm width bearing are presented in table 2, and table 3. The figure 7 includes polar load form applied to journal bearing and two different theoretical calculation of orbit of the shaft by using finite-volume and mobility methods [13, 14]. As seen from the figure 7, especially area of oil pressure start increases inside the oil film there is almost no difference for orbit calculation between the two methods. The area of surface fatigue on the bearing and shape of the fatigue cracks is shown in figure 8. In order to see the fatigue cracks more clearly, a penetrating fluid method was applied then pictures of the cracked areas were took. It is seen that the micro cracks start at the middle part of the bearing and then branch to the axial and circumferential directions and reach the edges of the bearing. The fatigue region of the test bearing was determined between circumferential angles of 105° and 165°. Considering the shaft orbit, the region of the fatigue crack was seen in a zone about 15° before the point where the minimum film thickness occurs depending on dynamic load forms. İnitial fatigue cracks on the test bearing approximately started at 74400 cycles. Furthermore, a study of comparison with the lubricating analysis of the bearing was carried out at each 10° crank angle.

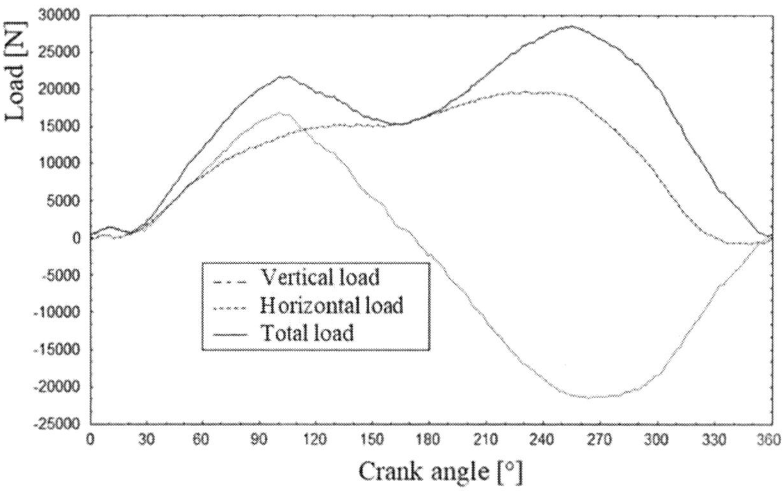

Figure 6. Variation of horizontal, vertical and total load as a function of crank angle.

The pressure distribution of the bearing at different angles was experimentally measured under dynamic loading conditions. Pressure distributions on the 10 mm width journal bearing for test-I are also given in figure 9. As seen from figure 9, pressure distribution on the journal bearing at the different crank angles shows same type of pressure distribution but different pressure values. However, the highest oil film pressure on the journal bearing appears at 273° crank angle with value of 1230 MPa and the force acting on this bearing at the same angle is approximately 26000 N. The highest oil film pressure obtained from analysis is within the fatigue region, which corresponds to the circumferential angles approximately between of 105° and 165°.

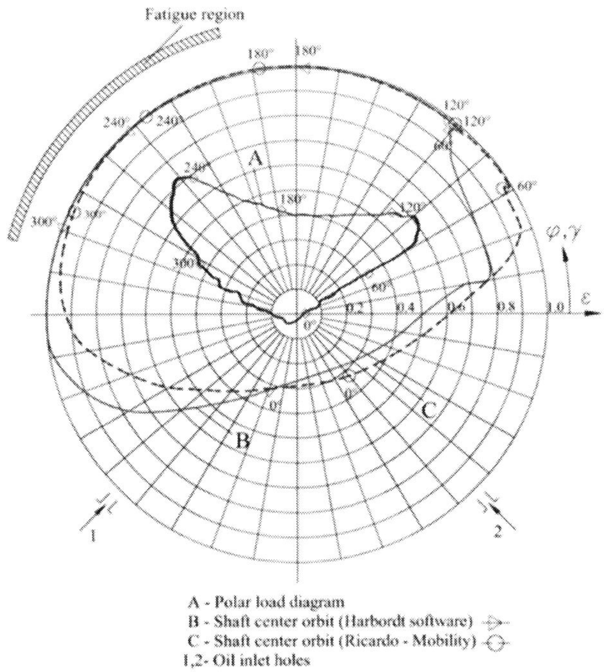

Figure 7. Journal orbits obtained from test bearing with 10 mm during test-I.

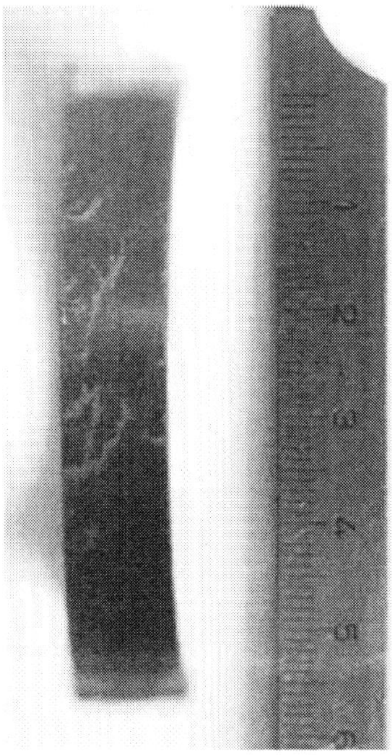

Figure 8. Surface fatigue of 10mm width test bearing.

4.2. Simulation results

Pressure distributions obtained from experiments (figure 9) used to calculate stresses occurring in the journal bearing by using finite element methods. Therefore, tangential, radial, and shear stresses distribution along the circumferential angle of the tested journal bearing surface calculated by using experimentally measured pressure distribution values and those stress distribution presented in figure 10 and figure 11, respectively. Tangential and radial stress distribution causes by radial pressure penetrating to journal bearing through the oil film, while the tangential stress includes compressive and tensile stresses components, the radial stress only includes compressive stress component. Of course, the tensile stress is the main stress causing initial micro-crack on the journal bearing and oil pressure helps to crack for penetrating everywhere in the fatigue region. After all, the maximum tensile stress will definitely appear at the point of maximum pressure appeared (273° crank angle) on the journal bearing that is calculated 51 MPa($\sigma_x = 0.041P_{max}$). Calculated tangential stresses in figure 10 caused first micro-crack and penetrated all around the failure region by helping radial pressure which clearly showed in figure 8. When the figure 8 carefully analysed, it can be seen that the first micro-crack appears and propagate all over the failure region from 105° to 120° and 145° to 165° circumferential angles. Can be seen from the figure 9, the highest pressure distribution values experimentally measured on the journal bearing are the same regions with the fatigue failure region.

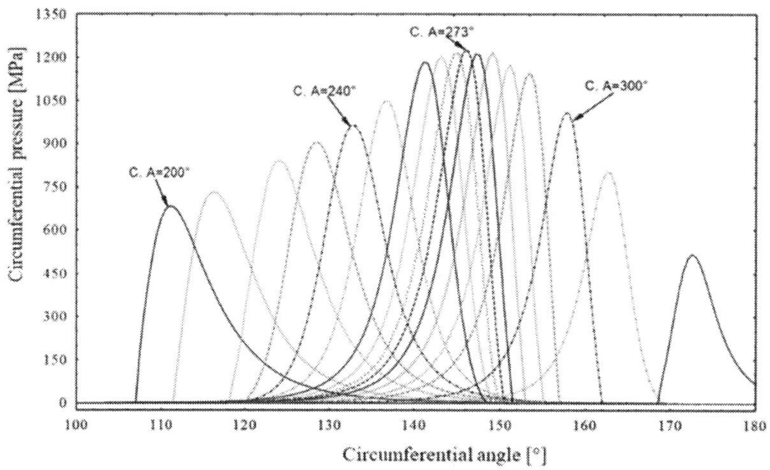

Figure 9. Pressure distribution of the test bearing along circumferential angle during Test-I.

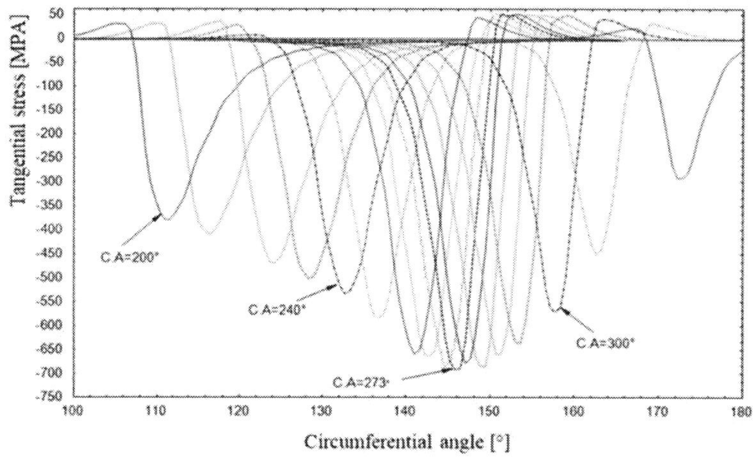

Figure 10. Tangential stress calculated in the fatigue regions at the bearing surface during test-I

Furthermore, the stresses and deformations occurring along the bearing thickness were calculated on the bearing modelled with two-dimensional finite elements due to investigate of relationship between stress and fatigue behaviour. The equivalent stress distributions calculated for journal bearing under the considered oil film pressure by using the Von Mises hypothesis shown in figure 12. As seen from the figure, the highest equivalent stress calculated as 971 MPa ($\sigma_{eq} = 0.79P_{max}$) 1 mm below the surface $h/s = 0.4$; h: depth from surface of bearing, s: depth of bearing) which is corresponding to the same point that peak pressure of the oil film appeared. It is important to know that the maximum equivalent stress applied point (1 mm blow the surface) will be the initial micro-crack starting point. Equivalent stress values is reduces above and below the maximum equivalent stress point such as 701 MPa ($\sigma_{eq} = 0.57P_{max}$) calculated just above the maximum equivalent stress value at the bottom of PTFE-Bronze layer and 486 MPa ($\sigma_{eq} = 0.4P_{max}$) calculated on the steel substrate. According to the figure 13, increasing pressure of the journal bearing creates deformation on the PTFE surface and maximum radial deformation (collapse) calculated 14.4 microns at the same point with maximum pressure appeared on the journal bearing. Can be call that the collapse value at that point is high value by considering the radial clearance and the oil film thickness formed at that point.

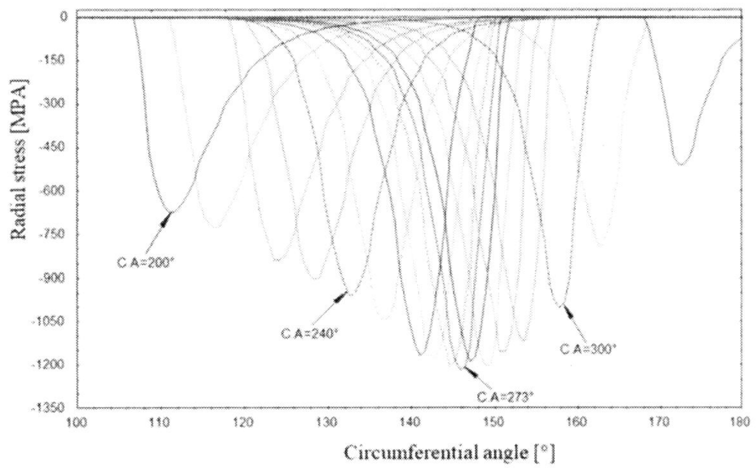

Figure 11. Radial stress calculated in the fatigue regions at the bearing surface during test-I

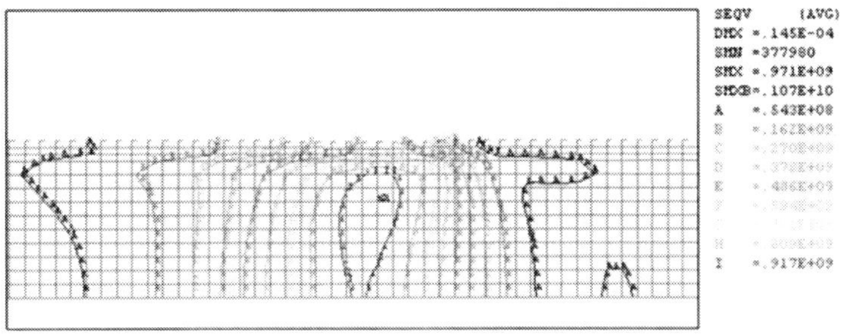

Figure 12. Von Mises equivalent stress distribution under the effect of pressure distribution at crank angle of 273°.

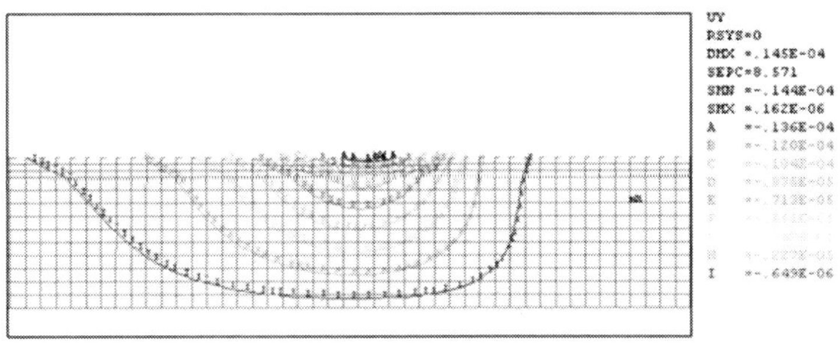

Figure 13. Von Mises equivalent stress distribution under the effect of pressure distribution at crank angle of 273°.

The experimental results of the test bearings according to bearing widths and type of test are illustrated in table 2, and table 3, respectively. According to table 2, the fatigue life increased with the increase in the loading area of bearing under the constant dynamic load; when the bearing area was increased by approximately 17%, the fatigue life increased by 2 times, when the bearing area was increased by approximately 40%, the fatigue life increased by 8 times until time to initial micro crack is observed at bearing surface. Fatigue cracks were observed in the most heavily loaded regions of the bearing surface at 273°-282° crank angles and also it appeared at between 105° and 165° circumferential angles. As seen from figure 14, PTFE-Bronze layered journal bearing showed better fatigue life than bearings include white metal alloy under real-time test conditions. B 0.25, B

0.34, and B 0, 42 are the test results of bearings include white metal. PTFE (0.2, 0.25, and 0.3) are the test results of the new test bearings.

Table 2. Experimental results of the test bearing with varying bearing widths under constant loading.

Bearing Widths (mm)	Load cycle at initial fatigue crack cycles (-)	Maximum Oil film Pressure (MPa) / crank angle (°)	Circumferential angle gaps at initial fatigue crack (°)
10	$7.4x10^4$	1230 / 273	105-165
12	$14.8x10^4$	866 / 282	130-160
14	$59.5x10^4$	656 / 277	125-165

Table 3. Experimental results of the test bearing under different loading with constant bearing width.

Test ID	Load cycle at initial fatigue crack cycles (-)	Maximum Oil film Pressure (MPa) / crank angle (°)	Circumferential angle gaps at initial fatigue crack (°)
Test-I	$7.4x10^4$	1230 / 273	105-165
Test-II	$12.6x10^5$	1154 / 252	140-220
Test-III	$20.8x10^5$	1014 / 212	185-220

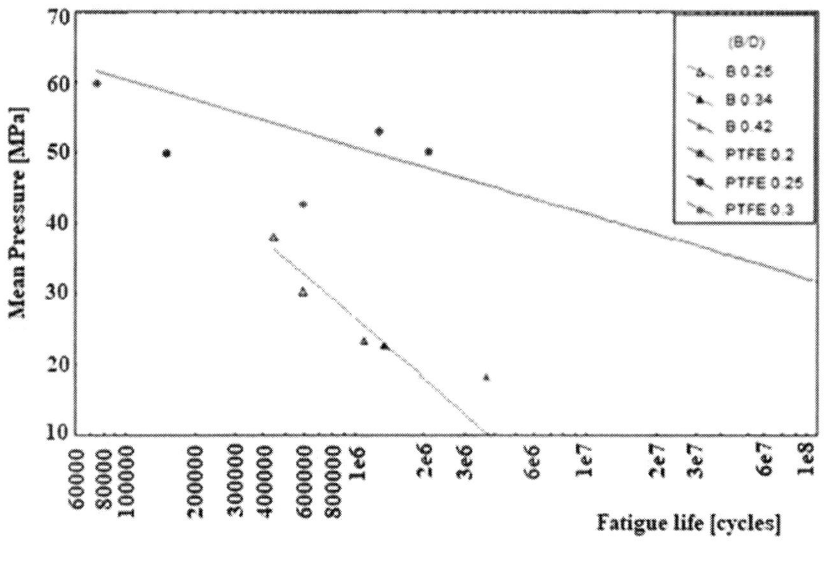

Figure 14. Comparison of mean pressure values and fatigue life (cycles) of the PTFE-Bronze layered journal bearings with the bearings include white-metal alloy.

Conclusion

PTFE-Bronze layered journal bearings were experimentally tested by according to bearing widths and type of test under real-time dynamic loading conditions. Fatigue regions, micro-crack dispersion and stress distributions occurred at the journal bearing were experimentally and theoretically investigated. It was observed that fatigue cracks occur in the main loading region in all experimental studies. The region of the fatigue crack was appeared in a zone about 10°-25° before the point where the minimum film thickness occurs depending on dynamic load forms. The fatigue life increased with the increase in the loading area of bearing under the constant dynamic loads and same operating conditions, i.e. when the bearing area was increased by 40%, the fatigue life increased by 8 times as seen Table 2. The fatigue life of the bearing increased 28 times with a 20% reduction in average pressure considering the average pressure determined according to the maximum bearing force (at constant journal width under different dynamic loads). The maximum value of the calculated oil film pressures was reduced by increasing the bearing area, i.e. when the bearing area was increased by 40 %, the maximum oil film pressure decreased by approximately 47%. The results of the finite element analysis are in good agreement with the experimental study. Results show that PTFE-Bronze layered journal bearings fatigue behaviour is better than the bearings include white metal alloy.

References

1. Bukovnik S., Dörr N., Caika V., Bartz W.J., Loibnegger B., Analysis of diverse simulation models for combustion engine journal bearing and the influence of oil condition, Tribology International, vol. 26, 820-826, 2006.
2. Mwanza M.C., Joyce M.R., Lee K.K., Syngellakis S., Reed P.A.S., Microstructural characterisation of fatigue crack initiation in Al-based bearing alloys, International Journal of Fatigue, vol. 25, 1135-1145, 2003.
3. Kalbande P.D., Dhande D.Y., Pande D.W., CFD analysis of carbon fiber reinforced polytetrafluoroethylene (PTFE) hydrodynamic journal bearings using optimization technique, International Conference on Automatic Control and Dynamic Optimization Techniques, Pune, India, 629-634, 2016.
4. Golchin A., Simmons G.F., Glavatskih, S.B., Break-away friction of PTFE materials in lubricated condition, Tribology International, vol. 48, 54-62, 2012.
5. McCarthy D.M.C., Glavatskih S.B., Assesment of polymer composites for hydro-dynamic journal bearing applications, Lubricant Science, vol. 21, 331-341, 2009.
6. Tevrüz T., Tribological behaviours of bronzed-filled polytetrafluoroethylene dry journal bearings, Wear, vol. 230, 61-69, 1999.
7. Iwai Y., Honda T., Onizuka T., Taniguchi Y., Kawabata M., Wear properties of a glass fiber filled PTFE composite under oil lubrication, Japanese Journal of Tribology, vol. 45, 389-401, 2000.
8. Lu, J., Qui, M., Li, Y., Numerical analysis of self-lubricating radial spherical plain bearings and investigations on fatigue damage mechanisms of the liner, Tribology International, vol. 96, 97-108, 2016.
9. INA Walzlager Schaeffler KG, Permaglide Plain Bearings-Catalogue 703, Nuremberg, Germany, 1991.
10. Bıyıklıoğlu A., Çuvalcı H., Adatepe H., Baş, H Duman M.S., A new test apparatus and method for friction force measurement in journal bearing under dynamic loading: part-I, Experimental Techniques, vol. 29, 22-24, 2005.
11. Bıyıklıoğlu A., Çuvalcı H., Adatepe H., Baş, H Duman M.S., A new test apparatus and method for friction force measurement in journal bearing under dynamic loading: part-II, Experimental Techniques, vol. 29, 33-36, 2005.
12. ANSYS, Users Manuel DN-R300:50-1, Houston 1992.

13. Harbordt, J., Klummp R., Computer-Programm zur Berechnung der Verlagerungsbahn, der Örtlichen Drücke und des Tatsachlichen Schmierspalts in Glleitlagern mit Beliebiger Geometrie unter Stasischer und dynamischer last, Forchungsberichte, heft 137, der FVV, 1972.

14. Journal Bearing Analysis-Orbit, Documentation/User Manual, Ricardo Software Version 1,2, IL, USA, 2000.

15. Duman M.S., Stress analysis and investigation of surface fatigue of PTFE-Bronze layered journal bearings under dynamic loading conditions, PhD Thesis, Karadeniz Technical University, Trabzon, 2002. (in Turkish)

Tribology Study of High-Technological Composite Coatings Applied Using High Velocity Oxy-Fuel

Kandeva M [1]***, Grozdanova T*** [2]***, Karastoyanov D*** [3]***, Ivanov Pl*** [1]***, Kalichin Zh*** [4]
[1]*Faculty of Industrial Engineering, Tribology Center, Technical University of Sofia, Sofia, Bulgaria,*
[2]*Space Research and Technologies Institute, Bulgarian Academy of Sciences, Sofia, Bulgaria,*
[3]*Institute for Information and Communication Technologies, Bulgarian Academy of Sciences, Sofia, Bulgaria,*
[4]*SciBulCom 2, Ltd., 7, Nezabravka Str., 1113 Sofia, Bulgaria*
kandevam@gmail.com[1],tinka_gr@abv.bg[2], dimikara@abv.bg[3], kalitchin@gmail.com[4]

Abstract: In the work are studied the differential parameters of wear and wear resistance of high-tech composite coatings of powder superalloys with nickel matrix, WC-12Co and mixed compositions. Coatings were created and applied to a substrate of steel with a different flame velocity - 700 m/s and 1000 m/s without preheating the substrate and with preheating the substrate to 650° C. The wear is carried out with a "thumb-disk" tribotester under dry surface friction with fixed black corundum abrasive particles. Comparative results were obtained for the microstructure and texture of the pre- and post- friction coating, the porosity, roughness, hardness, the dependence of mass wear, the speed and wear intensity and the wear resistance of the coatings on the number of friction cycles. Influence of the flame rate and substrate temperature on wear resistance and differential wear parameters has been determined.

Keywords: tribology, wear-resistance, coatings, HVOF

1. Introduction

The main task of the tribology is to increase the operating resource of the contact joints and work bodies in the machines by reducing the friction and wear of the contact surfaces. One of the modern methods for reducing the wear of the parts operating under extreme operating conditions - abrasion, erosion, corrosion, impact and vibration loads, vacuum and others, is associated with the application of HighVelocity Oxy-Fuel (HVOF) method [1÷9].

HVOF is a modern tribotechnology for obtaining high-tech composite coatings on surfaces in various areas of transport and industry: various types of transport, mining equipment, energy, gas and oil extraction, road machinery and others.

The essence of the HVOF method consists in the impact interaction between powder particles moving at supersonic velocity into a flame jet and the surface of the substrate on which the coating is deposited [1,3,6÷11].

Fuel (kerosene, acetylene, propylene and hydrogen) and oxygen are fed into a combustion chamber where the hot flame with a temperature > 3000°C flows under high pressure through a nozzle at supersonic speed. The powder composition is supplied axially in the combustion chamber under high pressure or in the nozzle in the side where the pressure is lower (Figure 1).

Powder compositions known in the literature as powder super alloys are micronized particle mixtures of various metals, alloys, ceramics, polymeric materials. The particles carried by the flame stream pass into a plastic and/or semi-plastic state in the form of a particle or a drop and after contact with the substrate due to the high kinetic energy they deform to form thin lamellae with different shape. When struck with irregularities, the droplet particles are cooled, forming adhesive and cohesive contact links with the base surface and with each other forming a laminar structure of the composite coating [12÷14].

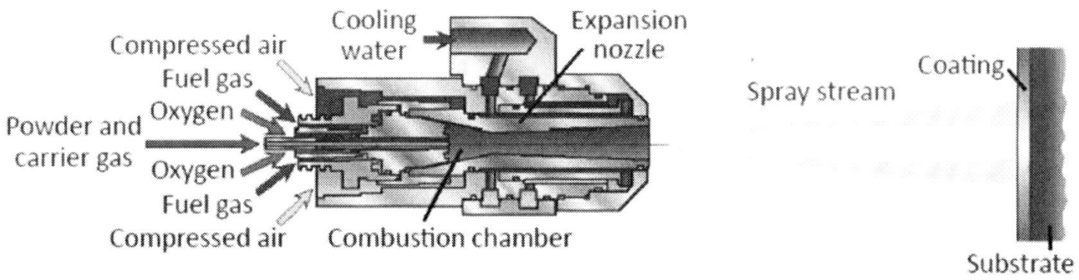

Figure 1. Schematic drawing of the high velocity oxygen fuel spraying (HVOF) process with gas-fuel gun (Reproduced by permission of Oerlikon Metco).

Gas-flame based coatings applied by the HVOF method are characterized by high adhesion hardiness/resiliency and low porosity due to the large kinetic energy of the particles. The extremely short contact time of the particles with oxygen from the environment creates favourable conditions for the absence of oxides in the coating, which in turn is of great importance for the wear resistance. As opposed to sonic gas-flame coatings, the HVOF method does not preheat the base surface to a high temperature, avoiding thermal deformations, and this makes it applicable to a number of responsible details in the industry. Certain research by the authors suggests that the combination of surface preheating and supersonic gas spray (HVOF method) leads to an increase of the wear resistance of coatings coated using powder compositions with lower hardness [13÷18].
The aim of the present study is to investigate the influence of three factors: particle velocity in the flame jet, preliminary heat treatment of the substrate and the presence of WC-12Co particles on the wear parameters of coatings applied with the 602P powder composition and HVOF method at the same surface abrasion conditions of rigidly bonded abrasive particles.

2. Experimental details

2.1. Materials and deposition conditions

Five types of HVOF (High Velocity Oxy-fuel) coatings were investigated. The coatings have been deposited on support (substrate) of one and the same material of chemical composition, represented in Table 1. The hardness of the substrate varies within the limits 193.6 ÷ 219.5 HV.

Table 1. Chemical composition (wt. %) of the coated material (medium-carbon steel substrate).

Element	C	Si	Mn	Ni	P	S	Cr	Fe
Percentage	0.4	0.2	0.55	0.3	0.45	0.045	0.3	Balance

The preliminary preparation of the substrate prior to the deposition of the coatings includes three stages: decontamination, blasting (erosion) and mechanical treatment.

The decontamination is done using lubricants and solvents, which remove the deposited mechanical components, moisture and organic molecules, adsorbed by the surface. Their extraction from the depth of the surface layer is accomplished by blazing in flame, i.e. increase of the temperature („sweating" of the surface) or by vapour-streaming device and once again purifying using solvents.

The blasting represents erosion of the surface by means of a system having definite technical parameters. The abrasive material "Grit" is used in accordance with the requirements of the standard ISO 11126 with granulomere composition of the abrasive in mm as percent ratio as follows: 3.15 ÷ 1.4 mm – 9.32%; 1.63 ÷ 0.5 mm – 16.4%; 1.4 ÷ 1.0 mm – 15.8%; 1.0 ÷ 0.63 mm – 39.6%; 0.5 ÷ 0.315 mm – 9.32%; 0.315 ÷ 0.16 mm – 9.32%; particles of size below 0.15 mm different fractions – up to 100% of the following chemical compounds: SiO_2 – 41% , combined in the form of silicates; AlO – 8.3%, MgO – 6.6%, CaO – 5.5% and MnO – 0.4%. The technical parameters of the mobile system for blasting are: input pressure 0.8 MPa; operating pressure in the nozzle – 0.4 MPa; diameter of the nozzle ϕ – 7 mm; distance between the nozzle and the surface – 30 mm; angle of interaction of the stream with the surface – 90°.

The coatings have been deposited applying a system model MICROJET POWDER using a set of technological parameters, listed in Table 2.

Table 2. Technological regime parameters for HVOF coating deposition.

№	Parameter	Technological regime	
1	Fuel/oxygen (C_3H_8/O_2) ratio, %	45/100	55/100
2	Particle velocity, m/s	700	1000
3	Spraying distance, mm	100	
4	Impact angle, grad	90	
5	Air pressure, MPa	0.5	
6	Nitrogen pressure, MPa	0.4	
7	Powder feed rate, g/min	22	

The coatings are applied with metal powdered compositions of nickel and tungsten. The designations and chemical composition of the coatings are presented in Table 3.

Table 3. Coating designation and appropriate powder chemical composition (wt. %)

Sample	Coating designation	Powder chemical composition, wt. %
1	602P:700	Cr: 14.5; Si: 4.61; B: 2.85; Fe: 4.42; Co: 0.05;
2	602P:700: PHS*	Cu: 2.65 ; Mn: 3.2
3	602P:1000	**Ni: Balance**
4	WC–12Co:1000	Co: 12; C: 5.4; Fe: < 0.1; Ni: < 0.1; **W: Balance**
5	602P/ WC–12Co:1000	1 : 1

With powder composition 602P, three types of coatings are obtained: Coating 602P: 700 – at particles velocity in the flame stream - 700 m/s without pre-heat treatment of the substrate; Coating 602P: 700: PHS * at a 700 m/s particle velocity in the flame stream with pre-heat treatment of the substrate in a thermal chamber at 650° C for 3 hours and a 602P: 1000 coating at 1000 m/s particle velocity in the flam stream without pre-heating the substrate.

Two types of coatings are obtained with a tungsten-based powder mixture: a WC-12Co: 1000 coating is applied at a particles speed of 1000 m/s and a 602P/WC-12Co: 1000 coating obtained by mixing the two powdered compositions 602P and WC-12Co in a ratio of 1:1. The average particle size range is $40 \div 45 \pm 2.5$ μm obtained by agglomeration process with the sintering.

Table 4 gives details of the thickness, porosity, roughness and hardness of tested coatings.

Table 4. As-deposited thickness, porosity, roughness and hardness of tested coatings.

Sample	Coating designation	Thickness μm	Porosity %	Roughness μm	Hardness HRC
1	602P:700	100	4.1	Ra=2,888; Rq=3,456	58-60
2	602P:700: PHS*	100	3.6	Ra=0,168; Rq=0,208	61-64
3	602P:1000	115	2.2	Ra=5,798; Rq=7,298	62-64
4	WC–12Co:1000	115	0.8	Ra=2,148; Rq=2,685	65-67
5	602P/ WC–12Co:1000	115	1.2	Ra=6,659; Rq=8,087	62-66

The roughness (Ra, Rq) of the coatings is measured with a TESA Rugosurf 10-10G profile meter (Figure 2) in two perpendicular directions.

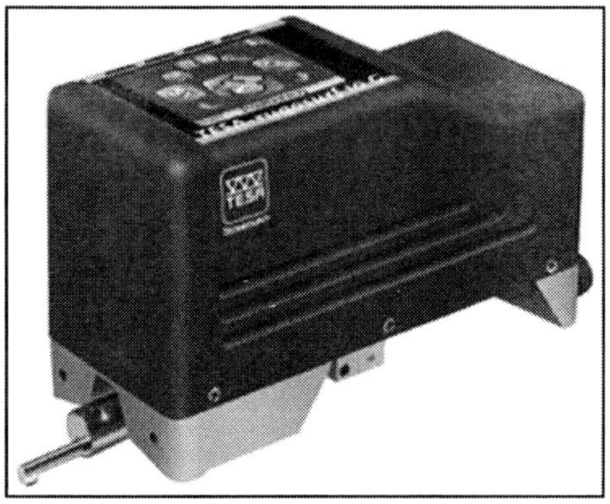

Figure 2. TESA Rugosurf 10G.

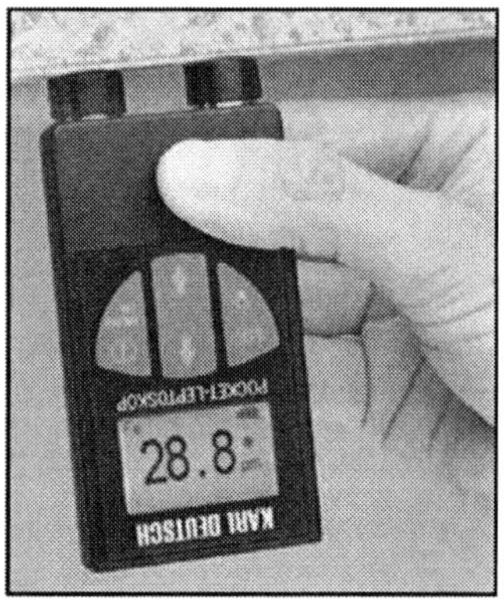

Figure 3. Pocket Leptoskop 2021Fe.

The thickness of the coatings is measured with the Pocket Leptoscope 2021 Fe (Fig. 3). Measurements were made at 10 points of the surface and the arithmetic average value was adopted.

2.2 Abrasive wear testing

The abrasive wear of the coatings is investigated with a "thumb-disk" tribotester under dry surface friction with hard-fixed abrasive particles. The functional diagram of the tribotester is presented in Figure 4.

The test specimen with coating (1) is fixed in the bed of the holder (2) placed in the loading head (8). The front surface of the sample (1) contacts the surface of the opponent body (3) fixed hard to the horizontal disc (4). The disc (4) is driven by an electric motor (6) and rotates about its vertical center axis with a constant angular velocity ω. The normal load P is applied to the center of gravity of the contact site between the sample and the surface and is set with a lever system in the loading head (8).

The friction path is set by the number of cycles with the tachometer 7.The device allows the sliding speed to vary from 0.55 m/s to 1 m s by varying the distance R between the axis of rotation of the disc (4) and the axis of the sample (1).

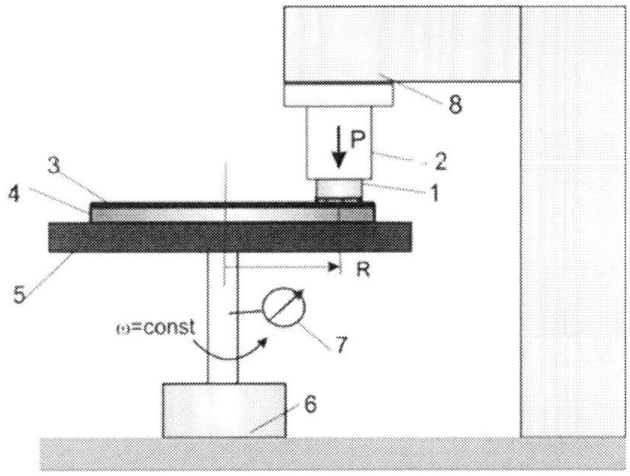

Figure 4. Schematic diagram of abrasive wear testing on pin-disc tribometer

The samples have a cylindrical shape with a diameter of 10 mm and a height of 25 mm. The test was carried out under the following constant friction conditions: normal load P = 4.6 N, nominal contact area - Aa = 0.785 cm^2; Sliding speed - V = 0.71 m/s; R = 0.032 m; abrasive surface - Corundum P 320, number of friction cycles - 100, 200, 300, 400. The friction path at N number of friction cycles is calculated by the formula $S = 2\pi RN$.

The wear characteristics test method consists in measuring the mass wear *m* of the samples for a specific friction path S (friction cycles) under constant conditions - load P and glide speed V. The method includes the following sequence: preparation of samples with the same dimensions; measure the mass *m* of the sample before friction with the electronic scale WPS 180 /C /2 to (accuracy 0,1 mg). Before each measurement of the scale the sample is cleaned of mechanical and organic particles, dried with ethyl alcohol to prevent the electrostatic effect. Calculated are the following wear characteristics:

Specific wear (i_s): Represents the ruptured mass of friction from the surface layer for normal load P = 1N per friction path S = 1m and nominal contact area A_a=1 mm^2:

$$i_s = \frac{m}{PSA_a}, \text{ mg/Nm.mm}^2 \tag{1}$$

Specific wear resistance (I_s): Expressed as the reciprocal value of the specific wear:

$$I_s = \frac{1}{i_s} = \frac{PSA_a}{m}, \text{ Nm.mm}^2\text{/mg} \tag{2}$$

Relative wear resistance(W_s): Relationship between the specific wear resistance of the test coating (I_s) and the wear resistance of the substrate (I_b) by equal friction conditions:

$$W_s = I_s/I_b \tag{3}$$

3. Results and discussion

Tables 5, 6, 7 and 8 represent experimental results for mass wear, specific wear, specific wear resistance, and relative wear resistance for test surfaces for four values of the number of friction cycles respectively.

Table 5. Abrasive wear of tested coatings

Sample	Coating designation	Number of cycles (N)			
		100	200	300	400
		Sliding distance, m			
		20	40	60	80
		Mass loss, mg			
1	602P:700	5.0	10.4	14.3	18.2
2	602P:700: PHS*	1.9	2.7	3.6	4.4
3	602P:1000	1.2	3.1	3.9	4.1
4	WC–12Co:1000	0.5	0.8	1.0	1.2
5	602P/ WC–12Co:1000	0.8	1.3	2.0	2.4
6	Substrate: Steel	12.4	16.6	19.8	22.1

Table 6. Abrasive specific wear of tested coatings

Sample	Coating designation	Number of cycles (N)			
		100	200	300	400
		Sliding distance, m			
		20	40	60	80
		Specific wear, $mg.mm^2/Nm$			
1	602P:700	6.9×10^{-4}	7.2×10^{-4}	6.6×10^{-4}	6.3×10^{-4}
2	602P:700: PHS*	2.7×10^{-4}	1.9×10^{-4}	1.7×10^{-4}	1.5×10^{-4}
3	602P:1000	1.7×10^{-4}	2.2×10^{-4}	1.8×10^{-4}	1.4×10^{-4}
4	WC–12Co:1000	0.7×10^{-4}	0.5×10^{-4}	0.3×10^{-4}	0.3×10^{-4}
5	602P/ WC–12Co:1000	1.1×10^{-4}	0.9×10^{-4}	0.9×10^{-4}	0.8×10^{-4}
6	Substrate: Steel	17.2×10^{-4}	11.5×10^{-4}	9.2×10^{-4}	7.6×10^{-4}

Table 7. Abrasive specific wear resistance of tested coatings

Sample	Coating designation	Number of cycles (N)			
		100	200	300	400
		Sliding distance, m			
		20	40	60	80
		Specific wear resistance, $Nm/mg.mm^2$			
1	602P:700	1.4×10^3	1.4×10^3	1.5×10^3	1.6×10^3
2	602P:700: PHS*	3.7×10^3	5.3×10^3	5.9×10^3	6.7×10^3
3	602P:1000	5.9×10^3	4.5×10^3	5.6×10^3	7.1×10^3
4	WC–12Co:1000	14.3×10^3	20×10^3	33.3×10^3	33.3×10^3
5	602P/ WC–12Co:1000	9.0×10^3	11.1×10^3	11.1×10^3	12.5×10^3
6	Substrate: Steel	0.6×10^3	0.9×10^3	0.9×10^3	0.8×10^3

Table 8. Relative wear resistance of tested coatings

Sample	**Coating designation**	**Wear resistance, Nm/mg.mm^2 (N=400)**	**Relative wear resistance (R)**		
			Influence of heat treatment	Influence of particle velocity	Both influence
1	602P:700	1.6 x 10^3	$R_{1,1} = 1$	$R_{1,1} = 1$	$R_{1,6} = 2.0$
2	602P:700: PHS*	6.7 x 10^3	$R_{2,1} = 4.19$	$R_{3,2} = 1.1$	$R_{2,6} = 8.4$
3	602P:1000	7.1 x 10^3	-	$R_{3,1} = 4.44$	$R_{3,6} = 8.9$
4	WC–12Co:1000	33.3 x 10^3	-	-	$R_{4,6} = 41.6$
5	602P/ WC–12Co:1000	12.5 x 10^3	-	-	$R_{5,6} = 15.6$
6	Substrate: Steel	0.8 x 10^3	-	-	$R_{6,6} = 1$

Figures from 5 to 12 show graphically the dependence of the wear from the friction path and diagrams on the influence of heat treatment of the substrate and the particles peed on the wear resistance of the tested coatings.

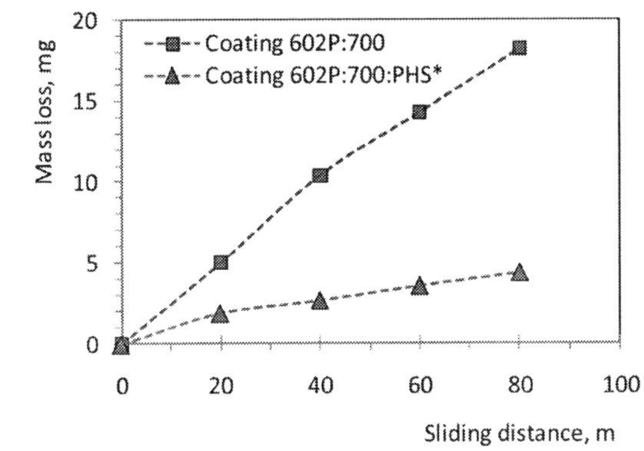

Figure 5. Mass loss vs. sliding distance for coatings on the without and with heat treatment of substrate at particle velocity 700 m/s.

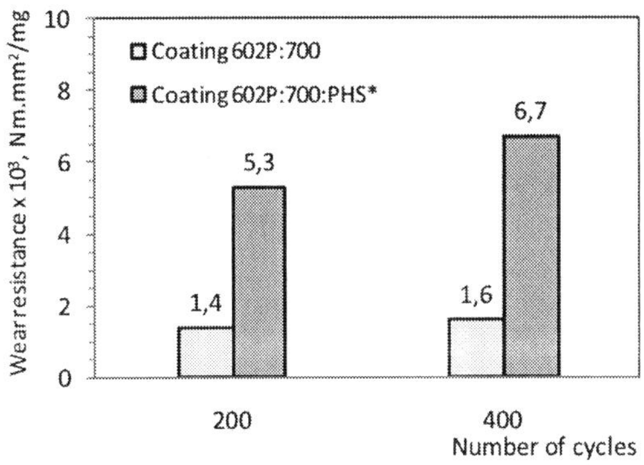

Figure 6. Influence of heat treatment of wear resistance of coatings at particle velocity 700 m/s.

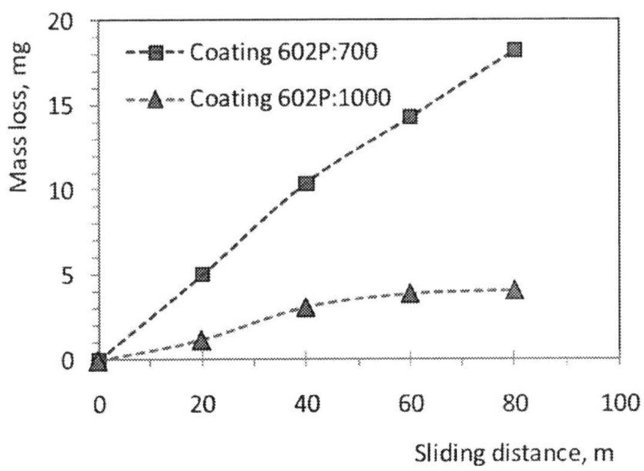

Figure 7. Mass loss vs. sliding distance for coatings at two particle velocity - 700 m/s and 1000 m/s without heat treatment of substrate.

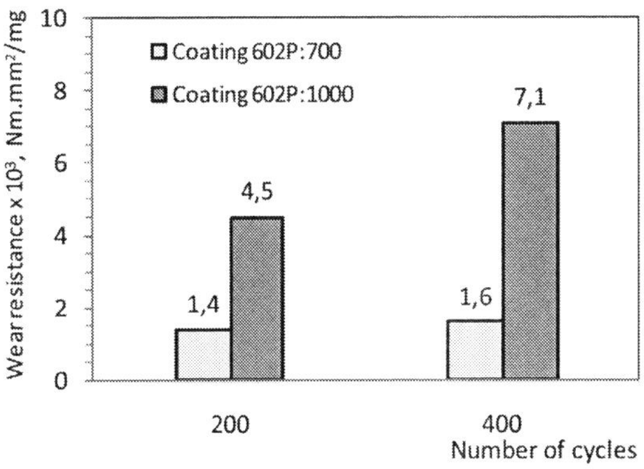

Figure 8. Wear resistance at two particle velocity 700 m/s and 1000 m/s of coatings without heat treatment of substrate.

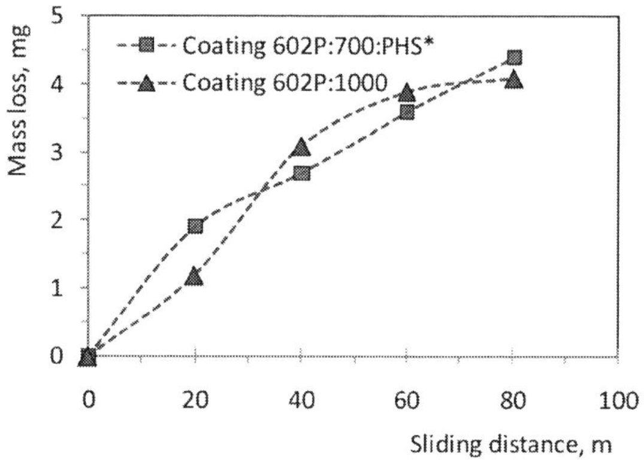

Figure 9. Mass loss vs. sliding distance for coatings at two particle velocity - 700 m/s and 1000 m/s with heat treatment of substrate.

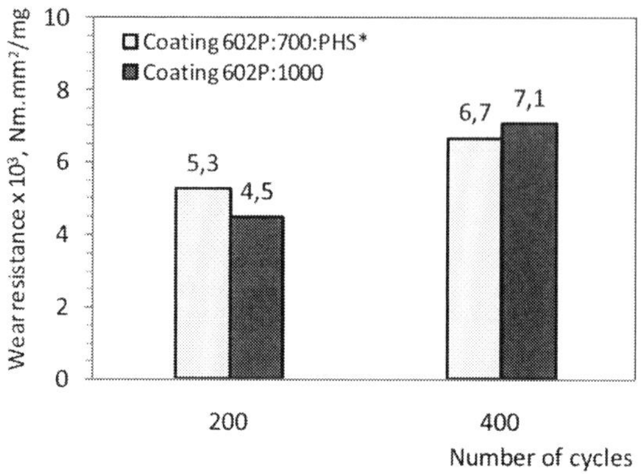

Figure 10. Wear resistance at two particle velocity 700 m/s and 1000 m/s of coatings with heat treatment of substrate.

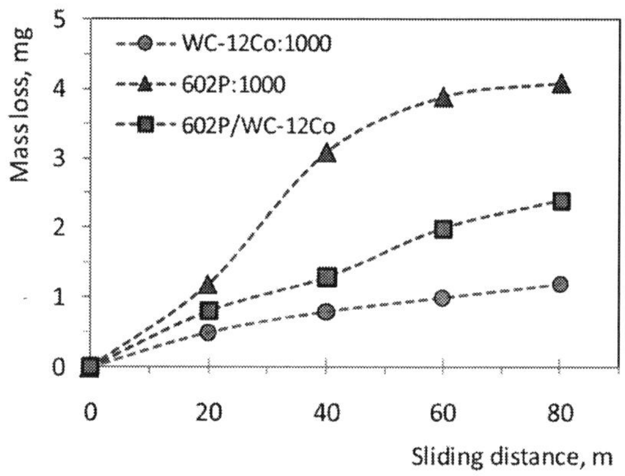

Figure 11. Mass loss vs. sliding distance for coatings 602P, WC-12Co and 602P/WC-12Co without heat treatment of substrate.

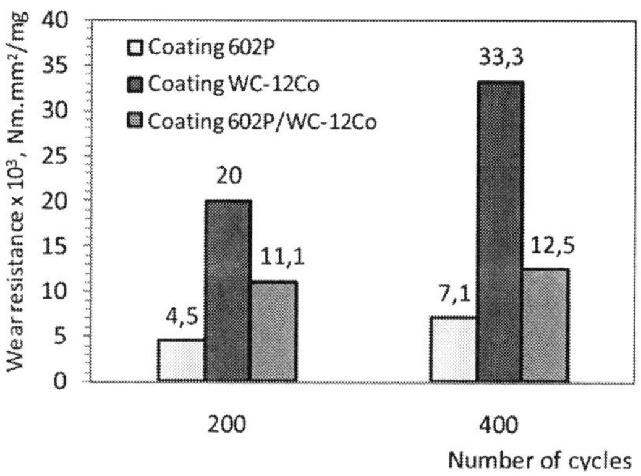

Figure 12. Wear resistance of coatings 602P, WC-12Co and 602P/WC-12Co without heat treatment of substrate.

Figures 13, 14 and 15give the microstructure and morphology of 602P:700 and WC-12Co coatings.

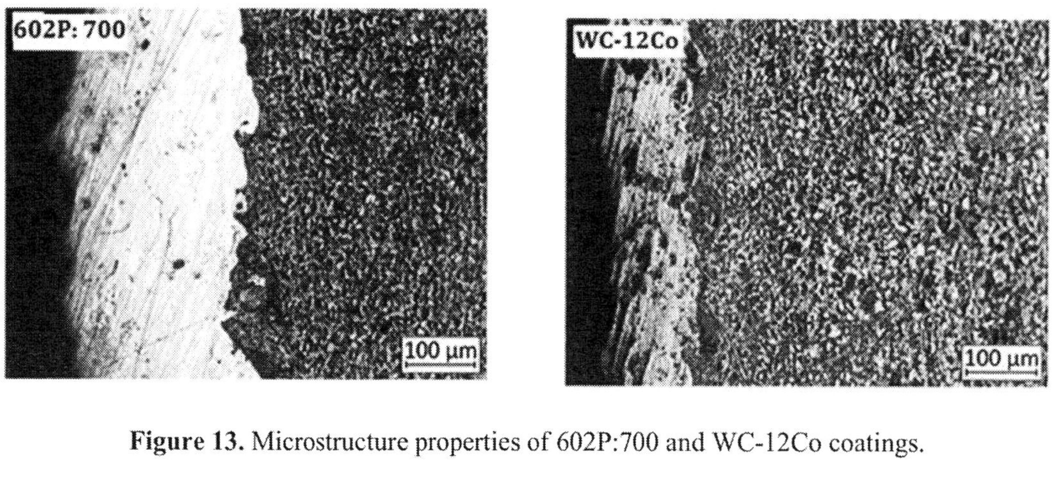

Figure 13. Microstructure properties of 602P:700 and WC-12Co coatings.

Figure 14. Morphology of 602P:700 coating at 400cycles.

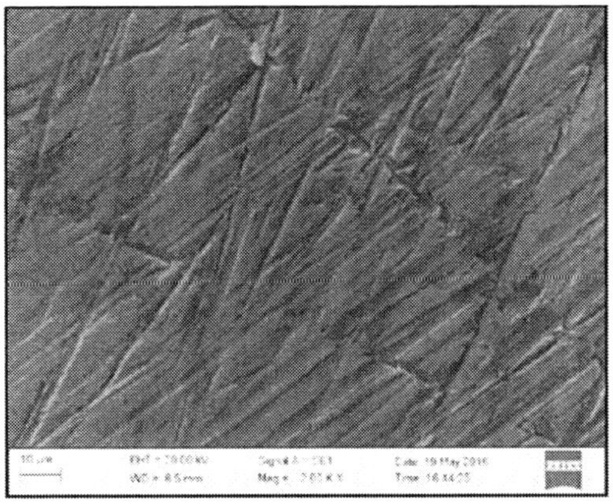

Figure 15. Morphology of 602P:700: PHS*coating at 400 cycles

The analysis of the results of the influence of heat treatment of the substrate and the particle speed on the wear characteristics show the following: In a 602P coating applied at a low supersonic particle velocity (700 m/s) on a pre-heat treated substrate at 650°C for 3 hours, the wear significantly decreases compared to the same coating without heat-heating the substrate. (Figure 5). The wear resistance of such coating increases 3.8 times for 200 cycles and 4.19 times for 400 cycles of friction path (Figure 5, Table 8). The reduction in wear rate and / or wear resistance from the friction path, respectively, is not proportional to the friction path due to the hardening of the coating.
The morphology of the worn coating without heat treatment of the substrate (Figure 14) after 400 cycles of friction is characterized by the presence of irregularly spaced fractal craters. In the surface layer of the worn surface of the heat treatment coating (602P:700:PHS*) the rare traces of the abrasive action of the abrasive particles without the presence of deep deformation (Figure 15). The heat factor influences the adhesion strength and coating density.
The velocity of particle velocity on wear only affects coatings without preliminary heat treatment of the substrate (Figure 7 and Figure 8). A thigh supersonic particle velocity (1000 m/s) the wear decreases, respectively, the wear resistance in creases 4.4 times over 400 cycles of friction path. This is explained by the high density of the coating, i. e. high cohesives trength in the coating volume. The latter is due, on the one hand, to the high kinetic energy of the particles and, on the other hand, to the absence of oxides in the coating.
The latter is the result of the short time of particles stay and their contact with The comparison of the two coatings 602P:700:PHS* and 602P:1000 shows that they have almost the same wear and wear resistance values (Figures 9 and 10).
The results of the new 602P/WC-12Co coating test, which was prepared by mixing a powder of 602P with a 1:1 powder WC-12Co and applied at a particle speed of 1000 m/s, showed that wear abruptly decreased takes twice as low as the 602P: 1000 without WC-12Co. Its wear resistance increases to 2.45 times for 200 cycles and 1.76 times for twice as many cycles (N = 400 cycles).

4. Conclusions

This paper presents results on the influence of particle velocity pre-heat treatment of the substrate and the presence of WC-12C particles on wear and wear resistance of coatings applied with HVOF powder 602P composition under the same surface friction conditions of fixed abrasive particles.
It has been found that the increase in the wear resistance of 602P coating can be achieved in three ways:

- Apply the coating at a particle speed of 700 m/s with pre-heat treatment of the substrate at 650°C for 3 hours. The wear resistance of the coating increases 4.19 times compared to the wear resistance of the coating without heat treatment.
- Apply the coating at a particle speed of 1000 m/s without preheating the surface. The wear resistance of the coating increases 4.44 times against the wear resistance of a coating applied at 700 m/s without heat treatment.
- Powder composition WC-12Co is added to the powder composition 602P in a ratio of 1:1 and the coating is applied at a particle speed of 1000 m/s. The wear resistance of the coating increases 1.76 times in comparison with the wear resistance of a coating applied at 1000 m/s without heat treatment.

Acknowledgements

This work has been performed as a part of activities within the following project Contract: DN 07/28-15.12.2016 "Research and creation of new wear-resistant coatings using composites and nano materials", funded by the National Science Fund of the Ministry of Education and Science, Bulgaria.

References

1. Wood, R. J., Roy K., Tribology of the thermal-sprayed coatings. – in: M. Roy (ed.), Surface engineering for enhanced performance against wear, Wien: Springer, 1-43, 2013.
2. Barbezat, G., A., Nicol R., Abrasion, erosion and scuffing resistance of carbide and oxide ceramic thermal sprayed coatings for different applications, Wear, vol. 162-164, part a, 1993.
3. Pawlowski, l., The science and engineering of thermal spray coatings. Chichester: John Wiley & Sons, 2008.
4. Oksa M., Turunen E., Suhonen T., Varis T., Hannula S., Optimization and characterization of high velocity oxy-fuel sprayed coatings: Techniques, Materials and Applications. – coatings, vol. 1, no 1, 17-52, 2011.
5. Schubert, J., Š. Houdková, M. Kašparová, Z. Česánek, Effect of co content on the properties of HVOF sprayed coatings based on tungsten carbide. – Proceedings of 21st international conference on metallurgy and materials – metal 2012, Brno (Czech Republic), 1086-1091, 2012.
6. Martin, P. M., Introduction to surface engineering and functionally engineered materials. Hoboken: john Wiley & Sons, 2011.
7. An introduction to thermal spray. Brochure, Oerlikon Metco, Winterthur, 2015, (http://www.oerlikon.com/metco).
8. About thermal spray, poster, flame spray technologies, Duiven, (http://www.fst.nl).
9. Mellor B. G., Welding surface treatment methods for protection against wear. – in: Mellor, B. G. (ed.), Surface coatings for protection against wear, Cambridge: Wood head publishing, 302-376, 2006.
10. Ponce de León C., C. Kerr, F. C. Walsh, Electroless plating for protection against wear. – in: mellor, B. G. (ed.), Surface coatings for protection against wear. Cambridge: Wood head publishing, 184-225, 2006.
11. Vencl Al., Optimization of the deposition parameters of thick atmospheric plasma spray coatings. – journal of the Balkan Tribological Association, vol. 18, 405-414, 2012.
12. Kandeva M., Al. Vencl, D. Karastoyanov, Advanced tribological coatings for Heavy-duty applications: Case studies, Academy publishing house, Sofia, Bulgaria, 147, 2016.

13. Chivavibul P., M. Watanabe, S. Kuroda, J. Kawakita, M. Komatsu, K. Sato, J. Kitamura, Effect of powder characteristics on properties of Warm-sprayed WC-Co coatings, journal of Thermal spray technology, vol. 19, no. 1, 81-88, 2010.
14. Kandeva M., Il. Peichev, N. Kostova, K. Stoichkov, Complex study of surface layers and coatings, J Balk Tribol Assoc, 17 (3), 387, 2011.
15. Kandeva M., B. Ivanova, V. Pozhidaeva, D. Karastoyanov, J. Javorova, Composite coatings toimprove durability of the working body of the drill, 5th World Tribology congress (WTC 2013), Turin (Italy), 2729-2732, 2013.
16. Kandeva M., D. Karastoyanov, Al. Vencl, Il. Peichev, B. Ivanova, E. Assenova. Wear-resistance of composite coatings deposited by gas-flame supersonic HVOF process, proceedings of XIV international scientific conference "Tribology and Reliability", Saint Petersburg (Russia), 307-321, 2014.
17. Kandeva M., B. Ivanova, D. Karastoyanov, T. Grozdanova, E. Assenova, Abrasive wear of High velocity oxygen fuel (HVOF) superalloy coatings under vibration load, IOP Conference Series: Materials Science and Engineering, Vol. 174, Issue 1, 1 March 2017, Article number 012010.
18. Kandeva M., T. Grozdanova, D. Karastoyanov, E. Assenova, Wear resistance of WC/Co HVOF-coatings and galvanic Cr coatings modified by Diamond nanoparticles, IOP Conference Series: Materials Science and Engineering, Vol. 174, Issue 1, 1 March 2017, Article number 012060.

IOP Conf. Series: Materials Science and Engineering **295** (2018) 012026 doi:10.1088/1757-899X/295/1/012026

Self-organization Effects on Tribosystems when Lubricated with a Metal-plating Additive „Valena"

***Kandeva M.*[1], *Karastoyanov D.*[2], *Grozdanova T.*[3], *Kalichin Zh.*[4], *Balabanov V.*[5], *Chikurtev D.*[2]**
[1]*Faculty of Industrial Engineering, Technical University of Sofia, Bulgaria*
[2]*Institute for Information and Communication Technologies, BAS, Sofia, Bulgaria*
[3]*Space Research and Technologies Institute, BAS,* Sofia, *Bulgaria*
[4]*SciBulCom 2, Ltd., 7, Nezabravka Str., 1113 Sofia, Bulgaria*
[5]*Faculty of Processes and Machines in Agrobusiness, Russian State Agrarian University – Moscow, Russian*
kandevam@gmail.com, dimikara@abv.bg, tinka_gr@abv.bg, kalitchin@gmail.com, vbalabanov@rgau-msha.ru, dchikurtev@abv.bg

Abstract: The external appearance of self-organization is very low friction and wear, and in some cases there is lack of wear, so a friction effect without wear is observed. In the present study are examined tribosystems of different materials - steel, spheroidal graphite cast iron micro alloyed with 0.051% of tin, bronze and other materials under conditions of border lubrication with grease "Litol 24" with a metal-plated additive "Valena". A unique multifunctional tribotester has been developed that allows varying load and slide speed across wide ranges and research different types of contact - point, ring, plane. Studies have been carried out on different friction modes and results have been obtained for friction coefficient and the parameters of wear of the specimens and the counter body depend of the friction times. Conditions have been established to produce a No-wear effect due to the formation of a metal-plated copper film on the surfaces of the body and the counter body.

Keywords: self-organization, tribosystems, friction, wear, No-wear effect

1. Introduction

Generally acknowledged self-organization is connected with the spontaneous transition of the open non-equilibrium systems from less complicated and more chaotic forms to more complicated and more aranged forms [1,2,3]. In the non-equilibrium thermodynamics, the Belgian physic-chemist, Nobel Prize winner Ilya Prigozhin has proved, that in the open natural systems independently of their nature: physical, chemical, biological and social there exist possibility for their sustaining in stable non-equilibrium and the latter takes the towards elaboration in their internal structure and their structural enrichment. This evolutionary state of the systems might be supported with the help of stationary energetically and mass streams. Similar conditions are accomplished for processes which are carried out spontaneously and in the human society to the spontaneous actions are added the impacts with set purpose from the side of the humans. The strongly non-equilibrium open system forms structures on the basis of the mechanism of self-organization. In the scientific literature these structures are known as dissipative structures. The term for the latter comes from its basic purpose and function to protect the system from outer powerful energetic flows by scattering them. Dissipative structures exist and they are developing themselves only in non-equilibrium state of the open systems.

The tribological systems are assigned to the open thermodynamically systems. They are characterized with exchange of substances, energy and information with the environment. In the tribology are well known many effects, which are assigned to the self-organization in the tribosystems: effect of anomalously friction, registered by Silin after the contact surfaces have been irradiated with alfa particles; effect of selective transfer of material from one to the other surfaces under friction; effect of processing of the roughness and the connected triboprocesses of friction; effect of Rebinder, which is expressed with the influence of special surfactants designed to decrease of the surface tension; auto vibrations in contact systems, etc.

Special attention of many scientists and basically of the authors of this publication is the phenomenon *no-wear effect under friction*. This effect has been discovered in 1956 in the Soviet union by a group of researchers under the scientific guidance of D. Garkunov [4÷8]. The effect is observed in cases of friction and it is a result of mechanical and physical-chemical processes, leading to auto-compensation of the wear and decrease of friction. Specific in the essence of this effect is the building and functioning of a protective film where is realized diffusion – vacational of deformations which are proceeding without accumulation of defects as opposed to the fatigue processes. As a result of the processes of self-organization during the friction are formed self-recoverable metal-polymeric structures – serving and servitic films with thickness of 100 nm, where the deformation-plastic phenomena are materialized without accumulation of defects in the crystal structure on the surface.

The processes of formation of such surfaces is carried out under stringent combination of certain conditions – composition and properties of the contacting surfaces; composition and properties of the lubricating material; characteristics of the deformation field; value of the relative pressure; friction rate; contact temperatures. Generally, the basic factor which has an influence for the simultaneous action of tribo-recovering and tribo-coordination processes leading of self-organization is the chaotic auto-vibration of copper containing products in the lubricating material [9÷12].

The self-organization is the basic mechanism for non-wear friction and it is a result of the evolution changes of the contact, which are conditioned by the achievement of stationary (stable) condition of the tribosystem with minimal friction and maximal wear resistance [13÷17].

Especially expressive for the self-organization are the strong non-equilibrium states of the system where the energetic and material flows from the outside are non-linear functions of the temperature gradients and the concentration of the substances.

The purpose of the present paper is to be studied the influence of the metal-plating additive Valena on the intensity of the wear of the tribosystems from different materials and different friction regimes.

2. Materials

The influence of metal-plating composite additive Valena on wear of different materials during friction in grease Litol 24 is studied. Possibilities for obtaining the selective transfer of materials effect are to be found.

Table 1. Designation of the tested contact systems

№	Contact systems	
	Designation	**Materials**
1	GJS 600-3 - C22	Spheroidal Graphite Cast Iron with 0.051%Sn – Steel
2	Ст4кп - C22	Steel – Steel
3	БрАЖ9-4 - C22	Bronze – Steel
Lubricant	Grease Litol 24 without Valena additive	
	Grease Litol 24 with Valena additive (4 mass %)	

Tables 2, 3, 4 and 5 give the data about the composition, hardness and roughness of the material.

Table 2. Composition and hardness of specimen GJS 600-3 (BDS-EN 1563-00).

Element	C	Sn	Mn	P	Si	Cr	S	Fe
wt. %	3.87	0.051	0.34	0.029	1.55	0.03	0.012	Balance
Hardness				277 HB				

Table 3. Composition and hardness of specimen Ст4кп (GOST 380 - 94).

Element	C	S	Mn	P	Si	Cr	Ni	Fe
wt. %	0.25	0.045	0.55	0.45	0.15	0.30	0.28	Balance
Hardness				135 HV 0.05				

Table 4. Composition and hardness of specimen БрАЖ9-4 (CuA19Fe3)(GOST 1628-78).

Element	Fe	Al	Mn	P	Si	Pb	Zn	Sn	Cu
wt. %	3.0	9.5	0.45	0.008	0.12	0.088	0.85	0.09	Balance
Hardness					180 HB				

Table 5. Composition and hardness of the counter body C22 (BDS-EN 10083-2/1966).

Element	C	Cu	Mn	P	Si	Cr	Ni	Fe
wt. %	0.19	0.15	0.55	0.05	0.12	0.07	0.03	Balance
Hardness				633 HV 0.05				

3. Experimental results and analysis

The materials are tested by means of the device shown in Figure 1. A multifunctional set of accessories is developed, which allows simultaneous testing of wear and seizure through 180° rotation of the body, with possibility to reconstruct different kinematic situations. The set is located on a common fundament (Figure 2).
The elements of the set are: 1 – fundament; 2 – vertical bearing frame of the chamber 4 for testing of wear; 3 – protective screen; 5 – mandrill; 6 – bearing partition; 7 – chamber for testing the seizure of specimens and oils with and without additive.
Three spheres 8 with diameter 11.4 mm are located in chamber 7, which are loaded with load P through the sphere with equal diameter. The spheres 8 are fixed and covered with lubricant. The test rig allows three cases of kinematic interactions: firstly, when the upper arbor rotates, and the spheres in the chamber are fixed; secondly, when the upper arbor is immovable, and chamber 8 with the spheres rotates; and the third case, simultaneous rotation of the arbor and the chamber with different revolutions and opposite directions. The chamber is driven by motor 14 through the belt drive 12, 11, 10. The motor is located in case 13.

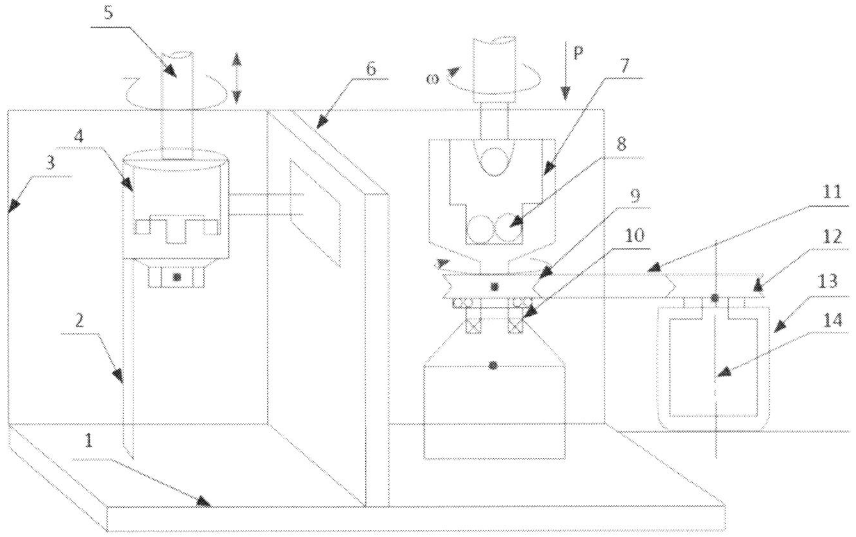

Figure 1. Set of accessories for testing of selective transfer at the specimens under normal and under seizure conditions.

Cylindrical specimens are prepared of the three materials, namely spheroidal graphite cast iron micro alloyed by 0.051% tin (GJS 600-3), steel (Ст4кп) and bronze (БрАЖ9-4). The specimens with their dimensions are shown in Figure 2*a*. The counter body is in the form of rectangular plate with dimensions 30.5 × 35 × 9 mm (Figure 2 *b*).

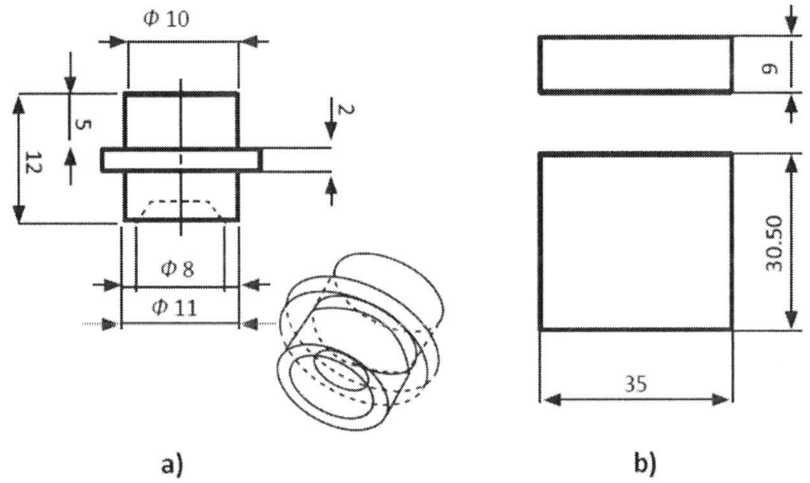

Figure 2. Dimensions of: the specimens (a); the counter body (b).

The interaction between specimen and counter body occurs in an annular contact with nominal area Aa = 44.75 × 10^{-6}m^2. The space above the annular contact is filled with lubricant, providing permanent penetration in the contact zone, and avoiding thus the regime of boundary and/or dry friction (Figure 2*a*).

The tests are carried out at two regimes of friction: the first one is realized at low load and high number of revolutions, i.e., big peripheral sliding speed, and the second one - at high load and small peripheral sliding speed. Wear is tested at every friction regime in two cases: lubrication with grease Litol 24 without Valena and lubrication with grease Litol 24 with Valena (Table 6). The time of friction for all tests is 40 min.

IOP Conf. Series: Materials Science and Engineering **295** (2018) 012026 doi:10.1088/1757-899X/295/1/012026

Table 6. Parameters of friction regimes in Litol 24 without and with Valena.

Regime of friction	Load N	Pressure MPa	Revolutions min^{-1}	Friction way m	Speed m/s
Regime I	10	0.22	2300	2744.40	1.33
Regime II	20	0.45	870	1038.00	0.5

Experimental results for the three contact systems at friction without and with Valena additive in both regimes of friction are presented below.

Contact system "Spheroidal graphite cast iron with 0.051% Sn – Steel"

(GJS 600-3 – C22)

Figures. 3 and 4 show wear variation for spheroidal graphite cast iron in both regimes of friction.

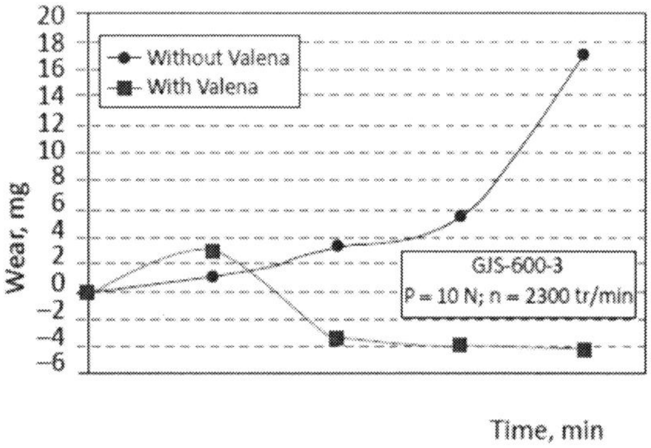

Figure 3. Influence of Valena additive on wear of GJS-600 at friction in Regime I

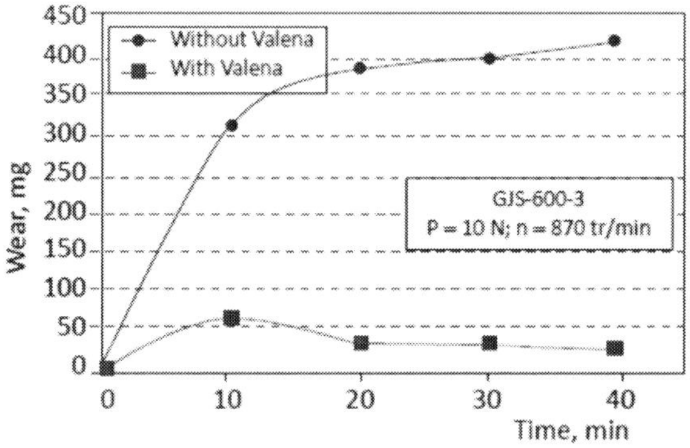

Figure 4. Influence of Valena additive on wear of GJS-600 at friction in Regime II

In the case of friction with high speed and low load (Regime I), the wear curve in the period of running-in up to the time of 10 min shows dissimilarities. The presence of Valena in grease Litol 24 leads to higher wear compared with that in grease without Valena additive. After that period, wear decreases, the specimen's mass increases, and that is due to the film deposited on the cast iron specimen's surface (Figure 3).

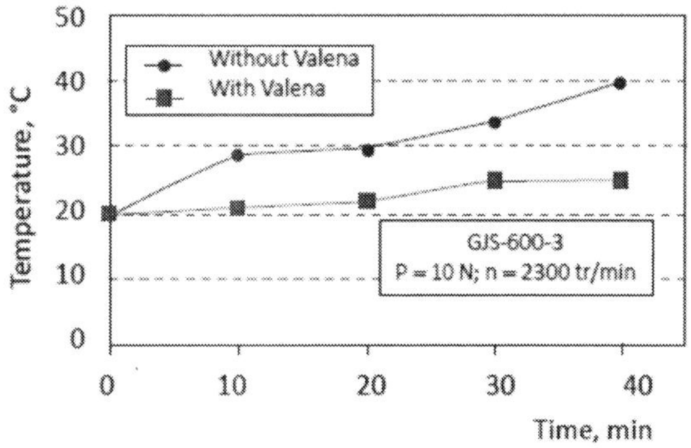

Figure 5. The temperature of the time friction in Regime I.

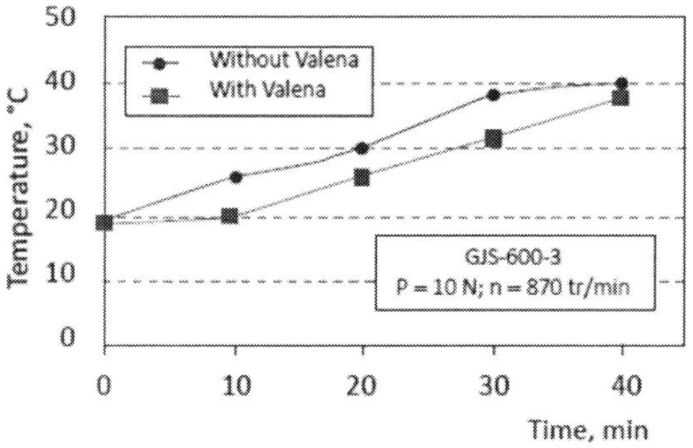

Figure 6. The temperature of the time friction in Regime II.

The temperature of lubricant material with Valena additive for the time of friction increases with 5° C only, whereas in grease without Valena – with 20° C (Figure 5). In the case of friction with twice higher load and 2.6 times higher speed (Regime II), specimen's wear occurs in both media, however, the wear in grease with Valena is 17 times lower than that in grease without Valena additive, which is

most probably the result of the protecting copper film formed on the working surfaces (Figure 3). The temperature in both media increases with 20° C (Figure 6).
Visual observation shows clearly metal-plating film with reddish colour on the surface of the spheroidal graphite cast iron specimen and on the surface of the steel counter body (Figure 7*a, b*).

a) b)

Figure 7. Specimens pictures of the spheroidal graphite cast iron (GJS) with 0.051% Sn (a); and of the counter body of steel (C22) (b) after friction duration 40 min at P = 10 N and n = 2300 min^{-1}.

The microstructure and the composition of the metal-plating film on specimens' surfaces are studied by scanning electronic microscope EVO MAX 10 ZEISS. Figure 8 shows the microstructure and the composition of the deposited metal-plating film on the surface of spheroidal graphite cast iron GJS 600-3 specimen.
The composition of the deposited film shows presence of the following elements: iron, copper, aluminum, nickel, chromium and chlorine.

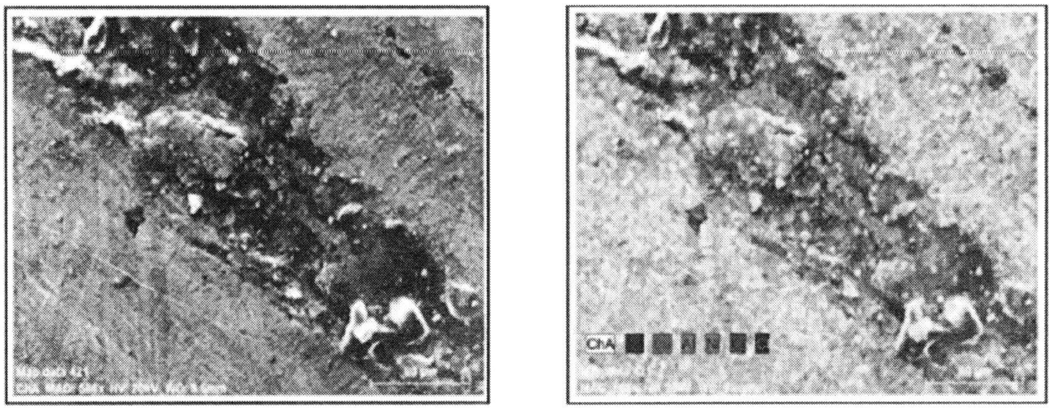

a) b)

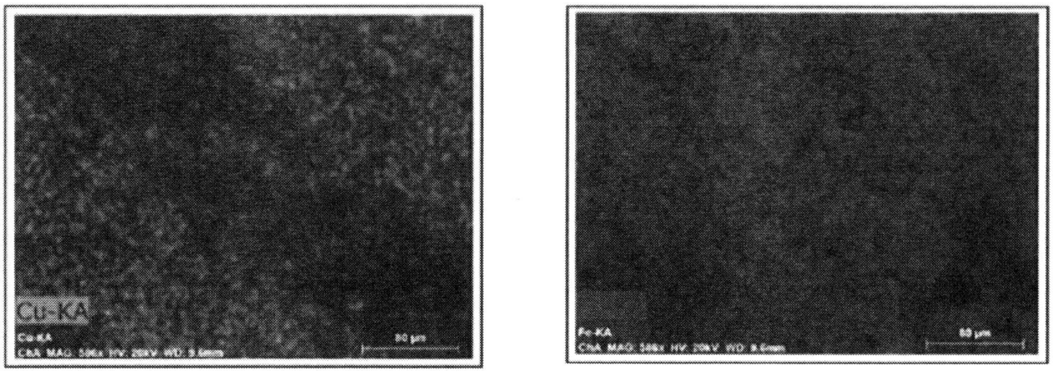

c) d)

Figure 8. Microstructure (a), composition (b), microstructure with Cu (c), and Fe (d) in metalplating film on the surface of spheroidal graphite cast iron GJS 600 – 3 with 0.051% Sn specimen, after friction process at P = 10 N; n = 2300 min^{-1}.

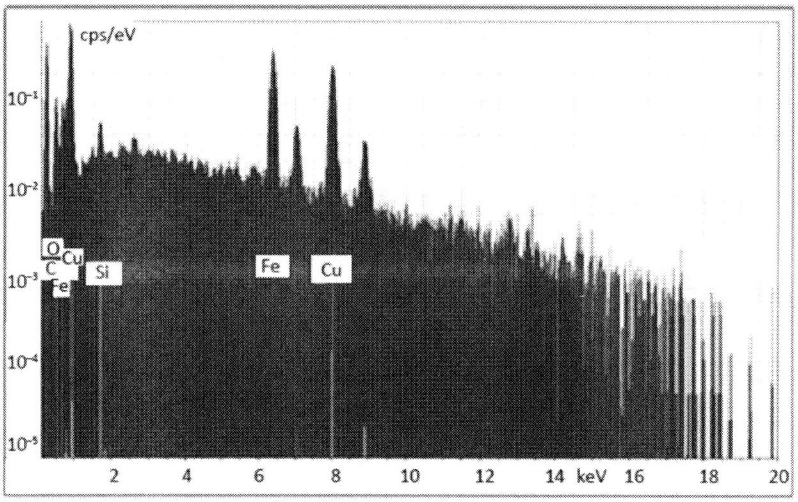

Figure 9. Chemical elements spectrum in metal-plating film on the surface of spheroidal graphite cast iron GJS 600 – 3 with 0.051% Sn, after friction at P = 10 N; n = 2300 min^{-1}.

Figure 9 shows the chemical elements spectrum in the composition of the metal-plating film. The wear curve of the counter body of steel C22 at friction in Regime I and the wear curve of the specimen of spheroidal graphite cast iron in the same regime of friction are similar (Figure 10).

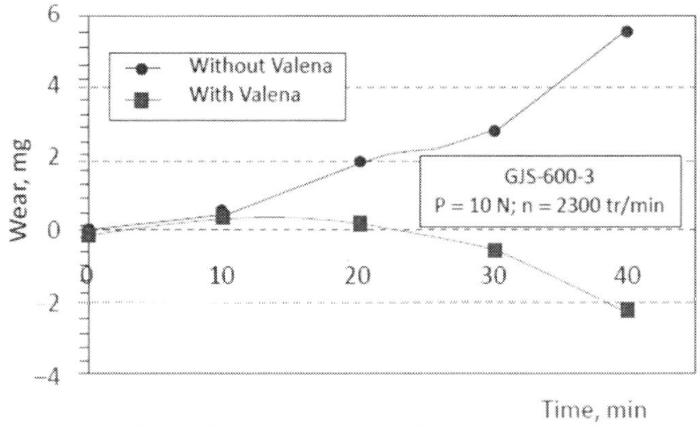

Figure 10. Influence Valena additive on the wear of the counter body of steel (C22).

Metal-plating copper films on the specimen of spheroidal graphite cast iron and on the counter body of steel form almost simultaneously. Figure 11 shows the microstructure and the composition of the metal-plating film on the surface of the counter body (C22).

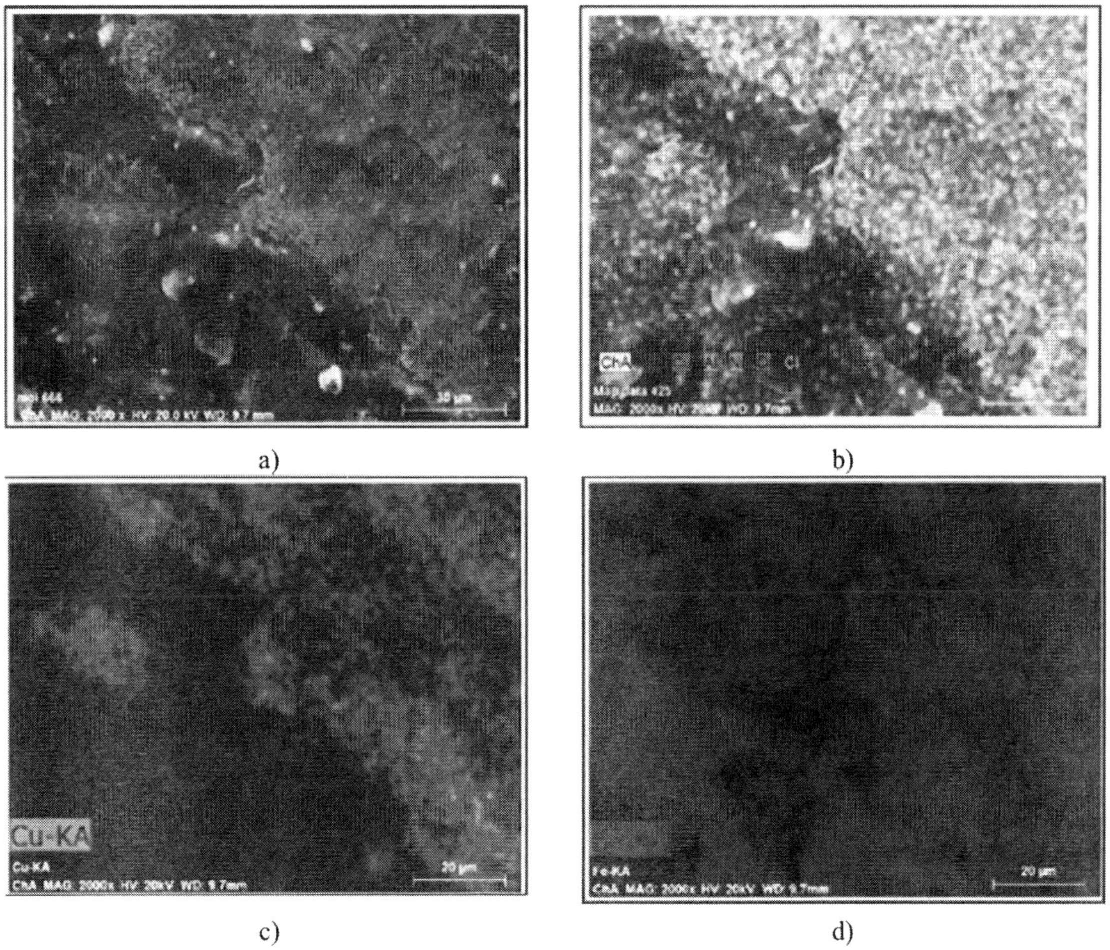

Figure 11. Microstructure (a), composition (b), microstructure with elements Cu (c), and Fe (d) in the metalplating film on the surface of the counter body (C22), after friction at P = 10 N; n = 2300 min^{-1}.

Roughness profiles of the surface before and after friction are obtained in radial direction of annular copper film (Figures 12, 13). The roughness of the surface before friction is Ra = 0.466 μm, and after friction the roughness of the deposited copper film is Ra = 0.196 μm.

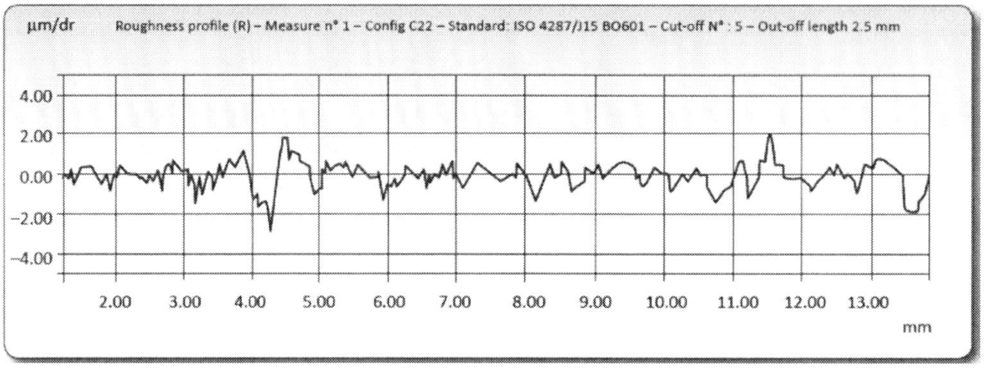

Figure 12. Profilogram of the counter body surface (C22) before wear: Ra = 0.466 μm.

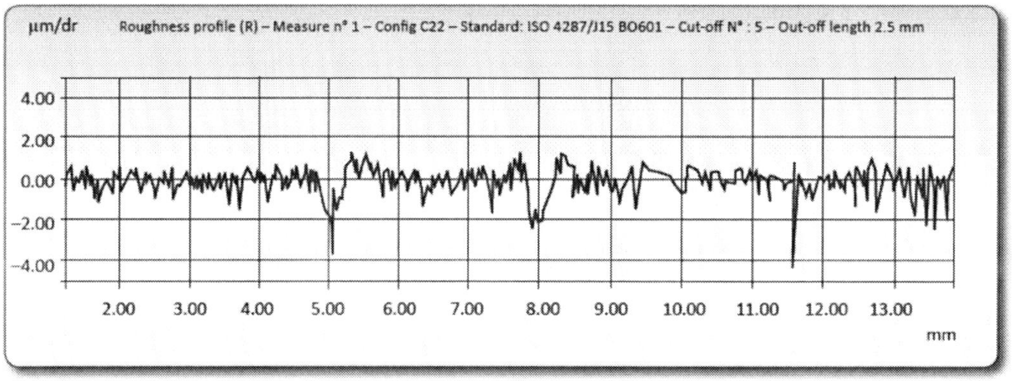

Figure 13. Profilogram of the counter body surface (C22) after wear: Ra = 0.196 μm at friction in regime P = 10 N; n = 2300 min^{-1}.

The width of the copper film is determined by the profilogram. It is 6 mm, i.e., twice larger than the width of the ring. The process of selective transfer occurs not only in the contact between the specimens, but also out of the contact zone. This fact is also found in other authors' studies [20, 30, 35, 38]. It is seen in the profilogram's indication from 5 to 11.5 mm.

Copper film thickness is no uniformly distributed both in radial direction and along the length of the trail. The average thickness film value is 3 μm.

Contact system "Steel – Steel" (Ст4кп – C22)

Figures 14 and 15 give the wear curves for specimen of steel Ст4кп in both regimes of friction. The influence of Valena on the character of variation is almost analogous with that in the case of spheroidal graphite cast iron micro alloyed by tin. In both regimes of friction is observed deposition of metal plating copper film upon the specimen and the counter body, which shows the presence of selective transfer process.

For the lighter regime of friction – P = 10 N; n = 2300 min^{-1}, effect of no-wear effect is observed, as it is in the case of spheroidal graphite cast iron specimens. The copper film thickness on the counter body is 2.5 μm.

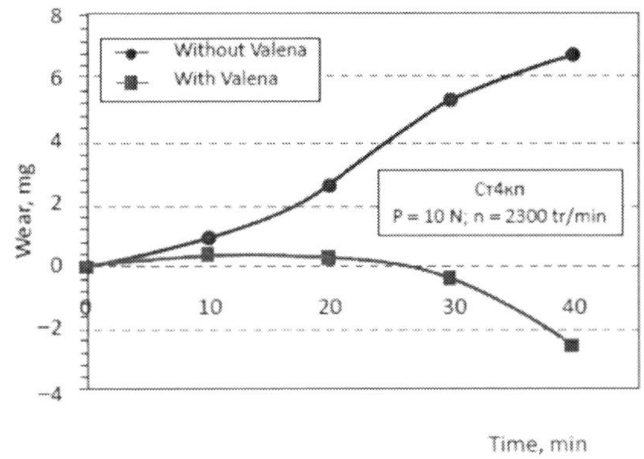

Figure 14. Influence of Valena additive on wear for Ст4кп under friction in Regime I.

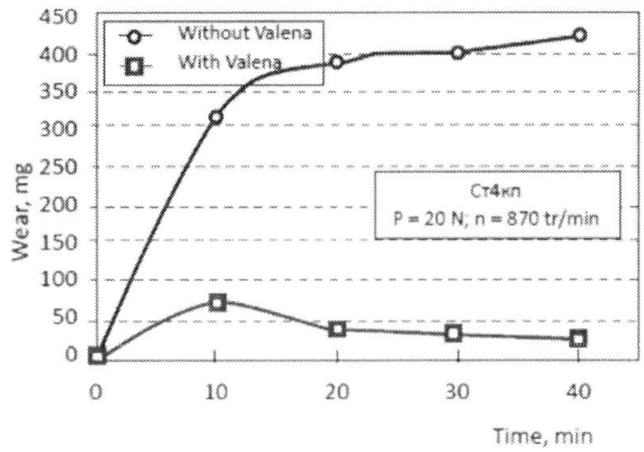

Figure 15. Influence of Valena additive on wear for Ст4кп under friction in Regime II.

The influence of Valena additive on lubricant temperature in both regimes of friction is shown in Figures 16 and 17. The increase of the temperature in both regimes decreases in the presence of Valena more than twice compared with the case without Valena additive.

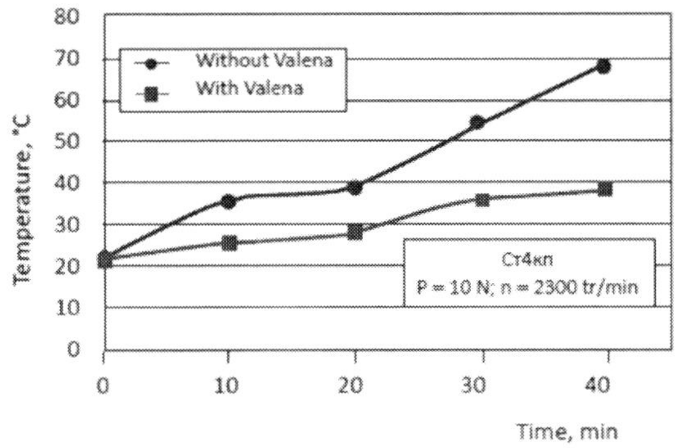

Figure 16. Variation of the temperature with time under friction in Regime I.

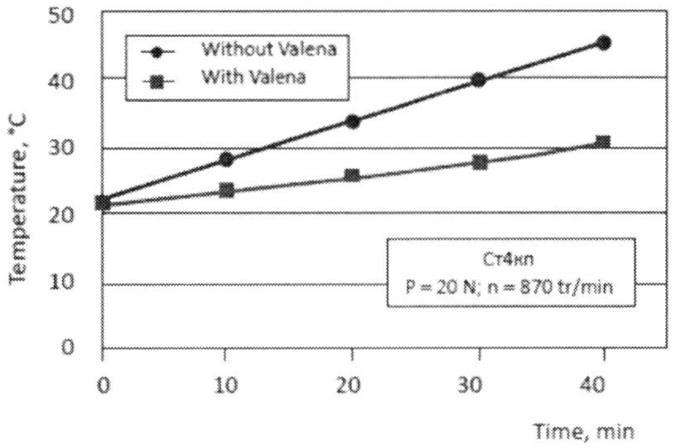

Figure 17. Variation of the temperature with time under friction in Regime II.

Contact system "Bronze – Steel" (БрАЖ9-4 – C22)

Significant wear decrease is observed during friction of bronze-steel couple in the first regime P = 10 N; n = 2300 min^{-1}, there is, however, no effect of specimen's mass increase as it was in the first two cases. Small wear of the bronze specimen is indicator that a selective transfer process. The thin copper film protects the surface from oxidation and destruction (Figure 18).

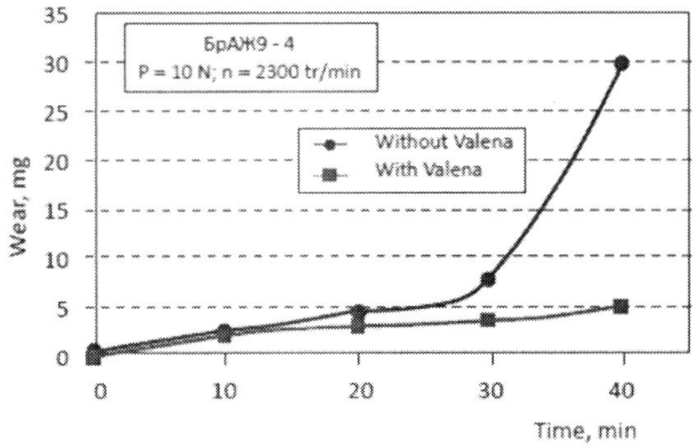

Fig. 18. Influence of Valena additive on wear of БрАЖ9-4 under friction in Regime I.

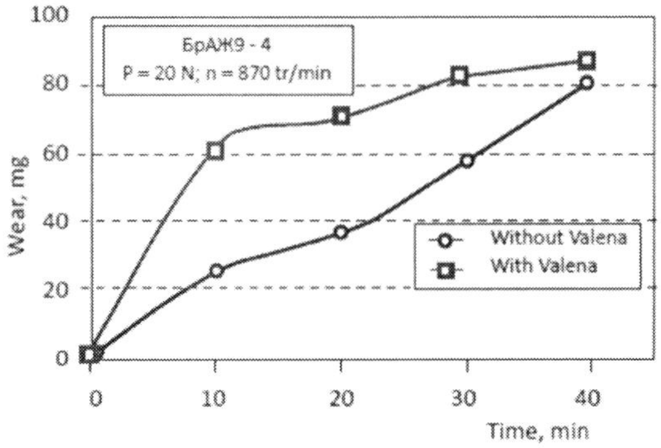

Fig. 19. Influence of Valena additive on wear of БрАЖ9-4 under friction in Regime II.

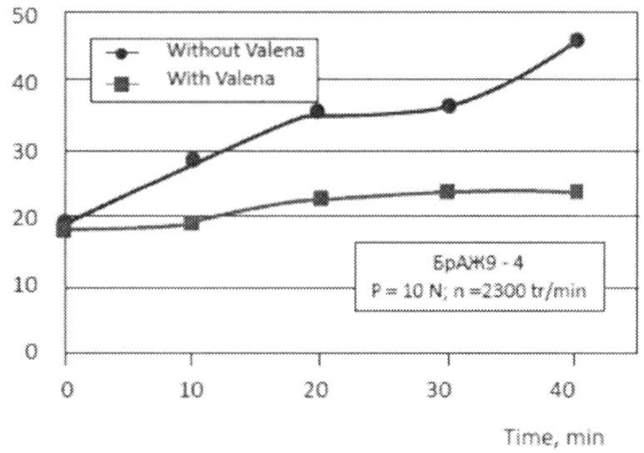

Figure 20. Variation of the temperature in time under friction in Regime I.

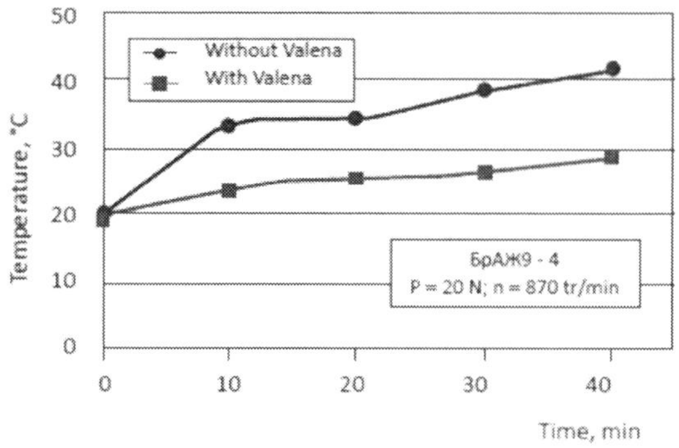

Figure 21. Variation of the temperature in time under friction in Regime II.

In the case of the second regime of friction P = 20 N; n = 870 min–1, quite high value of bronze wear is observed in the process of running-in when Valena additive available (Figure 19). In that case, there are no favourable conditions for selective transfer occurrence. The wear increase with load increase is explained with the character of the electrochemical contact processes. Microscopic study of the surface shows that cavities with liberated hydrogenare formed in some sections under the cathode copper, which causes detachment of copper flakes. As other authors' studies also show, saturation of the surface with hydrogen takes place under these conditions, which causes brittleness in some sections of the surface. The process of extraction of alloying elements from bronze and steel is more intensive at higher load. The extracted copper transferred to the steel surface does not succeed to crystallize, and forms crumbly copper film, which is not able to protect surfaces from wear [1, 3, 14].

One more feature affects the processes in the above considered case. The plastic deformation of the surface film increases under higher pressure, the real contact area increases, and respectively, the contact interstices decrease. Lubricant penetration in contact is hindered and direct mechanical interaction between asperities occurs. So, lubricant and additive lose their functional ability. Boundary friction and intensive wear take place.

The parameter wear intensity is announced below as criterion of comparative assessment for the tested materials. It is determined by the formula:

$$i = h / L, \qquad (1)$$

where $h = m/\rho.Aa$ is thickness of the worn film during friction way L = 1 m. Experimental results about wear intensity of the three materials are given in Table 7.

Table 7. Wear intensity [m/m] of materials in both regimes of friction without and with Valena.

Material	Regime I P = 10 N; P_a = 0.22 MPa n = 2300 min^{-1}		Regime II P = 20 N; P_a = 0.45 MPa n = 870 min^{-1}	
	Wear Intensity			
Steel Ст4кп	Litol 24	Litol 24 with Valena	Litol 24	Litol 24 with Valena
	$7x10^{-9}$	*No-wear* *Copper film*	$1.2x10^{-6}$	$6.6\ x10^{-9}$
Spheroidal graphite cast iron +0.051%Sn GJS-600-3	Litol 24 $17.8\ x10^{-9}$	Litol 24 with Valena *No-wear* *Copper film*	Litol 24 $0.11x10^{-6}$	Litol 24 with Valena $0.04\ x10^{-6}$
Bronze БрАЖ9-4	Litol 24 $32x10^{-9}$	Litol 24 with Valena $5.3x10^{-9}$	Litol 24 $0.2x10^{-6}$	Litol 24 with Valena $0.2\ x10^{-6}$

Figure 22 shows diagram of wear intensity for the materials at friction Regime I.

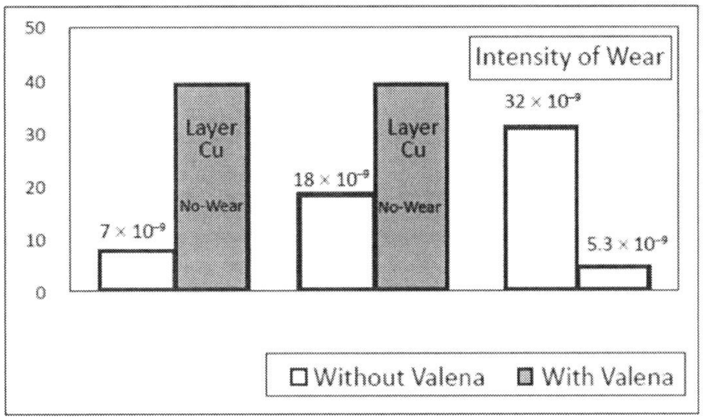

Figure 22. Wear intensity [m/m] of specimens at friction Regime I:
P = 10 N; n = 2300 min-1

4. Conclusions

The basic results of the present paper are:

1. Experimental data are gathered for the influence of metal-plating additive Valena on the contact characteristic of 3 types of: „GJS 600-3 - C22" (Spheroidal graphite Cast iron with 0.051%Sn – Steel), „Ст4кп - C22" (Steel - Steel) и „БрАЖ9-4 - C22" (Bronze – Steel) and their functioning at 2 friction regimes.
2. It has been found out that the additive Valena in grease Litol 24 in all three tribosystems the intensity of wear decreases and processes of selective transfer of the occur in both friction regimes.
3. In tribosystems „GJS 600-3 - C22" and „Ст4кп - C22" in friction regime at a contact pressure of P_a=0.22 MPa and revolutions of n = 2300 min^{-1} (Regime I) is observed a *no-wear effect under friction*. This effect is due to a proceeding mechanism of self-organization, where a dissipative structure is build up in the form of protective servovitic film with thickness of 2.5÷3 μm. The servovitic protective film is formed on the contact surfaces of the body and antibody of the tribosystem. The microstructure investigations of the microstructures and the chemical composition of the protective film with scanning electronic microscope shows the presence of the chemical element copper (Cu). The contact temperature decreased 2 to 4 times compared to the one in friction where there is no additive "Valena".

Acknowledgements

This study is connected with the completion and the financial support of the Project DUNK-01/3 "University R&D Complex for innovation and transfer of knowledge in micro/nano-technologies and materials, energy efficiency and virtual engineering" funded by the Bulgarian Ministry of Education and Science.

References

1. Glensdorf, P., I. Prigogine. Thermodynamic theory of structure, stability and fluctuations, Moscow: Mir, 280, 1973.
2. Nikolis G., I. Prigogine, Self-Organization in Nonequilibrium Systems, Moscow: Mir, 512, 1979.
3. Manolov, N., Contact Approach: Atomizing, Harmonization, Integration, Book Three, TEMTO, Sofia, 416, 2007 (in Bulgarian)

4. Garkunov, D. Tribotechnika. Wear and non-wear, Moscow: MCHA, 2001 (in Russian).
5. Garkunov, D. Tribotechnika. Designing, manufacturing, exploiting machines. Moscow, MCHA, 2002 (in Russian).
6. Garkunov, D. Scientific Discoveries in Tribotechnologies: No-Wear Effect Under Friction, Hydrogen Wear of Metals. Moscow, 2007.
7. Garkunov, D., I. Kragelsky, A. Polyakov. Elective transfer in friction nodes, Moscow: Transport, 100, 1969 (in Russian).
8. Polzer, G., Der Erfahrungsaustausch: Reibbeschichten und selective Übertragung. Publ. Bezirks-Neuerer-Zentrum, Gera, 1988.
9. Polyakov, A., Experience in the study of the dissipative structure of selective transport in a metal film in friction (review). - The effect of non-weariness and tribotechnology, № 3–4, 30–44, 1996 (in Russian).
10. Balabanov, V., V. Beklemyshev, A. Hamidov, S. Ishchenko, I. Makhonin, V. Filippov. An indiscriminate car service. Moscow: Izvestia, 135–159, 2007 (in Russian).
11. Balabanov, V. Self-organization in friction - a new trend in tribology. Moscow: MGAU, 8–15, 2002 (in Russian).
12. Balabanov, V., M. Kandeva, The effect of non-featurelessness and indiscriminate service of auto tractor technology in the process of exploitation, Agrarian science in conditions of modernization and innovative development of the agro industrial complex of Russia, Collection of materials of the All-Russian scientific and methodological conference with international participation, Anniversary of Academician DK Belyaev, volume 3, March 2, 2017, Ivanovo, FGBOU VO Ivanovskaya, 20-23, 2017 (in Russian).
13. Assenova, E., M. Kandeva. Self-organization in tribology: the role of synergism and selective transfer. – In: Proceedings of 7th International Conference on Tribology Balkantrib, pp. 305–310, Thessaloniki, 2011.
14. Kandeva, M., B. Ivanova, E. Assenova and A. Vencl. Influence of Additives and Selective Transfer on Wear Reduction in the Lubricated Contact. – Proc. of the Int. Conf. on Materials, Tribology, Recycling — MATRIB 2014, Vela Luka, Croatia, 197-206, 2014.
15. Kandeva, M., D. Karastoyanov, E. Assenova, K. Jakimovska, S. Simeonov, A. Vencl. Iinfluence of oil-soluble additive Valena on the friction in a "journal-bush" tribological system. – Journal of Friction and Wear, Vol. 37, No 2, 2016.
16. Metal-containing oil-soluble composition for lubricating materials: Pat. RF No. 2277579 / VG Babel, D.N. Garkunov, S.M. Mamykin, PI Kornik (in Russian).
17. Kandeva, M., Al. Vencl, D. Karastoyanov, Advanced tribological coatings for Heavy-duty applications: Case studies, Academy publishing house, Sofia, Bulgaria, 147, 2016.

Effects of Process Parameters on Copper Powder Compaction Process Using Multi-Particle Finite Element Method

***Güner F*[1], *Sofuoğlu H*[2]**
[1]*Giresun University, Department of Mechanical Engineering, 28200, Giresun, Turkey*
[2]*Karadeniz Technical University, Department of Mechanical Engineering, 61080, Trabzon, Turkey*
faruk.guner@giresun.edu.tr[1], *sofuoglu@ktu.edu.tr*[2]

Abstract: Powder metallurgy (PM) has been widely used in several industries; especially automotive and aerospace industries and powder metallurgy products grow up every year. The mechanical properties of the final product that is obtained by cold compaction and sintering in powder metallurgy are closely related to the final relative density of the process. The distribution of the relative density in the die is affected by parameters such as compaction velocity, friction coefficient and temperature. Moreover, most of the numerical studies utilizing finite element approaches treat the examined environment as a continuous media with uniformly homogeneous porosity whereas Multi-Particle Finite Element Method (MPFEM) treats every particles as an individual body. In MPFEM, each of the particles can be defined as an elastic- plastic deformable body, so the interactions of the particles with each other and the die wall can be investigated. In this study, each particle was modelled and analyzed as individual deformable body with 3D tetrahedral elements by using MPFEM approach. This study, therefore, was performed to investigate the effects of different temperatures and compaction velocities on stress distribution and deformations of copper powders of 200 μm-diameter in compaction process. Furthermore, 3-D MPFEM model utilized von Mises material model and constant coefficient of friction of μ=0.05. In addition to MPFEM approach, continuum modelling approach was also performed for comparison purposes.
Keywords: Multi-Particle Finite Element (MPFEM), Powder Metallurgy, Compaction Process, Compaction Velocity

1. Introduction

Optimization of the mass production parameters of Powder Metallurgy (PM) products, which are continuously increasing in industrial use, is very important in terms of reducing the production costs. Experimental determination of compaction process parameters is more costly than numerical methods.

Numerical methods are more efficient in terms of cost reduction, fast results, and close follow-up of parameters. Discrete Element Method (DEM), Finite Element Method (FEM) or Smoothed Particle Hydrodynamics (SPH) are some of the proper numerical methods to analyze compaction processes [1, 2].

Being an established model for the analysis of rigid particles, DEM is able to make elastic plastic analyses integrated with FEM as a result of the developments in time [3]. Mazor et al. [4] investigated roll compaction process with a combined Discrete and Finite Element method and figures out rolling process parameters.

However, the FEM shows more successful results. FEM analyses are also examined in two groups of approaches; continuum media approaches [5] and multi particle finite element method (MPFEM) [6]. The area to be investigated in the continuum approach is considered to be a solid body with a certain relative density. However, in MPFEM, each particle is defined as an elastic plastic independent body, and interactions between the particles can be examined [7]. Stresses, deformation, friction coefficients, particle interactions can be studied with more accuracy by the MPFEM approach [8].

The final relative density, punching force, particle distribution and pressing velocity are important in modelling the of compaction process. Yan et al. [9] studied compaction process of titanium powders to obtain process characterization. Moreover, Yan et al [10], investigated different particle size in order to get the size effect on compaction process.

In this study, copper powders in a diameter of 200 microns were compacted with temperature parameter of 20°C, 270°C, 400°C and 541°C in order to investigate temperature effect at 0.4mm/s, 0.8mm/s, 1.6mm/s, 4mm/s and 8mm/s compaction velocities with a proper material model. Work hardening elasto-plastic von Mises material model utilized to model copper in powder compaction process, similar to the previous studies given in literature [7, 11, 13]. Numerical results obtained from analyses were evaluated in terms of stress and particle distribution.

2. Numerical Studies

2.1. Finite Element Model

The geometrical details of numerical model is shown in Figure 1. The teeth angle of the die is 45° while ratio of the tooth width to its height is 0.5. The geometry of numerical model was taken from the study of Cora et al. [14]. 130 spherical particles with 200μm diameter was filled in the die. In aim to establish a realistic model, particles filled under effect of gravity using SolidWorks software (Solidworks Corp., Waltham, MA, USA). The pre-model meshed by 93000 tetrahedral elements with the capability of Patran software (MSC Software Corp., Santa Ana, CA, USA). Later on the model was imported into commercial Finite Element Analyses (FEA) package MSC Marc (MSC Software Corp., Santa Ana, CA, USA) and numerical analyses of compaction process was then performed by MSC Marc Mentat Solver.

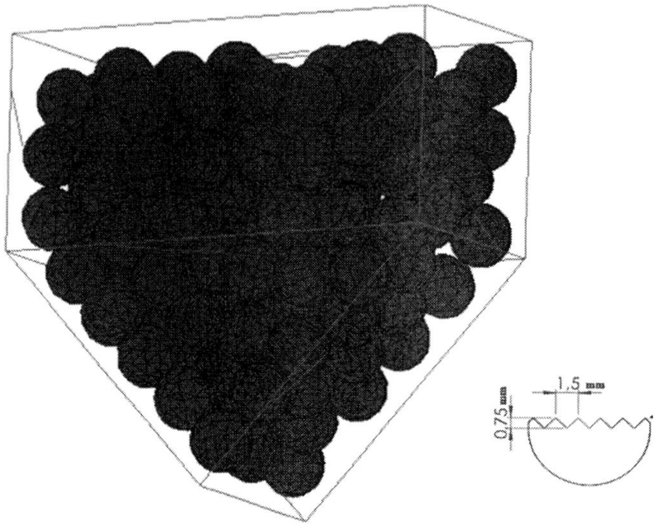

Figure 1 Numerical model (MPFM) of compaction process.

2.2. Material Model

In all FE analyses, von-Mises yield function along with power law was utilized in aim to represent work hardening issue [12]. The constitutive equation of von-Mises yield function is given in Eq. (1)

$$\sigma_Y = A(\varepsilon_0 + \bar{\varepsilon})^m + B\dot{\bar{\varepsilon}}^n \tag{1}$$

where σ_y is the yield stress, ε_o is initial yield strain, $\bar{\varepsilon}$ is equivalent strain, $\dot{\bar{\varepsilon}}$ is equivalent strain rate. The other material parameters were tabulated in Table 1. Parameter B takes very small value for 20°C and 270°C analyses so that it was ignored [15]. The parameters were taken from Güner at al. [11] or calculated by the similar method given in the mentioned reference where it is necessary.

Table 1. Material parameters used in numerical analyses

Material Parameter	20°C	270°C	400°C	541°C
A	451.6 (MPa)	296.7 (MPa)	64.56 (MPa)	42.12 (MPa)
B	0	0	202.16 (MPa)	151.91 (MPa)
m	0.324	0.3655	0.167	0.149
n	0	0	0.0079	0.0858
E	110 (GPa)	99.3 (GPa)	62 (GPa)	38.2 (GPa)
υ	0.35	0.35	0.35	0.35
σ_0	28.8 (MPa)	10.49 (MPa)	16.3 (MPa)	12.79 (MPa)

2.3. Friction Model

Amonton-Coulomb friction model which states friction stress as a portion of normal stress used for numerical analyses. A constant coefficient of friction that is shown by μ_c in Eq.2 expresses the proportionality of normal and friction stresses. $\boldsymbol{\tau}$ and $\boldsymbol{\sigma_n}$ stand for the shear and the normal stresses in Eq.2, respectively. A value of 0.05 was applied as a constant coefficient of friction to all MPFEM analyses [11].

$$\tau = \mu_c.\sigma_n \tag{2}$$

3. Results and Discussion

The numerical analyses was performed by Intel I7 2.20 GHz microprocessor along with 6 GB of RAM equipped PC. The mean analysis time was 43 hours and the study was completed with 20 successful analyses. Compaction processes under 0.4 mm/s, 0.8 mm/s, 1.6 mm/s, 4 mm/s and 8 mm/s punch velocities was analyzed at temperatures of 20°C, 270°C, 400°C and 541°C. 130 spherical copper powders were compacted in a die volume of 1.11 mm^3. Relative density (RD) which is the ratio of total particle volume to die volume changed from 49% to 79% throughout the compaction process.

Figure 2 shows the equivalent von Mises stress distribution of MPFEM analyses of copper powder compaction at 541°C. Maximum equivalent stress at 0.4 mm/s punch velocity is 205 MPa while it increases to 247 MPa at 8 mm/s punch velocity. The particles showed more homogenous stress distribution for the maximum value at 0.4 mm/s analyses. On the other hand, analyses at 8 mm/s, particles had less deformed shapes around the discontinuities of the die. The particles those are contacting with the punch and discontinuities of the die had more deformed shape although the stresses were not differ from the other particles. In their study of cold powder compaction by using the discrete element method Jerier et al [16] expressed that stress showed sudden increase after 80%RD

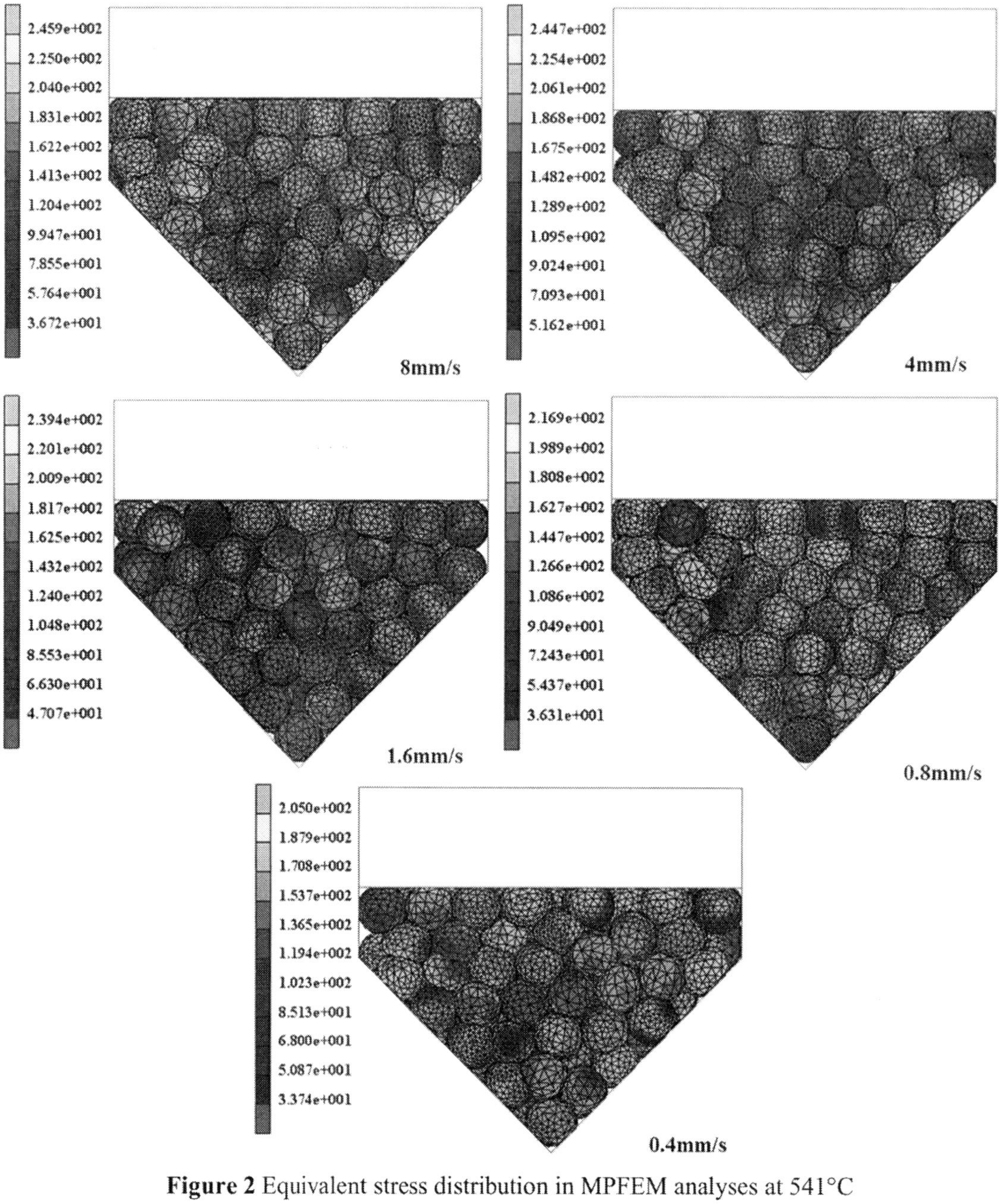

Figure 2 Equivalent stress distribution in MPFEM analyses at 541°C

Figure 3 shows equivalent stress distribution of MPFEM analyses at 270°C. In the analyses of 0.4 mm/s, the particles had more homogenous stress and deformation distribution. It can be seen that the particles were generally in blue or red in 0.4mm/s analyses. The particles especially contacting with the punch were exposed to excessive stresses on which maximum value of analyses were more apparent on them at punch velocities of 4 mm/s and 8 mm/s. In 8 mm/s punch velocity, the particles did not have enough time to rearrange and fill the gaps of die homogeneously.

The MPFEM analyses of copper powder compaction at 20°C and 400°C had similar stress distributions. The maximum equivalent von-Mises stress values of those analyses were given in Table 2. In all analyses maximum stress values increase in parallel with punch velocity increase. Also maximum value of equivalent von Mises decrease by temperature increase due to yield stress.

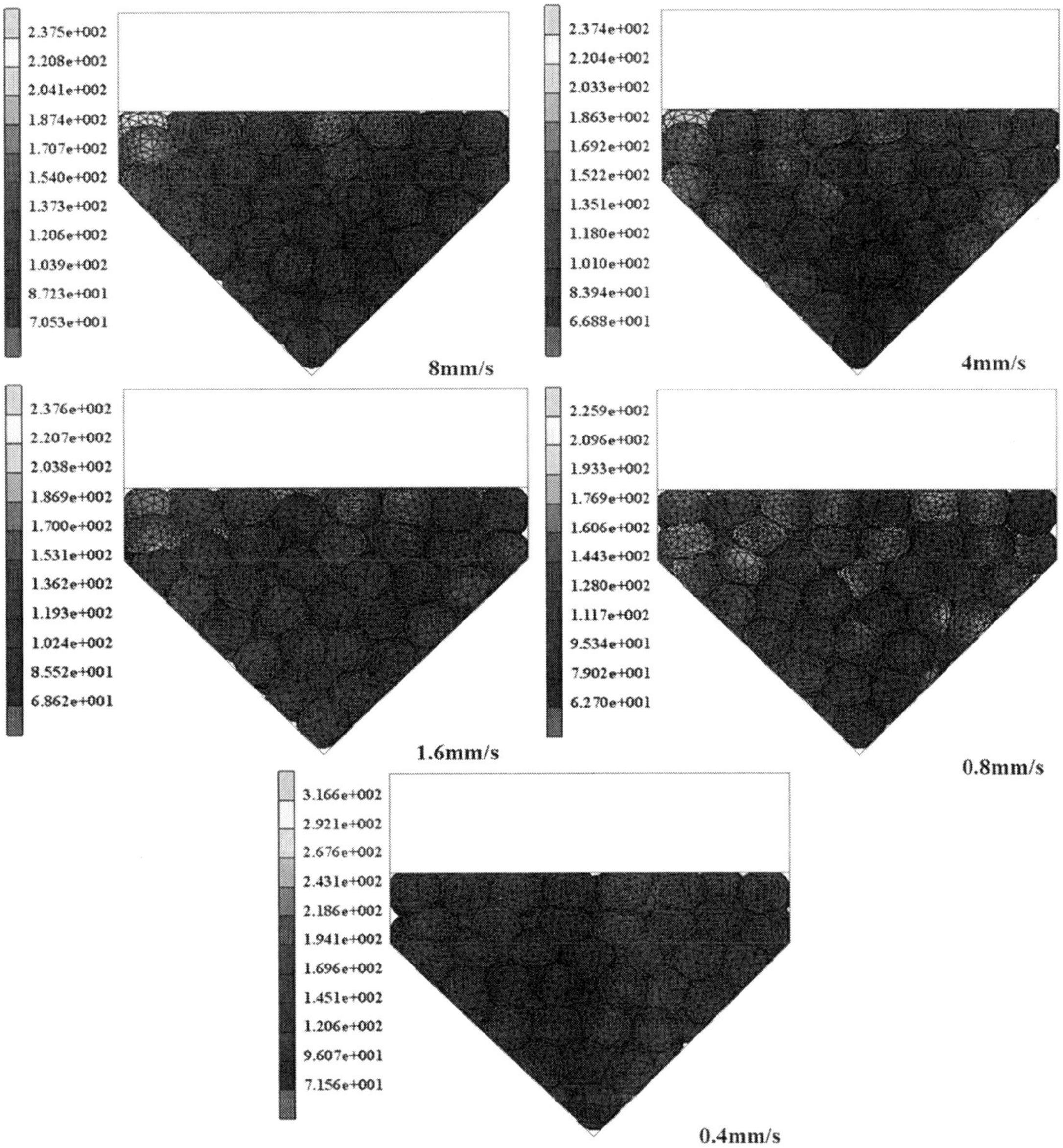

Figure 3 Equivalent stress distribution in MPFEM analyses at 270°C

Table 2. Maximum Equivalent von Mises stress values of MPFEM analyses

Max. Equivalent von Mises Stress (MPa)				
	20°C	270°C	400°C	541°C
0.4mm/s	366	316	282	205
0.8mm/s	354	225	283	216
1.6mm/s	402	237	291	239
4mm/s	343	237	315	244
8mm/s	372	237	314	247

Figure 4 shows the punch stress change via relative density. Except 1.6 mm/s analyses, punch stress arises after 61% RD cause of particle rearrangement. Particles were being rearranging until RD reached 61% and inter particle actions became more stable after this value of RD. The punch stress for 20°C MPFEM analyses varies between 160 MPa-180 MPa. The parabolic trend line of punch stress after 61% RD shows good agreement with the study of Jerier et al [16].

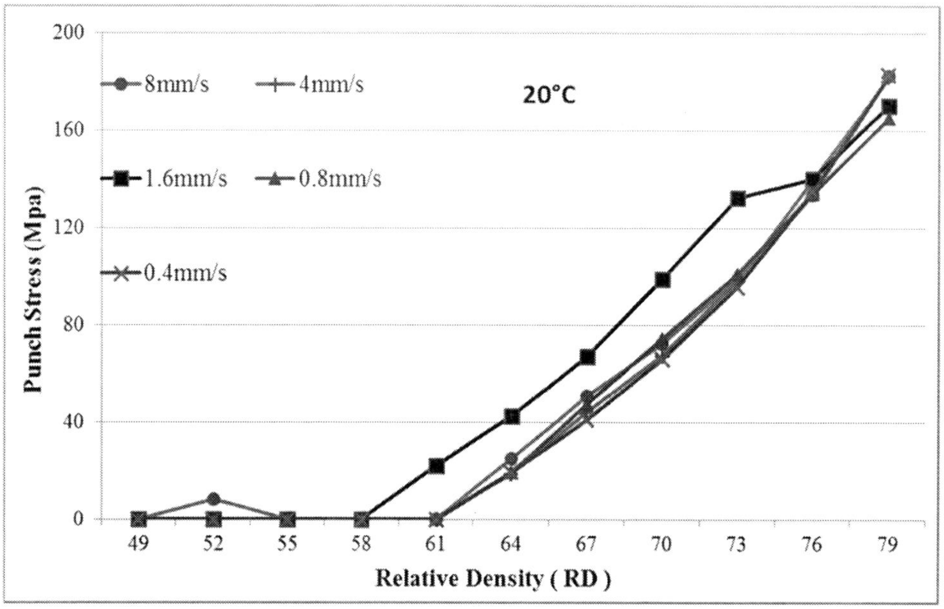

Figure 4 Punch stress of MPFEM analyses at 20°C

The punch stresses at the end of all MPFEM analyses are given in Figure 5. Although the stresses at the punch velocities of 8 mm/-s and 0.4 mm/s are close to each other, a sudden change in the stresses around the speed of 1.6mm/s has been achieved in Figure 5. The punch stresses differ from 110 MPa to 180 MPa due to varying temperature and punch velocity. Majzoobi and Jannesari [17] investigated copper powder compaction process experimentally in order to get cap material model parameters and figured out a punch stress varying between 120-160 MPa showing a good agreement with the present study.

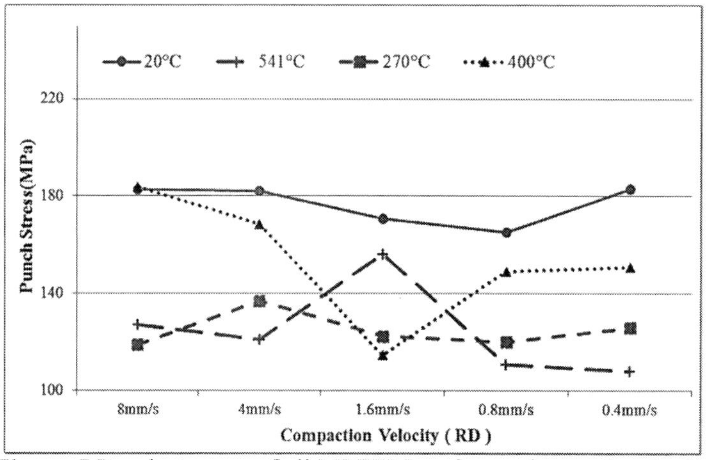

Figure 5 Punch stresses of all MPFEM analyses via punch velocity

4. Conclusion

MPFEM analyses of copper powder compaction process at 0.4 mm/s, 0.8 mm/s, 1.6 mm/s, 4 mm/s and 8 mm/s has performed in 20°C, 270°C, 400°C and 541°C in order to investigate effects of punch velocity and temperature. Within the limitation of study, followings are obtained:

1- The punch stresses of all MPFEM analyses varied from 100MPa-180MPa with a good agreement in the literature.

2- The punch stress showed sudden changes around 1.6 mm/s punch velocity so that this value of velocity can be named as a critical value.

3- If the punch velocity is low, particles can be rearrange in die so that a homogenous stress distribution can be observed in die.

4- High punch velocities can cause excessive shape deformation on particle especially around the punch and the zones where die has geometrical discontinuities.

References

1. Jerier J.F. and Molinari J.F., Normal Contact between Rough Surfaces by Discrete Element Method, Tribology International, 47 (2012) 1–8
2. Tan Y., Yang D. and Sheng Y., Study of Polycrystalline Al2O3Machining Cracks Using Discrete Element Method, International Journal of Machine Tools &Manufacture 48 (2008) 975–982
3. Gustafsson G., Häggblad H.A. and Oldenburg M., Smoothed Particle Hydrodynamic Simulation of Iron Ore Pellets Flow, AIP Conference Proceedings, 908, 1483, 10.1063/1.2741018.
4. Mazor A., Orefice L. and Michrafy A., A Combined DEM & FEM Approach for Modelling Roll Compaction Process, Powder Technology xxx (2017) xxx–xxx
5. Bier W., Dariel M.P., Frage N., Hartmann S. and Michailov O., Die Compaction of Copper Powder Designed For Material Parameter Identification, International Journal of Materials Sciences 49 2007 766-777
6. Dandekar C-R. and Shin Y-C., Modeling of Machining of Composite Materials: A Review, International Journal of Machine Tools &Manufacture, 57 (2012) 102–121
7. Guner F., Cora Ö.N. and Sofuoğlu H., Numerical Modeling of Cold Powder Compaction Using Multi Particle and Continuum Media Approaches, Powder Technology, 271 (2015) 238-247
8. Zhou Y., Wang Z.Y. and Ma X.L. Yu, Finite Element and Experimental Studies of The Formation Mechanism of Edge Defects During Machining of Sicp/Al Composites, International Journal of Machine Tools &Manufacture, 84 (2014) 9–16
9. Yan Z, Chen F. and Cai Y., High-Velocity Compaction of Titanium Powder And Process Characterization, Powder Technology, 208 (2011) 596–599
10. Yan Z, Chen F., Cai Y. And Yin J., Influence of Particle Size on Property of Ti−6Al−4V Alloy Prepared by High-Velocity Compaction, Trans. Nonferrous Met. Soc. China 23(2013) 361−365.

11. Guner F., Sofuoğlu H. and Cora Ö.N., An Investigation of Contact Interactions in Powder Compaction Process Through Variable Friction Models, Tribology International, 96 (2016) 1-10
12. Guner F., Cora Ö.N. and Sofuoğlu H., Numerical Modeling of Cold Powder Compaction Using Multi Particle and Continuum Media Approaches, Powder Technology, 271 (2015) 238-247
13. Han P., AnX., Zhang Y., Huang F., Yang T., Fu H., Yang X. and Zou Z., Particulate Scale MPFEM Modeling on Compaction of Fe and Al Composite Powders, Powder Technology 314 (2017) 69–77
14. Cora O.N., Usta Y., and Koc M., Micro-Manufacturing of Micro-Scale Porous Surface Structures for Enhanced Heat Transfer Application: An Experimental Process Optimization Study, Journal of Micromechanics Microengineering, 19 (2009) 045011
15. Kalpakjian S., Manufacturing Processes for Engineering Materials, 5th Ed., Pearson Education, New Jersey, 2010
16. Jerier J.-F., Hathong B., Richefeu V., Chareyre B., Imbault D., Donze F.-V. and Doremus P., Study of Cold Powder Compaction By Using The Discrete Element Method, Powder Technology 208 (2011) 537–541
17. Majzoobi G.H. and Jannesari S., Determination of The Constants of Cap Model For Compaction of Three Metal Powders, Advanced Powder Technology 26 (2015) 928–934

IOP Conf. Series: Materials Science and Engineering **295** (2018) 012028 doi:10.1088/1757-899X/295/1/012028

Synergy and Self-organization in Tribosystem's evolution. Energy Model of Friction.

***Fedorov S V*[1], *Assenova E*[2]**
[1]*Department of Theory of Mechanisms, Machines and Machine Parts, Kaliningrad State Technical University, Kaliningrad, Russia*
[2]*The Society of Bulgarian Tribologists, Sofia, Bulgaria*
fedorov@klgtu.ru [1], emiass@abv.bg[2]

Abstract: Different approaches are known to treat self-organization in tribosystems, related to the structural adaptation in the formation of dissipative surface structures and of frictional or tribo-films, using of synergistic modifying of layers and coatings, e.g. of the selective material transfer during friction, etc. Regarding tribological processes in contact systems, self-organization is observed as spontaneous creation of higher ordered structures during the contact interaction. The proposed paper considers friction as process of transformation and dissipation of energy and process of elasto-plastic deformation localized in thin surface layers of the interacting bodies. Energetic interpretation of friction is proposed. Based on the energy balance equations of friction, the evolution of tribosystems is followed in its adaptive-dissipative character. It reflects the variable friction surfaces compatibility and the nonlinear dynamics of friction evolution. Structural-energy relationships in the contacting surfaces evolution are obtained. Maximum of tribosystem's efficiency during the evolution is the stage of self-organzation of the friction surface layers, which is a state of abnormal low friction and wear.

Keywords: friction, energy, adaptation, dissipation, synergy and self-organization

1. Introduction

Friction is a global natural phenomenon of transformation and dissipation of the energy of external relative motion. Friction strictly obeys the energy balance equations. In the most general case it shows an adaptive-dissipative character. The deformable contact body evolves and adapts over time during friction showing well-defined stages in the evolution. Most characteristic is the property of self-organization [1], which is closely related to the synergy, i.e. the common mutual activities of the equilibrium states. When starting a self-organization process, the macro-synergy of the two opposite balance tendencies in the energy transformation, changes in an exclusive nano-synergy of frictional contact subsystems' states with properties of an attractor.

2.Thermodynamic analysis of friction process

Being a global phenomenon in nature, friction firmly obeys energy balance equations (figure 1). In terms of thermodynamics [2-4], friction results of two simultaneous interrelated and opposite tendencies: on the one hand, storage of the latent (potential) energy ΔU_e of various defects and structure damages in the contact volumes and, on the other hand, release (dissipation) of that energy Q due to various relaxation processes. The first trend defines the effect of deformation strengthening and integrally characterizes damage and fracture (a state parameter). The second trend defines the heat effect of friction; it is responsible for the quasi-viscous component of the process.

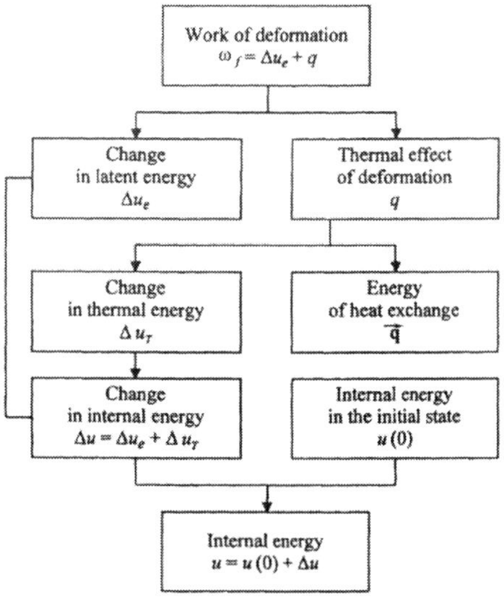

Figure 1. Scheme of energy balance of of solid bodies plastic deformation process [2, 3].

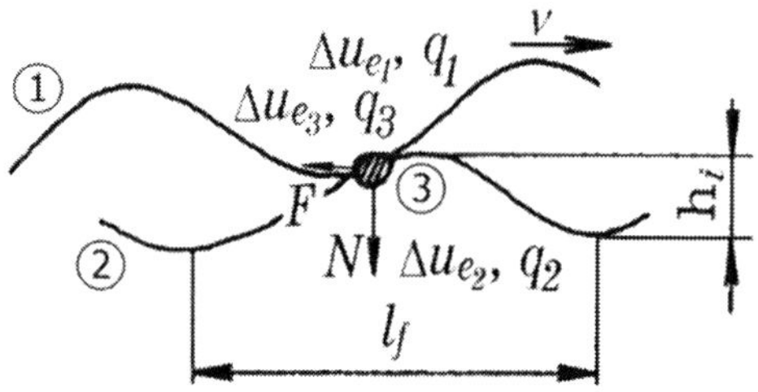

Figure 2. Conventional scheme of the elementary friction contact [2].

According to the energy balance scheme of plastic deformation and destruction processes (figure 1) [2, 3], we can present ω_f, i.e. the work of external forces needed for overcoming the friction F during the way l in a unit volume deformable contact (figure 1), as follows:

$$\omega_f = \Delta u_e + q \qquad \text{or} \qquad \dot{\omega}_f = \dot{u}_e + \dot{q} \tag{1}$$

where: $\dot{\omega}_f = d\omega_f/dt$ is the friction power (energy dissipation); $\dot{u}_e = du_e/dt$ is the rate of latent energy storage in the deformable (contact) volumes; $\dot{q} = dq/dt$ is the power of the heat effect of friction plastic deformation.

During friction, the contact volumes of both tribocouple materials undergo deformation, hence, we have to present equations (1) as generalized equations of friction surfaces compatibility [2], having in view that, in fact, they describe the common work (the deformation) of the friction surfaces:

$$\omega_f = \Delta u_{e1} + \Delta u_{e2} + q_1 + q_2\,; \tag{2}$$

$$\dot{\omega}_f = \dot{u}_{e1} + \dot{u}_{e2} + \dot{q}_1 + \dot{q}_2\,. \tag{3}$$

In general, under dry friction conditions, we should give the energy balance equations accounting the real (not unique) contact volume V_f :

$$W_f = \Delta U_e + Q = \Delta U_{e_1} + \Delta U_{e_2} + \Delta U_{T_1} + \Delta U_{T_2} + \vec{Q}_1 + \vec{Q}_2, \tag{4}$$

$$\dot{W}_f = \dot{U}_e + \dot{Q} = \dot{U}_{e_1} + \dot{U}_{e_2} + \dot{U}_{T_1} + \dot{U}_{T_2} + \dot{\vec{Q}}_1 + \dot{\vec{Q}}_2, \tag{5}$$

$$F_l = \frac{\Delta U_e}{l} + \frac{Q}{l} = \frac{\Delta U_{e_1} + \Delta U_{e_2}}{l} + \frac{Q_1 + Q_2}{l}, \tag{6}$$

$$F_v = \frac{\dot{U}_{e_1} + \dot{U}_{e_2}}{v} + \frac{\dot{Q}_1 + \dot{Q}_2}{v} = F_{mechanical} + F_{molecular}, \tag{7}$$

$$\mu_l = \frac{\Delta U_{e_1} + \Delta U_{e_2}}{Nl} + \frac{Q_1 + Q_2}{Nl} = \mu_{adapt} + \mu_{dis} = \mu_{adapt} + \mu_{T(dis)} + \mu_{\bar{Q}(dis)}, \tag{8}$$

$$\mu_v = \frac{\dot{U}_{e_1} + \dot{U}_{e_2}}{Nv} + \frac{\dot{Q}_1 + \dot{Q}_2}{Nv} = \mu_{deformation} + \mu_{adhesion} \tag{9}$$

Here μ is the friction coefficient; μ_{adapt} - the adaptive friction coefficient (of Amonton's Law); $\mu_{T(dis)}$ and $\mu_{\bar{Q}(dis)}$ - static and dynamic components of dissipative friction coefficient μ_{dis}; $\Delta U_e = \Delta u_e \cdot V_f$, ΔU_T - potential and heat components of internal energy; $\dot{U}_e = \dot{u}_e \cdot V_f$, ΔU_T - rates of change of the potential and heat components of internal energy; Q, $\dot{Q}$ - heat effect of friction and friction power, i.e.: $Q = \Delta U_T + \vec{Q}$, $\dot{Q} = \dot{U}_T + \dot{\vec{Q}}$; $\vec{Q}, \dot{\vec{Q}}$ - heat effect of the dynamic dissipation and its power; V_f - friction volume; N - normal load; l - friction way. The density of hidden energy Δu_e is an integral parameter of the tribo-state and fracture ability (of destruction (Δu_e^*) [3, 4]). Indices in equations (6) – (9) identify the friction parameters in time unit v and friction way l.

The correlations between energy balance components Δu_{e1} and Δu_{e2}, Δu_{T_1} and Δu_{T_2}, as well as $\vec{q}_1$ and $\vec{q}_2$ in equations (1) - (9) of the friction process vary widely depending on the physico-chemical properties and structure of contacting bodies materials, and on the friction conditions. The diversity of value ratios in the energy balance components actually determines the whole variety of particular (limit) displays of friction process [5, 6].

For the particular case of friction localization in third body volume (zone of compatibility (see below figure 6)), the equations (4) and (5) transform in:

$$W_T = \Delta U_{e3} + Q_3; \tag{10}$$

$$\dot{W}_T = \dot{U}_{e3} + \dot{Q}_3. \tag{11}$$

This way of recording the energy balance equations during friction is not inconsistent with above considerations, but complements them and has a larger physical validity. Here the index 3 marks energy balance components for the case of friction in the third body volume (the secondary structures), which are analogical to equations (4) and (5).

Thus, the thermodynamic analysis of the processes of plastic deformation and destruction of solid body contact volume allows us to get generalized (binomial) dependencies for friction.

3. Energetic interpretation of friction coefficient

For the case of low speed slip and small heat effect of friction $(Q \cong 0)$ energy balance equation (8) gives the Amontons' friction coefficient in a generalized form, as follows:

$$\mu = \frac{F}{N} = \frac{\Delta U_e}{Nl}; \quad F = \frac{\Delta U_e}{l}; \qquad Q \cong 0. \tag{12}$$

IOP Conf. Series: Materials Science and Engineering **295** (2018) 012028 doi:10.1088/1757-899X/295/1/012028

So, the friction coefficient of Leonardo da Vinci (G. Amontons) has a deep physical meaning. On the one hand, this is a parameter characterizing in general terms the resistance to the relative displacement (movement) of the surfaces, because it reflects the fraction of external motion energy that is "destroyed" by friction in the form of stored (potential) hidden energy ΔU_e. On the other hand, this is a generalized characteristic of the damageability, because the coefficient of friction is determined by the density of the latent energy Δu_e characterizing the degree of defectiveness of the structure and being a generalized parameter of damageability [3, 4]. Here we can also say that the coefficient of friction uniquely reflects in general terms the structural perfection (imperfection) of the deformable contact volume [2, 3], because the parameter Δu_e (ΔU_e) is determined by the energy of different types of defects and damages, being generalized characteristic of structure (structure parameter) in accordance with the main conclusion of the thermodynamic theory of strength [3].

4. Generalized experimental friction curves

The dependences obtained for the friction coefficient μ are in agreement with the experimental curves $\mu = \mu(N, v)$ (figures 3-5).

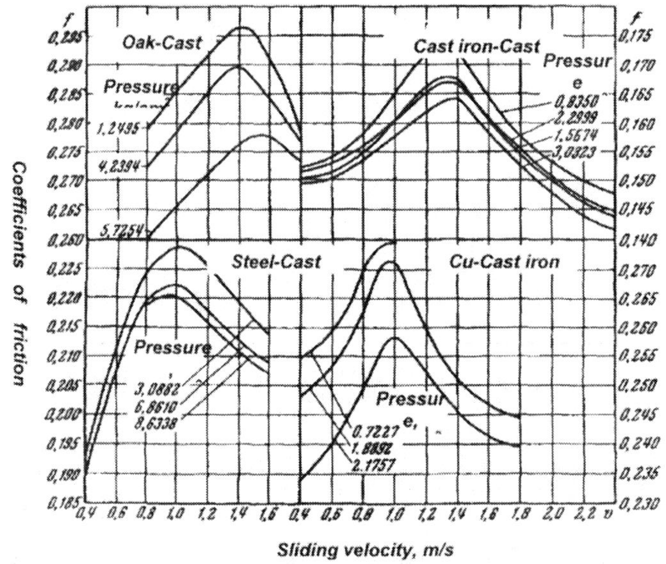

Figure 3. P. Conti's experimental results [7]

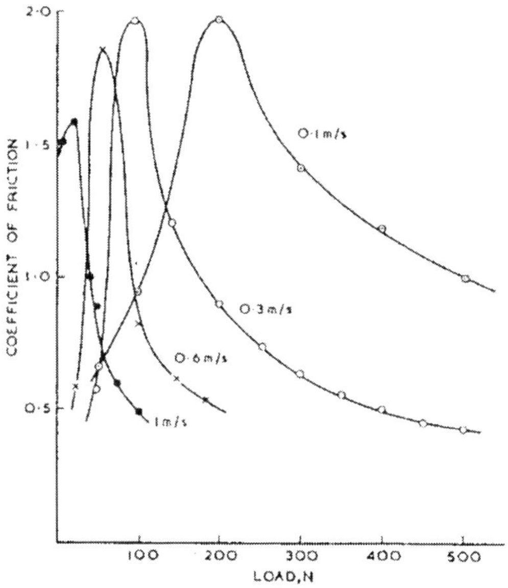

Figure 4. Watanabe's experimental results of Nylon 6 – Steel pair of friction [8]

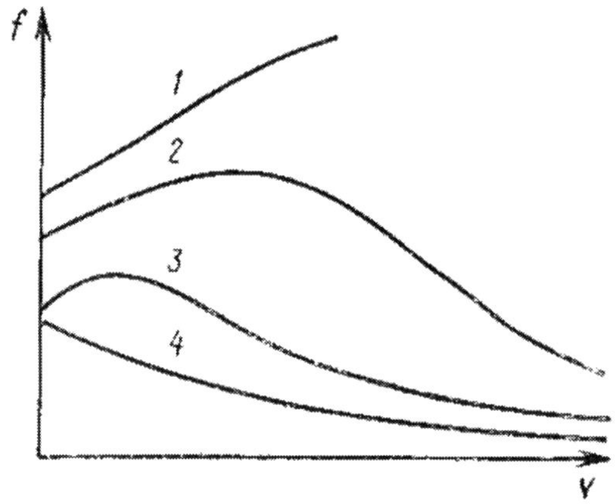

Figure 5. Generalized friction experiments in Kragelsky's interpretation [5]: v- sliding velocity (load: 1 - small; 2 and 3 – medium; 4 - considerable)

A subsequent analysis of modern experimental data using equations (1)-(12) has shown [2] that the experimental friction curves (figures 3-5) of type $\mu = \mu(N, v)$ are generalized friction curves that reflect the evolution (the change in the friction coefficient) of the tribosystem.

5. Regularities of tribosystems' evolution

Tribosystem's evolution given as a diagram (figure 6) is of adaptive-dissipative character (1)-(9) and reflects the competitive (dialectical) nature of friction.

The evolution curve has a number of principal points (1,2,3,4,5) of the transient stages of the tribosystem, which firmly obey the balance principle of friction. Between these points are located the most characteristic areas of tribosystem's behavior that reflect the general properties of its nonlinear dynamics.

So, we can see the following conditionally marked points and stages: 0-1- section of static friction and deformation strengthening; 1 – limit deformation strengthening point; 1-2 – stage of pumping of excessive energy; 2 – point of seizure and transition from external to internal friction (critical non-stability); 2-3 – stage of dissipative structures formation (formation of heat fluctuation in friction volume); 3 – minimum compatibility point; 1-2-3 – self-organization area; $3-4^1$- stage of compatibility; 4 - point of wearlessness (abnormal low friction); 5 – point of heat seizure.

.

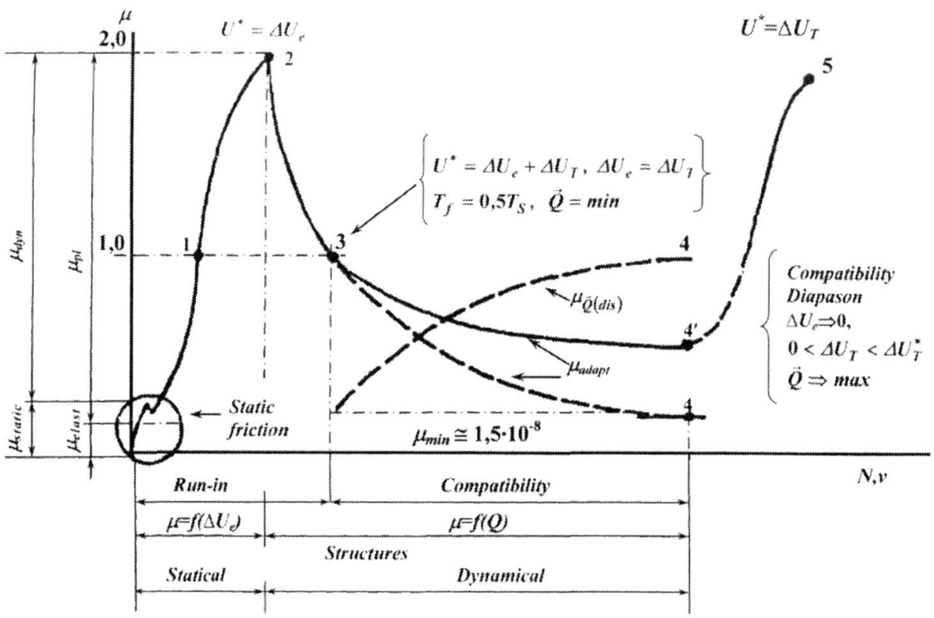

Figure 6. Structural-energy diagram of frictional surfaces evolution [2]. Designation on the axes: N, v - load and speed. $\mu_{st}, \mu_{dyn}, \mu_{elast}, \mu_{pl}$ - static, dynamic, elastic, plastic friction coefficients; T_f, T_S - the flash point in the friction contact volume in point 3 and the melting temperature.

The ideal evolution of the tribosystem is symmetrical. The process begins and ends in the areas of elastic behaviour. There is a plastic maximum of friction between them (a highly-excited state) as a condition for self-organization and adaptation.

In the most general case, the patterns of evolution (adaptation) of tribosystems can be represented in the form of two stages (figure 6). First stage (0-2) – stage of density increase of the latent energy Δu_e up to the limit value Δu_e^* in some critical volume of friction V_f^*. Second stage (2-4) – stage of the structural decomposition (transformation) into two parts of the critical volume of friction V_f^* (of the elementary tribosystem), namely: adaptive V_{adapt} and dissipative V_{dis} volumes. In the limit (point 4), this stage is characterized by a complete transformation of the adaptive critical volume V_{adapt}^* into dissipative volume V_{dis}^*.

Above volumes characterize different relationships of energy transformation of the external mechanical motion during friction. The adaptive volume V_{adapt} is related to an irreversible absorption of deformation energy. Latent energy of deformation Δu_e is accumulated in this volume, and then centers of fracture emerge. The dissipative volume V_{dis} is capable of reversible transforming (dissipation) the energy of the external motion. It does not accumulate latent energy of deformation due to a process of reversible elastic-viscoplastic deformation.

The proposed theoretical and calculated assessments [2] showed that the dissipative volume of friction produces a reversible elastic transformation of the energy of external mechanical motion with a density $\vec{q}^*$ equal to the critical density of the latent energy u_e^*.

The culmination of tribosystem's evolution is its final and limiting state, point 4 – the state of abnormal low friction and wearlessness (state of maximal performance and functional ability).

It has been shown [2] that the value of the minimum adaptive volume of friction $V_{adapt}^{\min}$ corresponding to zero value of the plastic friction component $\mu_{adapt}^{\min} = \mu_{elast}$ is not equal to zero, but is equal to the size of some minimal structural element of the deformable solid body.

It is of scientific interest to analyze this minimal structural composition during friction in the aspect of the general phenomenon of adaptation due to self-organization.

6. Structural-energy and physical patterns of self-organization and synergy

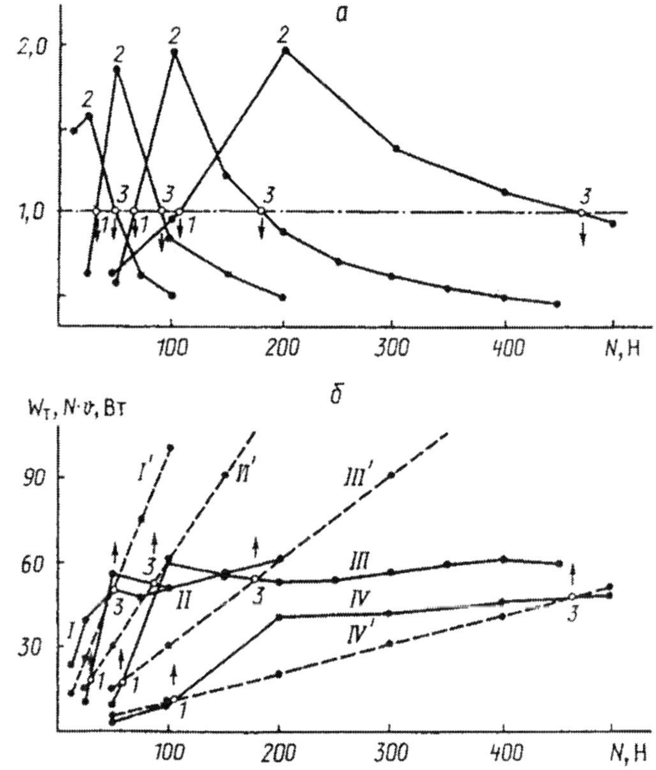

Figure 7. (a) Friction coefficient μ versus load [8], and (b) friction power $\dot{W}_T$ (solid line) and parameter $N \cdot v$ (dash line) versus load for the couple Nylon 6 – steel [2]: $I - v = 1\ m/s$; $II - 0{,}6$; $III - 0{,}3$; $IV - v = 0{,}1\ m/s$

An analysis of the laws of self-organization during friction will be carried out taking into account the equations of the energy balance of friction (1)-(11), the energy interpretation of friction coefficient (12), and the existence of generalized experimental friction curves (figures 3-5). It is particularly important to rearrange [2] the Wanatabe's [8] curve (figure 4) to the curves form of friction power W_f and external forces power $N \cdot v$ (figure 7). The curve in figure 4 is rearranged to three curves (figure 7). The first diagram (figure 7a) shows the friction coefficient versus load at different speeds; a horizontal line is given that crosses the friction coefficients curves at a unit value. Points 1 and 3 are denoted on the ascending and descending branches of these curves. The maximum of friction curves are points 2. Figure 7b shows that at points 1 and 3 there is an intersection of the friction work (power) W_f curves and the parameter Nv, which is the work (power) of external forces. Accordingly, between these points there is an excess of the values of friction powers W_f over the values of the work of external forces Nv. After points 3, on the contrary, the values of the friction power W_f are lower than the values of the Nv parameter.

Section 0-1 (figure 6) – preparatory (initial) stage. It is characterized by dislocation mechanism of plastic deformation, formation of the critical volume of friction V_f^* and accumulation of energy of dislocation strengthening of the limiting value ΔU_{eD}^*. The process of accumulation of latent (potential) energy of defects is miscoordinated with the process of its release. Up to point 1, the accumulated energy of friction is dissipated after wear particle detachment from the volume of the friction surface, i.e. there is no cooperative action or synergy.

Point 1 - the point of equilibrium. The friction coefficient has a balance value equal to unit (see figures 6, 7).

At point 1, the possibilities of plastic deformation according to the dislocation mechanism are exhausted; a critical volume of friction V_f^* is formed in point 1. This is the smallest contact volume that has accumulated the limiting density of dislocations energy $\Delta U_e = \Delta U_e^* = \Delta U_{eD}^*$. In fact, the volume of the elementary tribosystem was formed, i.e. the system capable of further transforming and dynamically dissipating energy. This is the point of the limiting dislocation (deformation) strengthening of the critical volume of friction V_f^*. Point 1 is the starting point of self-organization.

As seen from the experiment (figures 7a and b), we have equality of the power (work) of friction $\dot{W}_f$ and the power (work) of the external forces Nv:

$$\dot{W}_f = \mu N v = N v .$$

Hence we obtain the condition $\mu = \dfrac{\dot{W}_f}{Nv} = 1{,}0$, which corresponds to the experimental value of friction coefficient $\mu = 1{,}0$ (see figure 7a).

Accordingly, the energy interpretation of the friction coefficient (12) shows:

$$\mu = \frac{\dot{U}_e}{Nv} = 1{,}0 .$$

At point 1, equality takes place (see figure 7b) of the work of external forces Nv with the work of friction W_f and the accumulated potential energy ΔU_{eD}^* of dislocation strengthening in the critical volume of friction V_f^*:

$$Nv = W_f = \Delta U_{eD}^* .$$

Material Science aspect. The schematic diagram (figure 8a) shows how the critical volume of friction V_f^* is covered by a grid of dislocation slip lines with dislocations on them, which form "dislocation forests", i.e. the state of limiting dislocation strengthening. Such state of the critical volume of friction V_f^* can be represented by an equivalent scheme (figure 8b), which shows that only half of this volume is filled with dislocations (being large defects, dislocations cannot fill the entire crystal lattice of the volume of friction V_f^*).

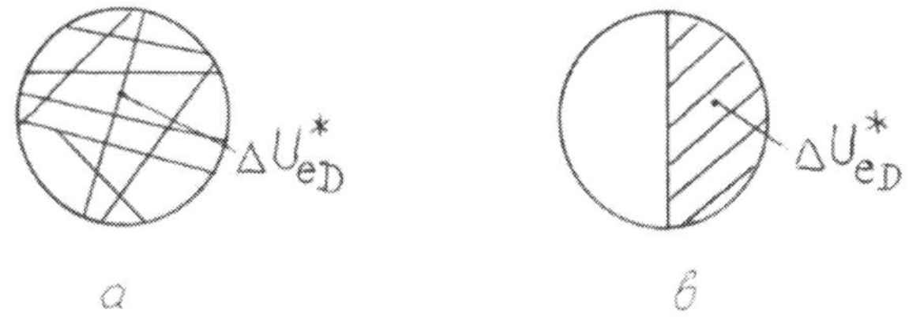

Figure 8. Schematic diagrams of potential energy ΔU_{eD}^* accumulation of dislocation slip lines in the critical volume of friction V_f^*

Synergy of the area 1-2. The work (transformation) of the internal energy fields of the limiting dislocation strengthening ΔU_{eD}^* is expended on the formation of internal energy fields of vacancies ΔU_{eV}^* in the critical volume of friction V_f^*. Self-organization began, doing it certainly as a self-dependent process! At point 1, the friction volume V_f^* possesses a stock of accumulated potential energy of dislocations ΔU_{eD}^* which is capable of performing the job of vacancies creation.
Actually, point 1 is the starting point of self-organization. Point 1 is in the zone of all-round, non-uniform compression $F = N$ and has the limiting reserve of accumulated energy ΔU_{eD}^*. But the system (figure 6) evolves further. Why? Self-organization! What does it mean? The system is further organized self-dependently.

Mechanism of the beginning of self-organization. At point 1, the centers of limiting deformation (dislocation) strengthening under conditions of all-round irregular compression (pressure plus shift (Bridgeman)) become sources of formation of nonequilibrium vacancies (work is done), which begin to accumulate in the regions of the crystal lattice (figure 9a), which are free from dislocations. Thus, work is done by internal forces on accumulation of energy of vacancies. Further, in the stage 1-2, the energy of vacancies ΔU_{eV}^* accumulates, equal to the energy of dislocation strengthening of point 1, i.e. $\Delta U_{eV}^* = \Delta U_{eD}^*$.

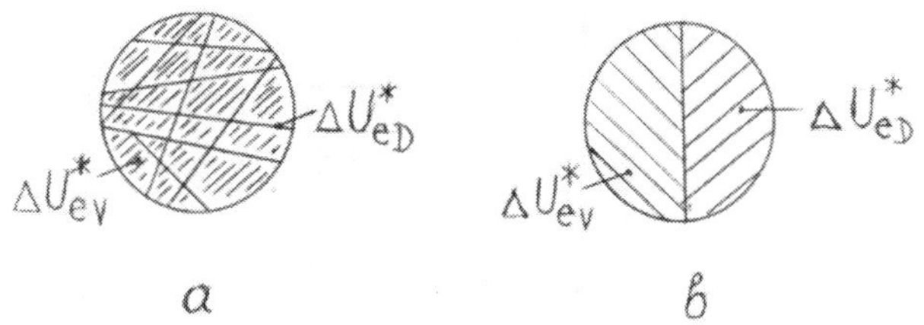

Figure 9. Schematic diagrams of potential energy of vacations ΔU_{eV}^* accumulation in crystal lattice areas free of dislocation strengthening regions of energy ΔU_{eD}^* in the critical volume of friction V_f^*

Material Science aspect. The schematic diagram (figure 9a) shows that vacancies fill-in those places in the crystal lattice, which are free from dislocation slip lines and dislocations on them. Sources of the vacancies become the centers of limiting deformation (dislocation). So, the work of external forces $Nv = W_f$ from point 1 and above maintains a level of constant energy density ΔU^*_{eD} of the limiting dislocation strengthening, and the centers of deformation supersaturation (of the strengthening) pump the energy of vacancies ΔU^*_{eV} into the regions of the crystal lattice that are free from dislocations. We have a joint action of two equal energy fields $-Nv = \Delta U^*_{eD}$ and $\Delta U^*_{eD} = \Delta U^*_{eV}$. Such a state of the critical volume of friction V^*_f can be represented by the equivalent scheme shown in figure 9b.

After point 1 the volume of friction V^*_f becomes an elementary tribosystem as a self-dependent energy transformer - it is capable not only of accumulating energy, but also of releasing it simultaneously! Synergy began working, namely simultaneous work of two opposing equalities - the accumulation of energy ΔU^*_{eD} and its release ΔU^*_{eV}.

An unique state exists in stage 1-2. The system releases accumulated potential energy ΔU^*_{eD} by the mechanism of simultaneous accumulation of potential energy ΔU^*_{eV}.

Point 2 - the point of equilibrium. This is the point of the superexcited state - the equilibrium state far from the state of the initial (zero) equilibrium. At point 2 (figure 9a) we have the equality of the densities of the accumulated potential energy of dislocation strengthening ΔU^*_{eD} and of the accumulated potential energy of vacancies ΔU^*_{eV} in the critical volume of friction V^*_f. This state of the critical volume of friction can be represented by the equivalent scheme shown in figure 9b. At point 2, the latent (potential) energy of the defects accumulates, equal to the double energy of the limiting deformation strengthening of point 1, i.e. $2 \cdot \Delta U^*_e = \Delta U^*_{eD} + \Delta U^*_{eV}$.

The state of point 2 is unique - characterized by a double energy balance. The coefficients of friction have a value of two (figure 6 and figure 7.) That's why, on the experimental friction curves (figure 7b), we see that at point 2 the friction work W_f is twice as high as the work of external forces Nv.

The principle of double energy balance of point 2 is a characteristic sign of self-organization!

Synergy of the area 2-3. Interaction of equal energy fields of dislocations ΔU^*_{eD} and vacancies ΔU^*_{eV} occurs in the critical volume of friction V^*_f with the development of the process of returning accumulated potential energy due to the annihilation of defects in the crystal structure. There is a transformation of half the latent potential energy $\Delta U^*_{eD} + \Delta U^*_{eV}$ of point 2 into thermal energy ΔU^*_T. We have an obvious synergy - the interaction of equal energy fields that determine a subsequent transformation.

Material Science aspect. Here, vacancies in the form of point defects (the smallest defects) of the crystal structure create the possibility for further development of the plastic deformation process. Dislocations begin to move due to the so-called mechanism of diffusion (vacancy) creep of dislocations. Vacancies are the most easily movable defects that accelerate all processes, in this case the process of plastic deformation. Dislocations go in passages (channels) – to the surface, grain boundaries, there are dislocations of opposite signs and the integrity of the crystal lattice is restored. A full return of accumulated potential energy is realized - the transformation of part of it into a related, heat component of internal energy ΔU^*_T. There is an effect of self-heating of the critical volume of friction $\Delta U_T^* = V_f^* \cdot \rho \cdot c_p \cdot dT_f$. This is the nature of the temperature flash T_f in the body and in the contact of friction.

Point 3 - the point of equilibrium. The coefficient of friction at point 3 is again unity (see the figures 6 and 7b). This is equality of the accumulated internal potential energy ΔU_e^* of defects, and the internal thermal energy ΔU_T^* (the kinetic component of internal energy):
$\Delta U_e^* = \Delta U_{eD} + \Delta U_{eV} = \Delta U_T^* = Q$ (figure10a). Such a state of the critical volume of friction V_f^* can be represented by the equivalent scheme shown in figure 10b, where one half of the critical volume of friction V_f^* has a reserve of potential energy of dislocations and vacancies $\Delta U_e^* = \Delta U_{eD} + \Delta U_{eV}$, and the second half possesses the energy of heat effect of friction $\Delta U_T^* = Q$.

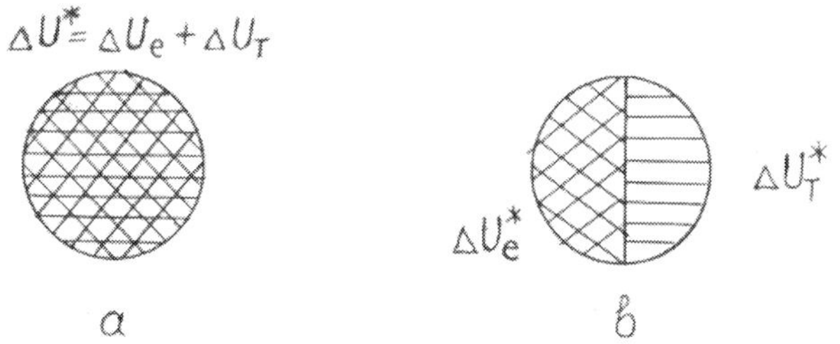

Figure 10. Schemes of the equilibrium state of the accumulated internal potential energy of dislocations and vacancies $\Delta U_e^* = \Delta U_{eD} + \Delta U_{eV}$ and the thermal (kinetic) component of internal energy $\Delta U_T^* = Q$ in the critical volume of friction V_f^*.

As a result, a static energy oscillator (a self-dependent transformer) was formed; it is able to dissipate part of the internal energy - the kinetic energy component ΔU_T^* by a thermal conductance mechanism, in depth perpendicular to the friction surface. However, being a slow dissipation channel for the energy Q - it takes time.

Self-organization area 1-2-3. The principle of point 2 with a double coefficient of friction and the principle of double energy balance generally characterize the synergy of self-organization. From the point of view of thermodynamics and tribology, there are, on the one hand, accumulation of potential energy ΔU_{eV}^* and change in the coefficient of friction from unity at point 1 to two at point 2, and on the other hand, release of the same part of the potential energy and its transformation into thermal energy ΔU_T^* with a decrease in the friction coefficient from two at point 2 to unity in point 3.
As a result, the evolution of the contact of friction, thanks to the self-organization, creates a state of point 3, which is more favorable than the states of point 1 and point 2. At point 3, the frictional contact (the elementary tribosystem) has a reserve of thermal energy $Q = \Delta U_T^*$ that can be released. The other part of the energy, the potential energy, can transform into a thermal one. As we see from the experiment (see figure 7), the frictional power and the work of external forces at point 3 are much higher (several times) than those in point 1.

Synergy of the area 3-4. The synergy in that case reflects the interaction (joint action) of two equal energy fields: of dislocations and vacancies $\Delta U_e^* = \Delta U_{eD} + \Delta U_{eV}$ (the potential component of internal energy) of the critical volume of friction V_f^* and of thermal energy of the self-heating

$\Delta U_T^* = Q$ (the kinetic component of internal energy). The result of this interaction will be the formation of dissipative friction structures, which dissipate the accumulated energy of the balance components $\vec{Q} = V_f^* \cdot \vec{q}^*$.

Material Science aspect. The temperature in the friction volume T_f of point 3 is equal to half the melting point ($0{,}5T_S$) of the most fusible material of the friction pair. This is a characteristic temperature, both for the recrystallization (the self-organization of a new, equiaxial structure in place of the extremely deformed one (distorted by defects)), and for the initiation of chemical reactions within the friction volume, which at the stage 2-3 communicated with the environment and realized an effect of self-doping. Defects came to the surface, and inside the bulk came substance of the environment, etc. The synergy of equal energy potentials of the activated surface (active centers) and the active substance of the environment was realized.

Dissipative structures at friction. As a result of the synergy effect, dissipative friction structures (third body [5] or secondary structures [9]) are formed in the region 3-4 (in figure 6). These dissipative friction structures possess the property of rapid dissipation of the energy $\vec{Q}$ of the thermal friction effect (see figures 6, 11). The main feature of these structures is that the structural dissipation principle is realized by them: the dissipation of accumulated energy is realized here; the energy is dissipated on the structural element (rather than inside it) in the form of elastic energy. Then the structural elements interact elastically and perform joint work (rotational modes of developed plastic deformation) of elastic visco-plastic flow (shift). The density of the energy of dynamic dissipative scattering $\vec{q}$ is equal to the critical accumulated density Δu_e^* of the potential energy. The system of the contact volume of friction V_f^* is under conditions of a flux of energy of critical density. In this case, the dissipation of energy by the dissipative structures takes place in the direction of motion, in contrast to the perpendicular direction of the energy scattering $Q = \Delta U_T^*$. The scattered energy of dissipation $\vec{Q}$ is a fast analog of thermal conductivity. A detailed physical model of dissipative structures is considered below in the model of mechanical (nano) quantum.

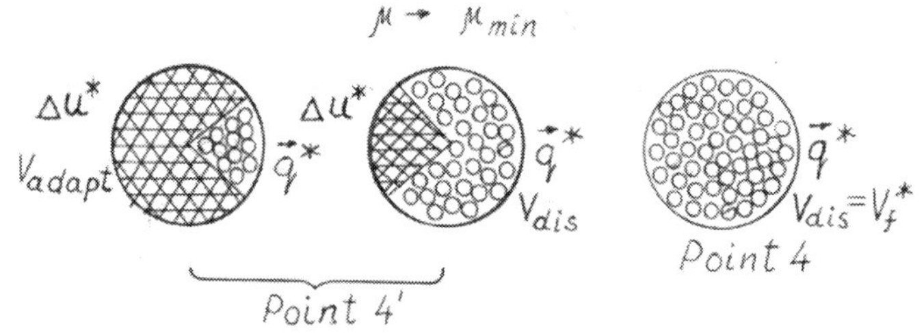

Figure 11. Schemes of the state of the critical friction volume V_f^* in the area of compatibility (section 3-4 in figure 6).

Point 4′ - the point of equilibrium (with reference to figure 11). This point is located at the level of friction coefficient equal to 0.5 (figure 6,7). Here, the state of equilibrium is determined by the equality of the work of friction W_f and the power of the thermal effect of dissipative friction $\vec{Q}$. In sum, both these balance components are equal to the work of external forces Nv (figure 7b.) The neighborhoods of this point in the region 3-4 (figure 6, 7b) can have a wide range of the structural arrangement of the critical volume of friction V_f^*, in which the fraction of the dissipative volume of friction V_{dis} will vary (figure 6) from the minimum (point 3) to the maximum (point 4). Accordingly, the proportion of the adaptive volume of friction V_{adapt} will vary inversely with the change of V_{dis}.

Point 4 - the point of ideal equilibrium (with reference to figure 6 and figure 11). The states and properties of this point are presented below in the idea of the mechanical (nano) quantum, as an attractor of the limit cycle for the process of plastic deformation at friction. This is the case of maximum realization of synergy as joint cooperative action.

7. The idea of mechanical (nano) quantum

Result of the ideal evolution of the elementary tribosystem is the formation of a unique nanostructure - a mechanical (nano) quantum. Strict concept about the mechanical quantum has been obtained [2, 10] considering the equation of a quasi-ideal solid body in point 4 of the friction evolution diagram, as follows:

$$\vec{Q}^* = \vec{S}_Q T = \mu_{dis}^* N l_f = U_e^* = V_f^* u_e^* = V_f^* \vec{q}_* = const \,, \tag{13}$$

which is a particular case of equation (8) solution of the energy balance of friction, with $\mu_{adapt} = 0$ and $\mu_{dis} = 1 = \mu_{dis}^*$. Here $\vec{S}_Q$ is the inertial entropy of the compatible friction volume; T - characteristic temperature of the friction contact volume; l_f - linear size of the friction volume.

Accordingly, under conditions of maximum compatibility (point 4), when the tribosystem realizes a complete evolutionary cycle of adaptation with formation of a perfect dissipative structure, its structure behaviour obeys the equation of quasi-ideal solid body state, i.e. it should be assumed that the interactions between the structure elements are minimized, namely as for the state of ideal elasticity in dynamics.

Taking into account the Planck-Boltzmann formula $S = k \ln W$ and the real number of atomic oscillators N_f in the volume of an elementary tribosystem V_f^*, equation (13) is reduced to the form:

$$\mu_{diss} = \frac{\vec{S}_Q T}{N l_f} = \frac{kTN_f \ln W}{N l_f}; \tag{14}$$

$$\mu_{adapt} = 1 - \mu_{diss} = 1 - \frac{kTN_f \ln W}{N l_f} = 1 - \frac{\vec{S}_Q T}{N l_f} = \frac{S_U T}{N l_f}, \tag{15}$$

explaining the friction laws from the point of view of the evolution of systems. The tribosystem always tends to some optimal state characterized by μ_{adapt}, i.e. to the most probable state $W' = N_f \ln W$ for these friction conditions. Here N_f is the number of atoms (oscillators) in the friction volume $V_{diss}^{\max} = V_f^*$; k - Boltzmann's constant; W - state probability; S_U - configurational entropy of the compatible friction volume.

Analysis and solution of these equations [2] allow us to show the probability value W constancy (the parameter of state (order) of the tribosystem) for the entire range of compatible friction, namely $\ln W = 3$ and $W = e^3 = 20{,}08553696$.

The value of thermodynamic probability W of the state equal to 20,085,53696 was interpreted [2] as the minimum number of linear, atomic oscillators in one of the three directions of the minimum adaptive friction volume $V_{adapt}^{\min}$ corresponding to the state of absolute elastic friction - abnormal low friction (the safe threshold of deformation). Accordingly, the number of atomic oscillators in this volume is $W^3 = 8103{,}083969...$.

On the other hand, taking the entropy S as Boltzmann entropy, a universal friction constant $R_f = kN_f$ [2] is obtained, which in physical sense characterizes the "energy dimension" of an elementary tribosystem (TS), containing under ideal conditions the same number of atomic oscillators N_f (mechanical quanta N_Q).

$$R_f = k \cdot N_f = k \cdot W^3 \cdot N_Q = R_{MQ} \cdot N_Q \ (\frac{J}{grad \cdot TS}); \quad (16)$$

$$R_{MQ} = k \cdot W^3 \ (\frac{J}{grad \cdot TS}). \quad (17)$$

As follows from the calculations [2, 10], the size of the minimum adaptive friction volume $V_{adapt}^{\min}$ coincides in magnitude with the size of the submicroscopic zone at the crack mouth, which is $(4 \div 9) \cdot 10^{-6} mm$ for metals, i. e. with the size of the critical volume responsible for the destruction. Thus, the size of the minimum adaptive friction volume $V_{adapt}^{\min} = V_{elast}$ can be presented as the size of a certain mechanical "Quantum". This mechanical quantum consists of the minimum number of atoms which are capable to provide such a configuration (structure) that has the property of reversibly receive and dissipate (return) the energy of the external mechanical motion (action). It also represents the smallest structural formation under conditions of plastic deformation and is formed during the transition of the tribosystem (the deformable volume) through the extremely activated (critical) state (figure 2), due to the development of self-organizing processes of tribosystem adaptation. Calculations show that in the volume of friction V_f^* the number of such mechanical quanta is equal to the number $0{,}63 \cdot 10^8$ [2], i.e. to the safe number of fatigue cycles.

The mechanical quantum itself is a dynamic oscillator of dissipative friction structures and its linear size is equal to the radius of a spherical ideal crystal:

$$D_Q = 2R_{MQ} = \sqrt[3]{\frac{3 \cdot W^3 \cdot \bar{d}_a^3}{4 \cdot \pi}} = 7{,}177 \, \text{nm}. \quad (18)$$

Here $\bar{d}_a$ is the average atom diameter.

The mechanical quantum should be regarded as an elementary nanostructure of a metallic solid. Figure 12 shows a constructed model of this atomically-rough theoretical crystal [2, 10-12], consisting of 8103 cubic atomic cells.

Figure 12. Ideal crystal model of the elementary nanostructure of friction contact [10-12]

The mutual rotational-vibrational motion of these mechanical quanta relative to each other inside the elementary tribosystem determines the state of the perfect dissipative structure of friction. Actually, such a state is described by the equation of state of a quasi-ideal solid body (13), a state where the interaction between the structure elements (the mechanical quanta) is minimized - the state of ideal elasticity of the quasi-viscous flow. The calculated friction coefficient between quanta is about 10^{-8} [2, 10-12].

8. Synergy of tribosystem's compatibility and optimality of state

The ideal, quasi-elastic state of friction contact at its complete evolution consists in the effect of the complete dissipation of external mechanical motion energy on the newly formed (by the mechanism of self-organization in the vicinity of the critical state) structural elements - mechanical quanta (dynamic oscillators). These oscillators realize their most complete rotation-vibrational behavior relative to each other in the volume of the elementary tribosystem. The resistance to their relative interaction is minimal, it is elastic and corresponds to the elasticity of ideal atomic (in thermodynamic equilibrium) interactions.
The universal constants of a mechanical quantum and an elementary tribosystem (a material point) are defined as quantum model of surface damping

$$\mu_{dis} = \frac{3R_{MQ}T\, n_i}{N\, l_f} = \frac{U_{1Q}n_i}{U_{1Q}n_*} = \frac{n_i}{n_*} = 1 - \mu_{adapt}; \qquad \mu = 1 - \frac{n_i}{n_*} = \frac{n_{dest}}{n_*}, \tag{19}$$

taking into account the destruction quanta n_{dest} (irreversible component of the process) and the damping quanta n_i (reversible, elastic component (number of fatigue)), as well as the probabilistic model of the evolution of the tribosystem towards the most ordered state

$$\mu_{adapt} = 1 - \mu_{dis} = 1 - \frac{R_f T \ln W_i}{N\, l_f} = 1 - \frac{\ln W_i}{\ln W_*}. \tag{20}$$

Here R_{MQ} is the constant of deformation (friction); $3R_{MQ}T = U_{1Q}$ - the energy of one mechanical quantum; W_i and W_* - current and limiting probabilities of states for the compatible tribosystems.
According to the model of quantum damping of surfaces at friction, under the conditions of complete evolution (adaptation) of the elementary tribosystem, all mechanical quanta except one transform elastically and reversibly the energy of external action (the mechanical motion). One mechanical quantum of radiation ($\cong 8103$ atoms) is a minimal loss (the essence of wearlessness or the standard of wear).

Actually, the principle of the mechanical quantum determines the nano quantum levels of all friction parameters of compatible friction surfaces [10-12], etc.

In such a way, one can speak of synergism in the evolution of friction, which manifests itself in the emergence of really new properties in the wholeness - the elementary tribosystem formed in point 2 - the center of self-organizing processes. These new properties are the essence of the collective, joint interaction of the assembly of mechanical (nano) quanta in point 4 of the ideal evolution of the tribosystem. The nanoquanta as tribo-subsystems in their elastic interaction realize the highest compatibility of tribomaterials and, accordingly, the maximum life of the elementary tribosystem.
The mechanical quantum, therefore, can be regarded as the smallest substance amount of a solid body, which has the sign of the latter. This quantum is really mechanical, because the next independent structural and material formation will be the atom. Thus, the mechanical quantum is precisely an asymptotically stable attractor of the limit cycle type for a deformable solid body (at friction). It really represents a limit in the structural evolution of a solid body and characterizes the ultimate durability of the working cycle of a rubbing pair.

9. Conclusions

1. Friction is a global process of joint action (work) of two surfaces in contact.
2. The synergy (joint action) of two rubbing surfaces activates the generalized mechanisms of transformation and dissipation of their external relative motion energy, which strictly obey the principle of energy balance.
3. The energy balance of friction characterizes in summarized form the adaptive, dual nature of a friction process - the competition of two simultaneously acting, interrelated and opposite tendencies: accumulation of internal energy by the deformable contact volume and its release.
4. The balance process of friction determines the essence of synergy principle - the equality of the joint action of two interrelated and opposite tendencies: accumulation of internal energy and its release.
5. Equality of the combined effect of the two opposite energy tendencies of the friction process activates the mechanisms of self-organization of the friction system (the deformable contact) as an evolution system.
6. In consequence of self-organization of the tribosystem (the contact volume), the latter evolves formation within itself of tribo-subsystems - structural units of an essentially new (unique) level. Thus, dissipative structures of friction are formed.
7. The basis of friction dissipative structures is a mechanical (nano) quantum - the smallest structural formation under conditions of plastic deformation, which has the property of maximal reversible receipt and return (dissipation) of the energy of external mechanical motion (action).
8. As a dynamic oscillator of friction dissipative structures the mechanical (nano) quantum is an attractor of the limit cycle for the process of plastic deformation (of friction). There is a maximum realization of synergy - a joint action of tribo-subsystems with minimal elastic interaction.
9. A maximum load capacity of the tribosystem is completed - the external load is received elastically at the internal level of the elementary tribosystem (contact). The tribosystem realizes abnormal low friction and minimal contact loss in the form of a single mechanical quantum (standard of wear) for the act of the contact loading. The highest fatigue strength and durability occur: "wearlessness" with the effect of a natural solid lubricant (the "servovite" film).

References

1. Assenova E., Complexity of Contact Systems: Self-organization and Synergetic Approach in Tribology, Proc. of the 15th international conference on Tribology "SERBIATRIB'17", 17-19 May 2017, Kragujevac, Serbia, 18-23, 2017.
2. Fedorov S. V., Fundamentals of Triboergodynamics and Physico-chemical Prerequisites of Compatibility Theory. KGTU, Kaliningrad, 2003 (in Russian).
3. Fedorov V. V., Thermodynamic aspects of strength and destruction of solids. Publ. House "FAN", Tashkent, 1979 (in Russian).
4. Fedorov V. V., Fundamentals of Ergodynamics and Synergetics of Deformable Bodies. Part 1. Basics of Physical Mechanics of Deformable Bodies (State of Problem). Edited by Fedorov S. V., KGTU, Kaliningrad, 2012 (in Russian).
5. Kragelskii, I.V., Dobychin, M.N., Kombalov, V.S., 1977, Friction and Wear Calculation Methods, Oxford: Pergamon Press, 1982.

6. Bowden, F.P., Tabor, D., The Friction and Lubrication of Solids, Oxford, 1964.
7. Conti, P., Sulla resistenza di attrito. Royal Accademia dei Lincei, v. II, 1875.
8. Lancaster, I.K., Basic Mechanisms of Friction and Wear of Polymers. Plastics and Polymers, 41, 296-306, 1973.
9. Surface Strength of Materials in Friction. Edited by Kostetzkii B.I. "Tehnika", Kiev, 1976 (in Russian).
10. Fedorov S.V., Fundamentals of Ergodynamics of Deformable Bodies and Triboergodynamics. Handbook. An Engineering Journal, Appl. № 8, pp. 1-24, Ed. House "Mashinostroienie", 2010. (in Russian).
11. Fedorov S.V. The Mechanical Quantum of Dissipative Friction Structures is the Elementary Tribonanostructure. Proceedings of the 4th World Tribology Congress (6-11 September 2009): Japanese Society of Tribologists. Kyoto, Japan, p. 926, 2009.
12. Fedorov, S.V., Generalized Energy Model of Sliding Friction Coefficient and Regularities of Tribosystem Evolution, 5th World Tribology Congress (8-13 September 2013), Turin, Italy, 2013, ISBN 9788890818509.

Cavitation resistance of surface composition "Steel-Ni-TiNi-TiNiZr-cBNCo", formed by High-Velocity Oxygen-Fuel spraying

Blednova Zh.M. [1], ***Dmitrenko D.V.*** [1], ***Balaev E.U.O.*** [1]
[1]*Kuban State Technological University, Department of dynamics and strength of machines, 350072, Moskovskaya 2, Krasnodar, Russia*
ddv-kk@yandex.ru, blednova@mail.ru, balaev1122@mail.ru

Abstract: The object of the study is a multilayered surface composition "Steel - a Multicomponent material with Shape Memory Effect - a wear-resistant layer" under conditions of cavitation effects in sea water. Multicomponent TiNi-based coatings with addition of alloying elements such as Zr in an amount up to 10% mass, allow to create a composite material with a gradient of properties at the interface of layers, which gives new properties to coatings and improves their performance significantly. The use of materials with shape memory effect (SME) as surface layers or in the composition of surface layered compositions allows to provide an effective reaction of materials to the influence of external factors and adaptation to external influences. The surface composite layer cBN-10%Co has high hardness and strength, which ensures its resistance to shock cyclic influences of collapsing caverns. The increased roughness of the surface of a solid surface composite in the form of strong columnar structures ensures the crushing of vacuum voids, redistributing their effect on the entire surface, and not concentrating them in certain zones. In addition, the gradient structure of the multilayer composite coating TiNi–$Ti_{33}Ni_{49}Zr_{18}$–cBN-10%Co Co makes it possible to create conditions for the relaxation of stresses created by the variable impact load of cavitation caverns and the manifestation of compensating internal forces due to thermo-elastic martensitic transformations of SME materials. The cavitation resistance of the coating TiNi–$Ti_{33}Ni_{49}Zr_{18}$–cBN-10%Co according to the criterion of mass wear is 15-20 times higher than that of the base material without coating and 10-12 times higher than that of the TiNi-TiNiZr coating. The proposed architecture of the multifunctional gradient composition, "steel-Ni-TiNi–$Ti_{33}Ni_{49}Zr_{18}$–cBN-10%Co", each layer of which has its functional purpose, allows to increase the service life of parts operating under conditions of cavitation-fatigue loading in corrosive environments.

Keywords: Cavitation, Surface, Multicomponent material, SME

1. Introduction

Increasing the reliability and durability of machine parts and mechanisms is primarily due to the elimination of wear caused by various types of impacts, in particular, on their surface layers. One of the types of such impact on the surface of parts working in a fluid environment is cavitation erosion of the metal, which is caused by the mechanical action of fast moving fluid particles on the surface of the metal, sand particles of solids, suspensions, gas bubbles, etc. Such influence is characterized by a high rate of loading, its small duration, locality and cyclicity. At the same time, stresses appear in the surface layers, they can be compared with the ultimate strength of the material and are concentrated in volumes close to the dimensions of its structural components. Effective ways to improve the strength properties and resistance to corrosion-erosive effects are thermal and thermomechanical hardening treatments and the deposition of protective coatings on the metal surface [1].

Traditional materials science formation methods of structure and properties have basically exhausted their possibilities. Under these conditions, taking into account the crucial role of the surface layer in the accumulation of damage and destruction, the task of providing multifunctionality, as well as increasing reliability and resource, can be successfully solved at the final stage of processing by means of technologies based on the principles of layer-by-layer synthesis using combined, functionally oriented Macro-, micro-, and nanotechnologies. [2,3]

A promising direction in the implementation of layered synthesis technologies is the intellectualization of products using shape memory effect (SME) materials. They have a wide range of functional and mechanical capabilities: effects of thermomechanical memory and superelasticity, high strength and damping properties, thermomechanical reliability and durability, wear and corrosion resistance [4]. It was shown in [5, 6] that multicomponent TiNi-based coatings with addition of extra alloying elements such as Zr in an amount of up to 10% by weight make allow to create a composite material with a gradient of properties at the interface of the layers. This helps to impart new properties to the coatings, and significantly improves their performance characteristics. The use of materials with shape memory effect (SME) as surface layers [7] or as a component of surface layered compositions [8] allows to provide an effective reaction of materials to the influence of external factors and adaptation to external influences.

The aim of this work is to obtain quantitative data on erosion resistance under cavitation of composition surface layers with a gradient distribution of properties from a multicomponent SME material $Ti_{49}Ni_{51}$-$Ti_{33}Ni_{49}Zr_{18}$ EPF and the wear-resistant material cNB-10% Co, which provides specified functional and mechanical properties.

2. Experiment

2.1 Technology of formation of surface composite layer

The formation technology of a surface composition with SME material is a complex multi-operation process involving the preparation of the surface and the deposited material (milling and mechanical activation), high-velocity oxygen-fuel spraying (HVOF), and subsequent thermo-mechanical (TM) treatment (Fig. 1). Characteristics of the layers are presented in Table 1.

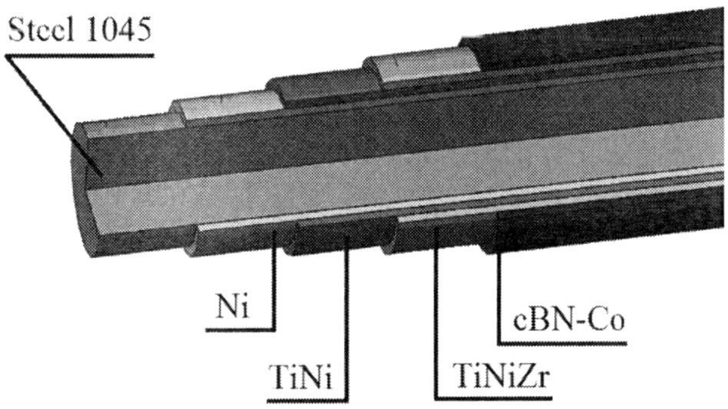

Figure 1. Scheme of the surface layer with a material with SME

Table 1. Layer characteristics

Layers	Material	Layer thickness, μm	T, °C	M_s, °C	M_f, °C	A_s, °C	A_f, °C	Microhardness, hPa	Phase state
base	Steel 1045	-	20					1,9-2,1	-
1	Ni	100±10	20					-	-
2	$Ti_{49}Ni_{51}$	600±50	20	23	55	13	31	10,2-11,9	austenite
3	$Ti_{33}Ni_{49}Zr_{18}$	1000±50	20	186	249	215	298	2,1-2,9 / 9,5-12,7	martensite / austenite
4	cBN-10%Co	200±10	20					34,6-35,6	-

M_f is the temperature of martensitic transformation expiry;
M_s is the temperature of martensitic transformation onset;
A_s - temperature of austenitic transformation onset;
A_f is the temperature of austenite transformation expiry.

The outer layer in contact with the working medium must have a high wear resistance with an increased bearing and damping ability, corrosion resistance with an optimum combination of strength and toughness. To meet these requirements, it is proposed to use a combination of layers of dissimilar materials. The outer layer is a hard and wear-resistant cNB-Co material, intermediate layers of $Ti_{33}Ni_{49}Zr_{18}$-$Ti_{49}Ni_{51}$, which are in austenitic-martensitic and austenite phase state, respectively. The purpose of the intermediate layer is to relax the stresses, damping the oscillations and blocking the cracks that appear in the outer layer and propagate deep into the material. An intermediate functional layer made of materials with an ESP in the martensitic state, which has an increased elasticity, helps to reduce the stress concentration at the tip of the crack and inhibits or blocks its movement. The gradient of properties in the outer contact and functional layers of materials with ESP is provided by a gradient of the temperatures of the phase transformations, which are set by the regulation of the chemical composition and thermomechanical processing. The intermediate layer of Ni is designed to increase the adhesion strength of the functional layer and the base material (Steel 1045), since Ni and the steel base are characterized by unlimited solubility of the basic elements. In this case, the crystallochemical affinity of the base material, the connecting and functional layers takes place, and, importantly, all the components have close values of the coefficient of thermal expansion.

A composite coating was deposited on the steel samples (Steel 1045) consisting of: a 100 μm adhesive layer of nickel, $Ti_{49}Ni_{51}$ 550-600 μm functional layers at room temperature in austenitic state, $Ti_{33}Ni_{49}Zr_{18}$ with a thickness of 950 ÷ 1000 μm, located at room temperature in a martensitic state, and a wear-resistant layer of cBN-10% Co whose thickness is 200-210 μm.

Grinding and mechanical activation can be performed using a variety of equipment. Milling and mechanical activation were carried out in the modernized ball mill GEPHEST-2 AGO-2U [9], in which mechanical effect is created by a series of successive mechanical impulses (strokes). Mechanical activation is not only an effective way to optimize the particle size distribution, but also due to the large plastic deformation it allows to pump energy into the processed material forming amorphous and nanostructural states [10]. This energy is released during the formation of the coating and activates the process of nanostructuring [11]. Since the formation of coatings was carried out by HVOF by imposing separate particles – splices, which move and harden at high speed, the phase composition, structure and properties of the coatings depend on the temperature, velocity of the particles colliding with the substrate and their cooling. HVOF is accompanied by large plastic deformations and the crystallization of particles occurs at high degrees of supercooling and is accompanied by the formation of crystalline nucleating centers whose size is determined by the degree of supercooling determined by the parameters of the process [12].

2.2 *Experimental*

In order to assess the quality of the composition functional layers, the sprayed samples were tested for adhesion, cavitation resistance, and fatigue strength. Adhesion was measured by a pin method on an Instron-8801 test machine. The micro-hardness test was performed on the Falcon-500 hardness tester. Cavitation tests were carried out in accordance with the American Standard ASTM G32-10 (Standard Test Method for Cavitation Erosion Using Vibratory Apparatus), using an ultrasonic vibrating unit and standard statistical processing of the experimental data (Fig. 2). [13]
Ultrasonic waves were used to create the cavitation zone. The signal from the generator is fed to a magnetostrictive transducer mechanically connected to a waveguide, at the end of which an experimental sample of a cylindrical shape with a diameter of 15 mm is rigidly mounted (Fig. 3), which provides an amplitude of oscillation of the sample end surface up to 50 ± 2 μm at a frequency of 20 kHz. The sample is immersed in a vessel with water. On the end surface of the sample, a zone with developed cavitation is formed. To assess the effect of corrosion processes on the magnitude of cavitation erosion, experiments were conducted in fresh and sea water.

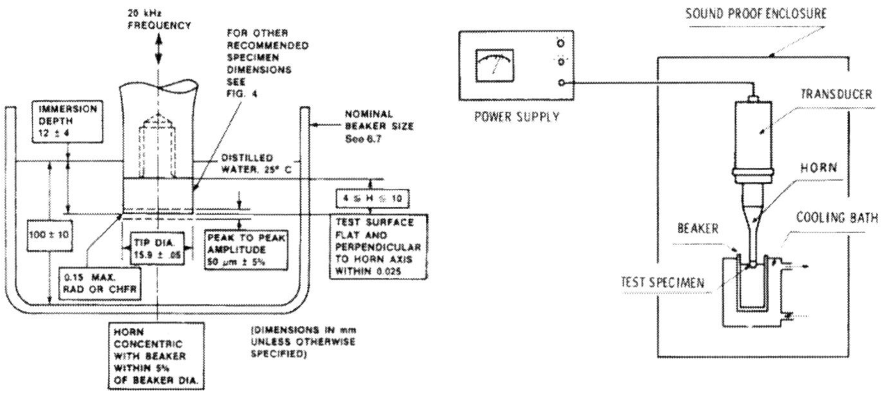

Figure 2. The test procedure for cavitation damage [ASTM G32-10].

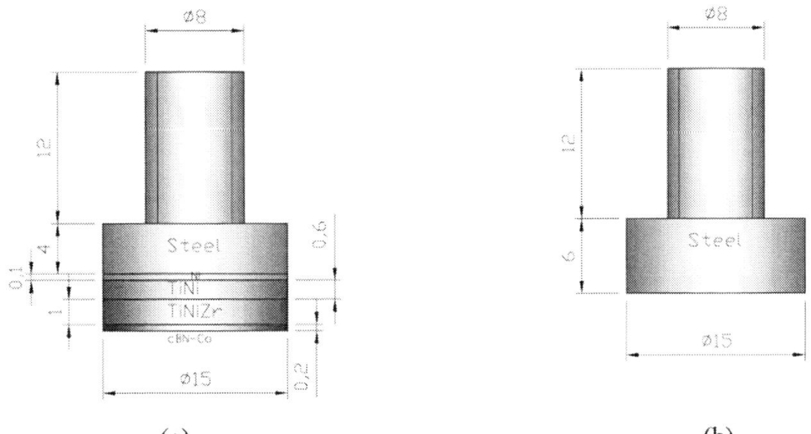

(a) (b)

Figure 3. Configurations and dimensions of cavitation test specimen for (a) with a composite layer, (b) steel (1045 or stainless 20Cr13).

The TiNiZr functional layer, which has a high-temperature SME [5], at normal temperature has increased relaxation properties and microhardness of 9.5 ÷ 12.7 GPa in the austenite state and 2.1 ÷ 2.9 GPa in the martensitic state. The microhardness of the wear-resistant cNB-Co layer varies depending on the cobalt content from 17.2 ÷ 22 GPa to 34.6 ÷ 35.6 GPa.

Damage accumulation and destruction of the samples surface was carried out with the visual inspection and area damage assessment on a stereoscopic zoom microscope MSP-1. Monitoring of the samples profile change, subjected to the cavitation fracture, as well as the damage depth was performed on the contour-tracing apparatus ABRIS PM7, the weight loss was measured on a DEMCON DL213 balance with an accuracy of ± 0.001 g. The dependence of the mass loss on the time of cavitation effect was measured; and from these data, the kinetic curves of the sample material destruction were constructed, which are characterized by the presence of the initial segment, when the destruction is small, and by the section with the maximum quasi-constant velocity.

2.3 Experimental results and discussion

The traditional criterion for resistance to material erosion is the loss of its mass over a period of time. Fig. 4 shows the study results of the cavitation resistance of uncoated samples and coated ones in fresh and sea water. In the process of cavitation in fresh water, the destruction of the surface is characterized by a classical curve of cavitation damage with an incubation period (30-60 min) and subsequent intensive loss of mass. The loss in mass of the composite coating $Ti_{49}Ni_{51}$-$Ti_{33}Ni_{49}Zr_{18}$ -cBN10% Co has a longer incubation period (more than 120 min), which is associated with an increased microhardness of the coating surface.

The presence of an aggressive component in the medium (salty sea water) somewhat smoothes the mass loss curve of the uncoated material and increases the intensity of the mass loss (Fig. 4). The materials used to form the coatings have an increased corrosion resistance, which is confirmed by the insignificant loss of weight of $Ti_{49}Ni_{51}$-$Ti_{33}Ni_{49}Zr_{18}$-cBN10% Co coated samples and has almost no difference when tested in fresh and sea water. The results of the tests show that the surface composition $Ti_{49}Ni_{51}$-$Ti_{33}Ni_{49}Zr_{18}$-cBN10% Co under consideration, each layer of which has its functional purpose, provides a set of properties such as high corrosion resistance, high microhardness, increased strength, and damping capacity. This contributes to the increase of stability of multicomponent coatings to cavitation erosion in corrosive environments.

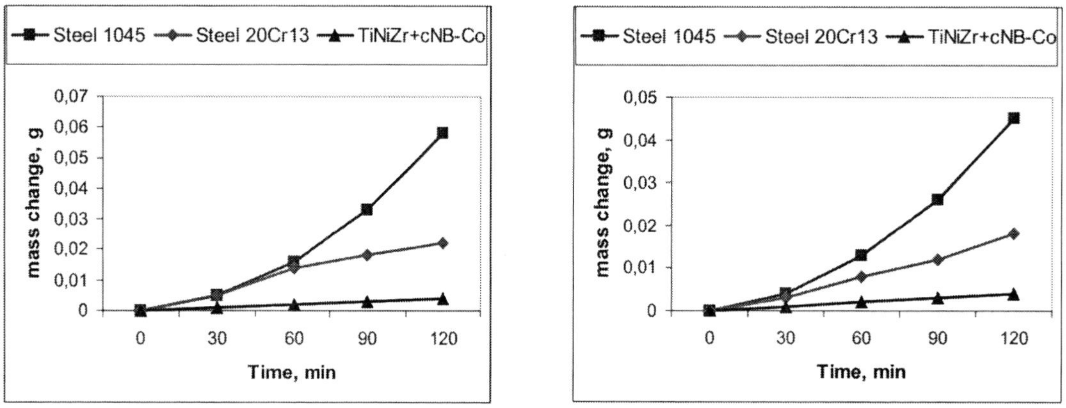

(a) (b)

Figure 4. Dynamics of the sample mass change as a result of cavitation influences: (a) Sea water; (b) Fresh water.

The cavitation resistance of materials can be evaluated both by weight loss and by the duration of the incubation period. However, neither method allows us to investigate the mechanisms of failure, which can differ significantly depending on the chemical composition of the material, heat treatment, design features of the product and operating conditions. Being an integral characteristic, the weight wear sums the damage within the limits of the entire cavitation "spot". However, a visual examination of the "spots" can be used to establish that the wear rate of the surface in different areas is different (Fig. 5).

Within the formed "spot" of erosive destruction, three characteristic regions can be distinguished. Firstly, it is the area in which the strong destruction of the surface are observed - zone 1. The next region corresponds to moderate disruptions and represents an annular area along the edge of the "spot" - zone 2. Finally, the region lying between zones 1 and 2, which has the smallest damage - zone 3.

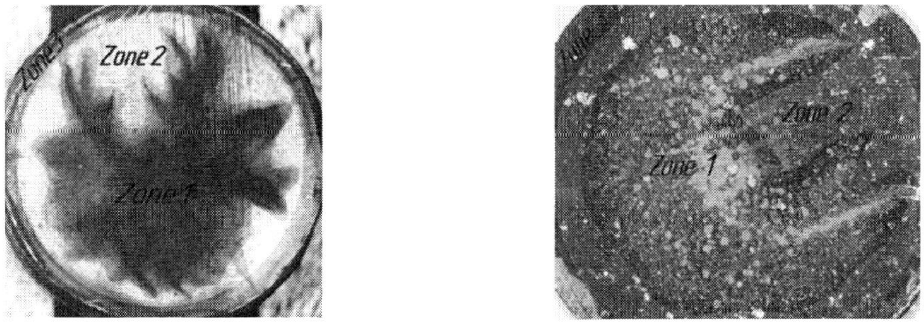

(a) (b)

Figure 5. Erosion spot of a cylindrical sample with a flat surface: (a) uncoated; (b) coated with TiNi–$Ti_{33}Ni_{49}Zr_{18}$

The surface without coating of Steel 1045, Steel 20Cr13 and the composite coating $Ti_{33}Ni_{49}Zr_{18}$–cBN10%Co before and after cavitation for 30 and 120 min in the sea water is shown in Fig. 6. It can be seen that the surface micro-relief practically does not change, i.e. the coating is practically not subjected to cavitation erosion.

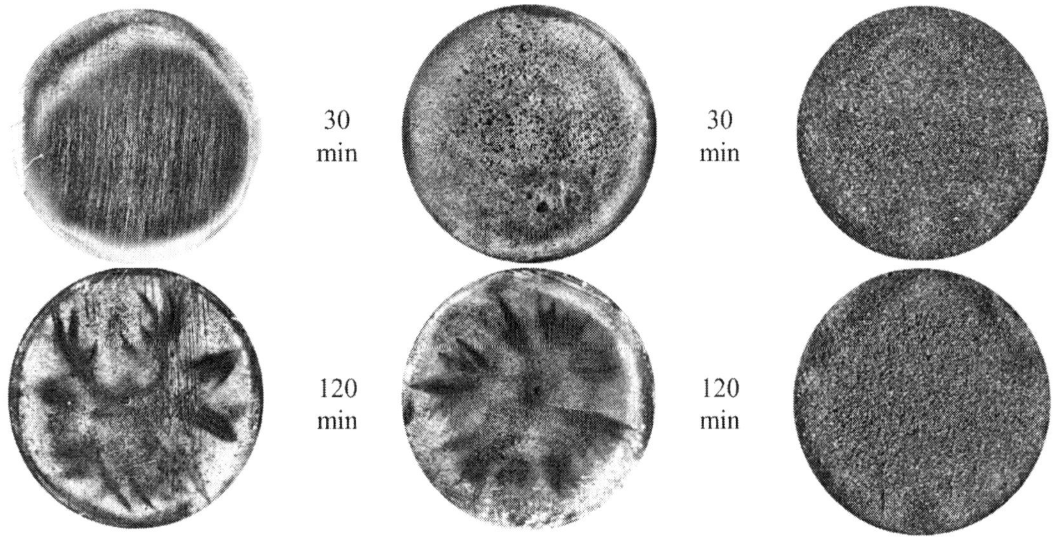

(a) (b) (c)

Figure 6. Destruction of the surface by cavitation for 30 and 120 min in the sea water: (a) Steel 1045; (b) Steel 20Cr13; (c) composite coating $Ti_{33}Ni_{49}Zr_{18}$–cBN10%Co

The composite coating $Ti_{49}Ni_{51}$-$Ti_{33}Ni_{49}Zr_{18}$-cBN10% Co in the process of cavitation for 120 min in the sea water is shown in Fig. 7. Cavitation damage of the composite layer is only visualized after 120 minutes of impact.

IOP Conf. Series: Materials Science and Engineering **295** (2018) 012029 doi:10.1088/1757-899X/295/1/012029

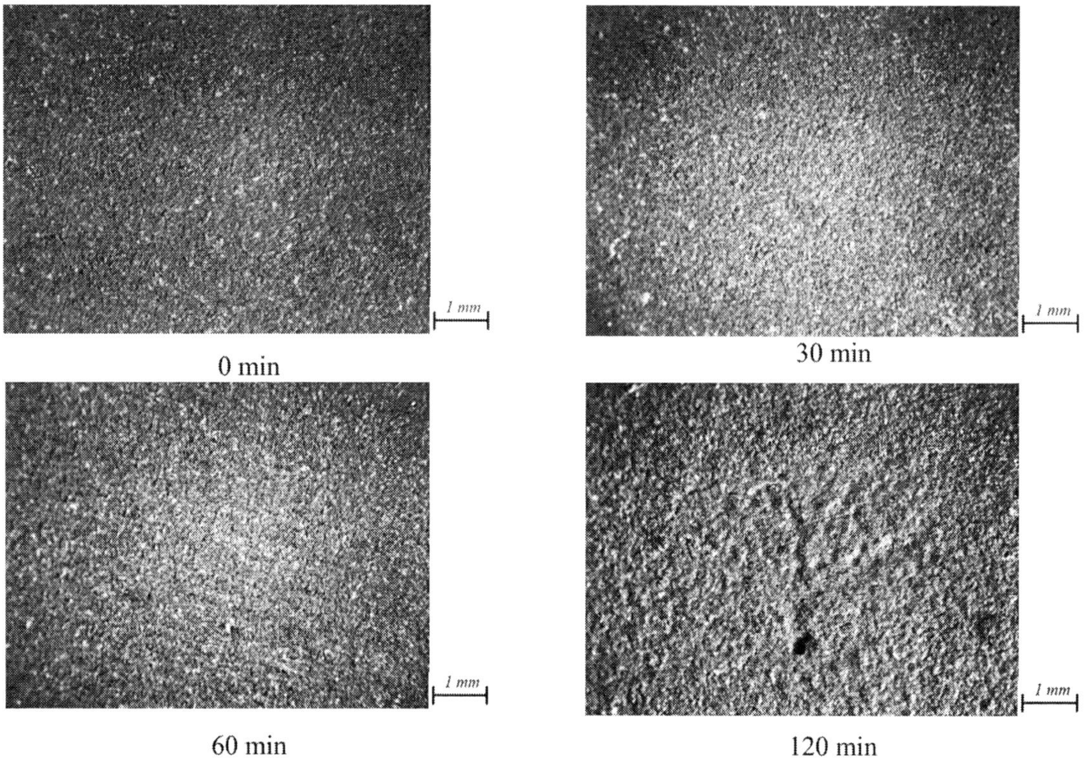

Figure 7. Cavitation destruction of the composite layer $Ti_{49}Ni_{51}$– $Ti_{33}Ni_{49}Zr_{18}$–cBN10%Co.

For a qualitative description of the cavitation fracture of a surface, it is expedient to investigate the parameters of the surface profile. The most characteristic factor for assessing the degree of surface cavitation wear is the following one: R_z is the height of the unevenness of the profile by ten points and R_{max} is the highest profile height at the base length of 7.5 mm.

Table 2 shows the dynamics of surface roughness variation within zone 1 on different samples during the cavitation effect in the sea water environment. The surface roughness of the samples with coatings is much higher than the initial roughness of the uncoated sample. This is due to the formation technology of surface coatings by the method of HVOF by the powder material particles, which forms a characteristic surface micro-relief.

Figure 8 shows the distribution of depths of the asperities from erosion fracture surface of tested samples for the three time values of cavitation (after 30 and 120 min, from top to bottom). The distribution was obtained by computer analysis of the measurement results of surface roughness in different areas of the samples. In the above diagrams the depth of erosion destruction of the surface was determined by the depth of asperities.

Table 2. Change in surface roughness (R_z / R_{max}), µm

	Process time, min				
	0	30	60	90	120
Steel 1045	0,27/1,56	5,23/18,54	9,56/28,14	13,88/44,48	16,25/47,2
Steel 20Cr13	0,35/1,72	1,05/4,18	1,43/12,65	4,87/21,42	8,48/32,18
Ni– $Ti_{49}Ni_{51}$–$Ti_{33}Ni_{49}Zr_{18}$ –cBN10%Co	61,76/79,36	63,14/80,12	65,86/80,67	67,24/81,04	69,21/85,60

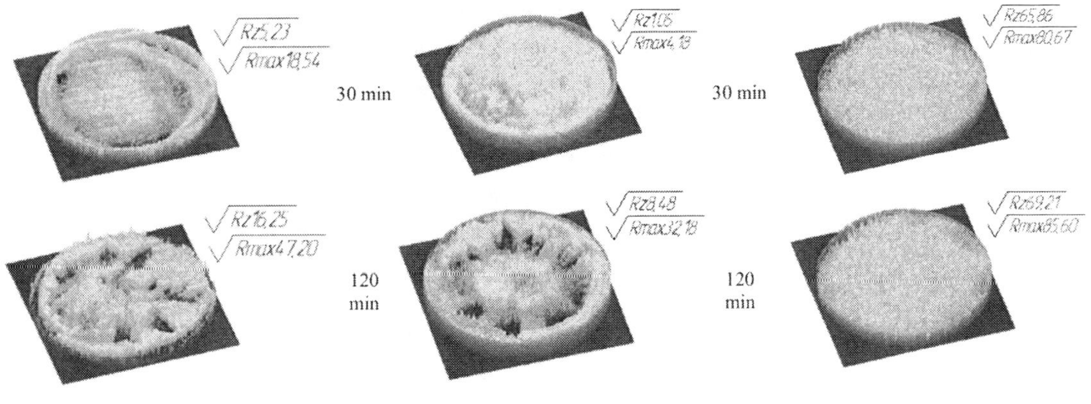

(a) (b) (c)

Figure 8. Destruction of the surface by cavitation for 30 and 120 min in the sea water: (a) Steel 1045; (b) Steel 20Cr13; (c) composite coating $Ti_{33}Ni_{49}Zr_{18}$–cBN10%Co

Analysis of Table 2 and Fig.8 shows that the change in the roughness of the samples with coatings is less than the change in the roughness of the sample without coating by 1.5-2 times. As we know, cavitation as a phenomenon, is associated with the discontinuity of the moving fluid, the formation of vacuum voids (caverns) in it, followed by their closing, accompanied by large hydrodynamic impacts. According to the ideas of a number of cavitation phenomena researchers [11,14,15,14], the cause of cavitation destruction of parts is the fatigue destruction of micro-volumes of a material due to repeated exposure of high-frequency pulses both as a detachment action (during formation and detachment of cavitation caverns) and shock action accompanying their clamping. In this case, the application site of the cavitation action pulses is so limited that the process of cavitation destruction should be considered from the point of the cyclic strength of individual structural components or local micro-volumes of the metal. Microvolumes of metal, damaged by cyclic loading and cavitation effect, weakened by loss of communication with the neighbouring micro-volumes, easily subject to erosion by the flow, forming deep relief cavitating surface, activating the cavitation processes to an even greater extent [14].

The surface composite layer cBN10%Co has high hardness and strength, which ensures its resistance to shock cyclic influences of collapsing caverns. The increased roughness of the surface of a solid surface composite in the form of strong columnar structures ensures the crushing of vacuum voids, redistributing their effect on the entire surface, and not concentrating them in certain zones. In addition, the gradient structure of the multilayer composite coating $Ti_{49}Ni_{51}$ –$Ti_{33}Ni_{49}Zr_{18}$–cBN10%Co makes it possible to create conditions for the relaxation of stresses created by the variable impact load of cavitation caverns and the manifestation of compensating internal forces due to thermo-elastic martensitic transformations of SME materials.

The increase in the performance characteristics of samples with a composite surface layer "steel – adhesive layer with a Ni – relaxing SME layer $Ti_{49}Ni_{51}$– a functional reinforcing SME layer $Ti_{33}Ni_{49}Zr_{18}$ – functional wear layer of cBN-10% Co" under conditions of cavitation erosion is explained by the special functional properties of SME materials, which are manifested in the ability to adapt to the loading conditions, which is provided by the formation technology and processing of the surface layer. The increase in the wear-and-fatigue characteristics of the surface composition is explained by the pseudoelasticity of the layer made of SME materials with the $Ti_{33}Ni_{49}Zr_{18}$ and the decrease in the stress concentration due to its relaxation capacity.

The field of use of the above surface composition can be expanded by controlling the functional properties of the layers composing the composition by varying the chemical composition and controlling the phase composition by thermal and thermomechanical processing.

3. Conclusion

According to the experimental data, the surface of a composite layered material, especially hardened by a hard wear-resistant composite, has a high resistance to surface defects as a result of cavitation processes. The cavitation resistance of the $Ti_{49}Ni_{51}$–$Ti_{33}Ni_{49}Zr_{18}$ coating according to the criterion of mass wear is 1,5-2 times higher than that of the base material without coating (steel 1045). The cavitation resistance of the coating $Ti_{49}Ni_{51}$–$Ti_{33}Ni_{49}Zr_{18}$–cBN10%Co according to the criterion of mass wear is 15-20 times higher than that of the base material without coating and 10-12 times higher than that of the $Ti_{49}Ni_{51}$-TiNiZr coating. The proposed architecture of the multifunctional gradient composition, "steel- Ni– $Ti_{49}Ni_{51}$–$Ti_{33}Ni_{49}Zr_{18}$–cBN10%Co", each layer of which has its functional purpose, allows to increase the service life of parts operating under conditions of cavitation-fatigue loading in corrosive environments.

Acknowledgements

The work was supported by the Russian Science Foundation (Agreement No. 15-19-00202).

References

1. Polovinkin V.N., Promising construction materials for special marine equipment, shipbuilding and military shipbuilding. [Electronic resource]. Http://www.proatom.ru/modules.php?name=News&file = article & sid = 5778
2. Mikhailov A.N., Mikhaylov DA, Grubka RM, Petrov MG. Increase of longevity of machine parts on the basis of functionally oriented coatings, High technology in engineering. - 2015.-7 (49). - P. 20-39.
3. Suslov, B.M. Bazrov, V.F., High technology in engineering. Without language, etc .; Ed. A.G. Suslova. M .: Mechanical engineering. 2012. 528 pp.
4. Blednova Zh.M, Rusinov P.O. Intellectualization of surface layers, working under cyclic loading and reversing friction, Applied Mechanics and Materials. Trans Tech Publications, Switzerland. 2015. Vol. 798 (2015). pp 440-446.
5. Rusinov P.O., Blednova Zh.M Structural and technological patterns of formation of surface nanostructured layers TiNiZr by high-speed flame spraying, Advanced Materials and Structural Engineering — 2016 (978), 21-25.
6. Rusinov P.O., Blednova Zh.M., Balaev E.Y., Dmitrenko D.V. Formation of composite layers TiNiZr-cBN-Co, working in conditions of cyclic loading and reverse friction // Procedia Structural Integrity 2. 2016. — P. 1506-1513.
7. Blednova Zh.M., Makhutov N.A., Rusinov P.O., Stepanenko M.A., Mechanical and tribological properties of the multifunctional composition "base - material with shape memory effect", formed under the conditions of high-energy effects, Factory laboratory. Diagnostics of materials - 2015. (81), No. 3.-P.41-49.
8. Blednova Zh. M., Rusinov P.O., Dmitrenko D.V. Failure analysis of screw propellers and increase of fail safety by surface modification with multicomponent materials with shape memory effect, Procedia Structural Integrity 2. 2016. P. 1497-1505.
9. Blednov, J. M., Dmitrenko D. V., M. E. Bezmogorychnaya. Planetary centrifugal mill, patent for useful model RUS 164078 20.10.2015.
10. Blednova Zh.M, Rusinov P.O., Bezmogorychnaya M.E. Influence of particle size distribution, energy condition and reactivity of the deposited material on nanopatterning of surface layers' made of shape memory materials, Key Engineering Materials. 2017, 730, pp.8-14.

11. Dmitrenko D.V., Blednova Zh.M., Balaev E.Yu. Cavitation resistance of products with a composite surface layer of SME materials. Intl. Conf. "Vitality and structural materials science" "ZhivKom-2016" .M .: IMASH RAS, 2016. P.232-235.

12. Blednova Zh.M., Rusinov P.O., The composite construction of the surface layers from materials with shape memory effect. – Krasnodar: Publishing House. KubSTU, 2017. – 280 c.

13. Petrov A.I., Skobelev M.M., Khanychev A.G.. Investigation of comparative resistance to cavitation erosion of materials samples and coatings of the flowing part of hydraulic machines - Vestnik MSTU. Ser. "Mechanical engineering". 2015. № 2 p. 128-137.

14. Vakulenko K.V., Biblak I.V., Cossack. I.B., Computer simulation of cavitation erosion of steel X20Cr13 with coatings. Open information and computer integrated technologies No. 72, 2016. - P. 242-248.

15. Razorenov S.V. Garkushin G.V., Kannel GI, Kashin O.A., Ratochka I.V., Behavior of nickel-titanium alloys with shape memory effect under conditions of shock wave loading // Physics of the Solid State, 2011, Vol. 4. - P. 768-773.

16. Wu S.K, Lin H.C., Yeh C.H., A comparison of the cavitation erosion resistance of TiNi alloys, SUS304 stainless steel and Ni-based self-fluxing alloy. Wear. 244 (2000), 85–93.

Tribological properties of multifunctional coatings with Shape Memory Effect in abrasive wear

Blednova Zh.M. [1]***, Dmitrenko D.V.*** [1]***, Balaev E.U.O.*** [1]
[1]*Kuban State Technological University, Department of dynamics and strength of machines, 350072, Moskovskaya 2, Krasnodar, Russia*
ddv-kk@yandex.ru, blednova@mail.ru, balaev1122@mail.ru

Abstract: The article gives research results of the abrasive wear process on samples made of Steel 1045, U10 and with applied composite surface layer "Nickel-Multicomponent material with Shape Memory Effect (SME) based on TiNi". For the tests we have chosen TiNiZr, which is in the martensite state and TiNiHfCu, which is in the austenitic state at the test temperature. The formation of the surface layer was carried out by high-speed oxygen-fuel deposition in a protective atmosphere of argon. In the wear test, Al_2O_3 corundum powder was used as an abrasive. It is shown that the wear rate of samples with a composite surface layer of multicomponent materials with SME is significantly reduced in comparison with the base, which is explained by reversible phase transformations of the surface layer with SME. After carrying out the additional surface plastic deformation (SPD), the resistance of the laminated composition to abrasion wear has greatly enhanced, due to the reinforcing effect of the SPD. It is recommended for products working in conditions of abrasive wear and high temperatures to use the complex formation technology of the surface composition "steel-nickel-material with high-temperature SME", including preparation of the substrate surface and the deposited material, high-speed spraying in the protective atmosphere of argon, followed by SPD.

Keywords: Abrasive wear, Surface, Multicomponent material, SME

1. Introduction

One of the most dangerous types of wear is abrasive wear [1-3]. The destruction of the workpiece surface as the result of abrasive wear occurs due to the interaction with solid particles which plastically and elastically deform the surface of the parts, as well as scratch, scrape and cut the material. When free abrasive particles get into places of details interface, they perceive the load transmitted through an interface pair and hit or squeeze into the surface details. The process of abrasive wear is an extremely complex phenomenon and is often accompanied by shock-droplet [4], erosion corrosion [5], the cavitation effect [6,7] and its distinctive feature is rapid development, leading to destruction of the surface. Currently, there are various ways to control abrasive wear: technological and operational ones. One of the promising control areas of abrasive wear is the introduction of new technologies of surface engineering, implemented in the integrated high-energy effects: laser, plasma, high-velocity flame ones[8,9].
High-energy impact allows to increase speed of heating and cooling that leads to the creation of the maximum non-equilibrium structures, which, under certain conditions, can provide the required complex of physical-mechanical and operational properties. Structure management of materials of a given composition, in conditions which are far from thermodynamic equilibrium, allows to ensure the

required durability of the product in extreme conditions when exposed to high contact and dynamic loads, abrasive and impact-abrasive wear.

It is known that the materials with high hardness are more resistant to abrasive wear. This is due to the fact that the higher the hardness is the smaller the depth of a cut or a crumple under the impact of the abrasive particles with the surface is. The interaction of abrasive particles with the surface is accompanied by surface deformation and causes the hardening that increases the hardness of the outer layers, and, as a consequence, increases the resistance to abrasive wear. However, further accumulation of defects, such as scratches and cracks, leads to intense destruction of the workpiece surface [3,10,11]. Studies proved the dependence of abrasive wear resistance and energy consumption of the material, which is defined as the product of tensile strength and relative residual contraction [12,13]. This allows us to conclude that abrasive wear resistance depends on the optimal combination of strength and plasticity.

In the present work in order to control abrasive wear, we proposed to use smart, intermetallic nanocrystalline and layered structures, which also include materials with reversible phase structure, in particular, materials with shape memory effect (SME). The use of SME materials [14], in order to form surface layers or layered structures [15] which are multi-functional, can provide an effective response of tribomaterials to active influence of external factors [16], including the conditions of abrasive wear. The special tribological properties of SME alloys based on TiNi in combination with high damping capacity and resistance to corrosion-erosion effects have determined the Ni alloy as a "frictional phenomenon" - "a new tribo-material" and "tribo-composite" [17,18]. High wear resistance of alloys with SME has got not only high hardness values, but also special functional-mechanical properties. On the workpiece surface made of SME materials, as a result of abrasive particle deformation at a voltage that exceeds the limit of elasticity, there is a nonlinear plastic deformation, which remains as a residual one and may accumulate in the case of conventional material after removing the tension. In the case of SME material, when elastic deformation is too high, the further deformation is non-linear and takes place due to thermoelastic martensitic transformation (austenite-martensite). This deformation is removed completely when the load is restored by the reverse martensitic transformation (martensite-austenite), in this case there is a phenomenon of pseudo elasticity [10,14-16,19]. Therefore, the damage accumulation in SME materials is much slower than that of the conventional materials. However, the production of workpieces from SME materials is not economically feasible. Therefore it seems promising to use SME materials in order to form the formation of the surface layers, resistant to this type of fracture.

In [26, 27] it is shown that multicomponent coatings on the basis of TiNi with additional alloying elements such as Zr in an amount of up to > 10% allow us to create a composite material with a gradient of properties at the interface between the layers, which allow to obtain new properties and to increase their performance. Use of materials with shape memory effect (SME) as the surface layers [25] or as a part of the layered compositions [27] allows to provide the effective response of materials to external factors and adaptation to external influences.

As a technology we selected High Velocity Oxygen Fuel spraying (HVOF) in the protective argon environment [15,20,21]. Surface layers formed by HVOF in a protective environment are characterized by high adhesive strength (90-100 MPa) and low porosity (less than 1%). As SME materials for the formation of surface compositions in the present work, we selected composite materials TiNiZr, TiNiHfCu, which have different structural-phase states [22-24]. As materials for results comparison, we applied structural steel 1045 and tool steel U10.

2. Experiment

2.1 Technology of formation of surface composite layer

As the basis were used Steel 1045, which is widespread in mechanical engineering, the transition layer is Nickel, which has unlimited solubility with iron and chemical affinity with the material of the transitional and functional layers, which provides increased adhesive strength. In order to for functional layers, we used SME materials, which have significantly different temperature phase transformations: $Ti_{33}Ni_{49}Zr_{18}$ alloy, which possesses high temperature memory effect and at temperature of 20 C is in the martensitic state [23]; $Ni_{44}Ti_{36}Hf_{15}Cu_5$ alloy at temperature of 20 C is in the austenitic state [28]. The characteristics of the materials are presented in table 1.

Table 1. Characteristics and structural-phase state of materials.

Material samples	Diameter, mm	Weight, g	T, °C	Microhardness, GPa	Phase state
Steel 1045	50,39	145,735	20	1,9-2,1	-
U10	50,24	158,943	20	5,7-6,5	-
Steel 1045 –Ni– $Ti_{33}Ni_{49}Zr_{18}$	50,53	154,176	20	2,1-2,9	martensite
Steel 1045–Ni– $Ni_{44}Ti_{36}Hf_{15}Cu_5$	51,12	155,214	20	9,5-11,9	austenite
Steel 1045 – Ni – $Ni_{44}Ti_{36}Hf_{15}Cu_5$ + (SPD)	50,87	155,663	20	11,7-14,2	austenite

Formation technology of surface composition involves several stages: preparation of the base and coating material, application of transitional and functional coatings by High-velocity oxygen-fuel spraying (HVOF), thermomechanical treatment (TMT).

Base preparation is a comprehensive machining, which includes creation of a surface with well-developed microstructure, followed by shotblasting, and chemical treatment, which consists of degreasing and surface etching with a mixture of hydrochloric and nitric acids.

Preparation of coating material is mechanical activation (MA) of a powder material, which provides necessary granulometric composition, energy state and reactivity of the material. Mechanical activation was carried out in the upgraded ball mill Gefest-2 of AGO-2U [30], in which the mechanical effect is carried out by a series of mechanical pulses (beats) of working bodies, that transmits a portion mechanical energy to the processed material.

The coating was carried out by HVOF at the upgraded device GLC-720 in a protective argon atmosphere [8, 29].

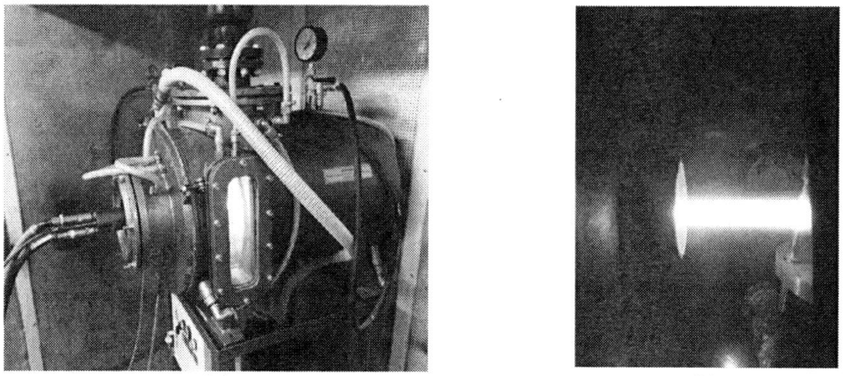

(a) (b)

Figure 1. Device for coating by HVOF in a protective atmosphere

Integrated TMT was carried out to give pseudo elastic properties or the properties of shape memory to a surface layer and to impart functional properties and increase the adhesion strength. Integrated treatment is assigned depending on the thickness of the composite layer. For coatings with thickness up to 1 mm we carried out comprehensive treatment consisting of surface plastic deformation (SPD) and ultrasonic treatment (UST). For coatings with thickness greater than 1 mm we carried out comprehensive treatment consisting of SPD and resistance spot welding [31].

IOP Conf. Series: Materials Science and Engineering **295** (2018) 012030 doi:10.1088/1757-899X/295/1/012030

Study of the deformation behavior influence of the samples (steel and composite surface layer of SME alloys) in conditions of abrasive wear was carried out on a modernized testing machine friction CMT-1-2070, which is equipped with a special chamber for abrasive. Figure 2 shows a diagram of abrasion testing. A device consisting of two disks of equal diameter, revolving in the opposite direction, is placed in hermetically sealed camera.

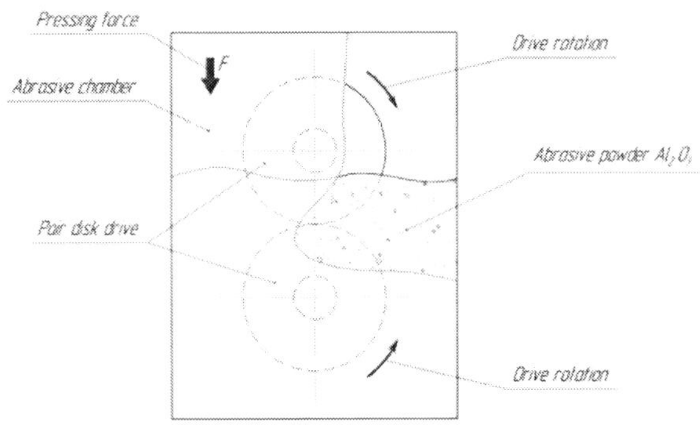

Figure 2. Scheme of test for abrasive wear

The chamber is filled with a powder of abrasive material to a level above the interface plane disks of 15 mm, which ensures a permanent contact of the powder particles in the contact zone of the two samples. We can see from the diagram that in the contact samples experience different effects: rolling friction, abrasive wear and sliding friction. This experiment may mimic the operation of a gear pair or crushing station rolls.
As the abrasive we used aluminium oxide Al_2O_3 (corundum) with a hardness of 25 GPa [1]. The average particle size of the Al_2O_3 powder is 120-180 μm.
The frequency of rotation of a pair is 300 min^{-1}, pressing force is 250 N, the test time is 2 hours, the interval measurement is every 15 minutes. The nucleation and growth dynamics of surface damage was observed using the stereomicroscope stamps LOMOM SP-2, with magnification 25-50, which is provided with a scale for measurement. Mass wear of the discs is determined by the gravimetric method (balance "DEMCOM DL") with a measurement accuracy of 0.001 g. The measurement of the diameters of the disks were performed with a micrometer with an accuracy of up to 0.005 mm., the wear rate was determined according to GOST 27674-88.
The study of coating hardening from existing forces in the process of cylindrical specimens rolling was carried out on microhardness tester Falcon 500.
To determine the resistance of the coating to the contact stresses we carried out their estimation based on the method described in [32].

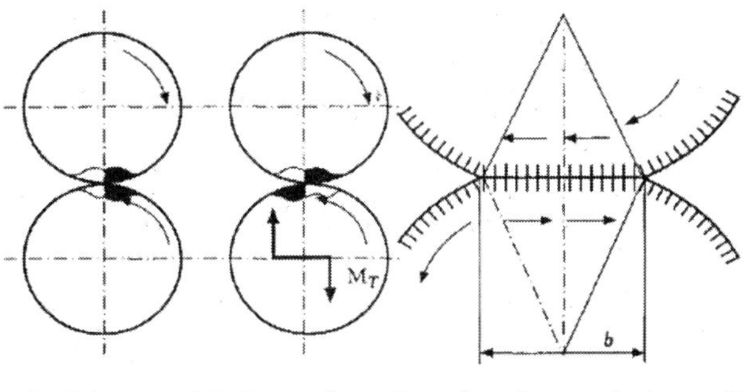

Figure 3. Scheme of deformation of surface layers during rolling of two cylinders

The contact patch has the form of a narrow rectangle whose length is equal to the width of the disk – 10 mm., and the width is determined by the formula:

$$b = 2{,}15\left(q\frac{\frac{1}{E_1}+\frac{1}{E_2}}{\frac{1}{R_1}+\frac{1}{R_2}} \right)^{1/2}$$

For alloy Ti33Ni49Zr18 Young's modulus in the martensitic condition are E=28 GPa, pressing force is 250 N on the length of 10 mm, therefore, a distributed load is equal to q=25000 N/m

$$b = 2{,}15\left(25000\frac{\frac{1}{28\times 10^9}+\frac{1}{28\times 10^9}}{\frac{1}{25\times 10^{-3}}+\frac{1}{25\times 10^{-3}}} \right)^{1/2} = 329{,}15\times 10^{-6}, \mathrm{m}$$

For alloy $Ni_{44}Ti_{36}Hf_{15}Cu_5$ Young's modulus in the martensitic condition is E=80 GPa.

$$b = 2{,}15\left(25000\frac{\frac{1}{80\times 10^9}+\frac{1}{80\times 10^9}}{\frac{1}{25\times 10^{-3}}+\frac{1}{25\times 10^{-3}}} \right)^{1/2} = 190{,}03\times 10^{-6}, \mathrm{m}$$

The greatest compressive stress acting at points of the axis of contact area, for alloy $Ti_{33}Ni_{49}Zr_{18}$ is

$$\sigma_{max} = 1{,}27\frac{q}{b} = 0{,}418\left(2\times q\frac{E_1\times E_2}{E_1+E_2}\times\frac{R_1+R_2}{R_1\times R_2}\right)^{1/2} = 98{,}9\ \mathrm{MPa}$$

The greatest compressive stress acting at points of the axis of contact area, for alloy $Ni_{44}Ti_{36}Hf_{15}Cu_5$ is

$$\sigma_{max} = 1{,}27\frac{q}{b} = 0{,}418\left(2\times q\frac{E_1\times E_2}{E_1+E_2}\times\frac{R_1+R_2}{R_1\times R_2}\right)^{1/2} = 167{,}2\ \mathrm{MPa}$$

Analysis of the stress state shows that the dangerous point is located at a depth equal to 0.4 of the width of the contact area. The principal stresses at this point have the following meanings:

контакта. Главные напряжения в этой точке имеют следующие значения:

$\sigma_1 = -0{,}180\times\sigma_{max}$ $\sigma_2 = -0{,}288\times\sigma_{max}$ $\sigma_3 = -0{,}780\times\sigma_{max}$

The main stress $Ti_{33}Ni_{49}Zr_{18}$

σ_1 =-17,802 MPa, σ_2 =-28,483 MPa, σ_3 =-77,142 MPa.

Shear stress

$\tau_{max} = 0{,}3\times\sigma_{max} = 29{,}67\ \mathrm{MPa}$

The main stress $Ni_{44}Ti_{36}Hf_{15}Cu_5$

σ_1 =-30,096 MPa σ_2 =-48,44 MPa σ_3 =-130,416 MPa

Shear stress is

$\tau_{max} = 0{,}3\times\sigma_{max} = 50{,}16\ \mathrm{MPa}$

According to the fourth theory of strength equivalent stress for the composition $Ti_{33}Ni_{49}Zr_{18}$ is

$\sigma_{equ} = (0{,}5[(\sigma_1-\sigma_2)^2+(\sigma_2-\sigma_3)^2+(\sigma_3-\sigma_1)^2])^{1/2} = 54{,}84\ \mathrm{MPa}$

According to the fourth theory of strength equivalent stress for the composition $Ni_{44}Ti_{36}Hf_{15}Cu_5$ is

$\sigma_{equ} = (0{,}5[(\sigma_1-\sigma_2)^2+(\sigma_2-\sigma_3)^2+(\sigma_3-\sigma_1)^2])^{1/2} = 95{,}54\ \mathrm{MPa}$

Contact stresses are dependent on the contact force of the rollers. For the selected contact force of the rollers of 250 N maximum contact stresses occur at the depth of the composition $Ti_{33}Ni_{49}Zr_{18}$ at 0,131 mm, $Ni_{44}Ti_{36}Hf_{15}Cu_5$ at 0,076 mm from the surface. Due to the fact that the adhesion strength of the coating layers is comparable to the magnitude of contact stresses, it is necessary to control the thickness of the top coat layer so that the interface between the layers was beyond the point of occurrence of maximum contact stresses.

2.2 Experimental results and discussion

During the test on abrasion wear, we monitored nucleation and accumulation of damages in the surface layers. Fig. 4 shows the pattern of damage accumulation through time for the studied materials, where more saturated colors correspond to the greater depth of the damage. Thus, on the Steel 1045 sample form the first minutes damages in the form of dings, tears, cuts and scratches appear on the surface. Areas with damage comparable to the size of Al_2O_3 particles, are highlighted in blue, and areas with damages that exceed the size of the abrasive particles are highlighted in red

(Fig.4). It should be noted that focal accumulation of damages (defects) under these test modes was observed only during the first 100-120 minutes, then the difference in depth of defects is reduced and intensive wear is observed throughout the sample surface. Similar tests on the abrasive wear of tool Steel U10 showed a decrease in intensity of the accumulation of defects and their depth compared with Steel 1045. The wear rate is reduced by about 30%. It is shown in figure of wear intensity (Fig. 5, a and b) and confirmed by other sources [1-7].

The deformation behavior of samples with a composite surface layer of a SME material with high temperature "Ni – $Ti_{33}Ni_{49}Zr_{18}$" (MF = 459 K, MS = 522 K, As = 488 K, AF = 571 K) [22-23] in the conditions of abrasive wear, showed that the emergence of deep damage is detected much later (after 60 minutes of testing); the speed of accumulation of defects is reduced as compared with the material without coating; with coating hardness of $Ti_{33}Ni_{49}Zr_{18}$ in the martensitic state two times less than the hardness of tool Steel U10, the intensity of wear is in 2 times less than that of Steel U10 (Fig. 5). Test temperature (20 C) corresponds to the martensitic condition of the coating material $Ti_{33}Ni_{49}Zr_{18}$.

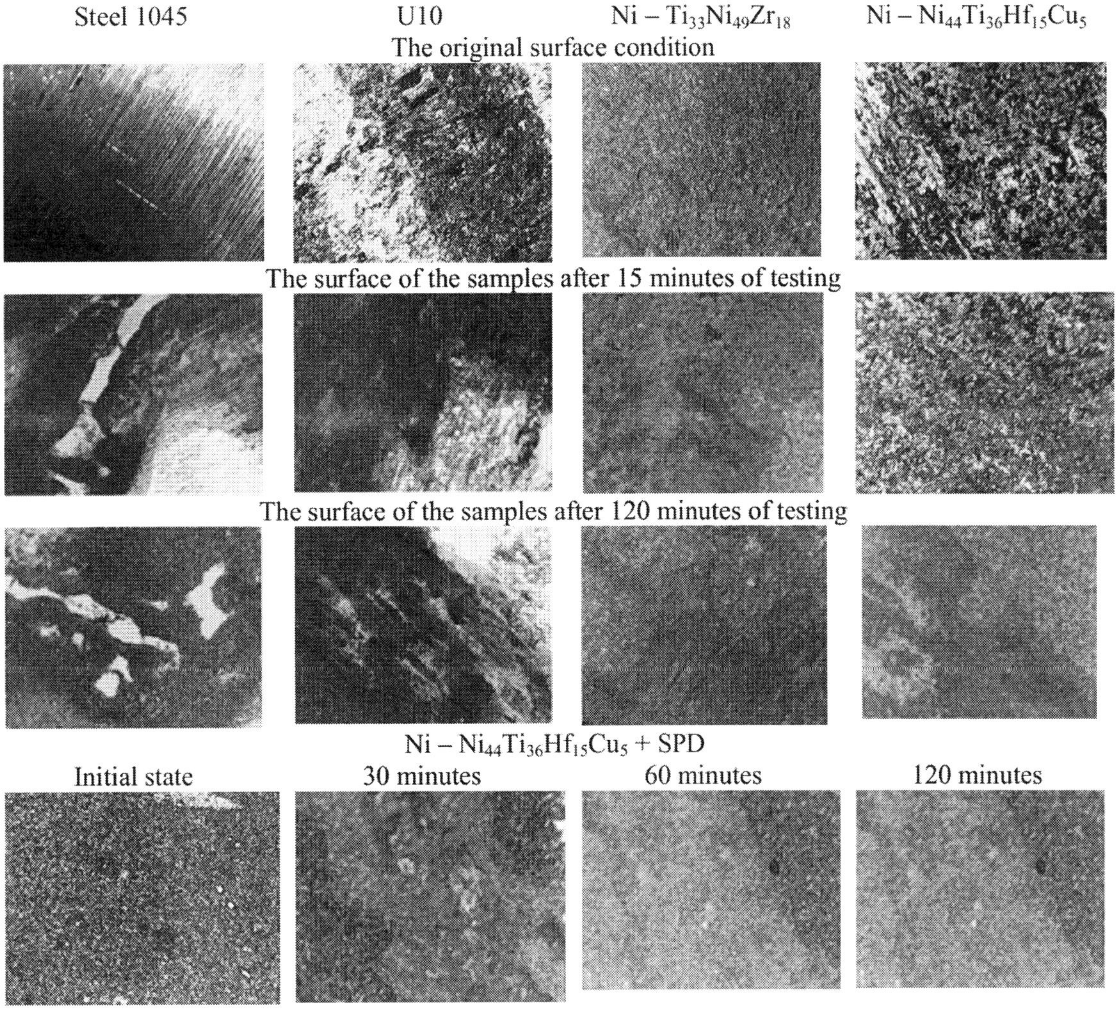

Figure 4. The accumulation of surface damage of the samples during the tests on the abrasive wear

To assess the effect of structural-phase state of SME materials on the wear rate at abrasive impact we carried out tests of samples with surface functional layer "Steel 1045–Ni–$Ni_{44}Ti_{36}Hf_{15}Cu_5$" after a full cycle of treatment: MA, VHOF, TMT and SPD in the test conditions in the austenitic state [24]. Analysis of the surface composition "Ni–$Ni_{44}Ti_{36}Hf_{15}Cu_5$" at various stages of the test (Fig. 4) showed that process of surface defects of considerable size formation slows down (after 90 minutes of the test). The wear rate is reduced by 50% compared to the wear of the coating "Ni–$Ti_{33}Ni_{49}Zr_{18}$" which is in the martensitic condition and more than 6 times than that of the the base material (Steel 1045).

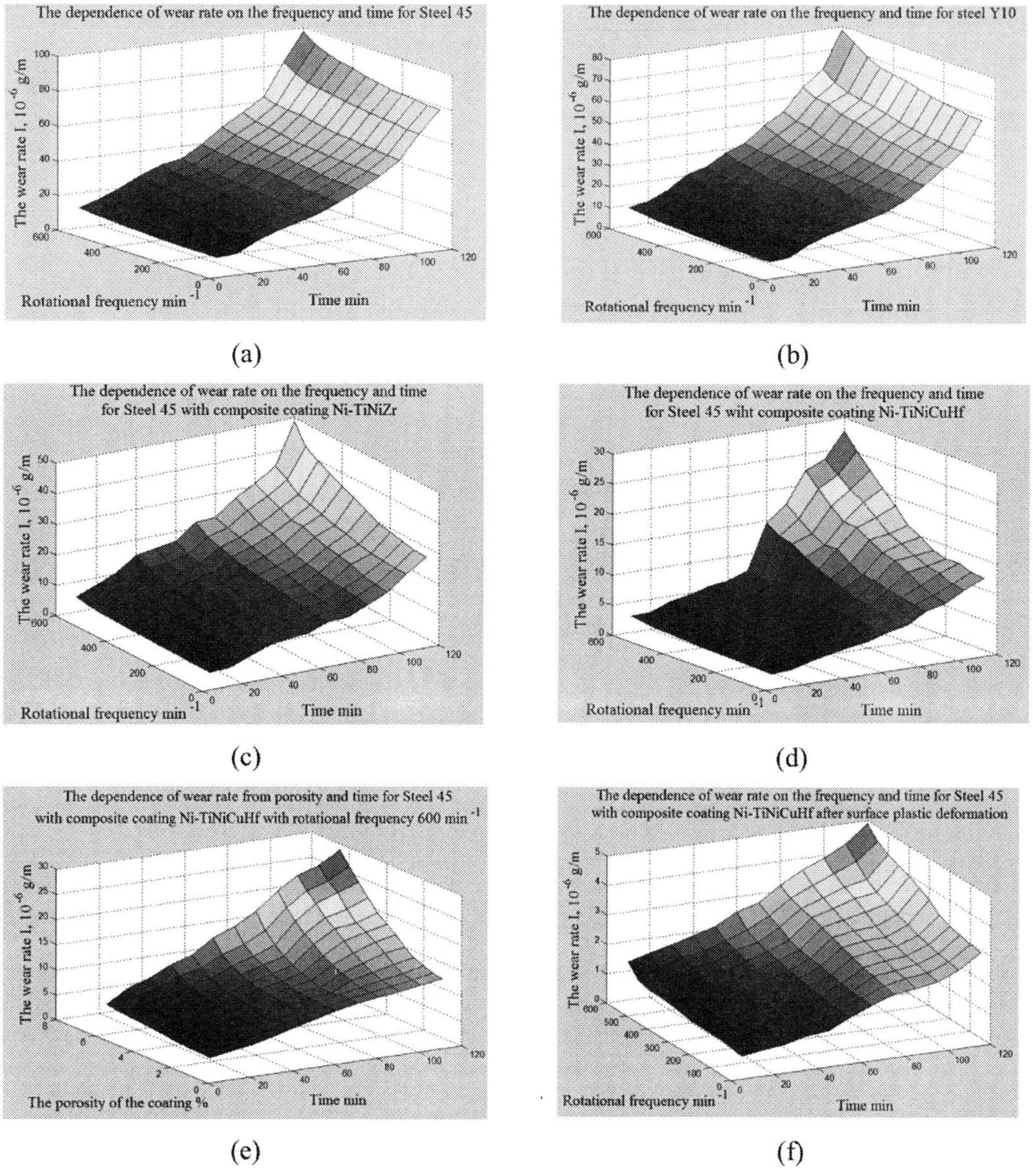

Figure 5. The wear intensity of specimens examined: (a) Steel 1045, (b) Steel У10, (c) coating Ni–$Ti_{33}Ni_{49}Zr_{18}$, (d) coating Ni–$Ni_{44}Ti_{36}Hf_{15}Cu_5$, (e) coating Ni–$Ni_{44}Ti_{36}Hf_{15}Cu_5$ at high velocity, (f) coating Ni–$Ni_{44}Ti_{36}Hf_{15}Cu_5$ + SPD

Full cycle processing of SME materials, including TMT and SPD is accompanied by strain hardening, which increases the hardness, creating a nonuniform stress-strain state (SSS) and is primarily focused on the formation of pseudo elasticity properties. The increased hardness of the surface layer, from one hand – increases the resistance to abrasive wear, and from another hand – leads to its embrittlement, and together with the heterogeneity of SSS, leads to the formation of defects at an early stage of testing of break type of and fractures. It is considered that the hardness of the SME materials is not the determining factor in the improvement of wear resistance, although it significantly affects abrasive wear (Fig. 5, c).

To assess the influence of abrasive wear on the distribution of hardness along the depth of the coatings, we conducted a study of samples' cross-section for microhardness after 120 min of testing (Fig. 5, a). From the diagrams of the distribution of hardness it can be seen (Fig. 5, c) that after 120 minutes of testing the test coating does not accumulate hardening. Microhardness on cross sections is distributed quite evenly. This suggests that after the load is removed the material returns to their original state.

The study of microstructure of the surface layer $Ni_{44}Ti_{36}Hf_{15}Cu_5$ + SPD on a raster electronic microscope of high resolution showed that the layer has a nanoscale structure with a grain size of 95 - 140 nm (Fig. 5, b). Nanoscale structure causes a decrease of critical transition temperature and an increase in the interval of temperatures within which it occurs [33].

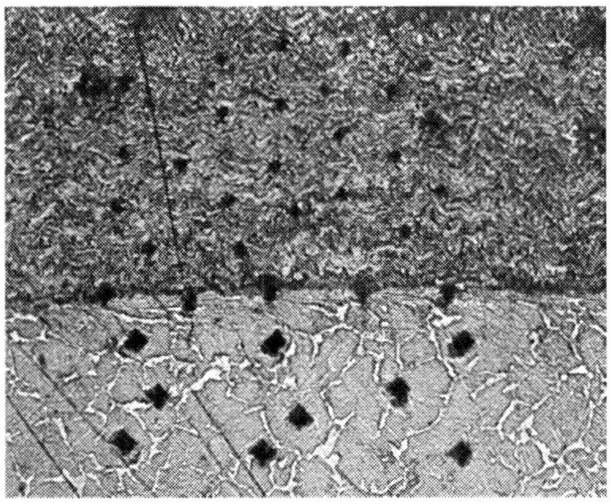
x10

(a)

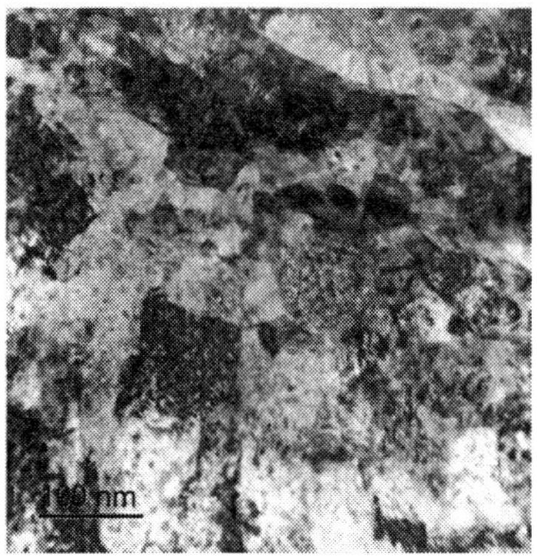

(b)

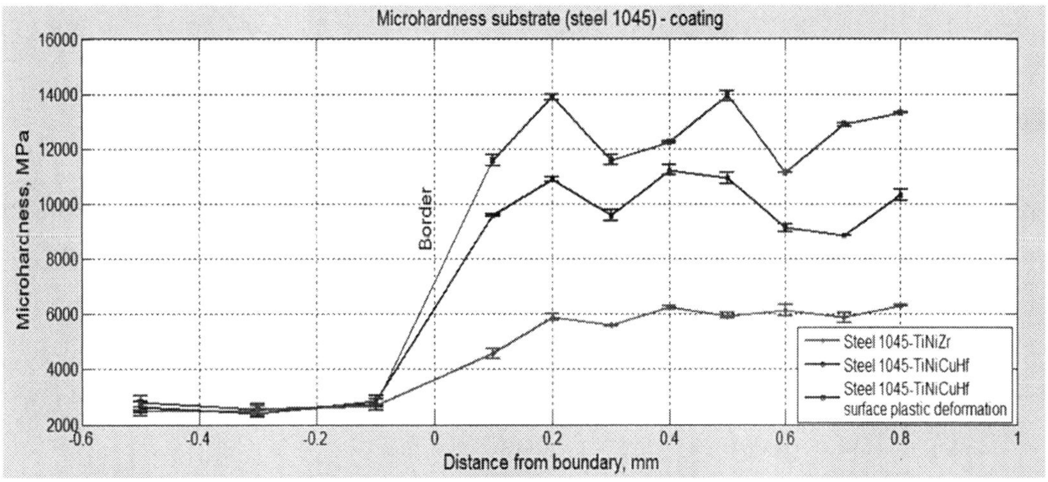

(c)

Figure 6. The distribution of hardness along the depth of the coating: (a) picture of microhardness measuring on the sample coated with Ni–$Ni_{44}Ti_{36}Hf_{15}Cu_5$ + SPD, (b) the microstructure of the coating Ni–$Ni_{44}Ti_{36}Hf_{15}Cu_5$ + SPD, (c) graphs of hardness distribution of the investigated coatings

Materials with SME, being the materials with reversible phase structure, have the unique ability to phase transformations under the action of load and temperature. When load is applied to the product, which is in the austenitic state, there is a direct austenitic-martensitic transformation, and at the termination of loading, the reverse martensite-austenite transformation takes place. The mechanism of abrasive wear of Ti-Ni alloy is considered in [11]. The TiNi alloy has a limited temperature interval of phase transformations, as a rule, not exceeding 373K. For articles exposed to abrasive wear in the conditions of higher temperatures, it is economically and technologically expedient to use the surface modification of multi-component materials with SME $Ti_{33}Ni_{49}Zr_{18}$, $Ni_{44}Ti_{36}Hf_{15}Cu_5$ providing a high temperature of SME . The most effective ways of increasing wear resistance and durability of the products in the conditions of abrasive wear is the formation of surface layered functionally graded compositions "the basis - transition layer – functional layer". As a transitional layer in the present work we used Nickel, which has a chemical affinity with material bases and the functional layer, providing a reliable adhesion. A transition layer of Nickel having a high ductility and toughness, is a relaxation layer that inhibits the movement of propagating defects, slowing down the process of destruction. Surface functional layer made of materials with SME on the basis of TiNi has a high abrasion resistance, sufficient hardness and ability to adapt to loading conditions by displaying pseudo elasticity properties during phase transformations.

3. Conclusion

Intensive development of smart materials have greatly increased modification and coating technologies with desired tribological and physical-mechanical properties used in the friction units. It was shown that the development of a new generation of coatings with SME materials, which is a multi-layered surface composition with alternating layers of different functional purposes, enhances operational properties of working surfaces of products. Study of deformation behavior of samples with a composite surface layer of multicomponent SME materials, formed under high-energy impact during abrasive wear showed a significant reduction of wear rate depending on the structural-phase state of the coatings and technology of their formation. We confirmed experimentally a decrease in the wear rate of the samples with composite surface SME layer: "Ni - $Ti_{33}Ni_{49}Zr_{18}$" which is in the martensitic state, by 3 times; "Ni–$Ni_{44}Ti_{36}Hf_{15}Cu_5$" which is in the austenitic state, by 6 times; "Ni–$Ni_{44}Ti_{36}Hf_{15}Cu_5$", after a full cycle TMT and rpm, by 30 times. The basis of improving the wear resistance of the surface composition of SME materials in the conditions of abrasive wear is the effect of pseudo elasticity and ability of SME materials to adapt to external influences due to reversible phase transformations.

Acknowledgements

The work is carried out with financial support of Russian Scientific Foundation (Agreement No. 15-19-00202).

References

1. Hrushhov M.M. Babichev M.A. Abrasion. G Science, 1970. 252 p.
2. Dobrovol'skij A.G. Abrasive wear resistance materials – K.: Tekhnika, 1989. 128 p.
3. Garkunov D.N. Tribotehnika. – M.: Mashinostroenie, 1985. – 424 p.
4. Varavka V.N., Kudrjakov O.V., Ryzhenkov A.V., Kachalin G.V., Zilova O.S. The use of nanocomposite coatings for the protection of power equipment from chipleaders erosion, Thermal Engineering, 2014, No. 11, Pp. 29–35.
5. Guilemany J.M., Cinca N., Dosta S., Benedetti A.V. Corrosion behaviour of thermal sprayed nitinol coatings Corrosion Science 51 (2009) 171–180.
6. Wu S.K., Lin H.C., Yeh C.H. A comparison of the cavitation erosion resistance of TiNi alloys, SUS304 stainless steel and Ni-based self-fluxing alloy. Wear 244 (2000) 85–93.

7. Dmitrenko D.V. Blednova Zh.M., Balaev E.U. Cavitation resistance of products with a composite surface layer of materials with SME. Tr. int. Conf. "Durability and structural materials science" "Ivkom-2016".M.: IMASH,2016. P. 232-235.

8. Polovinkin V.N Nanoengineering surfaces of mechanical engineering products. 2010, Electronic resource. www.proatom.ru/modules.php?name=News&file=print&sid=2730

9. Il'jushhenko A.F., Shevcov A.I., Okovityj V.A. The processes of formation of thermal spray coatings and their modeling. Minsk, Belarus. Nauka, 2011. — 357 p.

10. Potekaev A.I., Hohlov V.A., Galsanov S.V., Shulepov I.A. The study of fatigue wear of a material with shape memory. Bulletin of the Tomsk Polytechnic University. 2014. T. 324. No. 2. – Pp. 90-95.

11. Potekaev A.I., Hohlov V.A., Galsanov S.V. The nature and mechanisms of abrasive wear of materials with shape memory, for example, titanium nickelide Bulletin of the Tomsk Polytechnic University. 2012. T. 321 No. 2, pp. 107-111.

12. Mashkov, Ju.K., Poleshhenko K.N., Povoroznjuk S.N. The friction modifying materials tribosystem – M.: Nauka, 2000. 280 p.

13. Kosteckij, B.I., Nosovskij I.G., Karaulov A.K. and etc. Surface strength of materials under friction. K.: Tekhnika, 1976. 296 p.

14. Blednova Zh.M., Mahutov N.A., Chaevskij M.I. Surface modification of materials with shape memory effect– Krasnodar: Publishing house-South -2009.– 354 p.

15. Blednova Zh.M., Rusinov P.O., Balaev E.U.O. Increase of operational properties of products, the formation of the surface compositions of materials with EPF with a gradient of properties and temperatures of phase transformations. Modern high technologies. – 2016. No. 9-3. P. 378-384.

16. Kul'kov S.N., Rudaj V.V. The microstructure of composite materials TiC-TiNi with microgravity structurally unstable matrix, Izvestiya of higher educational institutions. Physics. 2012. No. 5/2. Pp. 167-169.

17. Li D.Y., Liu D.Y. The mechanism responsible for high wear resistance of Pseudo-elastic TiNi alloy - a novel tribo-material // J. Wear. 1999. Vol. 225-229. p. 777-283.

18. Li D.Y. Development of novel tribo-composites with TiNi shape memory alloy matrix // Wear. 2003. V. 255. P. 617–628.

19. Pushin V.G. Alloys of titanium nickelide with shape memory. Part 1. Structure, phase transformations and properties. Ekaterinburg: Uro ran, 2006 - p. 414

20. Rusinov P.O., Blednova Zh.M. Hardware-Technological Features of Layered Surface Composition Made of Shape Memory Materials in a Single Vacuum Cycle. Key Engineering Materials, 2017, Vol. 723, pp 497-502. doi:10.4028/www.scientific.net/KEM.723.497.

21. Blednova Zh.M., Rusinov P.O., Bezmogorychnaja M.E. The influence of granulometric composition, energy state and reactivity of the sprayed material on the formation of nanobacteria surface layers from materials with shape memory effect (SME) // Scientific journal of the Kuban state agrarian University [Electronic resource]. – Krasnodar: Kubsau, 2016. – №07(121). Pp. 484 – 498. Doi: 10.21515/1990-4665-121-021

22. Rusinov P.O., Blednova Zh.M. Structural and technological patterns of formation of surface nanostructured layers TiNiZr by high-speed flame spraying. Advanced Materials and Structural Engineering. Francis Group, London, 2016 p.21-24.

23. Rusinov P.O., Blednova Zh.M., Zirconium effect on the structure and properties of TiNi based alloy, Nauchnyj zhurnal KubGAU, No121(07), 2016 goda. Jelektronnyj resurs. http://ej.kubagro.ru/2016/07/pdf/20.pdf Doi: 10.21515/1990-4665-121-020
24. Rusinov P.O., Blednova Zh.M. Formation of nanostructured Ti-Ni-Hf-Cu coatings by high-speed flame spraying of mechanically activated powders. Advanced Materials, Structures and Mechanical Engineering. Taylor & Francis Group, London, 2016, Pp. 195-198.
25. Blednova Zh.M., Mahutov N.A., Rusinov P.O., Stepanenko M.A. Mechanical and tribological properties of multifunctional compositions "the basis - a material with shape memory effect", formed in conditions of high-energy effects, laboratory. diagnostics of materials— 2015.(81), № 3.–Pp. 41-49.
26. Rusinov P.O., Blednova Zh.M Structural and technological patterns of formation of surface nanostructured layers TiNiZr by high-speed flame spraying, Advanced Materials and Structural Engineering — 2016 (978), 21-25.
27. Rusinov P.O., Blednova Zh.M., Dmitrenko D.V. Formation of composite layers tinizr-cbn-co, working in conditions of cyclic loading and reverse friction. Procedia Structural Integrity. 2016. T. 2. Pp. 1506-1513.
28. M.B. Babanli, The parameters of martensitic transformations in $Ti_{32}Hf_{18}Ni_{50}$-xCux alloys, Fizika, Cild XIII №3, 2007
29. Rusinov P.O., Blednova Zh.M., Balaev E.U. A method of producing nanostructured coatings with shape memory effect on steel, patent for the invention RUS 2535432 16.08.2013.
30. Planetary centrifugal mill, Blednova Zh.M., Dmitrenko D.V., Bezmogorychnaja M.E. the patent for useful model RU 164078, 20.10.2015
31. Method of increasing the strength of the coated parts, Blednova Zh.M., Balaev E.U.O., Dmitrenko D.V., Plomodyalo R.L., the patent for the invention RU 2625508, 10.10.2016
32. Pisarenko G. S., Ageev V. A., Kvitka, A. L., Popkov, V. G., Usman, E. S., Strength of materials, Kiev, 1986, pp. 719-726
33. Malygin, G. A., Nanoscale effects in martensitic transformations in alloys with shape memory, solid state Physics, Vol. 8, 2008, Pp. 1480

Sliding contact on the interface of elastic body and rigid surface using a single block Burridge-Knopoff model

Amireghbali A[1], *Coker D*[1,2]
[1]*Department of Aerospace Eng., METU, Ankara, Turkey*
[1]*METU Center for Wind Energy (METUWIND), Ankara, Turkey*
aydin.amireghbali@gmail.com [1], coker@metu.edu.tr[2]

Abstract: Burridge and Knopoff proposed a mass-spring model to explore interface dynamics along a fault during an earthquake. The Burridge and Knopoff (BK) model is composed of a series of blocks of equal mass connected to each other by springs of same stiffness. The blocks also are attached to a rigid driver via another set of springs that pulls them at a constant velocity against a rigid substrate. They studied dynamics of interface for an especial case with ten blocks and a specific set of fault properties. In our study effects of Coulomb and rate-state dependent friction laws on the dynamics of a single block BK model is investigated. The model dynamics is formulated as a system of coupled nonlinear ordinary differential equations in state-space form which lends itself to numerical integration methods, e.g. Runge-Kutta procedure for solution. The results show that the rate and state dependent friction law has the potential of triggering dynamic patterns that are different from those under Coulomb law.

Keywords: Burridge and Knopoff model, Sliding contact, Rate and state dependent friction

1. Introduction

Burridge and Knopoff proposed a mass-spring model to explore interface dynamics along a fault during an earthquake [1]. The Burridge and Knopoff (BK) model is composed of a series of blocks of equal mass connected to each other by springs of same stiffness. The blocks also are attached to a rigid driver via another set of springs that pulls them at a constant velocity against a rigid substrate. In their study of interface dynamics an especial case with ten blocks and specific proper-ties is studied. A numerical model of the system is constructed to investigate statistics of the shocks on the fault. Energy of the system is also of interest as a measure of shock magnitude. The applied velocity weakening friction law, considers both seismic radiation and viscosity of the fault.

Stability of steady state slipping of a single block BK model is investigated within small perturbation theory by Rice and Ruina [2]. Using a one-state variable friction law, an analytic expression for minimum spring stiffness for stick-slip instability is obtained.

A detailed analysis of dynamics of a symmetric BK model with two mass blocks is performed in [3]. Later an asymmetric BK model with two mass blocks (with a different frictional force for each block due to velocity-weakening friction) was investigated in [4] also a cellular automata model is applied to investigate the symmetric BK model with two mass blocks [5].

Carlson and Langer [6] simulated a BK model including up to 200 mass blocks with a velocity weakening friction law by a coupled system of differential equations and noticed manifestation of the purely empirical Gutenberg-Richter law (known from 1950s). They proposed their continuum dynamical system as an alternative to the cellular automaton models for representing self-organized criticality (SOC) in earthquakes [7,8,9,10]. The concept of SOC argues that the statistically stationary state of dynamical many body systems is scale free [8], i.e. these systems drive themselves to a critical state in which no characteristic length, time or energy scale does exist [7] and avalanches occur in all sizes and scales. An analysis of criticality in BK models with velocity weakening friction law is also done in [11]. A study of dynamics of BK models (both single block and multi-block) is done in [12]. Transition from small scale events to large scale events is investigated for both velocity weakening and rate and state friction laws. Another study of dynamics of BK model using both discrete equations (ODEs) and continuum formulation (PDE) is given in [13].

2. Problem definition and the model

In our study a single block BK model is considered (Figure 1). Mass of the model is assumed to be 1 kg and the stiffness is 10 kN. Note that the spring with stiffness "k" does not have a physical length, i.e. it cannot be compressed. Therefore the spring experiences tension only and compression is not possible. Then, when the mass is right below the attachment point of the spring and the driver (AP), no spring force is exerted on it. Also the spring is not capable of pushing the block away. Often instead of a coil spring, the spring connecting each mass to the driver is shown as a leaf spring in the literature.

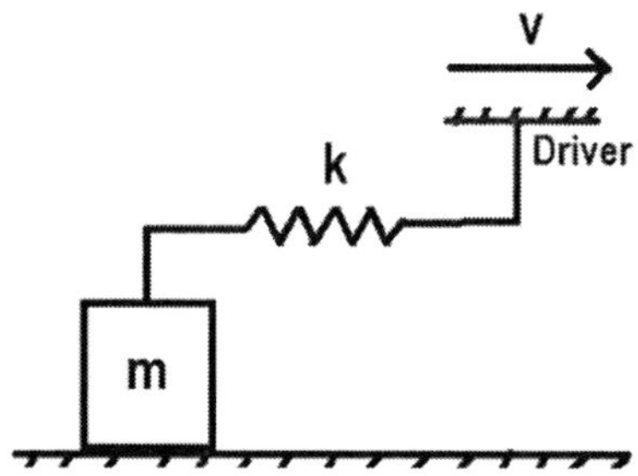

Figure 1: The single block BK model, mass of the block is represented by "m". "k" stands for stiffness of the coil spring and the driver moves with a speed "v" to the right.

The Coulomb friction law is very straightforward: when moving, dynamic coefficient of friction is constant; if not moving, there is a shear force threshold which when exceeded the block have to start moving. However, there are friction laws in the literature according to which the dynamic coefficient of friction is not a constant, but changes in rather complicated ways. The rate and state dependent friction law which is used here is adopted from the seminal works of Dieterich [14] and Rice and Ruina [2]. Dealing with test results for rocks sliding on each other, it has founded out that friction can be described more accurately by considering the effects of not only speed of sliding (rate) but also the history of evolution of this speed (state). It is assumed that the coefficient of friction is a multiplication of two functions $f = f(\dot{x})$ where $\dot{x}$ is velocity of the block (slip rate) and $g = g(\theta)$, where θ is called internal state variable:

$$\mu(\dot{x}, \theta) = f(\dot{x}).g(\theta) \tag{1}$$

where

$$f(\dot{x}) = \left(\frac{\dot{x}}{v_0} + 1\right)^{1/m} \tag{2}$$

and

$$g(\theta) = \frac{\mu_d + (\mu_s - \mu_d)\exp[-\left(\frac{L_0/\theta}{v_1}\right)^p]}{\left(\frac{L_0/\theta}{v_1} + 1\right)^{1/m}} \tag{3}$$

Note that the evolution of state variable is found from a first order ODE which is called evolution law and is given below:

$$\dot{\theta} = B\left(1 - \frac{\dot{x}\theta}{L_0}\right) \tag{4}$$

The evolution law can be analytically integrated to give an algebraic expression for the state variable to be substituted in the expression for $g(\theta)$:

$$\theta(t) = \left(\theta_n - \frac{L_0}{\dot{x}}\right)\exp\left[\left(\frac{\dot{x}}{L_0}\right)B(t_n - t)\right] + \frac{L_0}{\dot{x}} \tag{5}$$

The parameters of the above formula for sliding between rock plates are given in table 1.

Table 1. The adopted parameters of rate and state dependent friction law

Parameter	Value
μ_s	0.6
μ_d	0.5
v_0	100
v_1	0.001
p	1.2
m	0.001
L_0	2e-5
B	4.6

To have an idea about the coefficient of friction as it obtained from the above relations and how does it differ from Coulomb's coefficient of friction, the proposed model is plotted for an especial case of abrupt changes of sliding velocity. In figure 2, coefficient of friction is plotted as a function of slipped distance for two step changes in the speed of two sliding plates. Initially (in region A), the sliding speed is 1 mm/s then it suddenly jumps up to 10 mm/s (region B) and then again suddenly it drops down back to 1 mm/s (region C). The higher the sliding velocity is, the lower is the steady state dynamic friction coefficient. Note that the minimum steady state friction happens in region B (where the velocity is the highest among A, B and C). Note also that however after a jump to a higher velocity, a reduction in dynamic coefficient of friction (COF) is expected (due to what is called velocity weakening nature of the friction), the expected reduction starts only after a sudden rise (a spike) in the value of dynamic COF. In addition, when it is expected to increase, it decreases initially. Both spikes happen at the very beginning of the abrupt change in sliding velocity of the plates. The spike can even exceed the static coefficient of friction, but all of the dynamic steady state friction values are below the static coefficient of friction.

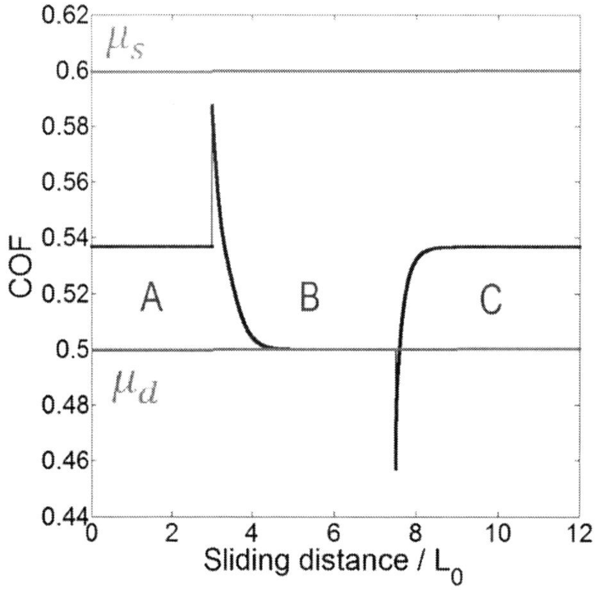

Figure 2: Coefficient of friction according to rate and state dependent friction law against nondimensionalized sliding distance. Sliding velocity of the plates are 1 mm/s in region A, 10 mm/s in region B and finally 1 mm/s again, in region C.

3. The equations of motion

Presence of the friction force makes the equation of motion (EOM) nonlinear. The first manifestation of this non-linearity is that direction of the friction must be opposite to that of the block velocity. Note that the velocity is obtained by solving the equation of motion and is not known a priori. Thus the solution for velocity is to be monitored continuously to switch the friction sign in opposition. It must happen any time the velocity passes through zero. The second manifestation is that when the block stops (sticks) the static friction (which is exactly equal and opposite to the spring force) starts to accumulate. Then the next stage of slipping is to be solved by considering motion of the driver. The block can stick and slip intermittently.

Using d'Alambert's principle:

$$m\ddot{x} = k(vt - x) - sign(\dot{x}) * f$$

or:

$$\ddot{x} = \frac{1}{m}[k(vt - x) - sign(\dot{x}) * f]$$

gives a rather incomplete picture of the dynamics of the mass under dry friction. It can also be misunderstood easily since according to the definition of the sign function: $\mathrm{sign}(0) = 0$, while it is not obviously the case for the stick state. Therefore the equation of motion is split to two cases; namely:

Case (1): motion towards right (positive velocity direction)

$$\ddot{x} = \frac{1}{m}[k(vt - x) - f]$$

Case (2): motion towards left (negative velocity direction)

$$\ddot{x} = \frac{1}{m}[k(vt - x) + f]$$

Putting into state space form by defining:

$$x_1 = x$$
$$x_2 = \dot{x}$$

(Which implies $\dot{x}_1 = x_2$)

In case (1):

$$\begin{bmatrix}\dot{x}_1\\ \dot{x}_2\end{bmatrix} = \begin{bmatrix}0 & 1\\ -\frac{k}{m} & 0\end{bmatrix}\begin{bmatrix}x_1\\ x_2\end{bmatrix} + \begin{bmatrix}0\\ \frac{1}{m}(kvt - f)\end{bmatrix}$$

In case (2):

$$\begin{bmatrix}\dot{x}_1\\ \dot{x}_2\end{bmatrix} = \begin{bmatrix}0 & 1\\ -\frac{k}{m} & 0\end{bmatrix}\begin{bmatrix}x_1\\ x_2\end{bmatrix} + \begin{bmatrix}0\\ \frac{1}{m}(kvt + f)\end{bmatrix}$$

Thus the strategy for developing the code is to make the algorithm choose between the different cases that may occur according to the outcome of the previous step and advance in a piecewise manner.

4. On the algorithm and the code

The mass can be either in stick state or in slip state (and slipping can occur in the right or left directions). The algorithm is to recognize whether the mass is stuck or it is slipping in every time step. It seems simple: to look at the velocity and if it zero then it is in stick state. Since it is hardly possible to detect the exact zero (if any) in the numerical solution of velocity, "zero velocity" is accepted to happen right at the beginning of the time increment in which the block changes sign

The loop syntax selected to perform this kind of piecewise solution of the EOM was at first written using a "for" command. But, the MATLAB ode23 function (which applies 4th order Runge-Kutta scheme for integration) automatically refines the given time step of the loop to even finer increments, it is quite possible that "the passing through zero" happen within the specified time step of the loop (and therefore missed). However, a "for" loop does not allow alternation of the loop variable (here, time) within itself. This fact makes it impossible to capture the "zero velocity" moment by a "for"

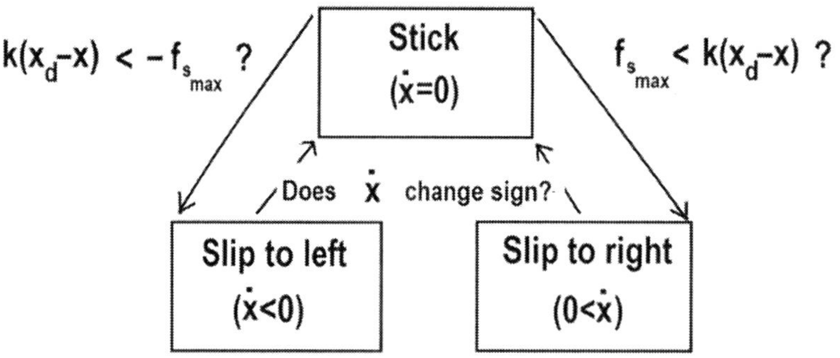

Figure 3: The part of the flowchart which determines the state of the block in each time step: Is it stuck? Does it continue being stuck? Will it slip to the right or to the left? Does it continue slipping?

loop. Then the proper loop command specified as "while". As a subsequence, the conditions for stick and slip became more straightforward. The flowchart for the code is shown in Figure 3.

As can be seen in figure 3, there is no bridge between slipping to the right and slipping to the left since the block has to pass through zero velocity point (i.e. stick state) and it is reasonable. In the case of zero initial conditions, the block is stuck at t=0. Figure 2 also specifies the very conditions for entering or exiting a stick or slip state. In each of these states the block is checked to see if it continues to remain in that state or it is to depart.

5. Results

Besides kinematics of the block motion, i.e. solution of the EOM (or what is called dynamic response of the block), kinetics of the motion (here we imply friction) is represented for a range of different μ_d values. Note that while for Coulomb law this quantity is the value of dynamic COF, it becomes a parameter in the rate and state dependent friction law. The other parameters of rate and state dependent friction law are as given in the table 1. The driver velocity is fixed to 1 mm/s. Mass of the block is sat to 1 kg, stiffness of the coil spring is 10 kN/m. The static COF is 0.6.

Effects of μ_d via the constitutive law of friction, eqn. (1), are investigated for three cases. In case (A) μ_d is sat to 0.1, in case (B) it is sat to 0.5 and in case (C) it is equal to 0.6, meaning that the dynamic COF is equal to the static threshold of friction.

The kinematics of case (A) is given in figure 4. The plots of position and velocity of the block (and the driver) against time are superimposed for both Coulomb and rate and state dependent friction laws are presented in figure 4. Note that the driver passes the block nearly at t=1s. Both of the friction laws trigger the stick-slip instability. Note that the driver passes the block nearly at t=1s. While the Coulomb law causes symmetric spikes of slip rate (the black line), under rate and state dependent law (the red line), those spikes are asymmetric, higher and sparser in time and start smoothly and end abruptly.

The kinetics of case (A) is represented in figure 5. The plots of shear divided by normal force (apparent COF) for both Coulomb and rate and state dependent friction laws are presented. While there is no transition from the stick state to the slip state under Coulomb law (the black line), there exist a smooth transition from stick to the slip for the rate and state dependent friction law (the red

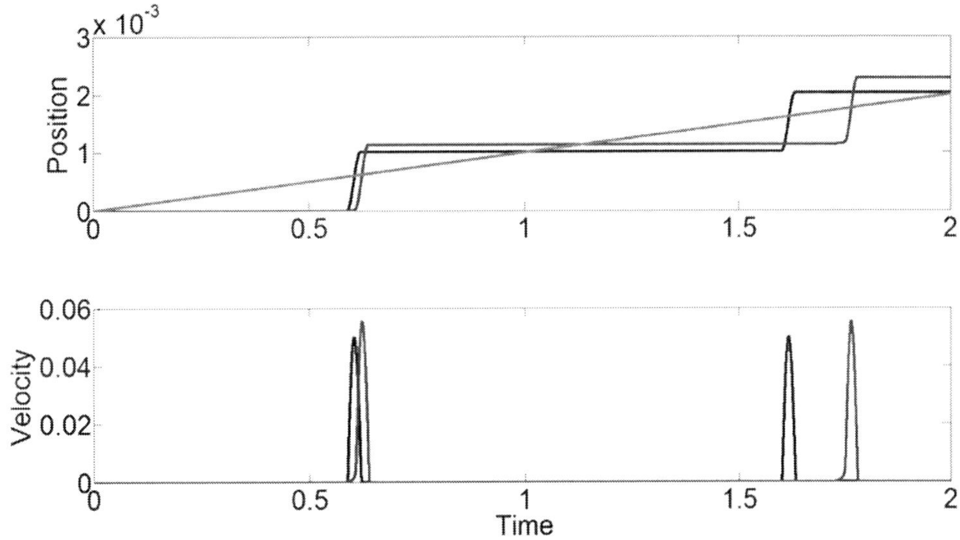

Figure 4: Case (A), for dynamic COF of 0.1, position (top) and velocity (bottom) of the mass block subjected to Coulomb law (in black), subjected to rate and state dependent friction law (in red) and the driver (in grey) against time for a constant driver velocity of 1 mm/s.

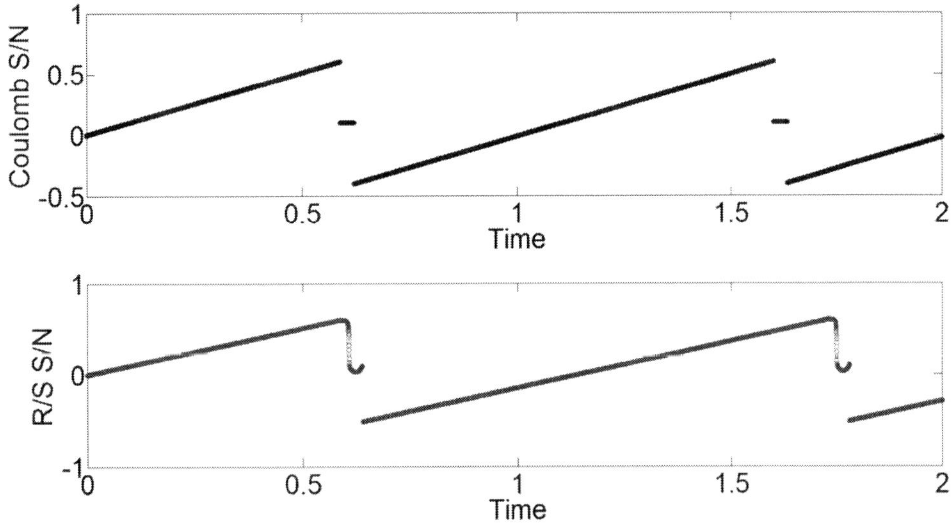

Figure 5: Case (A), for dynamic COF of 0.1, friction force (shear) divided by normal force (the block weight), conventionally called apparent COF is presented for Coulomb law (top) and rate- state dependent law (bottom) against time for a constant driver velocity of 1 mm/s.

line). The velocity weakening property of rate and state dependent friction law shows up in the concave shape of the COF in the slipping regions.

The kinematics of case (B) is given in figure 6. The plots of position and velocity of the block (and the driver) against time are superimposed for both Coulomb and rate and state dependent friction laws are

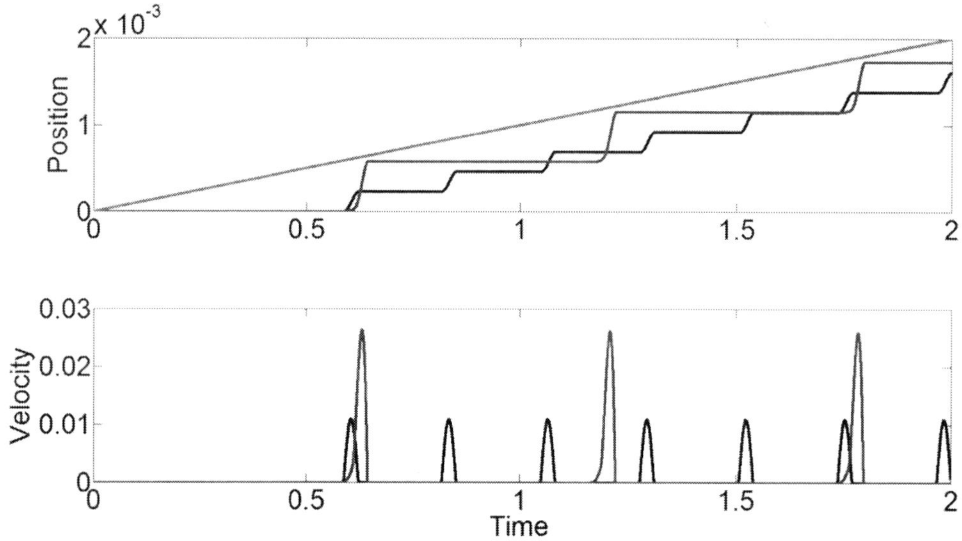

Figure 6: Case (B), for dynamic COF of 0.5, position (top) and velocity (bottom) of the mass block subjected to Coulomb law (in black), subjected to rate and state dependent friction law (in red) and the driver (in grey) against time for a constant driver velocity of 1 mm/s.

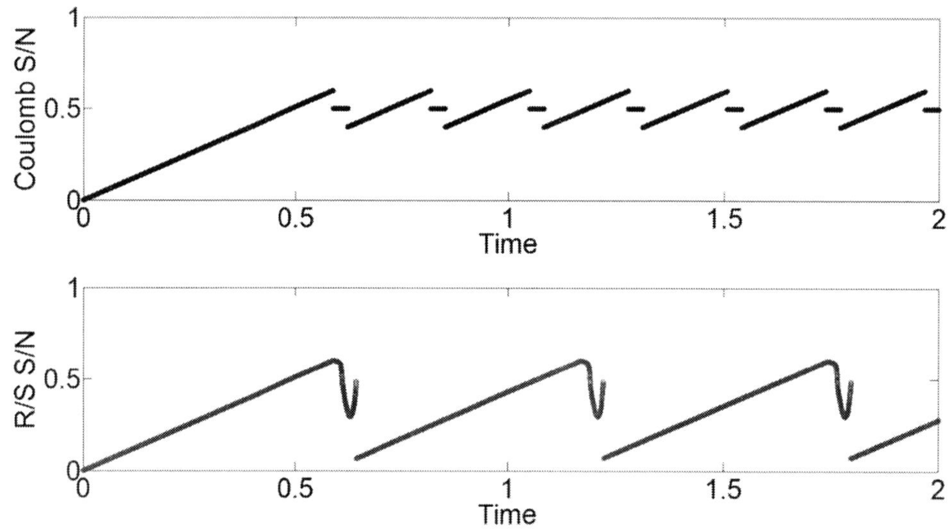

Figure 7: Case (B), for dynamic COF of 0.5, friction force (shear) divided by normal force (the block weight), conventionally called apparent COF is presented for Coulomb law (top) and rate- state dependent law (bottom) against time for a constant driver velocity of 1 mm/s.

presented in figure 6. Note that the driver never passes the block in this case, however, both of the friction laws trigger the stick-slip instability again. The spikes of slip rate have become more frequent and shorter it height.

The kinetics of case (B) is represented in figure 7. The plots of shear divided by normal force (apparent COF) for both of the laws are presented. The concave shape of drop in apparent COF has become more apparent, and well below the value of dynamic COF, 0.5.

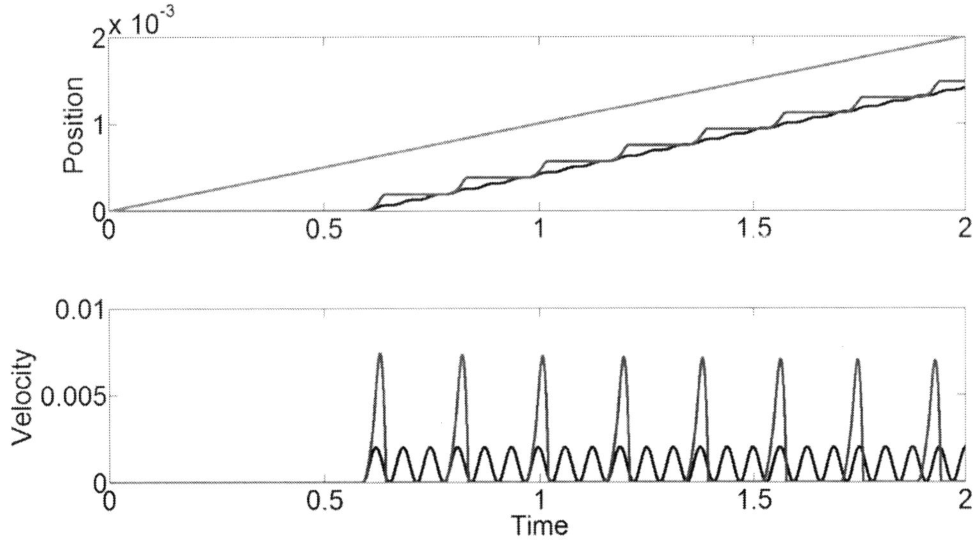

Figure 8: Case (C), for dynamic COF of 0.5999, position (top) and velocity (bottom) of the mass block subjected to Coulomb law (in black), subjected to rate and state dependent friction law (in red) and the driver (in grey) against time for a constant driver velocity of 1 mm/s.

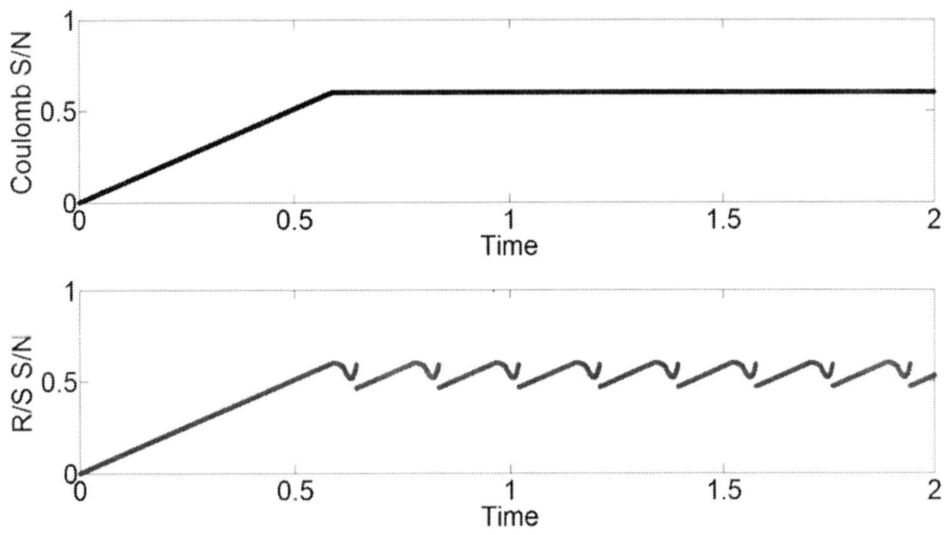

Figure 9: Case (C), for dynamic COF of 0.5999, friction force (shear) divided by normal force (the block weight), conventionally called apparent COF is presented for Coulomb law (top) and rate- state dependent law (bottom) against time for a constant driver velocity of 1 mm/s.

The kinematics of case (C) is given in figure 8. The plots of position and velocity of the block (and the driver) against time are superimposed for both Coulomb and rate and state dependent friction laws are presented in figure 8. Note that the driver does not pass the block. While the motion under Coulomb is oscillatory, stick-slip motion still happens for the rate and state- governed mass and the trend of increase in frequency and decrease in the height of slip rate spikes continues.

The kinetics of case (C) is represented in figure 9. The plots of shear divided by normal force (apparent COF) for both Coulomb and rate and state dependent friction laws are presented. While under Coulomb law (the black line), there exist a constant slip and therefore constant COF, rate and state dependent friction still causes decrease and increase of friction force during the slip period.

6. Conclusions

The rate and state dependent friction laws essentially proposed for modelling earthquakes and therefore rock plate sliding with abrupt change in their relative slip rate. Our study tried to detect differences in dynamics of a single block Burridge-Knopoff (BK) model under to Coulomb and rate and state dependent friction laws. During the slip period, instead of the step change of sliding velocity of the block an impulse of slip rate happen. This causes the apparent COF not to exactly imitate the behavior expected in figure 2. The velocity weakening nature of the law can be observed in the results. Increasing dynamic COF made slip spikes more frequent and less strong but could not eliminate stick-slip motion in the rate and state dependent law case. However, it caused a steady oscillatory motion in the Coulomb law - governed case.

References

1. Burridge R. and Knopoff L., Model and Theoretical Seismicity, Bulletin of the Seismological Society of America 57(3), 341-371, 1967.
2. Rice J. R. and Ruina A. L., Stability of Steady Frictional Slipping, Journal of Applied Mechanics 50(2), 343-349, 1983.
3. Nussbaum J. and Ruina A., A Two Degree-of-Freedom Earthquake Model with Static/Dynamic Friction, Pure and Applied Geophysics 125(4), 629–656, 1987.
4. Huang J. and Turcotte D. L., Chaotic Seismic Faulting with a Mass-Spring Model and Velocity-Weakening Friction, Pure and Applied Geophysics, 138(4), 569-589, 1992.
5. Narkounskaia G. and Turcotte D. L., A Cellular Automata Slider Block Model for Earthquakes I. Demonstration of Chaotic Behaviour for a Low Order System, Geophysical Journal International, 111(2), 250-258, 1992.
6. Carlson J. M. and Langer J. S., Properties of Earthquakes Generated by Fault Dynamics, Physical Review Letters 62(22), 2632-2635, 1989.
7. Bak P. and Tang C., Earthquakes as a Self-Organized Criticality, Journal of Geophysical Research, 94(B11), 15635-15637, 1989).
8. Olami Z., Feder H. J. S. and Christensen K., Self-Organized Criticality in a Continuous Nonconservative Cellular Automaton Modeling Earthquakes, Physical Review Letters, 68(8), 1244-1248, 1991.
9. Nakanishi H., Cellular Automaton Model of Earthquakes with Deterministic Dynamics, Physical Review A, 41(12), 7086-7089, 1990.

10. Huang J., Narkounskaia G. and Turcotte D. L., A Cellular Automata Slider Block Model for Earthquakes II. Demonstration of Self-Organized Criticality for a 2-D System, Geophysical Journal International, 111(2), 259-269, 1992.

11. Clancy I. and Corcoran D., Criticality in the Burridge-Knopoff Model, Physical Review E, 71(4), 046124, 2005.

12. Clancy I. and Corcoran D., State-variable Friction for the Burridge-Knopoff Model, Physical Review E, 80(1), 016113, 2009.

13. Erickson B. A., Birnir B. and Lavallee D., Periodicity, Chaos and Localization in a Burridge-Knopoff Model of an Earthquake with Rate and State Friction, Geophysical Journal International, 187(1), 178-198, 2011.

14. Deitrich, J. H., Modeling of Rock Friction: 1. Experimental Results and Constitutive Equations, Journal of Geophysical Research 84(NB5): 2161-2168, 1979.

Studies of worn surfaces by relocation profilometry

***Rîpă M*[1], *Iliuţă V*[2]**
[1] *„Dunărea de Jos" University of Galaţi, 800008, Galati, Romania*
[2] *„Dunărea de Jos" University of Galaţi, 800008, Galati, Romania*
minodora.ripa@ugal.ro[1], *virgil.iliuta@ugal.ro*[2]

Abstract: By relocation profilometry, a series of surface profiles can be recorded from the same track on a specimen. These techniques are used for monitoring specific particular points on the surface subjected to wear processes, in a more accurate manner as comparing to those involving average statistical information for surface. The method is providing a much more significant information about the surface, in a more efficient way, assuring that the same unworn investigated surface is studied after wear test. The studied roughness digital profiles were obtained before and after the testing of rolling/sliding line contacts, characteristic for spur gears, which has been simulated on SAE sets, with a two rollers test machine. The acquisition of the relocated profiles is performed on the same generatrix of the roller, before and after wear testing. To correlate the unworn and worn profiles, a spheroconical indentation was created on the circumferential surface of the disk, in the zone of the tested roller that remain unworn during the test. Measuring changes of the profiles by relocation techniques, two methods for wear assessment are presented: linear wear estimation by simulating the profile wearing and estimation of the volume wear.

Keywords: Relocation profilometry, wear tests, worn surface.

1. Introduction

In any experiment proposing to find a relationship between surface topography and tribosystem functioning, the objective is selecting the parameter that will better characterise the tribological function being under observation.
The interesting aspects of studying the running in process are information about surface characteristics and their changes during the running in, with or without experiment, the behaviour prediction starting from the unworn zones, for the efficient monitoring of the wear process to identify the wear types [1, 2, 3, 4].
The measurements of many surface parameters before, during and after testing rise immediate problems especially for isotropic surfaces. The measuring process of every parameter on a determined surface length has a statistically limited incertitude degree. And especially the summit height measurements are susceptible to incertitude.
There are a few ways of approaching this matter. One method is to perform measurements along the investigated surface and to average them, this one being not completely satisfactory as concerning economic and technical points of view as only a small percentage of asperities may be affected during the wear process. The consequence is that this averaging process may lead to loose the significance of severe local changes of surface topography.
The other method is by using relocation profilometry techniques. By relocation profilometry, a series of surface profiles can be recorded from the same track on a specimen [5]. These ones are monitoring

specific particular points on the surface, during the experiment, in a more accurate manner as comparing to those involving average statistical information for all over the surface. The advantage of the method is that of getting a much more significant information about the surface, in a more efficient way, assuring that the initial investigated surface is repeatedly studied. For the most experiments, the individual profiles of the surface are identified and consecutively monitored during the experiments usually using digital techniques. For different stages of the experiment, the profile is investigated and the signal offered by the amplification system is digitised and stored. Thus it is possible to evaluate and calculate the changes of the surface parameters. The example presented by Stout in the paper [1] indicated the changes of some surface parameters during wear tests with a rig of cross cylinders.

This kind of information is difficult to obtain without digital methods and devices. The relocation techniques were used for experiments involving many types of laboratory systems: cross cylinders, twin rollers testing machines and bearings testing rigs.

The major problem is to obtain a suitable solution for introducing the measuring device and the testing machine in the same investigating chain. Previous research demonstrated the necessity of two stages for applying the relocation technique, these being a certain repeated and accurate positioning both in the testing machine and in the measuring device of the specimen.

The results of applying the relocation techniques in tribology are presented in [1, 5, 6, 7, 8, 9, 10, 11, 12, 13]. These techniques require either an accurate positioning of the specimen in the complex measuring instrument [2, 14]. In the study presented in [11], a Hirth Coupling (Mitsubishi Heavy Ind.) was used for positioning a specimen on the measuring device. Placement, measurement, and removal were repeated 10 times and the deviation in the profile measurement direction was calculated using cross-correlation functions. In work [4], a procedure used to relocate and take profile measurements is presented.

Different gear tooth surface profile traces taken from standard steel FZG test gears have been used in [15]. The profiles are measured by mounting the test gear on a specially manufactured relocation rig for profilometer measurement, ensuring that the same area of the tooth is assessed at each load stage.

Metrology studies using indentation system, demanding the presence of control unworn zones on the inspected surfaces, are presented in papers [4, 7, 9, 16].

2. Studying wear by profilometric methods

2.1 Wear experiments

In rolling/sliding wear tests, studies on the surface profilometry can provide useful information about load-carrying capacity and some functional proprieties concerning lubricant retention.

The rolling/sliding line contacts, characteristic for spur gears, has been simulated on SAE sets, with a two rollers test machine [9]. An Amsler test machine was adapted for SAE test sets, adjusted and completed for a wear testing program. The test rig has been provided with a torque transducer (type T30FN200 - Hottinger Baldwin®) for measuring friction torque and rotational speed, a force transducer, temperature transducers and a data acquisition system (DMCPlus-Hottinger Baldwin®). The rollers were loaded on line contact using a hydraulic loading system (load range: 3-15 kN), at three values, producing Hertz pressure comparable to those in the studied spur gears.

For all tested pairs the running time was 10^5 cycles, corresponding to the early-stage wear, while the most significant tribolayer parameters changes occur. The tests were performed for the same rotational speed n= 525 rev/min. Because the rotational speed of each one of two rollers in running is the same, three different diameter combinations assure the diffrent rolling-sliding ratio, calculated as $s=(v_1-v_2)/(v_1+v_2)$.

Lubrication was achieved by splashing and the lubricant was paraffinic mineral base-oil (AGMA lubricant no. 4 EP).

The experimental research was done taking into account the following commanding factors influencing the wear process [9]:

- relative sliding speed characterised by sliding-rolling ratio $s=(v_1-v_2)/(v_1+v_2)$: $s_1=0,225$, $s_2=0,075$ and $s_3=0$ (pure rolling);
- normal loading on the line contact, by terms of Hertz pressure: $p_1=850$ MPa, $p_2=700$ MPa, $p_3=560$ MPa;
- different materials for rollers, C 45 (similar to 1045 SAE steel grade) and 40Cr10 (similar to 5140 SAE steel grade), being two common grades of steel for high pitch spur gears, through-hardened;
- different initial microgeometry pairs: roller 1 and roller 2 both turned or roller 1 turned and roller 2 grinded.

Based on relocation technique, there were done complex profilometry analysis of surfaces using statistical roughness parameters and Abbott-Firrestone curve, allowing to establish interdependent evaluation and characterisation of values and evolution of both some specific wear process parameters and tribolayer parameters.

2.2 Data acquisition equipment

The acquisition of the digital profiles was performed by the help of the profilometer Surtronic3+, widely used in tribological studies involving surface geometry evaluation and was processed with Talyprof (equipment and software Taylor Hobson®).

Some of the expert functions of the software were used for the studies. Figure 1 presents Talyprof screens, with "Compute area of the hole" feature, for the unworn profile (a) and for the worn profile (b), respectively.

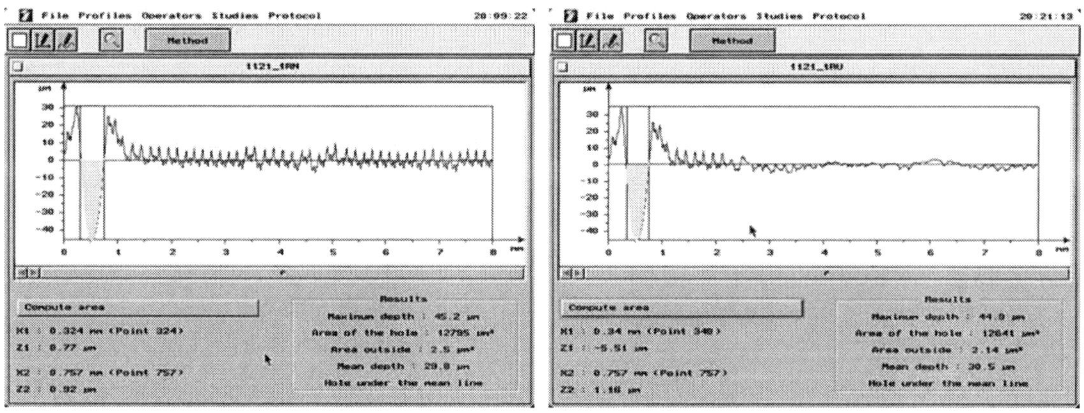

a) b)

Figure 1. Talyprof screens, with "*Compute area of the hole*" feature:
a) unworn profile; b) worn profile.

2.3 Methodology for studying the surface microgeometry based on the relocation technique

The authors used the relocation technique for evaluating rough profile, before and after a wear testing, without any intermediary investigation.

After 10^5 running cycles (rotations), the inspected roller (figure 3) has been designed with a larger width as compared to its mating roller. Two unworn zones, l_1 and l_2, border the worn zone (b_2), this one being the zone having contact with the mating roller, during the test.

The acquisition of the relocated profile is done on length $l_r=8$ mm and include the spheroconical indentation AC. The indentation is manufactured in the zone of the tested roller that will remain unworn during the test, using the indentor involved in measuring hardness in the HRC scale. The accurate positioning of the stylus is due to the AC indentation and to the V prismatic guide. The acquisition of the relocated profiles is performed on the same generatrix of the roller, both before and after wear testing. Figure 2 presents a profile with spheroconical indentation, used in relocation technique, represented with equal scales for both axes.

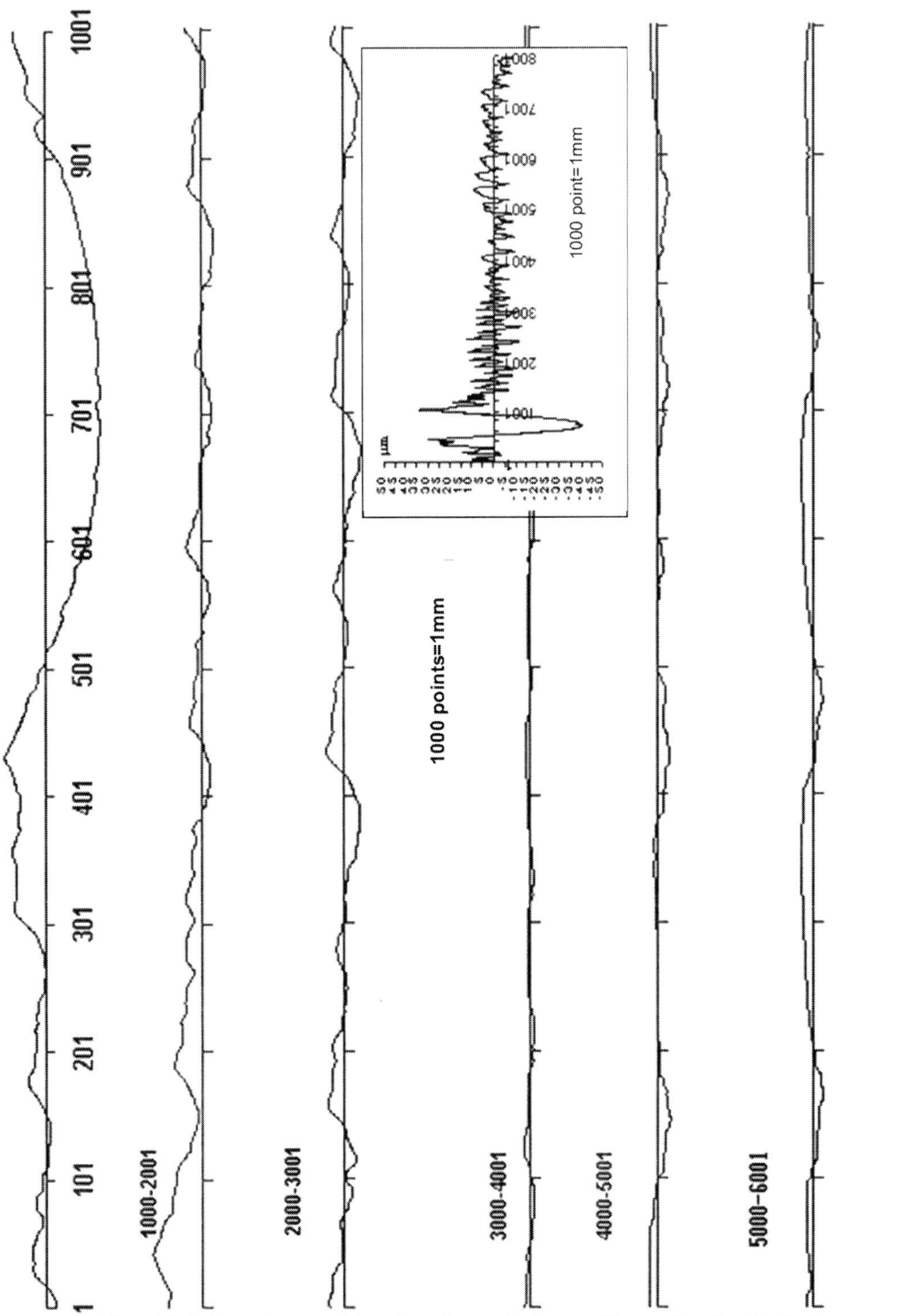

Figure 2. Profile with indentation, used in relocation technique, with equal scales for both axes. The profile is presented in 6 data sets (indentation between points 440÷1060). Regular presentation of the profile is shown in the detail box.

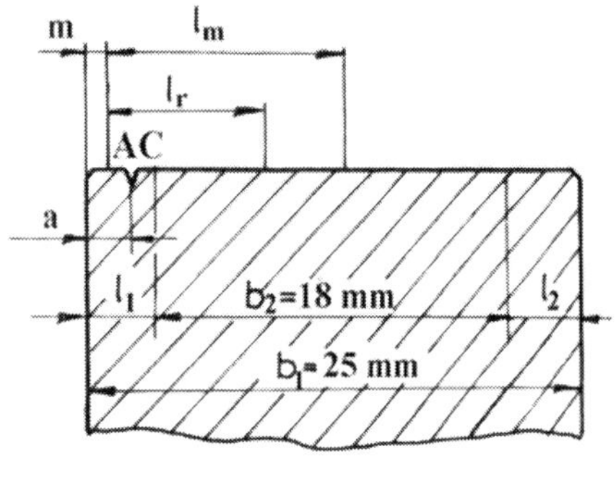

Figure 3. Indentation location on the roller, in order to apply the relocation technique.

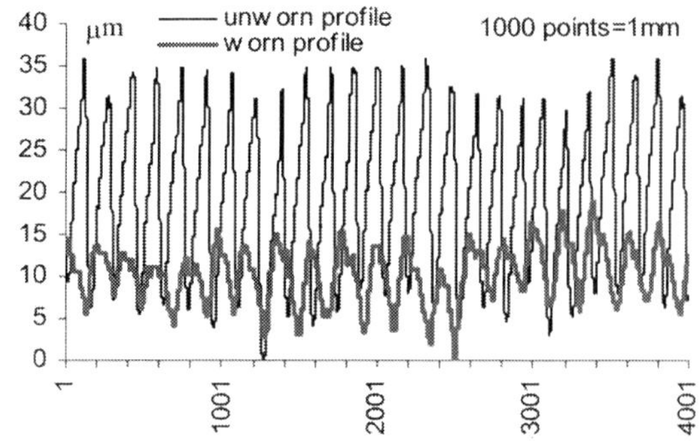

Figure 4. Unworn profile (U) and worn relocated profile (W), on the same generatrix of the roller.

The relocation of the device stylus on the worn surface was done on the same generatrix that was used for the acquisition of the unworn profile, in a transverse direction to the witness indentation. In order to achieve the relocation of the stylus, there were used two study operators of the Talyprof software: "Hole under the mean line" and "Compute area of the hole". Many successive tries make possible to obtained relocated worn profiles, characterised by very close values of these two operators as compared to those characterising the unworn profiles (U), meaning a difference of at most ±2% (for the profile presented in figure 1, this difference is only a percentage of 1,22%).
From the final right parts of the profiles (figure 1), the unworn profile (U) and the worn relocated profile (W) were extracted, their length being l_n=4mm, beginning at the same distance from the AC indentation (figure 4). For these two profiles, there were calculated two parameters usually used for studying wear by profilometric methods: worn area $\Delta S = S_u - S_w$ and the variation of the actual profile length $\Delta L_0 = L_{0u} - L_{0w}$ (Note; the parameter Actual profile length, Lo, describes how a real profile differs from a flat line is to determine how long the real profile is, compared to the horizontal evaluation length).

3. Wear evaluation by analysing relocated profiles

The shape and the dimensions of the rollers used for testing wear, allow obtaining the digital profiles by the help of the relocation technique. Based on these profiles, there are proposed two methods for wear evaluation.

3.1 Linear wear estimation by simulating the profile wearing

In order to estimate the linear wear, h, of the tribolayer, a truncation of the digital profile of the unworn surface was done using one of the operators available with Talyprof soft. The simulation criteria demand the simulated profile to have a value of roughness parameter R_q very close to the actual worn profile, meaning a difference of ±3% between the simulated profile and the actual worn one. Figure 5 presents an example with the three profiles involved in the estimation of linear wear: the unworn profile, the actual worn profile and the worn profile obtained by simulation (as a truncated profile). The high difference, h, is considered as linear wear.

3.2 Estimation of volume wear

Using Visual Basic applications, the worn volume could be calculated by the help of the two digital profiles, the unworn (U) one and the worn one (W) (obtained by relocation technique, figure 4). The following areas were calculated, S_u and S_w, both being limited downside by the reference line, characterising the maximum depth (on the basic length l=4mm) and upside by the line characterising the highest peak for the unworn and worn profile, respectively. The difference $\Delta S = S_u - S_w$ was calculated and also a specific worn area as $\Delta s=\Delta S/l$. The authors considered Δs as a mean value characterising the whole contact length (b_2=18 mm) and the roller circumference (taking into account the nominal value of the roller diameter, D). The specific mass of the worn material was also calculated, $\Delta m=\Delta s\cdot\pi D\cdot\rho$, with steel density ρ=7.9kg/ dm^3.

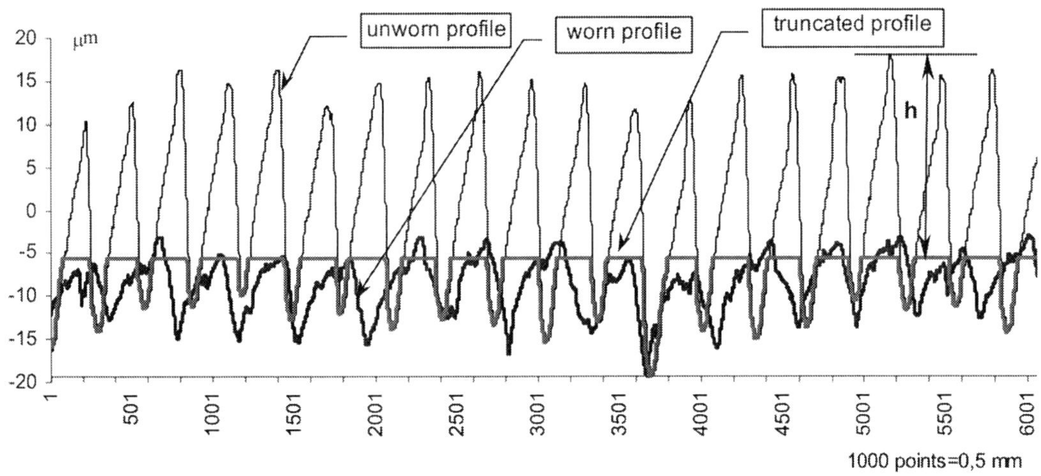

Figure 5. The profiles taken into consideration when estimating linear wear.

3.3 Discussions

Both methods presented at §3, may be applied only in the case of measurements performed using relocation techniques. Another condition for using these profilometric methods for wear evaluation, is that of having under study a wear process dominated by abrasive wear and characterised by an amount of material loss within the limit of the asperities'maximum height, R_y. This condition is necessary because calculations are based on the fact that a part of the voids patterns, initially noticed on the unworn profile, are still kept on the worn one. There is why these methods are useful in studies of microwearing process and for running in process, and only if the wear process does not have as consequence the completely taking-off of the initial microgeometry of the superficial layer.
The profilometric methods may be applied for studying wear processes characterised by a linear wear greater than maximum height of the profile only if the measured profiles contain witness (unworn) zones, bounding the wear trace (figure 6).

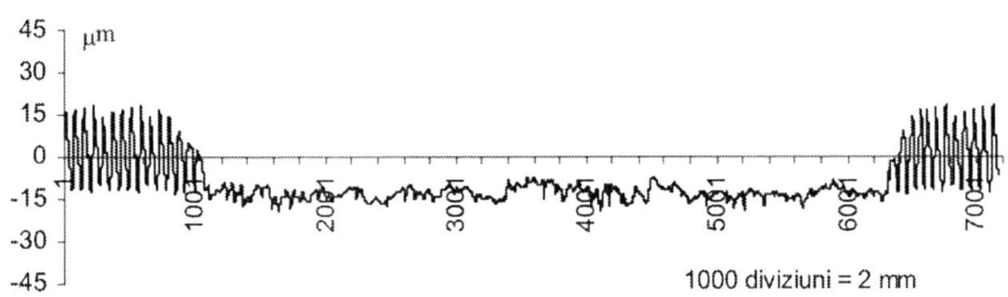

Figure 6. Worn profile having unworn bounding extremities.

4. Conclusions

Studying the microgeometry changes of a surface bearing a wearing process suppose to evaluate different roughness parameters and waviness, based on digital profiles, volume wear evaluation by the help of the wear trace and so on.
Making the average of many profile measurements on the same surface may lead to loose the significance of some severe local changes of the topography. The *relocation techniques* of the stylus

are more accurate as the roughness has been inspected on the same generatrix of the roller, before and after wear testing.
Quantitative evaluation of microwear cannot be made using the gravimetric method or other general measuring methods such as the dimensional method. It can be made by measuring changes of profiles. The wear study using profilometric methods impose the establishment of a methodology for analysing the surface microgeometry based on relocation technique, this one being recommended by the specialists particularly for tribological process research. There was also proposed a methodology for simulating a worn profile in order to evaluate variation of the wear rate for long time tests.

References

1. Stout K. J., Kin, T. G. and Whitehouse D. J., Analytical techniques in surface topography and their application to a running-in experiment, Wear, **43**, 99-115, 1997.
2. Bryant M. J., Running-in and residual stress: finite element contact analysis of as measured rough surfaces and comparison with experiment (Doctoral dissertation, Cardiff University, UK), 2013.
3. Sukumaran J., Vision assisted tribography of rolling-sliding contact of polymer-steel pairs (Doctoral dissertation, Ghent University), 2014.
4. Weeks I., An experimental investigation into the mixed lubrication of steel surfaces (Doctoral dissertation, Cardiff University), 2014.
5. Williamson J. B. P. and Hunt R. T., Relocation profilometry, J Phys E Sci Instrum., **1** (7), 749-752, 1968.
6. Snidle R. W., Alanou M. P., Evans H. P., Krantz T. L., Surface fatigue lives of case-carburized gears with an improved surface finish, J Tribol-T Asme, **123** (4), 709-716, 2001.
7. Newman P. T., Radcliffe S. J. and Skinner J., The accuracy of profilometric wear volume measurement on the rough LC1B-coated surfaces of an articulating joint, Surface Topography **2**, 59-77, 1989.
8. Raharijaona, F. et al., Determination de parametres rugosimetriques fonctionnels-Application au cas du frottement multipasse sur acier et aluminium en geometrie cylindre/cylindre, JIFT 97, Journ. Inter. Francophones de Tribologie, "Tribologie et Precision", 51-52, 1997.
9. Rîpă M., Studiul pe tribomodel al tribostratului in procesul de uzură specific roţilor dinţate [Study of the tribolayer for wear process characterising gears in rolling/sliding wear tests]. Ph.D. Thesis, Dunărea de Jos University of Galaţi, 2002.
10. Roizard, X. et al., Action d' un decapage chimique sur la topographie de surface d'un substrat composite en vue d'ameliorer l'adherence d'un revetement metallique, JIFT 97, Journ. Inter. Francophones de Tribologie, "Tribologie et Precision", 57-60, 1997.
11. Sasajima K., Naoi K., Tsukada T., A software-based relocation technique for surface asperity profiles and its application to calculate volume changes in running-in wear, Wear, **240** (1-2), 152-163, 2000.
12. Yetim A. F., Kovacı H., Aslan M., Çelik A., The effect of magnetic field on the wear properties of a ferromagnetic steel. Wear, **301** (1), 636-640, 2013.
13. Jeng Y. R., Impact of plateaued surfaces on tribological performance. Tribol T, **39** (2), 354-361, 1996.
14. Gee M. G., Nunn J. W. Gant A. J., Relocation profilometry of worn coated surfaces, National Physical Laboratory, 2014, available at http://www.npl.co.uk/upload/pdf/10787%20-%20M%20Gee.pdf (23.04.2017).

15. Snidle R. W., Hopkinson N. A., Hughes T. G., Tao J., Elastohydrodynamic lubrication analysis of gear tooth surfaces from micropitting tests, J Tribol-T Asme, **125** (2), 267-274, 2003.

16. Balcon M., Marinello F., Carmignato S., Savio E., Wear analysis through surface relocation. Journal of Physics: Conf. Series, 311 (1), 012020, IOP Publishing, 2011.

Investigating Tribological Characteristics of HVOF Sprayed AISI 316 Stainless Steel Coating by Pulsed Plasma Nitriding

Mindivan H
Bilecik Seyh Edebali University, Department of Mechanical Engineering, 11210, Bilecik, Turkey
harun.mindivan@bilecik.edu.tr

Abstract: In this study, surface modification of aluminum alloy using High-Velocity Oxygen Fuel (HVOF) thermal spray and pulsed plasma nitriding processes was investigated. AISI 316 stainless steel coating on 1050 aluminum alloy substrate by HVOF process was pulsed plasma nitrided at 793 K under 0.00025 MPa pressure for 43200 s in a gas mixture of 75 % N_2 and 25 % H_2. The results showed that the pulse plasma nitriding process produced a surface layer with CrN, iron nitrides (Fe_3N, Fe_4N) and expanded austenite (γ_N). The pulsed plasma nitrided HVOF-sprayed coating showed higher surface hardness, lower wear rate and coefficient of friction than the untreated HVOF-sprayed one.

Keywords: Duplex surface treatment, HVOF process, pulsed plasma nitriding, stainless steel coating, wear.

1. Introduction

Aluminum alloys have numerous industrial applications due to their attractive properties such as light weight, high specific strength and favorable mechanical formability. Unfortunately, hardness and tribological properties of pure and unprotected aluminum with high pitting corrosion tendency are poor comparing with cast iron and steel [1, 2]. The surface modification of aluminum can be classified into three categories: (1) coating with hard materials [3]; (2) forming alloyed or composite layers [4 - 6]; (3) hardening with ion or gas nitriding [7]. Until now, electrodeposition or electroless codeposition [8], reactive spraying [5], and nitriding [9] have been used as the process, although some of these techniques cause difficulties in matching aluminum substrates.

Duplex surface treatment can improve the anti-wear property of aluminum alloy against severe environmental uses by application of thermally sprayed steel coating plus pulsed plasma nitriding. An effective coating system combined a diffusion process and a coating has been recently developed in an attempt to further improve the tribological and tribocorrosion performances [10]. It is well known that since the thin oxide layer forming the deposition process is hard and brittle, the abrasive particles are occurred during wear and wear rate increases. Therefore, it might be required to apply a diffusion process on thermal sprayed surfaces for improving the wear resistance properties of the resulting coating due to its porosity, oxide content and lamellar boundaries in the weakest link in most thermally sprayed coating itself or along the interface, which would lead to cracking and delamination.

In the present research, a duplex coating can be divided into two steps, namely depositing AISI 316 stainless steel coating on 1050 aluminum alloy substrate by High-Velocity Oxygen Fuel (HVOF) thermal spray process, and pulsed plasma nitriding of HVOF-sprayed AISI 316 stainless steel coating in a mixture atmosphere made of nitrogen and hydrogen. The microstructure and wear property of duplex coatings were also investigated

2. Experimental Procedure

1050 aluminum alloy plates were cut to the dimensions of 20 mm×20 mm×4 mm. Initially, the aluminum plates were coated with an intermediary Ni-Al layer that increases the adhesion of coating. The duplex coating was prepared on 1050 aluminum alloy substrate with two procedures (deposition/nitriding). AISI 316 stainless steel coatings were applied by HVOF thermal spray process on the aluminum plates. Before pulsed plasma nitriding, the surfaces of the HVOF-sprayed AISI 316 stainless steel coating were ground using 1200 grit SiC paper and mechanically polished with a fine grade Al_2O_3 paste to achieve a certain surface uniformity. Finally, the surfaces were ultrasonically cleaned in an acetone bath for 10 min. The coatings were then pulsed plasma nitrided at 793 K under 0.00025 MPa pressure for 43200 s in an industrial furnace (Er-Mir Textile and Machine Ltd) in a gas mixture of 75 % N_2 and 25 % H_2. The coatings were cross sectioned and ground with successive SiC papers (grit 320–1200) and polished mechanically for metallographic examinations.

Microstructural characterization of the treated and untreated HVOF-sprayed AISI 316 stainless steel coatings was made by microscopic examinations, X-ray diffraction (XRD) and microhardness measurements. Microscopic examinations were performed on the cross-sections of the coatings by utilizing Nikon Eclipse LV150 Light Optical Microscope (LOM). XRD analysis was carried out by utilizing CuKα radiation with a Panalytical Empyrean diffractometer in order to identify the phase structures present in the surface layers. The cross-sectional microhardness measurements were carried out using a Knoop microhardness tester with at the indentation load 10 g for 15 s.

Dry sliding wear tests of the coatings were performed on a reciprocating wear tester operating in ball-on-disc configuration. In this configuration, wear tests were carried out for testing distance of 50 m by rubbing an Al_2O_3 ball with a diameter of 10 mm to the untreated and treated surfaces under normal loads of 5 N and 10 N, respectively. The stroke and the sliding speed of the ball were 10 mm and 1.7 cm s^{-1}, respectively. During the wear tests, the temperature and the relative humidity were maintained as 25 ± 1 °C and 30 ± 1 %, respectively. The coefficient of friction (COF) was recorded automatically by the sensing cell mounted in the tester. Width and depth of the wear tracks were measured by a surface profilometer (Mitutoyo Surftest SJ-400) to calculate wear rate of the coatings. Following the wear tests, wear tracks were examined using Zeiss Supra Field Emission Gun Scanning Electron Microscope (FEG-SEM).

3. Results & Discussion

Figure 1 gives the low and high magnification cross-sectional LOM micrographs of AISI 316 stainless steel coated 1050 aluminum alloy before and after pulsed plasma nitriding. The HVOF-sprayed AISI 316 stainless steel coating had a dense microstructure but oxide inclusions between the particle interfaces can be clearly observed. It should be noted that black spots in the micrographs represent the porosity while dark grey regions represent the oxidation. Pulsed plasma nitriding produced a layer with a thickness of about 10 μm on the surface of the HVOF-sprayed AISI 316 stainless steel coating. The cross-section morphology of pulsed plasma nitrided HVOF-sprayed AISI 316 stainless steel coating consisted of an outer compound zone and an inner diffusion zone. Generally, the compound zone contains nitrides formed from nitride-forming elements in the material.

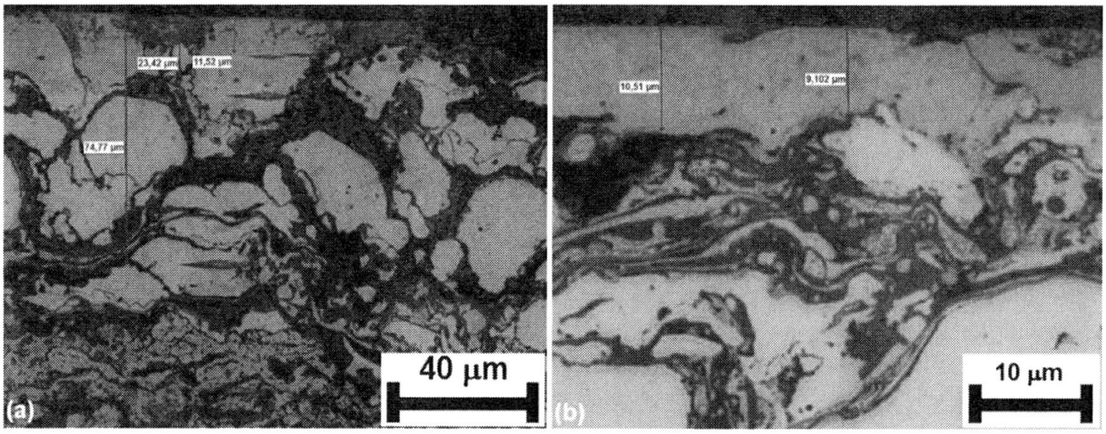

Figure 1. (a) Low and (b) high magnification cross-sectional LOM micrographs of the pulsed plasma nitrided HVOF-sprayed AISI 316 stainless steel coating.

The XRD examinations shown in Figure 2 demonstrate the transformation of the phases present in the treated and untreated HVOF-sprayed AISI 316 stainless steel coatings. The results show that the dominant phases in the surface layer of the untreated HVOF-sprayed coating were α-Fe, γ-Fe and Fe_3O_4CrO. However, the pulsed plasma nitriding treatment induced new phases corresponding to γ_N, CrN, Fe_3N, and Fe_4N peaks.

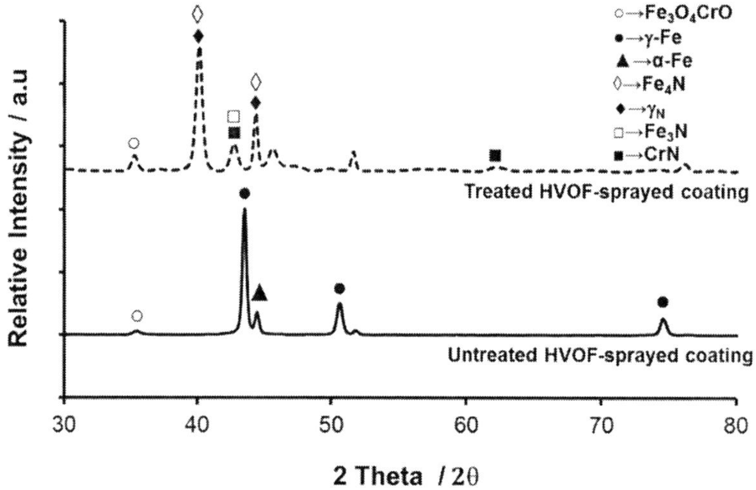

Figure 2. XRD patterns of the untreated and treated HVOF-sprayed AISI 316 stainless steel coatings.

Figure 3 shows the cross-sectional microhardness values of the untreated and treated HVOF-sprayed AISI 316 stainless steel coating layers. It is clearly observed that the surface microhardness of the coating subjected to pulsed plasma nitriding was above 1250 $HK_{0.01}$, three times greater than that of the coating produced without nitriding (430 $HK_{0.01}$). Microhardness values of the untreated HVOF-sprayed coating strongly depend on porosity, oxidized, unmelted and semimelted particles, and inclusions. Pulsed plasma nitrided coating maintained a high microhardness above 1250 $HK_{0.01}$ from the surface to a depth of ~10 µm due to the presence of the CrN, Fe_3N, and Fe_4N hard phases in the compound layer. Moreover, the microhardness of the pulsed plasma nitrided HVOF spray coating decreased depending on the depth from the surface. The decrease of the microhardness on the case depth profile of the pulsed plasma nitrided coating can be attributed to the reduction of the solubility of nitrogen in the diffusion zone [11].

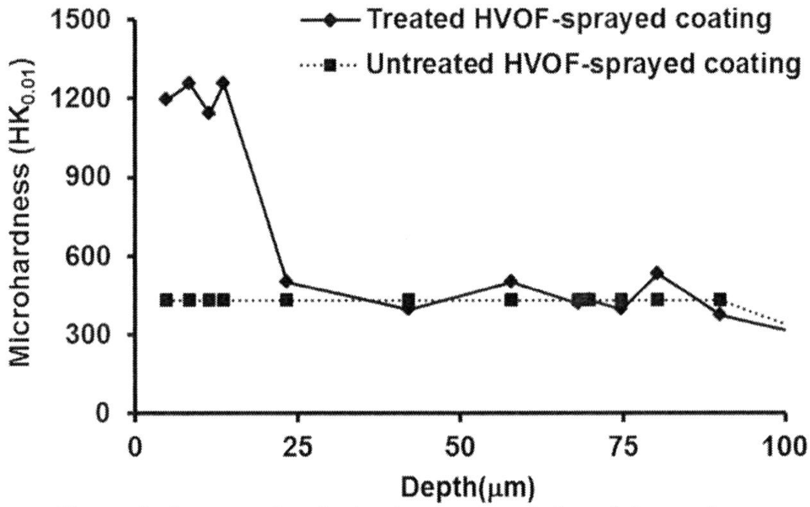

Figure 3. Cross-sectional microhardness variation of the coatings.

The friction coefficients vs sliding distance curves of the treated and untreated HVOF-sprayed coatings are shown in Figure 4. The stable friction coefficients of the tested coatings were found to be similar (1.06 for the untreated one, 0.97 for the nitrided one). However, the sliding curves of the untreated HVOF-sprayed coating exhibited larger fluctuation than the treated one. This large fluctuation in the sliding curves resulted from the lower surface hardness of the untreated one.

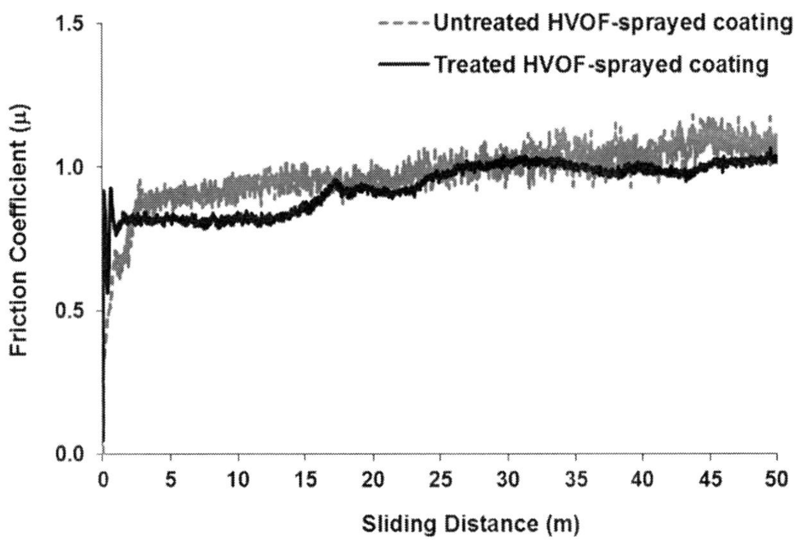

Figure 4. Friction curves of treated and untreated HVOF-sprayed coatings.

2D profiles of the wear tracks developed on the treated and untreated HVOF-sprayed coating surfaces by Al_2O_3 ball are presented in Figure 5. On the treated HVOF-sprayed coating surface, very small or no wear tracks has been formed when compared to untreated HVOF-sprayed coating surface. The wear rates of the untreated and treated HVOF-sprayed coatings were calculated as 9.35 x 10^{-4} mm^3/Nm and 0.5 x 10^{-4} mm^3/Nm, respectively. When compared to the untreated HVOF-sprayed coating, the wear rate of the treated HVOF-sprayed coating is almost negligible.

IOP Conf. Series: Materials Science and Engineering **295** (2018) 012033 doi:10.1088/1757-899X/295/1/012033

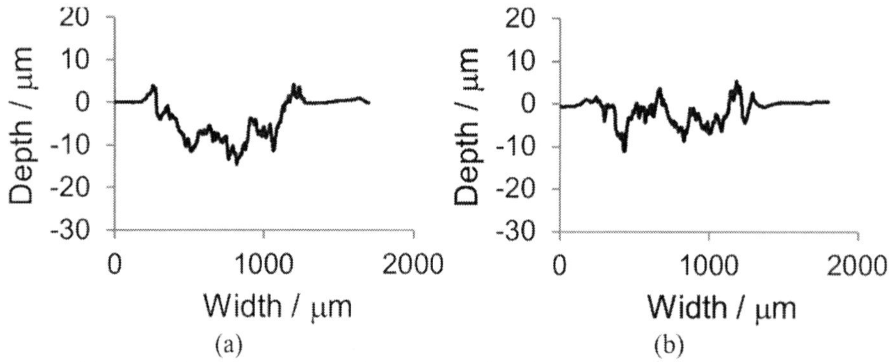

Figure 5. Wear track profiles of the (a) untreated and (b) treated HVOF-sprayed coatings.

Figure 6 presents low and high magnification SEM micrographs of the worn surfaces of the untreated and treated HVOF-sprayed coatings. It can be seen from the micrographs that, on the untreated surfaces wear progressed by extensive shear deformation due to the ploughing action of the ball under the test load of 5 N. On the worn surfaces of treated coating, there is no indication of nitrided layer failure; it still has capability to support the test load of 10 N by remaining largely intact. When compared to the untreated surface, it is possible to see the change in wear mode from the heavy ploughing to the combination of limited ploughing and polishing.

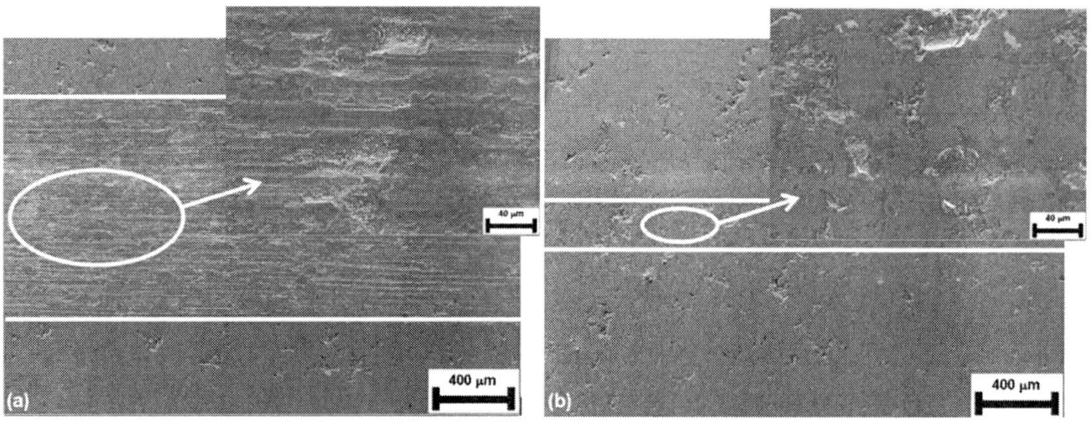

Figure 6. Low and high magnification SEM micrographs of wear tracks generated on the (a) untreated, (b) treated HVOF-sprayed coatings.

Figure 7 shows the morphologies of the wear scars of Al_2O_3 balls mating with the untreated and treated HVOF-sprayed coatings at applied loads of 5 N and 10 N, respectively. It was clearly to see that the contact surface of Al_2O_3 ball sliding against the untreated HVOF-sprayed coating was rough and appeared as dark region within the wear scar, while the wear scar surface of Al_2O_3 ball sliding against the treated HVOF-sprayed coating exhibited very smooth, light and flat. The reduced incidence of delamination phenomena in the treated HVOF-sprayed coating might also be explained by a somewhat lower tendency to adhere to the Al_2O_3 counterpart, witnessed by the lower friction coefficient with minor fluctuation (Figure 4).

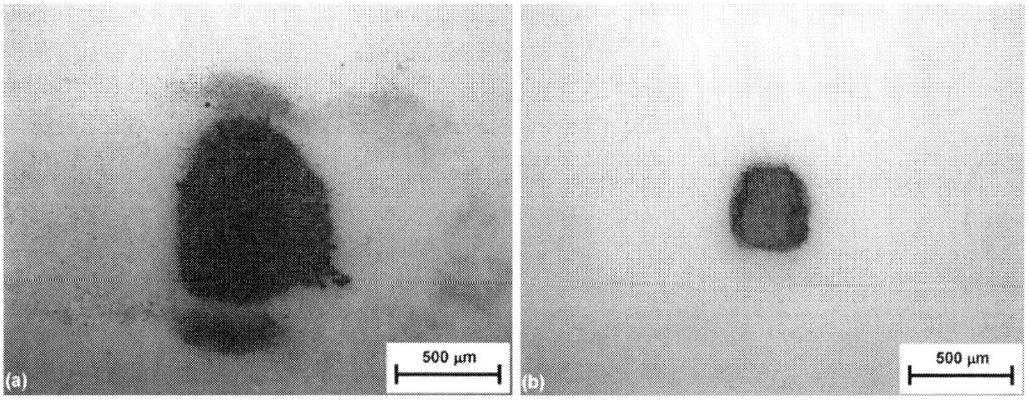

Figure 7. Morphology of the worn scars of Al_2O_3 balls mating with the (a) untreated and (b) treated HVOF-sprayed coatings at applied loads of 5 N and 10 N, respectively

4. Conclusions

Pulsed plasma nitriding process was performed on HVOF-sprayed AISI 316 stainless steel coating. The coatings were characterized by microstructure analysis, microhardness measurements and dry sliding wear tests, and the most relevant conclusions can be summarized as follows:

(1) A novel duplex coating was fabricated on the surface of 1050 aluminum alloy by pulsed plasma nitriding.

(2) Pulsed plasma nitriding formed a surface layer with a thickness of about 10 µm. This layer is composed of CrN, iron nitrides (Fe_3N, Fe_4N) and expanded austenite (γ_N) phases and exhibits extremely high hardness (more than 1250 $HK_{0.01}$) compared to the untreated HVOF-sprayed coating (430$HK_{0.01}$).

(3) The wear rate of the treated HVOF-sprayed AISI 316 stainless steel coating decreased remarkably compared with that of the untreated one.

Acknowledgements

The financial support of the research foundation of Bilecik S.E. University (Project no.: 2015-01.BSEU.03-04) is gratefully acknowledged. I would also like to express our thanks to Dr. Ersin. E. KORKMAZ for the support given in nitriding process.

References

1. Saglam U., Baydogan M., Mindivan H., Kayali E.S., Cimenoglu H., Influence of retrogression and reageing on mechanical and corrosion properties of 7039 aluminium alloy, Zeitschrift Fur Metallkunde, vol. 95, 14-17, 2004.
2. Mindivan H., Baydogan M., Kayali E.S., Cimenoglu H., Wear behaviour of 7039 aluminum alloy", Materials Characterization, vol. 54, 263-269, 2005.
3. Mindivan H.,Wear behavior of plasma and HVOF sprayed WC-12Co+6 % ETFE coatings on AA2024-T6 aluminum alloy, Surface & Coatings Technology, vol. 204 (12-13), 1870-1874, 2010.
4. Akın U., Mindivan H., Samur R., Kayali E.S., Cimenoglu H., Tribological behaviour of combined coatings deposited on aluminium, Key Engineering Materials, vol. 280-283, 1453-1458, 2005.

5. Mindivan H., Tekmen C., Dikici B, Tsunekawa Y., Gavgali M., Wear behavior of plasma sprayed composite coatings with in situ formed Al_2O_3, Materials & Design, vol. 30, 4516-4520, 2009.
6. Mindivan H., A study on wear and corrosion behaviors of thermally sprayed alumina titania composite coatings on aluminum alloys, Kovove Materialy-Metallic Materials, vol. 48, 203-210, 2010.
7. Richter E., Gunzel R., Parasacandola S., Telbizova T., Kruse O., Moller W., Nitriding of stainless steel and aluminium alloys by plasma immersion ion implantation, Surface & Coatings Technology, vol. 128-129, 21-27, 2000.
8. Sudagar J., Venkateswarlu K., Lian J., Dry sliding wear properties of a 7075-T6 aluminum alloy coated with Ni-P (h) in different pretreatment conditions, Journal of Materials Engineering and Performance, vol. 19 (6), 810-818, 2010.
9. Okumiya M., Tsunekawa Y., Murayama T., Surface modification of aluminum using ion nitriding and fluidized bed, Surface & Coatings Technology, vol. 142-144, 235-240, 2001.
10. Mindivan F., Mindivan H., Surface properties and tribocorrosion behaviour of a thermal sprayed martensitic stainless steel coating after pulsed plasma nitriding process, Advances In Materials and Processing Technologies, vol. 2 (4), 514-526, 2016.
11. Mindivan F., Mindivan H., Comparisons of wear performance of hardened Inconel 600 by different nitriding processes, Procedia Engineering, vol. 68, 730-735, 2013.

Wear Resistance of Austempered Ductile Iron with Nanosized Additives

Kaleicheva J.K. [1] ***, Mishev V.*** [1]
[1] *Technical University of Sofia, Department of Materials Science and Technology, 1000, Sofia, Bulgaria*
jkaleich@tu-sofia.bg , v_mishev@tu-sofia.bg

Abstract: The wear resistance, microstructure and mechanical properties of austempered ductile iron (ADI) with nanosized additives of cubic boron nitride cBN are investigated. Samples of ductile iron are put under austhempering at the following conditions: heating at 900°C, 1 h and isothermal retention at 280°C, 2 h and 380°C, 2 h with the aim to achieve a lower bainitic structure and an upper bainitic structure. The experimental wear testing of austempered ductile irons is performed in friction conditions of a fixed abrasive by a cinematic scheme „pin - disc" using an accelerated testing method and device. The microstructure of the ADI is investigated by metallographic and X-Ray analyses. The Vickers hardness testing and impact strength examination are carried out. The influence of the nanosized additives of cBN on the wear resistance, microstructure, impact strength and hardness of the ADI is investigated.

Keywords: nanoadditives, austempered ductile iron, hardness, wear resistance, microstructure.

1. Introduction

Ductile cast iron is a structural material, characterized with high strength, toughness and wear resistance. The graphite presence into the ductile iron provides resistance to mechanical wear and improves its machining. It is possible to influence graphite morphology, also the material base structure using alloying, heat treatment or trough a combination of these two processes to achieve the optimum in the cast irons properties. The austempering of the cast irons is a heat treatment leading to a formation of a bainitic structure in the metal base. This transition is a reason for the internal stresses and deformations reduction and details impact strength enhancement. During this heat treatment the lower or upper bainitic structure is formed, which leads to material high strength, increased toughness and a wide practical use in the treatment of structural steels and ductile cast irons [1,2]. Incomplete austempering is also applicable to the heat treatment of some hipper eutectoid and ledeburite steel [3,4].

The possibility for wider practical application of this type of heat treatment requires an additional data in the bainitic transformation of iron – carbon alloys of different composition, considering alloys within nanoadditives [5-9].

The aim of this study is to investigate the wear resistance, microstructure and mechanical properties of austempered ductile iron, containing the nanosized additives of cubic boron nitride cBN.

2. Materials and Investigation Methods

The composition of the spherical graphite iron samples is: Fe-3, 55C-2, 67 Si-0, 31 Mn-0,009 S-0,027 P-0,040 Cu-0,025 Cr-0, 08 Ni-0,06Mg wt%. The cBN nanosized particles are coated by electroless nickel method EFTTOM-NICKEL [10] prior to the edition to the melt. The nickel coating improves the particles wetting into the melt and their uniformity distribution into the casting volume.
The spherical graphite iron samples undergo to austempering, including heating at 900°C for 1 hour, after that isothermal retention at 280°C, 2 h and 380°C, 2h. The austempered ductile iron samples' microstructure is observed by means of an optical metallographic microscope GX41 OLIMPUS. The samples surface is treated with 2 vol. % HNO_3 - C_2H_5OH solution.
The hardness testing is performed by Vickers method. The impact strength testing with Charpy impact test is carried out.
The austempered ductile iron samples are examined by X-Ray diffraction analysis to find out the retained austenite quantity in the structure before and after tribological testing. X-ray powder diffraction patterns for phase identification were recorded in the angle interval 35-115° (2θ), on a Philips PW 1050 diffractometer, equipped with Cu Kα tube and scintillation detector. Data for cell refinements and quantitative analysis were collected in θ-2θ, step-scan mode in the angle interval from 35 to 115° (2θ), at steps of 0.03° (2θ), and counting time of 3 s/step. Quantitative analysis was carried out by BRASS - Bremen Rietveld Analysis and Structure Suite [11].
The experimental wear testing of austempered ductile iron is performed in friction conditions of a fixed abrasive by a cinematic scheme „pin - disc" using an accelerated testing method and device [7]. The impregnate material Corundum 220 is used as an abrasive surface during the test. The basic experimental parameters are: Nominal contact load, Pa=0, 4. 10^6[Pa]; Average sliding speed = 24,5 [cm/s]; Nominal contact surface, Aa = 50,24 [mm^2]; Density, $\rho = 7{,}80.10^3$ [kg/m^3] .

3. Experimental results and analysis

The iron structure after austempering at 280°C, 2 h consists of lower bainite (Figure 1) and at 380°C, 2 h – of upper bainite (Figure 2). Bainite is an oriented structure, consist of α – phase needles (bainitic ferrite), of carbides and untransformed austenite. The α – phase is formed from austenitic areas (containing lower carbon) by martensitic mechanism and is carbon over-saturated α – solid solution. Part of the untransformed austenite undergoes martensitic transformation and part of it remains in the structure as retained austenite A upon cooling from the isotherm to ambient temperature [1, 2]. The nanosized additives change the bainitic transformation kinetic and accelerate $\gamma \rightarrow \alpha$ transformation. The quantity of the retained austenite A in the samples with nanoadditives of cBN is 21% for the lower bainitic structure and 30,2% for the upper bainitic structure. For the samples without nanoadditives the quantity is respectively 30, 6% and 40, 4% (Table 1). The retained austenite quantity is more for the samples with upper bainitic structure than this one for the samples with lower bainitic structure.

Table 1. Wear resistance, hardness, impact strength and retained austenite.

№ of sample		Nanosized additive	Hardness *HV10*	Impact strength KC MJ/m²	Wear resistance *I*	Retained austenite A, %	
						before wear test	after wear test
1	lower bainite	-	388	0,771	$7{,}13.10^6$	30,6	25,0
2		*cBN*	422	0,884	$9{,}56.10^6$	21,0	11,8
3	upper bainite	-	314	1,137	$7{,}67.10^6$	40,4	31,3
4		*cBN*	312	1,387	$8{,}03.10^6$	30,2	25,3

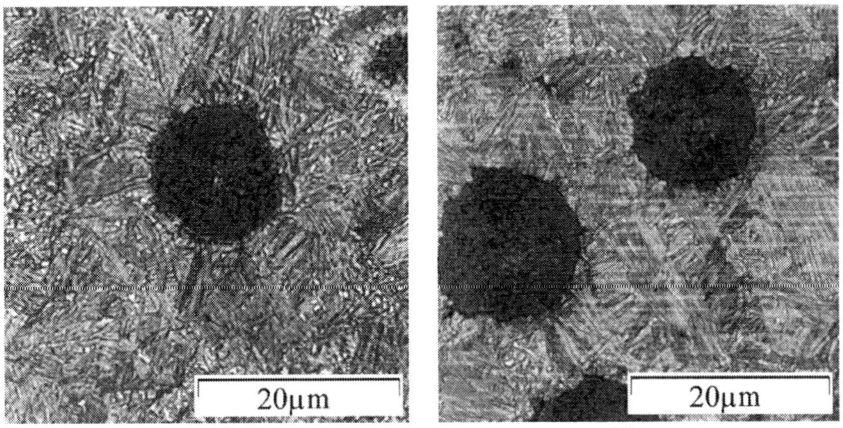

a) **b)**

Figure 1. Lower bainitic microstructures: a- sample 1; b- sample 2.

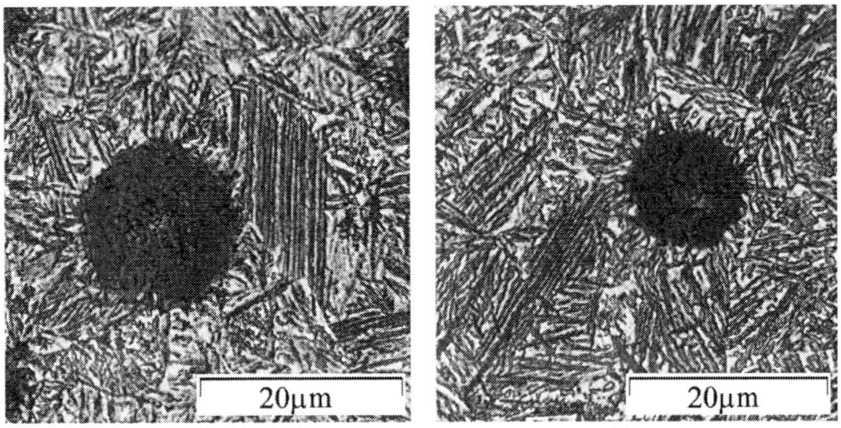

a) **b)**

Figure 2. Upper bainitic microstructures: a- sample 3; b- sample 4.

This is associated to the mechanism peculiarities of the bainitic transformation in the upper and lower interval of the bainitic region. The lower bainitic hardness (388-422 *HV10*) is higher than this one of the upper bainite (314-312 *HV10*) (Figure 3). α – phase carbon over-saturation in the lower bainite is more than this one in the upper bainite, which is the answer for its higher hardness. The retained austenite quantity is more for the samples with upper bainitic structure than this one for the samples with lower bainitic structure, which defines their higher toughness (Figure 4). The wear resistance of the samples with nanosized cBN additives and lower bainitic structure is about 34% higher than this one of the samples without nanoadditives. The addition of nanosized cBN increases the wear resistance for the samples with upper bainitic structure with 5% (Figure 5). The quantity of the retained austenite in the tested samples before and after tribological testing is defined by X-Ray analysis (Table 1). It is concluded that the retained austenite quantity decreases in all samples after teribological testing (Figure 6).

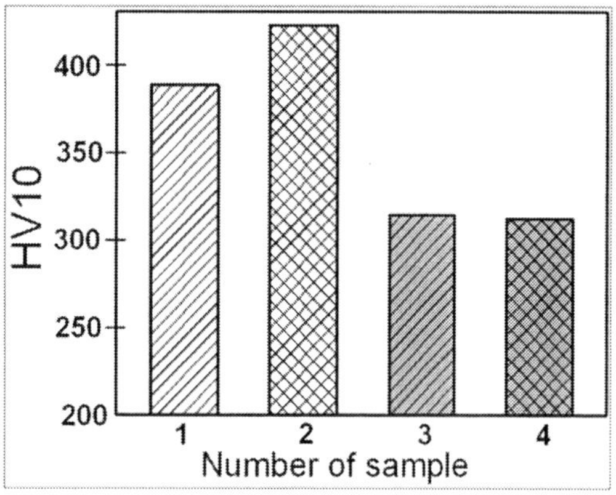

Figure 3. Hardness *HV10* of austempered ductile iron samples without (1,3) and with (2,4) nanoadditive *cBN*. 1,2 - lower bainite; 3,4 - upper bainite.

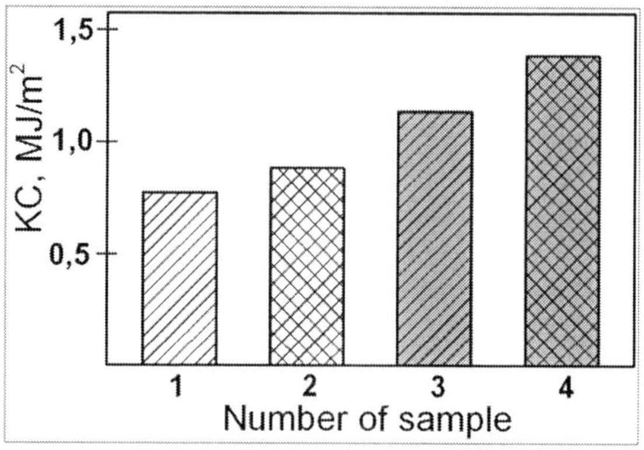

Figure 4. Impact strength *KC* of austempered ductile iron samples without (1,3) and with (2,4) nanoadditive *cBN*. 1,2 - lower bainite; 3,4 - upper bainite.

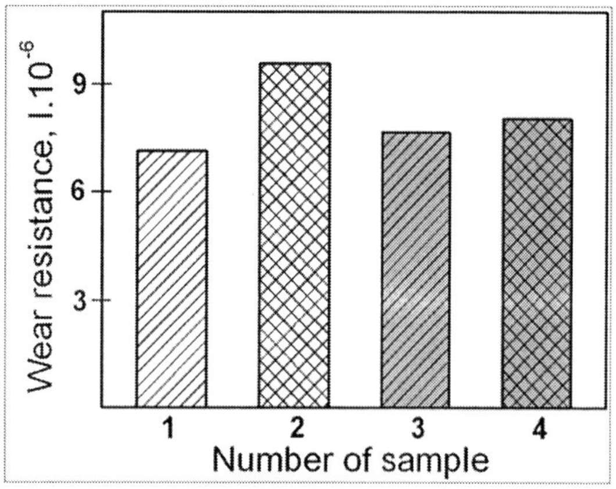

Figure 5. Wear resistance *I* of austempered ductile iron samples without (1,3) and with (2,4) nanoadditive *cBN*. 1,2 - lower bainite; 3,4 - upper bainite.

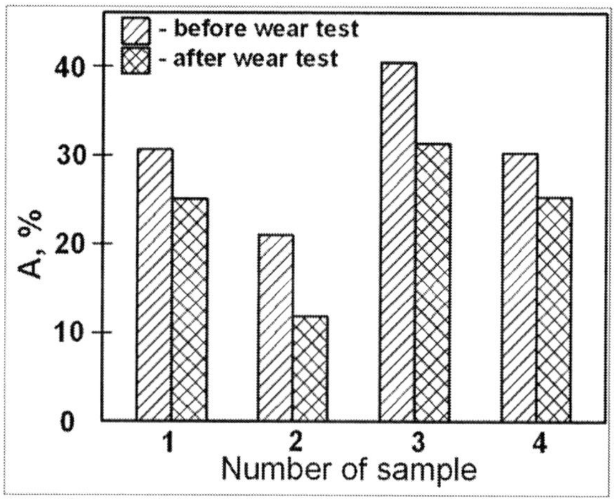

Figure 6. Retained austenite quantity *A* of austempered ductile iron samples without (1,3) and with (2,4) nanoadditives *cBN* before and after wear test. 1,2 - lower bainite; 3,4 - upper bainite.

The standard measured properties (hardness, etc.) not always are reliable criteria for the steel and cast iron wear resistance. The tribological properties of the metallic materials significantly depend on the structural condition, forming on the contact surface during the friction. When the material is exposed to an intensive plastic deformation the structural transformation goes off in the frictional contact area in the metastable structures (retained austenite, martensite, bainite). They greatly influence on the effective surface strength and respectively on the material tribological properties. The metastable retained austenite may be subjected to $\gamma \rightarrow \alpha$ martensitic deformation. During friction the retained austenite transforms partly into deformation martensite with the same amount of carbon as in high carbon austenite. This deformation martensite is untempered martensite characterized with high hardness and ability for intensive strengthening by wear [8,9]. The quantity of retained austenite for the lower bainitic structure decreases from 30,6% to 25% during friction for the samples without nanoadditives (Table 1) and from 21% to 11,8% for the samples with nanoadditives (Table 1, Figure 7). The formation of deformation martensite from the metastable retained austenite in the frictional area probably is one of the reasons for the wear resistance increase of these materials.

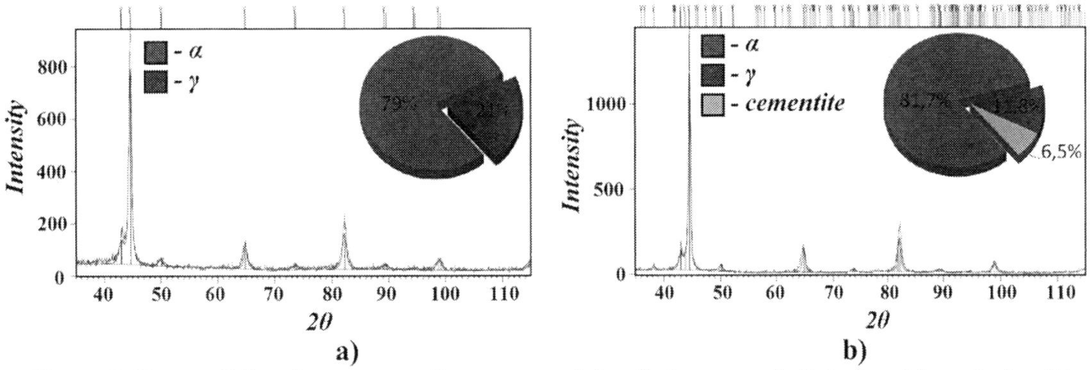

Figure 7. X– ray diffraction pattern of austempered ductile iron sample 2 before (a) and after (b) wear test.

4. Conclusions

The wear resistance, hardness, impact toughness and microstructure of austempered ductile irons without and with nanosized additives of cBN are investigated. Nanosized additives change the bainitic transformation kinetic and accelerate austenite transformation in bainite. Austempered ductile irons with nanosized cBN additives possess higher wear resistance (5-34%) and higher impact strength (15-22 %) in comparison to the samples without nanoadditives. The ADI with nanoadditives and lower bainitic structure have highest wear resistance (I = 9,56.10^{6}). It is found that the partial transformation of the metastable retained austenite to deformation martensite in the greatest extent proceeds in these materials during friction load, which influences on their wear resistance.

Acknowledgements

The work presented is carried out and financed under the project "University complex for research and development of innovations and knowledge transfer in the field of micro/nanotechnologies and materials, power effectiveness and virtual engineering", contract DUNK-01/3 between National Science Fund, MES of RB and R&D Sector of TU-Sofia.

References

1. Bhadeshia, H.K.D.H. Bainite in Steels, 2nd ed., Inst. of Materials, Cambridge, London, 2001.
2. Dorasil E. High-strenght Bainitic Nodular Cast Iron, Academia, Prague, 1985.
3. Kaleicheva, J.A. Structure and properties of high-speed steels after austempering, Int. J. Microstructure and Materials Properties, Vol. 2, 16–23, 2007.
4. Vatavuk J., Totten G.E., Nucci J.E., Albano L.L.M., Canate, I.C.F. Comparative impact behavior of high-C steel after conventional quenching, tempering, and austempering, Materials Performance and Characterization, Vol. 5, 227-238, 2016.
5. Li J., Chen M., Gao H., Zhao Y. Structures and Properties of Cast Irons Reinforced by Trace Addition of Modified SiC Nanopowders, Chinese Journal of Chemical Physics, Vol.20, 625-631, 2007.
6. Wang Y., Pan Z., Wang Z., Sun X., Wang L. Sliding wear behavior of Cr–Mo–Cu alloy cast irons with and without nanoadditives, Wear,Vol. 271, 2953– 2962, 2011.

7. Kaleicheva J., Kandeva M., Mishev V., Karaguiozova Z. Wear Behavior of Ductile Cast Irons with Nanoparticle Additives, J. Chem. Chem. Eng.,Vol.7, pp. 1044–1049, 2013.
8. Kaleicheva J. Wear Behavior of Austempered Ductile Iron with Nanosized Additives, Tribology in Industry, Vol. 36, 74-78, 2014.
9. Kaleicheva J., Mishev V., Avdeev G., Karaguiozova Z., Dineva B. Influence of Nanoadditives on the Structure and Properties of Austempered Ductile Irons, European Conference on Heat Treatment and 21st IFHTSE Congress, Munich, Germany, 537-543, 2014.
10. Gavrilov G., Nicolov C.. Electroless Nickel and Composite Coatings, Tehnika, Sofia, 1985.
11. Birkenstock J., Fischer R., Messner T. The Bremen Rietveld Analysis and Structure Suite. Ber. DMG, Beih. z. Eur. J. Mineral. 15(1), 21, 2003.

Investigation of Selective Laser Melting Surface Alloyed Aluminium Metal Matrix Dispersive Reinforced Layers

***Kamburov V.V.*[1], *Dimitrova R.B.*[1], *Kandeva M.K.*[2], *Sofronov Y.P.*[1]**
[1]*Technical University of Sofia, Department of Materials Science and Technology, 1000, Sofia, Bulgaria*
[1]*Technical University of Sofia, Laboratory of Tribology, Faculty of Industrial Technologies, 1000, Sofia, Bulgaria*
vvk@tu-sofia.bg[1], *r_dimitrova@tu-sofia.bg*[1], *kandevam@gmail.com*[2]

Abstract: The aim of the paper is to investigate the improvement of mechanical properties and in particular wear resistance of laser surface alloyed dispersive reinforced thin layers produced by selective laser melting (SLM) technology. The wear resistance investigation of aluminium matrix composite layers in the conditions of dry friction surface with abrasive particles and nanoindentation tests were carried out. The process parameters (as scan speed) and their impact on the wear resistant layers have been evaluated. The alloyed layers containing metalized SiC particles were studied by Optical and Scanning Electron Microscopy (SEM) and Energy Dispersive X-ray microanalysis (EDX). The obtained experimental results of the laser alloyed thin layers show significant development of their wear resistance and nanohardness due to the incorporated reinforced phase of electroless nickel coated SiC particles.

Keywords: wear resistant layers, selective laser melting (SLM), laser surface alloying, SiC particles.

1. Introduction

Lasers provide a powerful, high-speed source with precise control of the applied amount of energy. On their basis, the selective laser melting technology (SLM) is developed, and is used for the local melting of powder mixtures for the purpose of layer construction of monolithic details, as well as for cladding of individual layers [1].
The purpose of this work is to investigate the characteristics of wear of laser surface alloyed dispersive reinforced thin layers of composite material on aluminium alloy matrix produced by selective laser melting (SLM) technology. Two types of powder mixtures with different content of metalized microparticles of silicon carbide have been used. The introduction of the reinforcing phase of metalized SiC microparticles is accomplished by pre-deposition and subsequent laser melting of a powder mixture to produce an alloyed layer of composite material on an aluminum alloy matrix.

2. Materials and equipment

The following materials were used for the realization of the study: substrate (basic alloy) of aluminium alloy EN AW-2017; electroless nickel coated silicon carbide microparticles, fraction 7-10 μm. To improve wettability of the reinforcement, a surface electroless nickel plating (Ni-P) on SiC particles in alkaline bath using two nickel salts ($NiSO_4$ and $NiCl_2$) is conducted [2, 3]. The electroless nickel coated particles are heat treated in order to increase the hardness of the electroless nickel-phosphorus (Ni-P) coating deposited on their surface. Applying a temperature between 250°C and 400°C leads to microhardness increase due to the formation of the Ni_3P compound in the crystallization of the amorphous phase [4].

The alloying of layers with dispersive reinforced material on a metal matrix is performed by the SLM 125 HL system for selective laser melting in the Prototype Lab at the Faculty of Industrial Technologies of the Technical University of Sofia. The main parameters of the laser installation are: YLR-Faser-Laser emitting source; maximum power of the laser beam 100/200 W; dimensions of the workspace in the camera 125 x 125 x 75 (125) mm; (Ar / N_2) inert gas consumption 0.5 l/min.

The mechanical properties of the basic aluminium alloy EN AW-2017A are as follows: yield strength $Rp_{0.2}$ = 279 MPa; ultimate tensile strength R_m = 437 MPa.

Laser surface alloying is performed on a substrate (base metal) of aluminum alloy EN AW-2017A with dimensions 125 x 125 x 10 mm (figure 1a) at several different SLM scanning speeds in the range from 400 mm/s to 600 mm/s. Two types of powder mixtures were deposited on the substrate (figure 1), namely: - substrate SLM2 – electroless nickel coated silicon carbide microparticles and aluminum powder in a 1: 1 ratio - for 1L and 1R samples; - substrate SLM2 – electroless nickel coated silicon carbide microparticles only for 2L and 2R samples.

The L-specimens are scanned only with longitudinally strokes and the R-specimens are scanned both with transversely and longitudinally strokes.

a) drawing of the aluminum substrate *b) substrate SLM2 with alloyed zone – for specimens 1L and 1R* *c) substrate SLM3 with alloyed zone – for specimens 2L and 2R*

Figure 1. Substrate of aluminium alloy EN AW-2017A with zones under different scanning speeds

Five strips of 16 mm width and 80 mm lengths are laser surface alloyed on each substrate at SLM speeds in the range 400 to 600 mm/s through 50 mm/s (figure 1b and figure 1c). For the purpose of the study, five specimens of the same size are prepared, as well as a reference specimen of the base alloy, without coating.

The experimental study of wear resistance is carried out under the guidance of Prof. Dr. M. Kandeva, applying dry friction study according to a procedure developed in the Applied Laboratory "Tribology" at the department "Material Sciences and Material Technologies" in Technical University of Sofia. Abrasive wear at dry friction sliding according to pin-on-disc kinematic scheme is investigated in laboratory conditions [5-7]. The abrasive surface is modelled by impregnated corundum (E) with 9.0 Mohs hardness, which meets the requirement of the standard for at least 60% higher hardness than that of the surface layer of the tested materials.

3. Experimental results and analysis

3.1. Experiment conducting

The reinforcing phase is introduced according to the technology of laser surface alloying through selective laser melting (SLM). A mixture of metal powder and reinforcing phase of nickel coated SiC microparticles are melted locally on the aluminum alloy surface through the laser beam into the SLM 125 HL camera.
The reinforcing phase of metalized SiC microparticles is introduced in the SLM 125 HL camera by means of preliminary deposition and mechanical flattening of powder layer with thickness of about 100 μm on the substrate. The next melting by laser beam scanning results in obtaining an alloyed laser cladding track of dispersive reinforced layer (figure 2b).

a) Removing the oxide layer from the surface *b) Deposition of strips by different scanning speeds* *c) Repeating the scanning process (SLM)*

Figure 2. SLM 125 HL camera view during laser surface alloying.

Laser surface alloying is considered to be successful in obtaining a solid homogeneous black laser surface alloyed layer with clearly outlined contours on the field of the aluminum substrate (figure 2c). The technological parameters of the laser beam, which produce a stable surface alloyed layer with nickel coated SiC microparticles, are: scanning speed - from 400 to 600 mm/s; laser spot diameter in the working plane - 200 μm; distance between the tracks (distance between axes) 140 - 200 μm.
Upon completion of the alloying, the substrate is partially cooled in the chamber and removed to its final cooling in atmospheric air. It is followed by mechanical cleaning of the surface layer of soot and uncharged particles, washing and drying. Before the test of nanoindentation, specimens are cut out of the aluminum substrate and metallographic sections transverse to the direction of scanning are prepared.

3.2. Structure and chemical composition of the alloyed laser cladding tracks

The alloyed laser cladding tracks represent alternating strokes of absorbed SiC microparticles with a width of about 160 - 220 μm. Figure 3 gives top view of the alloyed tracks on the aluminium substrate obtained by optical macrostructure analysis (figure 3a) and scanning electron microscopy (figure 3b) without further surface treatment.

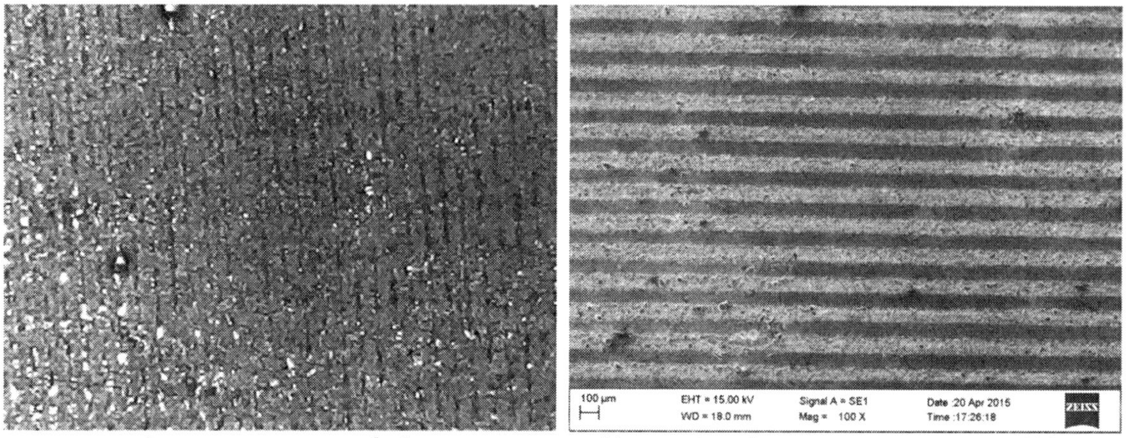

a) macrostructure analysis *b) scanning electron microscopy Mag 100 X*

Figure 3. Alloyed laser cladding surface view.

Figure 4 shows different magnifications of the surface of the alloyed layers made again by scanning electron microscopy. At higher magnifications, it is clearly shown the morphology of the individual alloy strokes, as well as the common areas with attached microparticles of silicon carbide.

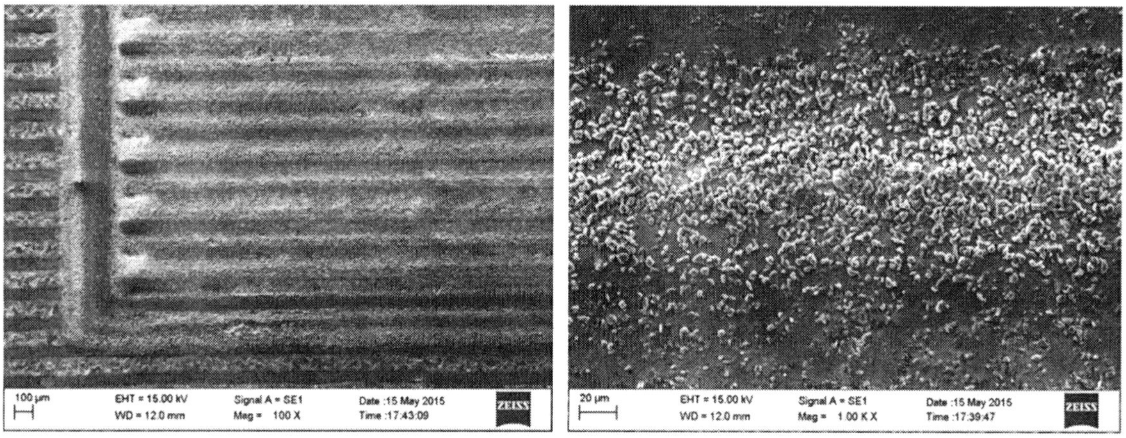

a) scanning electron microscopy Mag 100 X *b) scanning electron microscopy Mag 1.00 KX*

c) scanning electron microscopy Mag 3.00 KX *d) scanning electron microscopy Mag 5.00 KX*

Figure 4. Specimen 2L-4 – SLM3 scanning electron microscopy.

The conducted energy dispersive X-ray spectroscopy microanalysis confirms the presence of silicon carbide particles in the alloyed layers. The presence of elements C, Si, Ni, Al in the range of the scanned areas is established. The additional scanning electron mapping (with higher magnifications) confirms the conclusions drawn about the content and chemical composition of the alloyed layers (figure 5).

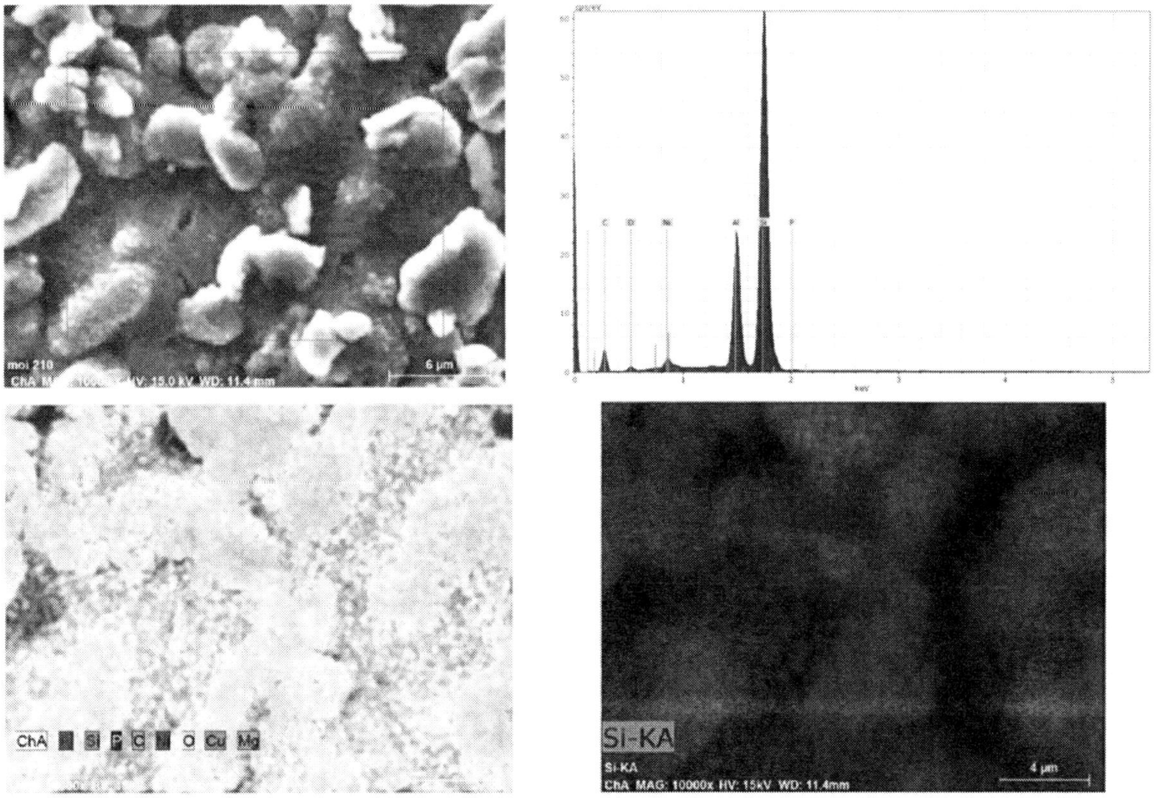

Figure 5. Energy dispersive X-ray spectroscopy microanalysis of specimen 1L-4 – SLM2

In the prepared transverse metallographic sections (figure 6) of the specimens, the alloyed layers stand out clearly enough, so that their geometrical dimensions can be determined.

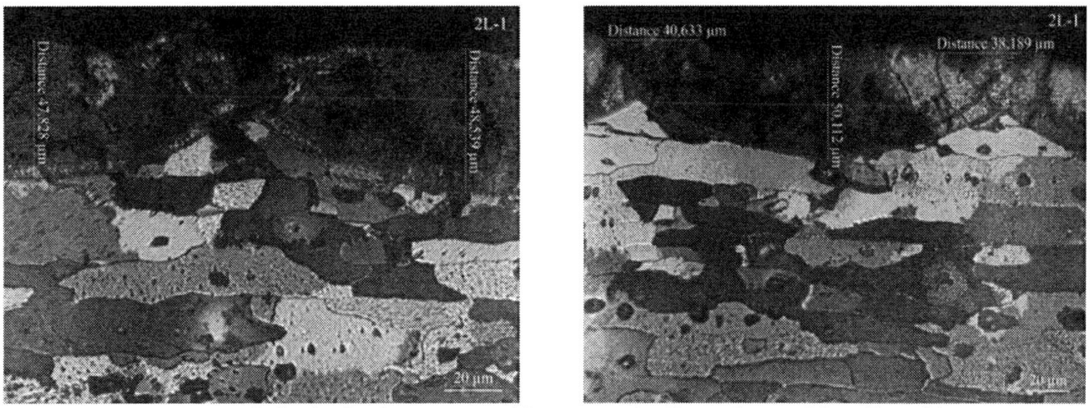

Figure 6. Optical microscopy with geometric dimensions of the alloyed layers.

The data from the optical microscopy show the following geometric parameters of the alloyed layer: average alloy depth about 40 - 50 μm at the average track width about 200-220 μm. Overlapping of the alloy layers is observed in an area of about 10 - 40 μm (figure 6).

3.3. Results of the nanoindentation test

The modulus of elasticity and hardness are determined by indentation of the aluminum alloy substrate and the alloyed layer (obtained by selective laser melting) of the reinforcing phase (SiC microparticles). The nano- indentation tests are performed on metallographic sections in depth of the cross-sections covering the alloyed zone and the transition to the basic metal [8].

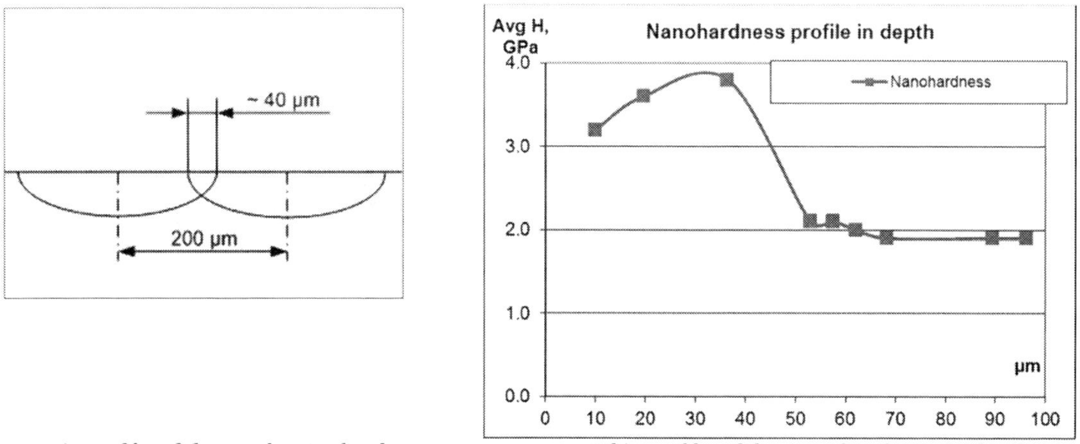

a) profile of the strokes in depth *b) profile of the nanohardness in depth*

Figure 7. Profiles in depth of the alloyed layer.

The observed increase in nanohardness (figure 7b) in the alloyed layer area (up to about 40 μm) is 1.5 to 2.0 times higher than that of the aluminium alloy due to the absorbed reinforcing phase of SiC nickel coated microparticles The measured nanohardness confirms the results for the geometry of the alloyed layer obtained by the optical metallographic analysis. This characteristic changes, i.e. the hardness decreases to values typical for the basic metal (after about 50 μm), with no clear heat impact area. The elasticity modulus shows the same trend, namely, relatively high values in the area of the alloyed layer (up to about 40 μm), a decrease after about 70 μm, and a slight increase in depth of the values typical of the base alloy.

3.4. Results of the wear resistance tests of laser cladded layers of dispersive reinforced material

The tribological parameters of the basic aluminium alloy (EN AW-2017A) are used as reference parameters of the wear resistance tests. The wear characteristics for the base metal specimens under the given test conditions indicate a deformation strengthening and an almost linear increase in the relative intensity of wear at the increasing of cycles number.

Figure 8 and figure 9 show the results of the tests for the wear resistance of the 1R (transversely and longitudinally scanned) sample coating and the 2L (longitudinally scanned) sample coating (electroless nickel coated silicon carbide microparticles and aluminum powder in a 1: 1 ratio). The characteristics of the basic metal are represented by a dashed line.

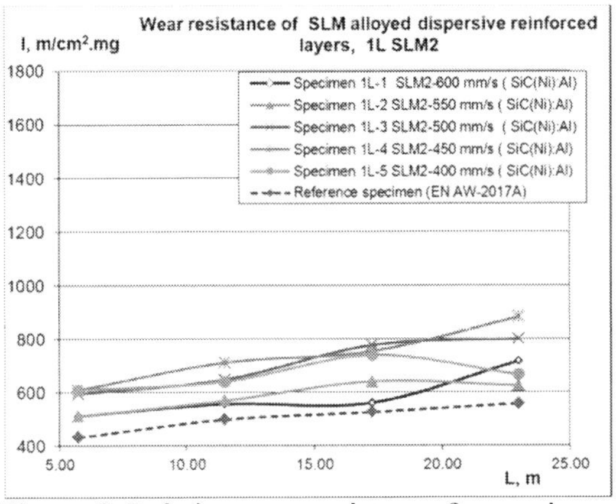

Figure 8. Relative wear resistance for specimens 1L at different scanning speeds (SLM)

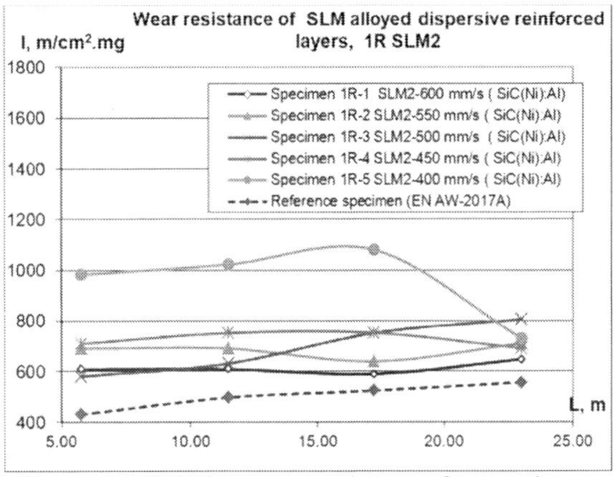

Figure 9. Relative wear resistance for specimens 1R at different scanning speeds (SLM)

Figure 10 and figure 11 show the results of the tests for the wear resistance of the 2R (transversely and longitudinally scanned) sample coating and the 2L (longitudinally scanned) sample coating (only of nickel coated silicon carbide microparticles). The characteristics of the basic metal are represented by a dashed line.

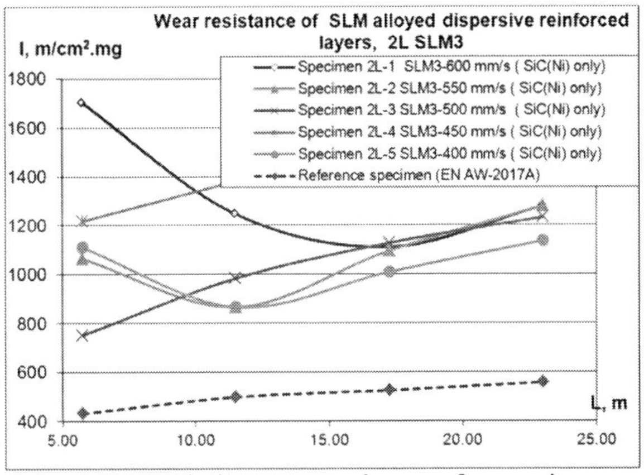

Figure 10. Relative wear resistance for specimens 2L at different scanning speeds (SLM)

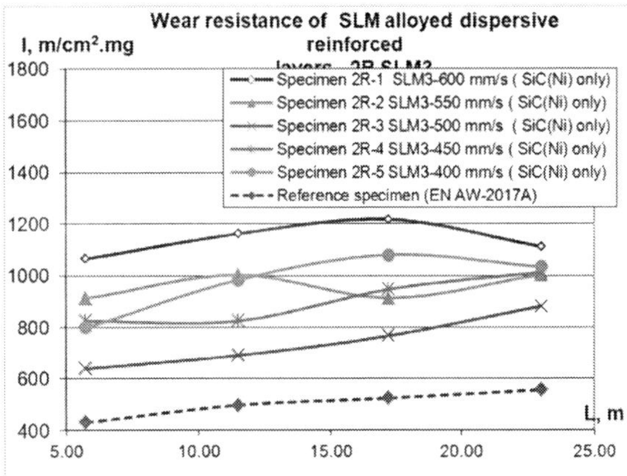

Figure 11. Relative wear resistance for specimens 2R at different scanning speeds (SLM)

The 2L and 2R samples coatings (only of nickel coated silicon carbide microparticles) show the lowest wear rate and respectively the highest relative wear resistance at a higher scanning speed (in the range of 600 - 550 mm/s). The results for the change of relative wear resistance tests at different scanning speeds for the coating on 2L samples are given in figure 12.

The average relative change in the wear resistance of coatings on 2L specimens (only of nickel coated silicon carbide microparticles) in figure 10 is high - from 37% to 96%, with a maximum values of 172% at scanning speed of 600 mm/s. The average relative change in the wear resistance of coatings on 2R specimens (only of nickel coated silicon carbide microparticles) in figure 11 is also high - from 25% to 54%, with a maximum values of 128% at scanning speed (SLM) of 600 mm/s. The maximum values of the relative wear resistance for SLM layers with metalized SiC microparticles on the aluminium alloy matrix are obtained at a scanning speed of 600 mm/s. This shows a significant increase in wear resistance compared to the base alloy and indicates the obtaining a high percent alloyed and absorbed phase of nickel coated silicon carbide microparticles at higher scanning speeds.

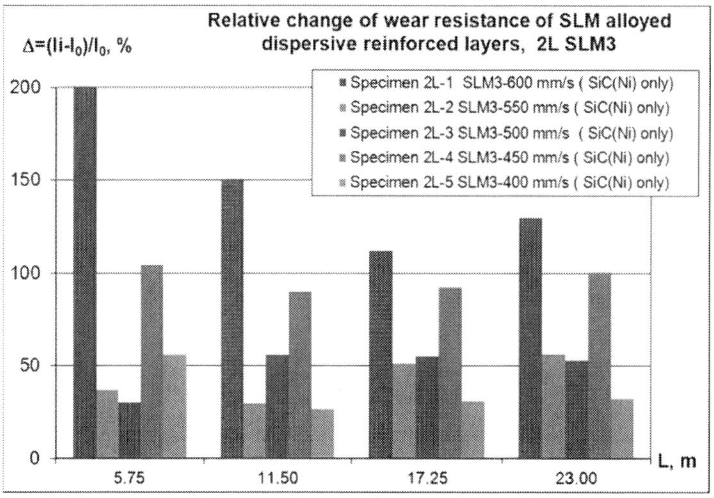

Figure 12. Relative wear resistance change for specimens 2L under different scanning speeds (SLM)

Figure 12 shows the results of the tests for the relative wear resistance of the 2L (longitudinally scanned) sample coating (only of nickel coated silicon carbide microparticles). Figure 13 shows the results of the tests for the relative wear resistance of the 2R (transversely and longitudinally scanned) sample coating (only of nickel coated silicon carbide microparticles).

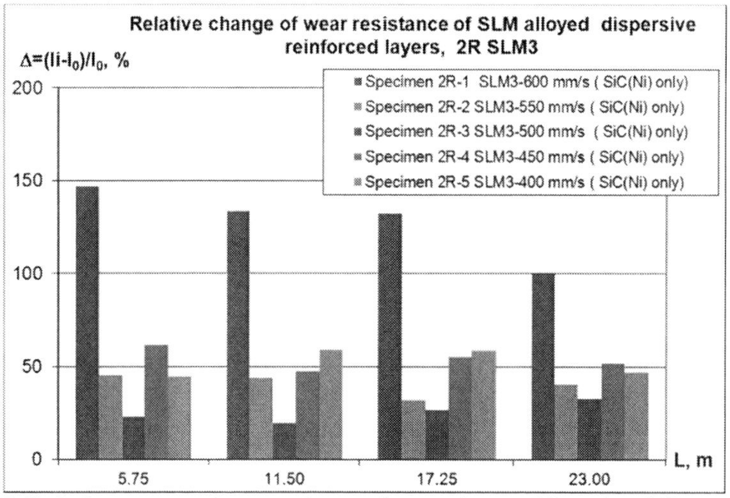

Figure 13. Relative wear resistance change for specimens 2R under different scanning speeds (SLM)

The average relative change in the wear resistance of coatings on 1L samples (of nickel coated silicon carbide microparticles and aluminium powder) is relatively low – from 14% to 33%, with a maximum values of 34% at scanning speed (SLM) of 450 mm/s. The average relative change in the wear resistance of coatings on 1R samples (of nickel coated silicon carbide microparticles and aluminium powder) is also low – from 23% to 34%, with a maximum values of 61% at scanning speed (SLM) of 400 mm/s. This shows a relatively low increase in the wear resistance compared to the basic metal, due to the obtaining of coatings of low percent alloyed and absorbed phase of nickel coated silicon carbide microparticles.

3.5. Analysis and comparison of the obtained results

Comparison of the different coatings (figure 14 and figure 15) is performed with respect to the relative wear resistance intensity of the coating at different scanning speeds (SLM) ranging from 400 - 600 mm/s obtained from the nearly stable test data for 75 and 100 cycles. The summarized data from the test are given in Figure 14 and Figure 15.
The characteristics of the basic metal are represented by a two-point dotted line. The characteristics of the coating of metalized microparticles of silicon carbide and aluminum powder (1L and 1R) are given with a dotted line and of the coating only with metalized particles (2L and 2R) - with a solid line.

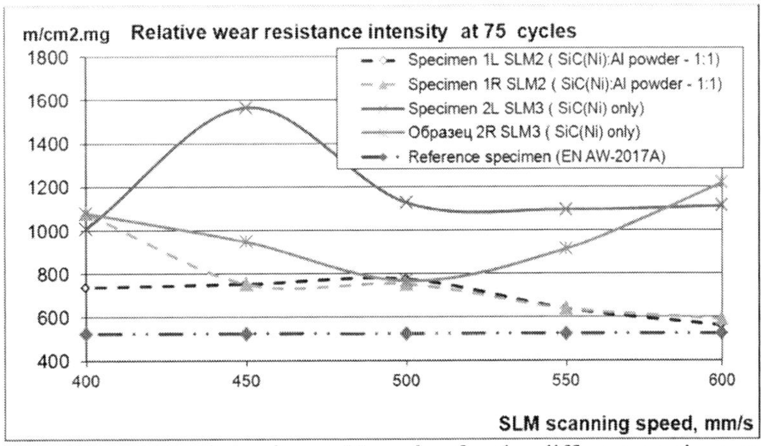

Figure 14. Relative wear resistance intensity at 75 cycles for the different coatings ranged in 400 – 600 mm/s scanning speed (SLM)

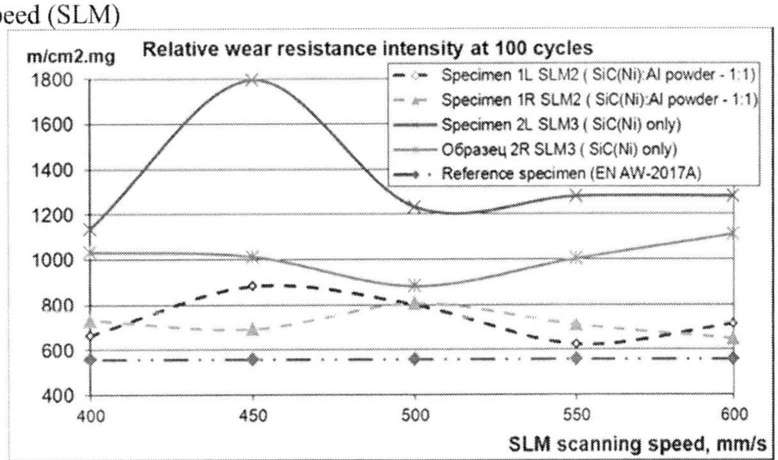

Figure 15. Relative wear resistance intensity at 100 cycles for the different coatings ranged in 400 – 600 mm/s scanning speed (SLM))

Laser surface alloyed dispersive reinforced thin layers with only metalized silicon carbide microparticles (2L and 2R samples) showed an 1.9 to 3.0-fold higher average relative wear resistance intensity within the whole range of speeds, with the maximum wear resistance at scanning speed (SLM) of 450 mm/s. Increasing of the wear resistance intensity of the coating to 3.0 times compared to the basic metal is due to the realization of a laser cladded layer with a high content of alloyed and absorbed reinforcing phase.
Laser surface alloyed dispersive reinforced thin layers with metalized silicon carbide microparticles and aluminium powder in 1:1 ratio (1Land 1R samples) show relatively weak coating wear resistance intensity increase (from 1.2 to 1.5 times), which decreases to the values of the basic aluminium alloy at the higher scanning speeds (SLM).

4. Conclusions

1. A surface layer of a dispersive reinforced composite material with an aluminum alloy matrix containing metalized SiC microparticles is obtained by selective laser melting (SLM) technology at scanning speeds ranging from 400 mm/s to 600 mm/s.
2. Nanohardness in the alloyed zone of dispersive reinforced composite material containing nickel coated SiC microparticles increases from 1.5 to 2.0 times, compared to that of the basic aluminium alloy.
3. The maximum values of the relative wear resistance for SLM layers with metalized SiC microparticles on the aluminium alloy matrix are obtained at a scanning speed of 600 mm/s, however their mixing with aluminum powder reduces and removes its maximum values toward a lower scanning speed.
4. Alloying of the reinforcing phase of metalized SiC microparticles provides up to 3 times higher relative wear resistance intensity of the SLM layers than the basic alloy, however their mixing with aluminium powder significantly reduces its tribological parameters.

Acknowledgements

This study is done with the financial support of the Project **DUNK-01/3** "University R&D Complex for innovation and transfer of knowledge in micro-/nano- technologies and materials, energy efficiency and virtual engineering" funded by the Bulgarian Ministry of Education and Science.
The authors wish to thank to Assoc. Prof. S. Vylkanov, Department of Materials Science and Technology at the TU – Sofia, for his invaluable help in carrying out the SEM and EDX.

References

1. Yadroitsev I., Smurov I., Selective laser melting technology: From the single laser melted track stability to 3D parts of complex shape, Physics Procedia, vol. 5, part B, 551-560, 2010.
2. Dimitrova R., Electroless nickel coatings on silicon carbide particles using two nickel salts in alkaline solution, International Working Conference ''Total Quality Management – Advanced and Intelligent Approaches'', 2–5 June, 2015, Belgrade, Serbia.
3. Dimitrova R., Rangelov R., Kamburov V., Nikolov A., Obtaining a composite material reinforced by SiC particles based on aluminium matrix, International Conference Foundry, April 2014, Pleven, Bulgaria.
4. Dimitrova R., Investigation of creation possibilities of dispersive reinforced aluminium metal matrix for hardfacing, PhD Thesis, Technical University of Sofia, 2016, Sofia, Bulgaria.
5. Dimitrova R., Kandeva M., Sofronov Y., Chervenkov L., Wear resistance investigation of composite layers cladded over aluminium alloy matrix by selective laser melting, Bultrib 2015" 11-13 September, 2015, Sozopol, Bulgaria.
6. Dimitrova R., Kandeva M., Kamburov V., Jordanov M., Mechanical and tribological characteristics of hardfaced dispersive reinforced aluminium metal matrix layer, Journal of Balkan Tribological Association (JBTA) book 4, Vol. 22, 2016.
7. Dimitrova R., Kandeva M., Kamburov V., Mechanical and tribological characteristics of TIG hardfaced dispersive layer by reinforced with particles extruded aluminium, Tribology in Industry Journal, vol. 38, number 1, March 2017.
8. Datcheva M., Dimitrova R., Iankov R., Sofronov Y., Kamburov V., Hardness characterization of laser cladding microsurfaced SiCp layers using instrumented indentation technique, International Scientific Conference „70 years FIT", 11-13 September, 2015, Sozopol, Bulgaria.

Improvement of the Wear Resistance of Ferrous Alloys by Electroless Plating of Nickel

Kaleicheva J.K. [1] , ***Karaguiozova Z.*** [2]
[1] *Technical University of Sofia, Department of Materials Science and Technology, 1000, Sofia, Bulgaria*
[2]*Bulgarian Academy of Sciences, Space Research and Technology Institute, 1113, Sofia, Bulgaria*
jkaleich@tu-sofia.bg , karazuzi@yahoo.com

Abstract: The electroless nickel (Ni) and composite nickel – nanodiamond (Ni+DND) coatings are investigated in this study. The method EFTTOM-NICKEL for electroless nickel plating with nanosized strengthening particles (DND 4-6 nm) is applied for the coating deposition. The coatings are deposited on ferrous alloys samples. The wear resistance of the coatings is performed by friction wear tests under 50-400 MPa loading conditions – in accordance with a Polish Standard PN-83/H-04302. The microstructure observations are made by optic metallographic microscope GX41 OLIMPUS and the microhardness is determined by Vickers Method. Tests for wear resistance, thickness and microhardness measurements of the coatings without heat treatment and heat treatment are performed. The heat treatment regime is investigated with the aim to optimize the thermal process control of the coated samples without excessive tempering of the substrate material. The surface fatigue failure is determined by contact fatigue test with the purpose to establish suitable conditions for production of high performance materials.

Keywords: electroless coating, wear resistance, microhardness, contact fatigue, nanodiamond.

1. Introduction

The market demand for high reliability, effective, with environment protection working machines requires development of new materials and innovative methods for production of high performance parts and details. The involvement of nanotechnology in different fields of the industry open a wide perspective for the following application-related advantages to be expected: improved efficiency, increased component life, which is the expectation on the side of automotive, aircraft and airspace industry. The reduced costs, increased quality of the production makes the European enterprises a serious competitor in these sectors.

There are many studies proving the materials surface properties improvement by incorporation of super hard micro and nanosized particles by electroless metal deposition technique. Electroless method for nickel based composite coating deposition is especially important seeing the expectation to replace the toxic Cr-containing coatings [1]. The characteristics of the electroless plating method ensure the uniform dispersion of the particles within the composite coating. This makes the coating consistent and regenerative in use and offers a practical and economical opportunity to provide the properties of super hard particles to the substrate surface.

Authors in [2] study the effect of co-deposition of PTFE and/or MoS2 particles on the morphology, wear, and corrosion properties of electroless nickel coating and achieve increase of the wear resistance of the coating by about 30% and a reduction of the average value of friction coefficient to 0.25 from 0.65. About corrosion studies the co-deposition of the PTFE and MoS2 particles led to reduction in corrosion resistance by 10 and 5 times. Authors also prove the heat treatment influence on the coatings hardness and its maximum is gained at 400°C.

[3] investige the possibility for reduction of the corrosion and enhancement of the wear resistance of ferrous based bearings used for water lubricated applications by application of electroless nickel coating with addition of silicon nitride (Si3N4) which possesses high hardness and chemical inertness. The amorphous to crystalline transition with precipitation of Ni3P and Ni around 340 °C even when 2.9 wt.% Si3N4 is present in the composite is observed by Differential scanning calorimetry (DSC) and X-ray diffraction (XRD). Scanning electron microscopic (SEM) photographs of the cross-section and surface of the coated specimen reveal most desired uniformity at 2.9 wt.% from 5 g/l bath loading of the strengthening particles. Almost similar corrosion rate as that of carbon steel in demineralized water adjusted to pH 10 by addition of LiOH and satisfactory performances in terms of weight loss, pitting levels and coefficient of friction and life time of about 9 years under the actual application environment of coated AISI 52100 ball-bearings and races coated with 4 µm thick ENC are achieved.

In [4] Electroless nickel (EN), EN-silicon carbide, and EN-alumina composite coatings are put under testing by the two-body abrasive wear using a scratch test with a diamond indenter and are also heat treated at temperatures of 100–500° C. The coatings hardness increase with heat treatment temperature from 500 HV100 for the as-deposited condition to 1008 HV100 when fully hardened is determined. Scratch testing show that the as-deposited coating have scratch tracks with a high degree of plasticity, signs of micro ploughing and tensile cracking and is characterized as a ductile failure. On the other hand, the heat-treated coatings show chipping and cracking on the edge of the scratch tracks, failing in a brittle manner. The heat-treated EN-silicon carbide coatings, however, exhibit neither cracking nor chipping, believed to be due to its higher fracture toughness than the other heat-treated coatings, attributable to its lower phosphorus content. The volume of material removed from the silicon carbide scratch track is 1/3 of the volume removed from the steel substrate at a 20 N load, and show the best wear/ scratch resistance of any of the coatings tested.

A sliding wear tests performed at both room temperature and high temperature (100 and 300°C) on silicon carbide/nickel composite coatings deposited industrially on AISI 1020 steel discs by using a proprietary self-catalyzing chemical reduction process (Hardex) and annealed at 400°C in argon show nearly 23 and 11·6 times larger wear rate for the heat treated coatings tested at 100 and 300°C, respectively when compared with the wear resistance of the heat treated coatings tested at 25°C [5].

In the study [6] a prediction model for the wear depth of the electroless Ni-P deposits under lubricated conditions with varying normal load, sliding speed and sliding time using multiple regression analysis and fuzzy logic, which is a very efficient artificial intelligence technique for modeling and monitoring systems is compared with experiments carried out according to Taguchi's L27 orthogonal array of experiments and it is found a good agreement between the two approaches. The authors find that the wear mechanism is mild abrasive in nature.

[7] examine electroless nickel based alloy deposition as an excellent alternative to hard chromium and nickel-boron coating as a result of its superior hardness and outstanding wear resistance in abrasive environments. Therefore, this study reveals the inherent and enhanced mechanical properties that composition and heat treatment have on multi-alloy, and that duplex nickel based coating has on mild steel substrate. The coating is a supersaturated alloy in its as-plated state but became strengthened as a result of controlled moderate annealing by precipitation of nickel phosphide crystallites which effectually enhance its overall mechanical properties within a specified heat treatment temperature range. However, the entire mechanical properties of this coating decrease with excessive heat treatment above 450oC due to grain coarsening. The results here show that the optimum wear resistance of Ni-W-1.5% P and Ni-W-9.5%P coatings correspond to peak hardness at 350oC to 450oC respectively after one hour annealing in argon environment.

The authors in [7, 8] receive a uniform and compact surface morphology in Ni-P/nano-TiO2 and Ni-P/nano-carbon nanotube composite coatings, ensuring better corrosion resistance than that in the 5083

aluminum alloy substrate and in the electroless Ni-P coating. The investigation of the influence of different concentration of the added particles shows that the corrosion resistance of the nanocomposite coatings is substantially increased to a maximum at 10 g/L nano-TiO2 and nano-CNT introduction. The superior corrosion resistance exhibits nano-CNT codeposited coating compared to the nano-TiO2 codeposited coating. [9].
In [10] electroless Ni–P coatings on magnesium composite reinforced with multiwall carbon nanotube under specific coating conditions are studied. In general, it is observed that the surface finish, micro hardness, specific wear rate and friction of the EN-coated layers improve significantly with the addition of nano additives. Nano additives to the EN bath such as Al2O3, SiO2 and ZnO are used.
[11] proves the influence of the presence of the surfactant in the EN bath, its type and concentration on the process of electroless deposition of Ni–P–nano-ZrO2 composite coatings. The most stable bath is obtained with the addition of dodecyl trimethyl ammonium bromide (DTAB). In the presence of DTAB, the Ni–P–nano-ZrO2 layers change the surface topography and become smoother. The surfactant can increase the possibility for the particle being embedded changing the composition of obtained coatings and influencing the P/Ni ratio in the coatings produced without and with surfactant.
The purpose of the present study is to improve the ferrous materials surface properties by addition of nanosized diamond particles (DND) with the aim to increase fatigue lives of gears using a novel technology for electroless coating deposition. The electroless coating selected for this investigation is composite nickel-nanodiamond (Ni-DND) coating.

2. Materials and Investigation Methods

2.1. Manufacture of test samples

The samples are manufactured using 17CrNiMo6 steel and a new air hardening steel suitable for gears manufacturing OVATECTM 677L.
The 17CrNiMo6 steel samples are carburised, hardened and tempered and finish ground. The thermal treatment process cycle conditions are: soak for 30 minutes at 930°C; gas carburise for 120 minutes at 930°C and 1.2% C diffuse for 60 minutes at 930°C; fast cool to 830°C and hold for 60 minutes at 0.75% C; direct oil quench (standard speed 22 grade quench oil); temper for120 minutes at 180°C.

2.2. Coating procedure

The nanocomposite nickel coatings are obtained by electroless nickel plating method EFTTOM-NICKEL. This method is developed at Technical University of Sofia. The ultra dispersed diamond powder (DND) is used as a strengthening material. The particle size is between 4-6 nm. The nanosized diamond powder is produced by a detonation method developed at SRTI-BAS. The technology for nanostructured coating deposition is developed at BAS-SRTI. The electroless nickel coatings (Ni) are composed of one layer. The composite nickel coatings consist of two layers: electroless nickel followed by a second layer of Ni-DND. The coatings are heat treated at the following temperature modes:
1. 200°C for 10 hours
2. 220°C, 9 hours
3. 250°C, 7.5 hours
4. 290°C for 6 hours.

2.3. Test procedure

The microstructure observations are carried out by means of the optical metallographic microscope GX41 OLIMPUS at magnifications x100 before and after coating layering. The samples are treated with 3% HNO_3-C_2H_5OH solution before the examination. The microhardness of the coating is determined by Vickers Method. The used loading is 300gf.

The resistance to wear of samples is estimated by means of the friction test in 3-rollers-cone system (Polish Standard PN-83/H-04302). Testing procedure is carried out under the 4 various loading conditions: 50 MPa, 100 MPa, 200 MPa and 400 MPa. Tests are conducted with drop lubrication at a rate of 30 drops of Lux-10 oil per minute. Toughened 45 grade steel (Polish standard) with a hardness of 28-30HRC is used as a counter piece.
The contact fatigue tests are carried out using the 60mm centre distance disc rig. Two tests are performed on each coating and a summary of the operating conditions is shown in Table 1. The discs are run for $1x10^6$ cycles at each of the load stages shown in Table 2 until the coating is removed. It is noted that a Hertzian contact stress of 1200MPa is selected for load stage 1 as the base steel would be expected to perform without concern under such conditions.
All studies are conducted with the presumption that the results will be used in the manufacture of gears.

Table 1. Operating conditions for contact fatigue tests on coated disc samples.

Lubricant:	100cSt mineral oil + 4% Anglamol A99 EP additive
Lubricant temperature:	70°C
Lubricant flow rate:	1.4l/min (into and out of mesh)
Speed:	1100rpm
Slide roll ratio:	0.33
Load cycles:	$1x10^6$ (per load stage)

Table 2. Loading conditions for contact fatigue tests on coated disc samples

Load stage	Applied load (N)	Hertzian contact stress (MPa)
1	143	1200
2	182	1300
3	227	1400
4	280	1500
5	339	1600
6	407	1700
7	483	1800
8	568	1900

3. Experimental results and analysis

3.1. Metallographic analysis

Micro-hardness measurements are carried out on the discs to determine if the coating process has adversely affected the properties of the substrate material. The micro-hardness profiles measured on each disc are plotted in Figure 1. The micro-hardness profile measured on the uncoated disc in the carburised and finish ground condition is also included for comparison. The Ni - DND coated discs that are thermally processed at temperatures of 200 and 290°C give a reduction in case hardness of approximately 5 to 65Hv relative to the carburised and finish ground reference condition. However, these reductions in case hardness are not unexpected given that the thermal processing is carried out at temperatures of between 20 and 110°C higher than the typical tempering temperature of the steel from which the discs are manufactured.

IOP Conf. Series: Materials Science and Engineering **295** (2018) 012036 doi:10.1088/1757-899X/295/1/012036

Despite the observed reductions in case hardness, the values measured on the Ni - DND coated discs that are thermally processed at temperatures of 200 and 290°C would not be deemed unacceptable for carburised gears. In each case, the hardness values measured at a depth of 0.05mm are greater than the minimum case hardness of 660Hv recommended for carburised gears.

The coatings applied show good adhesion to the surface of the substrate material and there is no evidence to suggest that the microstructure of the substrate material has been modified during the coating process. Ni- DND coatings can also be applied without modifying the gear geometry and surface finish present after finish grinding.

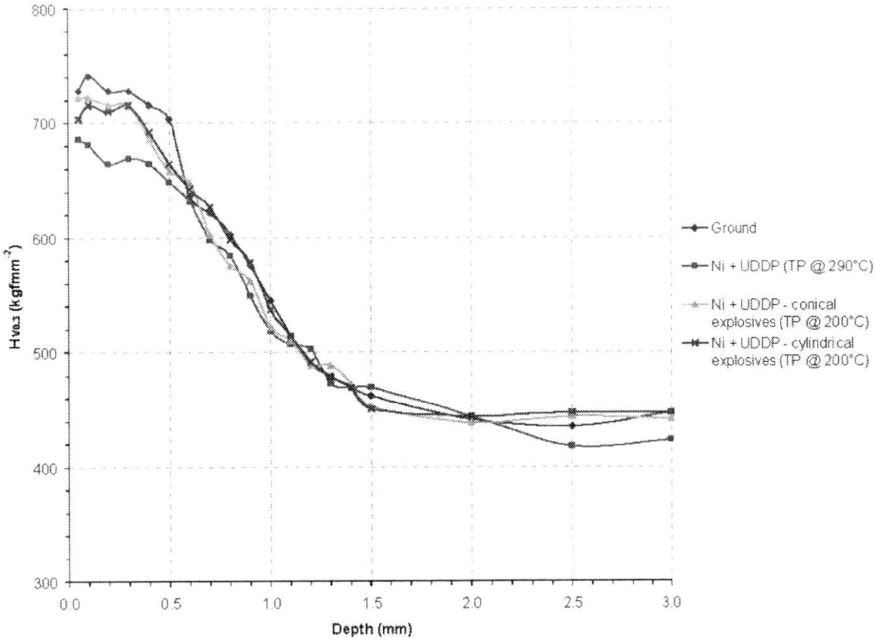

Figure 1. Figure 1: Microhardness profile measured on coated samples

3.2. Metallographic observations

A comparison of the near-surface microstructure present before and after coating is shown in Figure 3. The Ni-DND coating is found to be approximately 8μm in thickness and show good adhesion to the substrate. There was no evidence to suggest that the microstructure of the substrate material is modified during the coating process.

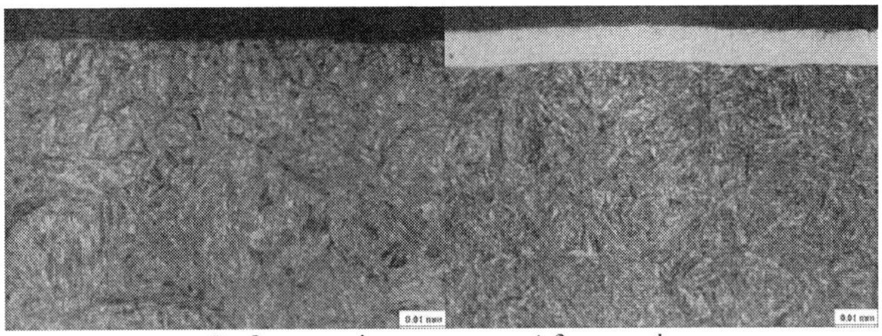

Before coating After coating

Figure 2. Near-surface microstructure before and after plating

3.3. Resistance to wear test

Friction tests are performed for evaluating the coating behaviour obtained by different technologies. Several unit pressure of loading are applied during the test. Low and medium range of the load are: 50 MPa, 100 MPa and 200 MPa. For high loads the pressure value of 400 MPa is used.
During tests carried out under low and medium range load, it is observed that all of the coating layers tested don't undergo galling or the wear rate during the whole testing cycle don't exceed the critical value of wear (maximum of the measured wear range is 25μm).
The baseline of the resistance to wear friction test is the wear resistance of carburized specimen treated without any additional surface processing. The results of wear test (Figure 2) show relatively high resistance to wear under the load of 50 and 100 MPa. Under the load of 100 MPa more intensive wear is observed, while under load of 400 MPa specimens quickly (20 minutes of the test) undergo galling. While the tests are worked out under low and medium loading conditions of unit pressures, resistance to wear of all coated variants are comparable to wear resistance of the carburized only specimens. Among all of the coated specimens, the best wear resistance result is obtained by Ni – DND with next thermal processing coating variant of the surface treatment. Under the high loading conditions all of the chemical method specimens undergo galling. Galling occur after 80 minutes of test of thermally processed Ni -DND coating, while for all other coating variants the galling take place much quicker (50-60minutes of friction test).

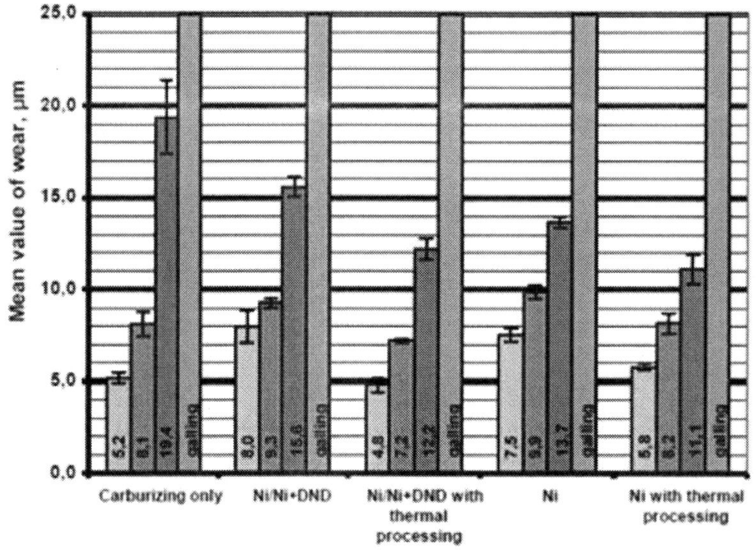

Figure 3. Wear test of the samples

3.3. Contact fatigue test

The contact fatigue tests are carried out on the coatings showing the best performance during previous test procedure. The results are summarized and the minimum Hertzian contact stresses at which the coatings are seen to be removed during testing are shown in Table 3.
The results described indicate that the contact fatigue performance of the Ni - DND coatings that are thermally processed at 200° C is limited by delaminating of the Ni plate. Therefore, the heat treatment regime is optimized with the aim the Ni-plate adhesion to the substrate to be improved by increasing the thermal processing temperature without causing excessive tempering of the substrate material.

As four different regimes of thermal process of the coatings are investigated - at 200°C, 10 h; 220°C, 9h; 250°C, 7,5h and 290°C, 6h and the coated samples are tested for contact fatigue examination. The work outlined is demonstrated that thermal process at 290°C can give the greatest improvement in contact fatigue performance without adversely affecting the properties of the substrate material.

Table 3. Overall results of contact fatigue tests on coated disc samples

ID	Coating	Thermal processing	Material of Twin Discs	Hertzian contact stress required to remove coating, σ_H [MPa]		
				Test 1	Test 2	Test 3
1	Ni - DND (conical explosives)	200°C, 6 hours	17CrNiMo6	≤ 1200	≤ 1200	-
2	Ni - DND (cylindrical explosives)	200°C, 6 hours	17CrNiMo6	≤ 1200	≤ 1200	-
3	Ni - DND	220°C, 9 hours	17CrNiMo6	1200...1300	1500...1600	-
4	Ni - DND	250°C, 7.5 hours	17CrNiMo6	1500...1600	1400...1500	-
5	Ni - DND	290°C, 6 hours	17CrNiMo6	1700...1800	1200...1300	1200...1300

There are noticeable differences in the results obtained. We suppose they are due to the bed thermal control, which is proved during the next experiments on real details. The contact fatigue tests are performed on three pairs of through hardened and finish ground OvatecTM 677L spur gears coated with Ni - DND and thermal processed for 6 hours at 290°C under controlled conditions. They all show the same result: Ni - DND coating give 14% increase in maximum Hertzian contact stress required to generate failure (Table 4).

Table 4. Contact fatigue test results

Material	Surface condition	Maximum Hertzian contact stress required to generate failure (MPa)	Maximum profile deviation at end of test (μm)
OvatecTM 677L	Ground	1404	13μm
	Ni - DND	1601	15μm

4. Conclusions

Metallographic observations, microhardness measurements, wear and contact fatigue tests of the electroless Ni and Ni-DND coatings plated on 17CrNiMo6 and OvatecTM 677L samples in as-plated condition and aftyer heat treatment show that the temperature at which the Ni - DND coatings are thermally processed can be increased to 290°C without causing excessive tempering of the substrate material. The contact fatigue performance of the Ni - DND coatings can be significantly improved by increasing the temperature at which thermal processing is carried out. Improved control of the thermal processing temperature is required to reduce the variability in the contact fatigue performance of the Ni - DND coatings. 14% increase in maximum Hertzian contact stress required to generate coatings failure is obtained for Ni-DND coating after heat treatment.

Acknowledgements

The work presented is carried out and financed under the project "University complex for research and development of innovations and knowledge transfer in the field of micro/nanotechnologies and materials, power effectiveness and virtual engineering", contract DUNK-01/3 between National Science Fund, MES of RB and R&D Sector of TU-Sofia.
Exceptional acknowledgements to colleagues from the Institute of Precision Mechanics- Warsaw, Poland and from Design Unit – Newcastle, UK for their support in performing wear test and contact fatigue tests.

References

1. Feldstein M., The Environmental Benefits of Composite Electroless Nickel Coatings, Product Finishing, 01.08.200, Available: http://www.pfonline.com/articles/the-environmental-benefits-of-composite-electroless-nickel-coatings.
2. Mohammadi M. And Ghorbani M., Wear and corrosion properties of electroless nickel composite coatings with PTFE and/or MoS2 particles, Journal of Coatings Technology and Research, vol. 8, 527–533, 2011.
3. Das C.M., Limaye P.K., Grover A.K., Suri A.K., Preparation and characterization of silicon nitride codeposited electroless nickel composite coatings, Journal of Alloys and Compounds, vol.436, 328-334, 2007.
4. Subramanian C. and Pallotta E., Two-body abrasive wear of electroless nickel composite coatings, Tribology Letters, vol. 2, 133–150, 1996.
5. Staia M.H., A. Conzoño A., Cruz M.R., Roman A., Lesage J., Chicot D, Wear Behaviour of Silicon Carbide/Electroless Nickel Composite Coatings at High Temperature, Journal Surface Engineering, vol.18, 265-269,2002.
6. Mukhopadhyay A., Duari S., Barman T.K., Sahoo P., Wear Analysis of Electroless Ni-P Coating Under Lubricated Condition Using Fuzzy Logic, Portugaliae Electrochimica Acta, vol. 34, 63-83,2016.
7. Oloyede O.R.,, Liu Z., Liu H., Effect of Heat Treatment on the Microhardness and Wear Resistance of Electroless Ni-P/Ni-W-P Duplex Coatings, Petroleum Technology Development Journal: An International Journal,vol. 6, 49-55, 2016.,
8. Lee Ch., Comparative Corrosion Resistance of Electroless Ni-P/nano-TiO2 and Ni-P/nano-CNT Composite Coatings on 5083 Aluminum Alloy, Int. J., Electrochem. Sci., vol.7,12941 – 12954, 2012.
9. A. Miteva, Microstructure and tribology of nickel and nickel alloys, Tribological Journal BULTRIB, vol.III,371-375, 2013.
10. Dhinakaran R., Elansezhian R., Lalitha A.A., Effect of Nanoadditives with Surfactant on the Surface Characteristics of Electroless Nickel Coating on Magnesium-Based Composites Reinforced with MWCNT, Advances in Tribology, vol. 2013, Article ID 315965, 10 pages, 2013, Available: http://dx.doi.org/10.1155/2013/315965.
11. Kurkowska M., Bieliński J., Sałacińska A., Kozera R., Boczkowska A., Electroless deposition of Ni-P/nano-SiO2 composite coatings on PET and carbon fiber substrates, Composites Theory and practice, 14: 1,54-59, 2014.

Theoretical and experimental research on the phenomenon of stick-slip at traction railway vehicles

Ioan SEBESAN[1], Ion MANEA[2], Marius Adrian SPIROIU[3], Sorin ARSENE[4]
[1,3,4]Polytechnic University of Bucharest, Transport Faculty, Department of Railway Rolling Stock, Splaiul Independentei no. 313, Sector 6, Bucharest, Romania
[2]SC Softronic SRL, R&D Department, Craiova, No.40, Dolj, Romania,
ioan_sebesan@yahoo.com[1], ion.manea@softronic.ro[2], marius_spiroiu@yahoo.com[3],
sorin.arsene@upb.ro[4],

Abstract: The stick-slip phenomenon may occur when the limit adhesion force is exceeded at one of the motor axles of the vehicle. In the present paper is analysed the physical phenomenon stick-slip, the conditions for its occurrence and the mechanical model recommended for the study of the phenomenon. Regarding the experimental research, it is presented the stand built in the laboratory of the Rolling Stock Department at Politehnica University of Bucharest.

Keywords: stick-slip phenomenon, rail vehicle, experimental results

1. General considerations

The motion of the drive axle may be accompanied by some intermittents or jerks, a phenomenon known in the speciality literature as stick-slip. As a consequence of the stick - slip, there are important dynamic overloads both in the axle and in its drive system, as well as variations in traction forces at the periphery of the wheels, which cause disturbance of the ride and thus reduce the traction performances of the vehicle.

Issues related to the stick-slip phenomenon will be analyzed for the case of the axle drive system with traction electric motor, at which the motor moment is transmitted to the traction gear by means of a torsion elastic shaft.

The stick - slip vibrations, the primary cause of which lies in the form of the wheel – rail friction and traction force characteristics, occur at low slide speeds, generally at the start of the vehicle, when it is possible the axle slip due to exceeding by the traction force of the limit force given by adhesion.

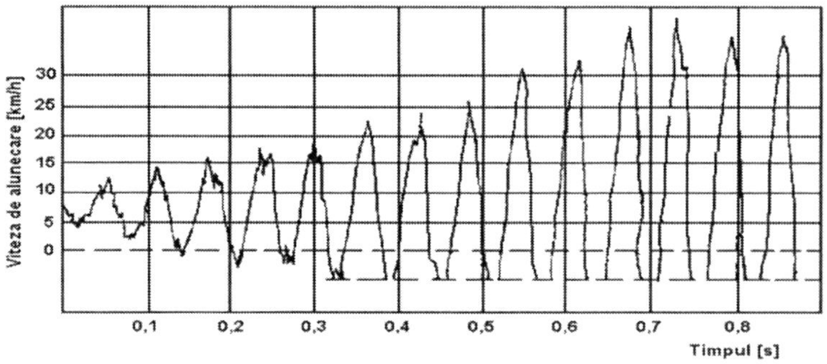

Figure 1. The time variation of the sliding speed of the wheel.

The time variation of the sliding speed of the wheel when the stick – slip occurs, experimentally determined by Schröter and Schönenberger [1], can be seen in fig.1.

Generally, vibrations that occur under the influence of dry friction on the wheel - rail contact are strongly depending on the sliding speed. Thus, at low slip speeds (Fig.2 a), the motion has an adherent phase (stick) and a sliding phase (slip), i.e. an intermittent sliding occurs when the friction force varies between a maximum level T_{max} limited by adhesion and a minimum level T_{min} corresponding to the friction coefficient at the maximum slip speed. At high slip speeds (Fig.2, b) autovibrations are produced, the amplitude of which is much lower than in the case of stick-slip.

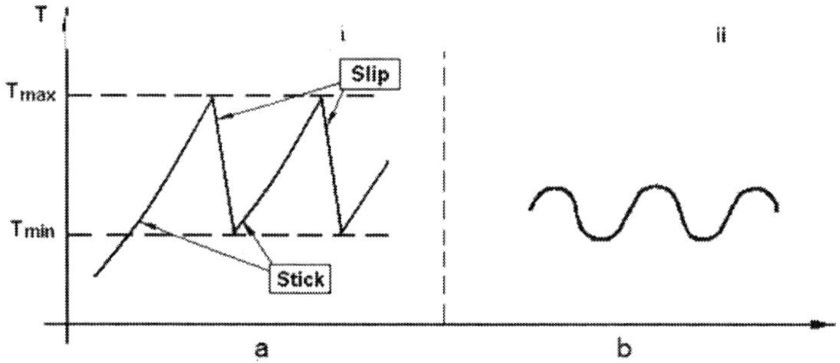

Figure 2. Vibrations produced under the influence of friction:

a – stick-slip; b – vibrations produced at high slip speeds.

The occurrence and development of the stick - slip is dependent on the variation law of the wheel - rail friction coefficient versus the slip speed.

In the case of axle sliding there is intense mechanical action between the particles of the wheel-rail contact surfaces, with a significant heat energy generation. The important modifications of the contact surfaces make the friction coefficient vary with the slide speed.

2. Factors that influence the stick-slip phenomenon

As is was highlighted before, the occurrence and time development of stick-slip depend on the variation of wheel-rail friction coefficient as a function of sliding speed.

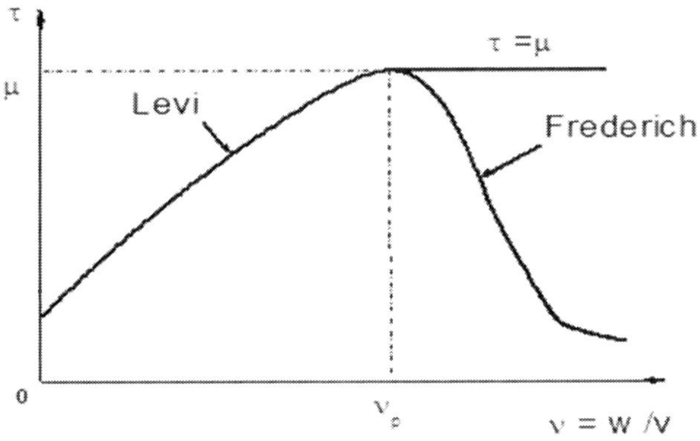

Figure 3. Characteristics of friction coefficient.

The friction coefficient τ depends on the rail-wheel creep ν, which is given by the ratio between the slip velocity w the wheelset forward speed (the vehicle speed, practically) v, as demonstrated by F. Carter. In Levi's law, the characteristic of the coefficient of friction has an ascending branch up to the creep value ν_p, corresponding to the maximum coefficient of friction (adhesion coefficient). As shown by Frederick F. [2], this characteristic has a descending branch for $\nu>\nu_p$ (see Fig. 1).

The stick-slip oscillations occur when the drive force F and the friction force T=τQ, (Q being the wheel load), are decreasing characteristics function of slip velocity w, for a given value of speed v- see Fig. 2.

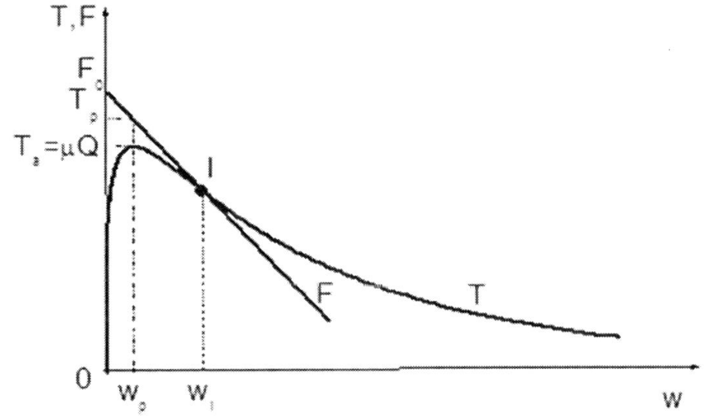

Figure 4. F(w) and T(w) characteristics for forward speed v .

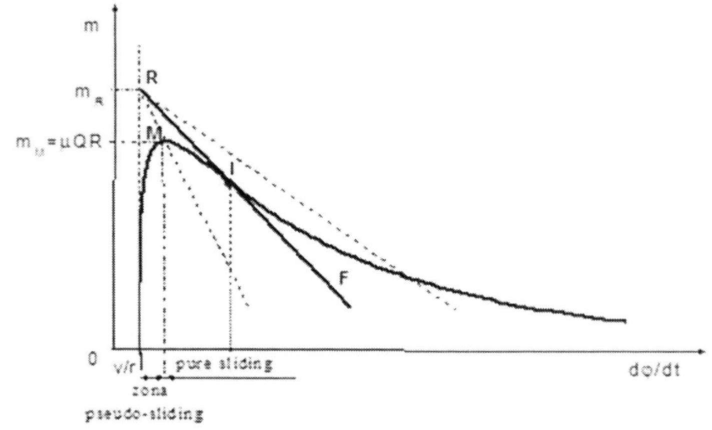

Figure 5.

The tangent in I intersect vertical dφ/dt =v/r in point R. Ordinate m_g of this point defines the minimum value of wheelset torque. Straight line RM defines the possible limit (with maximum slope) of reduced torque characteristic.

Observations:

- The point R resulted as intersection of the tangent in inflection point I with the vertical in dφ/dt=v/r. Any line starting from a point with the ordinate less than m_g will intersect the decreasing branch characteristic of friction coefficient corresponding to the pure sliding in one or two points. If an intersection point is in the portion MI of the characteristic, then appears and the second point with ordinate less than m_I. Any line taken from a point with ordinate higher than m_g intersect the characteristic in a single point.
- The line MR is characteristic limit of the torque reduced to the wheel (minimum slope)

3. Mechanical model for the study of stick-slip phenomenon

In Fig. 6 is shown an equivalent model of the driving system of motor wheelset. The occurrence of the stick-slip phenomenon can be explained as follows: in the slip velocity range (0, w_p), the torque of the traction motor causes torsion of its elastic shaft and of the wheelset axle. Due to the fact that at the slip velocity w_p the traction force $F_p>T_a$, T_a being the adhesion force, the wheels start to slip with the acceleration of the drive system. At the same time, there is a tension release of the elastic elements from drive system and a reduction of the forces acting on the wheel circumference up to the point I where F=T. Further, the motion slows down, the adhesion is restored, and in the same time the engine torque increases until it exceeds again the adhesion limit and the motion repeats itself.

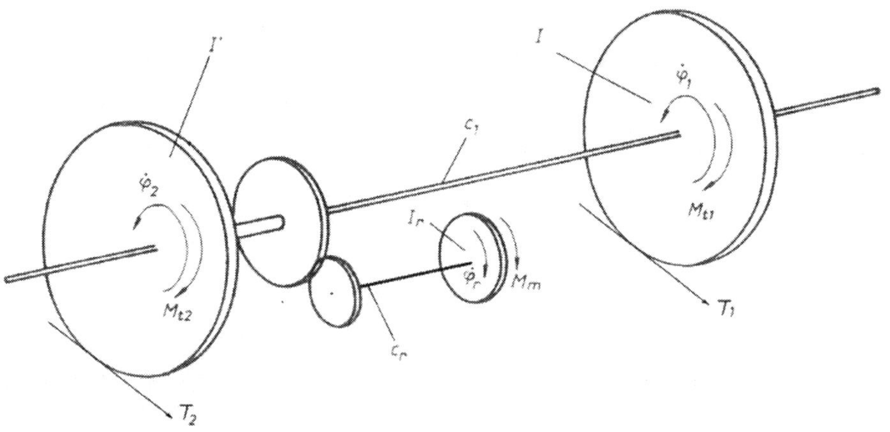

Figure 6. Mechanical model for the study of stick-slip.

The system includes: the engine shaft on which acts the drive moment M_m, a torque shaft of the motor with a stiffness k_c, the assembly consisting of two-wheel gear with the transmission ratio u, transmitting the motion from the engine shaft to the wheelset axle having a stiffness k_1 and the two wheels of the vehicle on which act the moments M_{t1}, M_{t2} of friction forces T_1, T_2.
I_r – is the moment of inertia of the engine shaft;
I - is the moment of inertia of vehicle wheel;
$I'=I+u^2I_r$ - is the moment of inertia of the wheel, including reduced moment of inertia of the two-gear assembly drive axle, axle stiffness portion of it and the axle of the vehicle considering the infinite value;
u is the gear transmission ratio;
φ_r - is the angular displacement of electric engine shaft;
φ_1, φ_2 - angular displacements of the vehicle wheels.

4. Experimental model for the study of stick-slip phenomenon

In the laboratory of the Rolling Stock Department at Politehnica University of Bucharest has been developed a functional model of a motor vehicle bogie on which can be to modelled and experimented the wheel-rail interaction phenomena that occur when a fully equipped bogie runs on the rail (fig. 7,8,9).

Figure 7. Stand for the study of stick-slip phenomenon – general view.

Figure 8. Location of braking rail and force transducers.

Figure 9. Stand for the study of stick-slip phenomenon.

The experimental stand for dynamic performance determination consists of a bogie model fitted with an axle and two wheels driven by an electric motor by means of an elastic shaft. In the lower part of the wheels there are provided two rails operated in the vertical direction by means of a rectangular metal frame provided with a helical screw (Fig. 9) in order to tighten the rails on the wheels. Each frame arm is equipped with four tensiometric marks, the force transducer thus realized having the possibility of measuring the traction-compression force from the bar driving the rails.

The electric motor is of the asynchronous type and has the possibility of modifying the speed, depending on the traction force we want to achieve. Depending on the braking force, the speed of the electric motor will change, even reaching the wheel lock. A tachometric probe with an infrared emitter - receiver system is used to measure the instantaneous speed.

There were determined the accelerations in the vertical and horizontal - longitudinal directions at the left and right journals of the axle by means of piezoelectric accelerometers. The measuring transducer chain is connected to an acquisition system that performs simultaneous 12-channel acquisition.

5. Conclusion

The stick-slip phenomenon, which occurs generally at the start of traction vehicles, has a negative impact on both the mechanical resistance of the elements of the axle drive system and the traction performance of the vehicle.

The mechanical model proposed for studying the stick-slip phenomenon was validated by the experimental results obtained on the stand in the laboratory.

It is worth mentioning that experimental determinations have also been made on a locomotive in starting regime. The measurements were consistent with those obtained in the laboratory, thus confirming the correctness of the mechanical model.

Acknowledgements

This work was supported by a grant of the Romanian National Authority for Scientific Research and Innovation, CNCS–UEFISCDI, project number PN-III-P2-2.1-PTE-2016-0008, within PNCDI III.

References

1. Sebeşan I, Mitu A-M, Sireteanu T., On the stick-slip phenomena in traction railway vehicles, Proceedings of the Romanian Academy, series A, vol. 16, nr. 2, 2015
2. Frederich F., Beitrag zur Untersuchung der Kraftschluss beins pruchurgen an schrögrollenden Schienefahrryeugrödern, Disertation TH Braunshweig, 1969
3. Kalker, J. J., Same new results in rolling contact, Vehicle System Dynamics,18, 1989
4. A. HAC, Adaptive control of vehicle suspension, Vehicle System Dynamics, 16, pp. 57–74, 1987
5. D. Stăncioiu, T. Sireteanu, M. Giuclea, A.M. Mitu, Simulation of random road profiles with specified spectral density, Proceedings of the Annual Symposium of the Institute of Solid Mechanics, Bucharest, Romania, pp. 331-338, 2003
6. Tyan, F., Hong, Y.-F., Tu, S.-H., and Jent, W. S., Generation of random road profiles, Journal of Advanced Engineering, pp. 1373 - 1378, 2009
7. S. Turkay, H. Akcay, A study of random vibration characteristics of the quarter-car model, Journal of Sound and Vibration, 282, pp. 111-124, 2005.
8. G. Verros, S. Natsiavas, C. Papadimitriou, Design optimization of quarter-car models with passive and semi-active suspensions under random road excitation, Journal of Vibration and Control, 11, pp. 581-606, 2005.

IOP Conf. Series: Materials Science and Engineering **295** (2018) 012038 doi:10.1088/1757-899X/295/1/012038

Physical and Constructive (Limiting) Criterions of Gear Wheels Wear

Fedorov S V

Kaliningrad State Technical University ,Department of Theory of Mechanisms and Machines and Machine Parts, 236022, Kaliningrad, Russia fedorov@klgtu.ru

Abstract: We suggest using a generalized model of friction - the model of elastic-plastic deformation of the body element, which is located on the surface of the friction pairs. This model is based on our new engineering approach to the problem of friction-triboergodynamics. Friction is examined as transformative and dissipative process. Structural-energetic interpretation of friction as a process of elasto-plastic deformation and fracture contact volumes is proposed. The model of Hertzian (heavy-loaded) friction contact evolution is considered. The least wear particle principle is formulated. It is mechanical (nano) quantum. Mechanical quantum represents the least structural form of solid material body in conditions of friction. It is dynamic oscillator of dissipative friction structure and it can be examined as the elementary nanostructure of metal's solid body. At friction in state of most complete evolution of elementary tribosystem (tribocontact) all mechanical quanta (subtribosystems) with the exception of one, elasticity and reversibly transform energy of outer impact (mechanic movement). In these terms only one mechanical quantum is the lost – standard of wear. From this position we can consider the physical criterion of wear and the constructive (limiting) criterion of gear teeth and other practical examples of tribosystems efficiency with new tribology notion – mechanical (nano) quantum.

Keywords: friction, energy, evolution, dissipation, nanostructure, wear standard.

1. Introduction

We suggest using a generalized model of friction the model of elastic-plastic deformation of the body element, which is located on the surface of the friction pairs. This model is based on our new engineering approach to the problem of friction - triboergodynamics.

Triboergodynamics [1] is an extension (one of its parts) of general Ergodynamics of deformable bodies [2-4]. Ergodynamics is a synthesis to the problem of deformation most general laws of thermodynamics for non-reversable processes, molecular kinetics and dislocation theory in their mutual, dialectical tie on the basis of a most general law of nature – the law of energy conservation at its transformations.

Triboergodynamics is based on modern knowledge of friction too: 1. Friction is a phenomenon of resistance to relative motion between two bodies, originating at their surfaces contact area; 2. Friction is the process of transformation and dissipation of energy of external movement into other kinds of energy; 3. Friction is the process of elasto-plastic deformation localized in thin surface layers of rubbing materials.

Thus, within the framework of triboergodynamics the model of elastic-plastic deformation of contact volumes is examined as a generalized mechanism of transformation and dissipation energy and determines essence of resistance to surfaces displacement.

The major distinction of triboergodynamics from general Ergodynamics of deformed solids is «scale factor» which exhibits itself in existence of critical friction volume. This volume determines the limit friction parameters and separate, in essence, the surface deformation from the traditional volume deformation.

2. Short fundamentals of ergodynamics of deformed solids

2.1. Structural model

Deformable body is considered as an open, multicomponent, substantially nonhomo-geneous and nonequilibrium system with hierarchy of different levels (from submicro- to macrolevel) of metastable structural elements (defects and damages) which are statistically uniformly distributed in the volume. Some of these elements are virtual sources and sinks of elementary defects (vacancies, dislocations, etc.), the others are a barrier to their motion.

The structure state is defined by the basic parameters [4]: γ_σ is overstress factor of interatomic bonds which evaluates nonuniformity of external stresses σ distribution in the bonds $\sigma^0\left(\gamma_\sigma = \sigma^0/\sigma \geq 1\right)$; u_e is the density of latent (free) energy of defects and damages; v is the coefficient of nonuniformity of latent energy distribution in volume which is equal to ratio between latent energy density in local volume u_e^0 and average value of $u_e\left(v = u_e^0/u_e\right)$. A complex structural parameter $k = \gamma_\sigma / v^{0,5} = \sigma_* / S_*$ specifies a relationship between theoretical σ_* and actual S_* strength of a solid body.

2.2. Physical model and structural-energetic interpretation of the process

Macroscopic phenomenon - plastic deformation and fracture of the body element is considered as a cooperation of a huge number of microscopic elementary acts of atomic-molecular regroupings under external force field (mechanical, thermal, electrical, etc.) which are activated by the thermal energy fluctuations. From the thermodynamic point of view, all the mechanisms and structural levels of the process are divided into two most characteristic groups of adaptive and dissipative (relaxation) types. They differ in physical nature and kinetic behavior. The simple acts controlling generation and accumulation of unit defects in deformed body element (damage) are classified as the first group. The specific (referred to unit volume) pumping power of excessive (latent) energy $\dot{u}_e$ is an overall characteristic of the processes rate

$$\dot{u}_e = \frac{du_e}{dt} = A \sinh\left[\left(\alpha\sigma_i^2 - v\,u_e\right)/2kT\right]. \tag{1}$$

The mechanisms and simple acts controlling relaxation (dissipative) processes of plastic deformation are classified as the second group. The specific power of thermal effect $\dot{q}$ of plastic deformation is overall characteristic of the processes

$$\dot{q} = \frac{dq}{dt} = B \sinh\left[\left(\alpha\sigma_i^2 + v\,u_e\right)/2kT\right]. \tag{2}$$

Here A and B are the kinetic coefficient

$$A = \frac{2kT}{hV_0}\sum_1^n U_i'(\sigma_0,T)\exp\left[-\frac{U'(\sigma_0,T)_i}{kT}\right], \tag{3}$$

$$B = \frac{2kT}{hV_0}\sum_1^n U_i''(\sigma_0,T)\exp\left[-\frac{U''(\sigma_0,T)_i}{kT}\right], \tag{4}$$

$$U_i'(\sigma_0,T) = U_{0i}' + \Delta U'(T) \pm \beta\sigma_0^2, \qquad U_i''(\sigma_0,T) = U_{0i}'' + \Delta U''(T) \pm \beta\sigma_0^2, \tag{5}$$

$$\alpha = \frac{\gamma_{\sigma}^2 V_0}{6G}, \qquad \beta = \frac{\gamma_{\sigma}^2 V_0}{2K}, \tag{6}$$

U'_{0i}, U''_{0i} — activation energy of formation and diffusion of the i-th defect; σ_0, σ_i — hydrostatic stress and stress intensity; V_0 — atomic volume; k — Boltzmann constant; h — Planck constant; T — absolute temperature; G, K — shear and bulk modules.

2.3. *Thermodynamic analysis of interrelation between deformation and fracture*

From thermodynamic point of view, the plastic deformation and the fracture are defined by a competition of two opposite interrelated and simultaneous trends (figure 1): growth of latent energy density u_e of various defects and damages which are generated and accumulated in the material due to work done by the external forces ω_p and reduction (release) of the density as a result of relaxation processes in deformed body element. The first trend is concerned with strain hardening and damage of material, the second — with dynamic recovery and dissipation of the strain energy which govern the thermal effect of plastic deformation q.

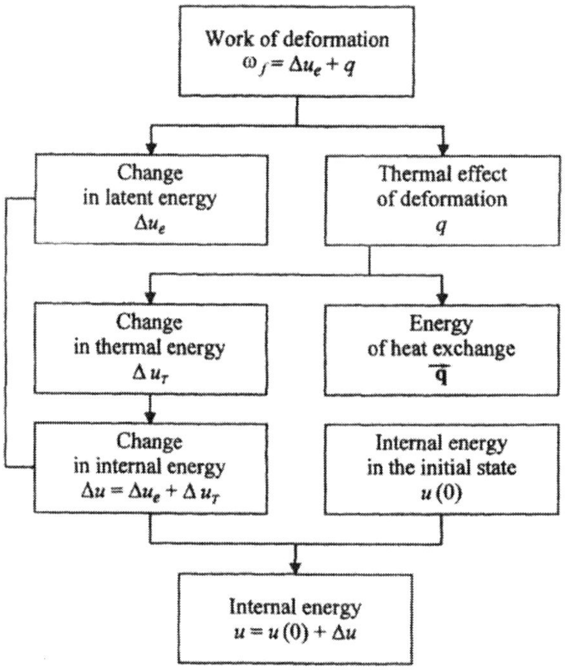

Figure 1. Scheme of the energy balance for the plastic deformation of a solid body [2-4].

A significant portion of the dissipative energy q is not retained in the deformed element, but passes through it and is dissipated in the environment due to heat exchange $\vec{q}$. Only insignificant portion of the energy q is accumulated in deformed element as a heat component of internal energy $\Delta u_T = q - \vec{q}$ increasing its temperature (self-heating effect).
According to conservation law:

$$\omega_p = \Delta u_e + q \quad \text{and} \quad \dot{\omega}_p = \dot{u}_e + \dot{q}. \tag{7}$$

In mechanics of deformable solids the irreversible work ω_p and power $\dot{\omega}_p$ of deformation are related to stress-strain state of the element by

$$d\omega_p = \sigma_i d\varepsilon_i^p, \qquad \dot{\omega}_p = \sigma_i \dot{\varepsilon}_i^p, \tag{8}$$

From (7) and (8) a one-to-one relation follows between stress-strain and thermodynamic states of the element

$$\dot{\varepsilon}_i^p = \frac{\dot{\omega}_p}{\sigma_t} = \frac{1}{\sigma_i}(\dot{u}_e + \dot{q}) = \dot{\varepsilon}_i^e + \dot{\varepsilon}_i^q. \tag{9}$$

Consequently, from thermodynamic point of view, the total values of work ω_p and irreversible strain ε_i^p as well as their rates ($\dot{\omega}_p, \varepsilon_i^p$) may be presented as a sum of two components related to strain hardening and damage $\left(\varepsilon_i^e = \dot{u}_e / \sigma_i\right)$, and dynamic recovery $\left(\dot{\varepsilon}_i^q = \dot{q} / \sigma_i\right)$ controlling quasi-viscous flow of the body element, respectively.

This deduction is of important value in analyzing interrelation between deformation and fracture processes. Only a portion of plastic (irreversible) strain ε_i^e which is controlled by microscopic processes related to strain hardening and accumulation of latent energy of defects and damages is responsible for damage and fracture of the body element. The significant portion of the irreversible strain ε_i^q controlled by relaxation (dissipative) processes does not effect the damage and fracture of the body element and only causes quasi-viscous flow (steady state creep). The relationship between work and extent of irreversible deformation and their components varies in a very wide range and depends on the structure and deformation conditions of the material [2].

2.4. Thermodynamic condition of local fracture

As a parameter of damage (scattered fracture) we shall take the density of internal energy stored in the deformed volume. The energy is defined as a sum of two components: potential (latent) energy u_e and kinetic (thermal) energy u_T that is,

$$\Delta u = \Delta u_e + \Delta u_T, \qquad \dot{u} = \dot{u}_e + \dot{u}_T. \tag{10}$$

The energy is related to static (Δu_e) and dynamic (Δu_T) damages and distortions of crystal lattice in deformed body. Consequently, it is responsible for scattered fracture (damage).

The body element is looked upon as fractured if at least in one local volume responsible for fracture the internal energy density reaches the critical (ultimate) value u_*. This value corresponds to the loss of stability «in great» by crystal lattice. At this instant the cracks of critical size (after Griffith-Orowan-Irwin) and sharp localization of the process at the crack tip occur in a local volume. The thermodynamic condition of local fracture is written as

$$u(\vec{r}_*, t_*) = u(\vec{r}_*, 0) + \int_0^{t_*} \dot{u}(\vec{r}_*, t)dt = u_* = const. \tag{11}$$

Here $u(\vec{r}_*, 0)$— density of internal energy of the material in initial (before deformation, $t = 0$) state; $\dot{u}(\vec{r}_*, t)$— specific power of internal energy sources in local macrovolume of the material responsible for fracture; $\vec{r}_*$— parameter characterizing coordinates (x_*, y_*, z_*) - of the local volume responsible for fracture.

2.5. Thermodynamic criterion of fracture

According to structural-energetic analogy between mechanical fracture and melting of metals and alloys [5] and experimental data [2], the critical value of internal energy u_* in the local volume responsible for fracture agrees with known thermodynamic characteristic of material ΔH_S (enthalpy of melting)

$$u_{*} = \Delta H_{S} = \int_{0}^{T_S} c_p dT + L_S \,. \tag{12}$$

Here T_S — melting temperature; c_p - specific heat; L_S - latent melting heat.

2.6. Relationship between force and energy criteria of local fracture

The analysis of kinetic equation of state (damage) (1) indicates that the real solid body approachs the stationary (stable) state under constant action of external fields ($\sigma_0 = const$, $\sigma_i = const$, $T = const$) if

$$\Delta u_e = const \qquad \text{and} \qquad \dot{u}_e = 0\,. \tag{13}$$

From the kinetic Equation (1) under condition (13) an important consequence follows

$$\sigma_i = \sigma_S, \quad \sigma_S = \left(\frac{\nu\, u_e}{\alpha}\right)^{1/2} = \frac{1}{K_a}\left(6Gu_{e^*}\right)^{1/2}, \tag{14}$$

according to which the structure state of material σ_S (α, u_e, ν) adapts (shakes down) in a stable stage to external conditions, which are defined unambiguously by deviatory component of the stress tensor $\sigma_i(\sigma_S = \sigma_i)$. Relationship (14) generalizes the known proposition of dislocation theory on mutual relation between the flow stress σ_S and the density of latent (stored) energy u_e [6] for the case of combined stress state. The material damage u_e in the local volume responsible for fracture becomes critical, therefore, relationship (14) makes it possible to estimate the actual strength S of the material.

$$S = \left(\frac{\nu\, u_{e^*}}{\alpha}\right)^{1/2} = \frac{1}{K_a}\left(6Gu_{e^*}\right)^{1/2}. \tag{15}$$

Here

$$u_{e^*} = u_e - u_{T_0} = \Delta H_S - \int_{0}^{T_S} c_p dT\,. \tag{16}$$

From (15) and (16) under $k_\sigma = 1.0$ and $T = 0$ the theoretical shear strength is:

$$\sigma_{*} = \left(6G\Delta H_S\right)^{1/2} = \left(\frac{3E\Delta H_S}{1+\mu}\right)^{1/2}. \tag{17}$$

Here E, μ — elasticity modulus and Poisson's ratio.

3. Triboergodynamics method

3.1. Structural-energetic interpretation of friction process

It is known friction is characterized a product of frictional forces F by friction distance ℓ, that is, the work ω_f, expended on overcoming frictional force:

$$\omega_f = F\ell\,, \tag{18}$$

$$\omega_f = \Delta u_e + q\,, \tag{19}$$

$$\dot{\omega}_f = \dot{u}_e + \dot{q}\,. \tag{20}$$

Here $\omega_f = d\omega_f / dt$ is a power of friction dissipation of energy; $\dot{u}_e = du_e / dt$ is the rate of storing latent energy in deformed (contact) volumes; $\dot{q} = dq / dt$ the power of thermal effect of plastic deformation (friction).

Since the contact volumes of both materials of the friction pair (figure 1) are deformed, Equations (19) and (20) should be rewritten as:

$$\omega_f = \Delta u_{e1} + \Delta u_{e2} + q_1 + q_2\,, \tag{21}$$

$$\dot{\omega}_f = \dot{u}_{e1} + \dot{u}_{e2} + \dot{q}_1 + \dot{q}_2. \tag{22}$$

These equations show, that from thermodynamic point of view, the work ω_f of friction forces, (friction power $\dot{\omega}_f$) is related to plastic deformation of the contact volumes. The work ω_f may be divided conventionally into two specific parts.

The first part is related to variation of the latent (potential) energy (Δu_{e1} and Δu_{e2}) in deformed (contact) volumes. This is the energy of various simple defects and damages which are generated and accumulated in the bulk. This energy is unique and the total characteristic of submicro- and microstructural variations occurring in plastically deformed volumes [2, 3].This is a measure of strain hardening and damage of material.

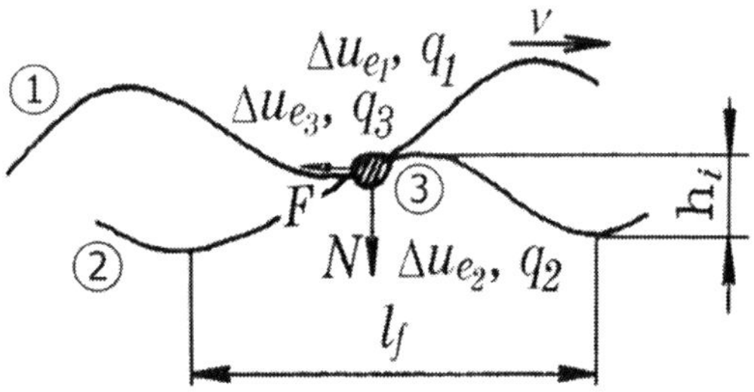

Figure 2. Schematic view of friction's contact [1].

The second part of the friction work ω_f is related to dynamic recovery which is accompanied by releasing latent energy and thermal effect of friction (q_1, q_2).This energy involves displacement and annihilation of various simple defects of opposite sign terminating at the free surface, healing reversible submicroscopic discontinuities, etc.

The relations between Δu_{e1} and Δu_{e2}, as well as q_1 and q_2 are defined by physico-chemical properties of the materials of the friction pair, their structure and friction conditions.

Since the contact volumes (not unit sizes) of the materials forming a friction couple become strained by friction (figure 2), equations (1) and (2) can be rewritten in the form

$$W_f = \Delta U_e + Q = \Delta U_{e_1} + \Delta U_{e_2} + Q_1 + Q_2, \tag{23}$$

$$\dot{W}_f = \dot{U}_e + \dot{Q} = \dot{U}_{e_1} + \dot{U}_{e_2} + \dot{Q}_1 + \dot{Q}_2, \tag{24}$$

where $\Delta U_e = V_f \Delta u_e$; $\dot{U}_e = V_f \dot{u}_e$; V_f - is the deformable (friction) volume.

Solving equations (23) and (24) for the frictional force F, one obtains:

$$F_l = \frac{\Delta U_{e_1} + \Delta U_{e_2}}{l} + \frac{Q_1 + Q_2}{l}, \tag{25}$$

$$F_v = \frac{\dot{U}_{e_1} + \dot{U}_{e_2}}{v} + \frac{\dot{Q}_1 + \dot{Q}_2}{v}, \tag{26}$$

where l and v are the friction path and the slip velocity.

Dividing equations (25) and (26) by the normal force N gives generalized equations for the friction coefficient:

$$\mu_l = \frac{\Delta U_{e_1} + \Delta U_{e_2}}{Nl} + \frac{Q_1 + Q_2}{Nl}, \tag{27}$$

$$\mu_v = \frac{\dot{U}_{e_1} + \dot{U}_{e_2}}{Nv} + \frac{\dot{Q}_1 + \dot{Q}_2}{Nv}. \tag{28}$$

Therefore, the friction is generally described by the energy balance equation and with thermodynamical point of view [1-3] is the process of two interrelated, oppositely directed and concurrent trends operating in a strained contact. According to the energy balance scheme (figure 1) for plastic deformation and fracture [2] presented above, equations for friction work W_f, frictional force F and friction coefficient μ (without lubrication) has view:

$$W_f = \Delta U_e + Q = \Delta U_{e_1} + \Delta U_{e_2} + \Delta U_{T_1} + \Delta U_{T_2} + \vec{Q}_1 + \vec{Q}_2, \tag{29}$$

$$\dot{W}_f = \dot{U}_e + \dot{Q} = \dot{U}_{e_1} + \dot{U}_{e_2} + \dot{U}_{T_1} + \dot{U}_{T_2} + \dot{\vec{Q}}_1 + \dot{\vec{Q}}_2, \tag{30}$$

$$F_l = \frac{\Delta U_e}{l} + \frac{Q}{l} = \frac{\Delta U_{e_1} + \Delta U_{e_2}}{l} + \frac{Q_1 + Q_2}{l}, \tag{31}$$

$$F_v = \frac{\dot{U}_{e_1} + \dot{U}_{e_2}}{v} + \frac{\dot{Q}_1 + \dot{Q}_2}{v} = F_{mechanical} + F_{molecular}, \tag{32}$$

$$\mu_l = \frac{\Delta U_{e_1} + \Delta U_{e_2}}{Nl} + \frac{Q_1 + Q_2}{Nl} = \mu_{adapt} + \mu_{dis} = \mu_{adapt} + \mu_{T(dis)} + \mu_{\vec{Q}(dis)}, \tag{33}$$

$$\mu_v = \frac{\dot{U}_{e_1} + \dot{U}_{e_2}}{Nv} + \frac{\dot{Q}_1 + \dot{Q}_2}{Nv} = \mu_{deformation} + \mu_{adhesion}, \tag{34}$$

where $\Delta U_e = V_f \Delta u_e$; $Q = V_f q$; $\vec{Q} = V_f \vec{q}$; $\dot{U}_e = V_f \dot{u}_e$; $\dot{u}_e = \mathrm{d}u_e/\mathrm{d}t$ - is the rate of latent energy density change in the contact volumes; V_f - is the deformable (friction) volume; μ - friction coefficient; μ_{adapt} - adaptive friction coefficient; $\mu_{T(dis)}$ and $\mu_{\vec{Q}(dis)}$ - statical and dynamical components of dissipative friction coefficient; ΔU_T - thermal component of internal energy; N - normal load; l - distance of friction; v -sliding velosity. The latent energy density Δu_e is an integral parameter of tribostate and damageability (failure (Δu_e^*)).

Thus, viewed thermodynamically, the work done by friction forces W_f (the friction power $\dot{W}_f$), the friction force F and the friction coefficient μ may be classified conventionally into two specific components with different kinetic behavior [3]. The first component is associated with microscopic mechanisms of adaptive type and relates to the change of latent (potential) energy ($\Delta u_{e_1}, \Delta u_{e_2}$) of various elementary defects and damages that are generated and accumulate in the deformable volumes of materials friction pair (figure 1). This energy is a unique and integral characteristic of the submicro- and microstructural transformations that occur in plastically strained materials [2-4]. This energy is a measure of strain hardening and damageability of materials. The second component is associated with microscopic mechanisms of dissipative type and relates to dynamic recovery processes in which latent energy is released and heat effect of friction (q_1, q_2) take place. This energy originates in the motion and destruction of various elementary defects of opposite signs, the egress of these defects to the surface, the healing of reversible submicroscopic discontinuities, etc. The ratios of the components Δu_{e_1} and Δu_{e_2} as well as $q_1 q_2$ of the balance vary over a wide range, depending on the physical, chemical, and structural properties of the materials that comprise the friction couple and the friction process conditions.

Thus, the thermodynamic analysis of friction (plastic deformation and fracture) has led to generalized (two-term) relations for the force F and coefficient of friction μ, which agrees with current concepts of the nature of friction [7,8] – molecular-mechanical theory (32) and deformable-adhesion theory (34). But it is more correct to speak about the adaptive-dissipative nature (model) of friction (33).

Relationships (21)-(28) which generalize the mechanism of energy dissipation at friction allow to classify the tribosystem states. According to ergodynamics of deformed solids (relationships $\Delta u = \Delta u_e + \Delta u_T$ and $q = \Delta u_T + \vec{q}$) and equations (23)-(24) may be transformed to:

$$W_f = \Delta U_{e1} + \Delta U_{e2} + \Delta U_{T1} + \Delta U_{T2} + \vec{Q}_1 + \vec{Q}_2, \tag{35}$$

$$\dot{W}_f = \dot{U}_{e1} + \dot{U}_{e2} + \dot{U}_{T1} + \dot{U}_{T2} + \dot{\vec{Q}}_1 + \dot{\vec{Q}}_2. \tag{36}$$

As follows from equations of energy balance (35), (36), all exhibitions of friction and wear may be reduced conventionally at least to two basically different states: the first state defines all types of damage and wear, the second — the so-called "wearless" condition.

The state of damage and wear is characterized by the components of energy balance (35), (36), which are responsible for accumulation of internal energy in deformed volumes $\Delta u = \Delta u_{e1} + \Delta u_{e2} + \Delta u_{T1} + \Delta u_{T2}$, i.e. the process is irreversible. The "wearless" state is characterized by the components responsible for dynamic dissipation (reversibility) of strain energy into elastic and structural dissipated energy of friction contact $\vec{q} = \vec{q}_1 + \vec{q}_2$.

In its turn, the first state may be classified depending on the relation between potential Δu_e and kinetic Δu_T components of internal energy. It is subdivided conventionally into mechanical damage and wear (due to so-called structure activation) and thermal damage and wear (due to thermal activation). For instance, let the thermal component of internal energy Δu_T be equal to zero ($\Delta u_T = 0$) and the internal energy variation at damage and wear be defined only by variation of the potential component $\Delta u_e (\Delta u = \Delta u_e)$. Then, the mechanical damage and wear with brittle fracture of surfaces take place. On the contrary, if we have $\Delta u_e = 0$ ($\Delta u = \Delta u_T$), then the thermal damage and wear with ductile fracture of surfaces take place. All the intermediate values of the components are associated with quasi-brittle or quasi-ductile fracture of solids.

In the most general case, the energy balance at dry friction (23) should be written as:

$$W_f = \Delta U_{e1} + \Delta U_{e2} + \Delta U_{e3} + Q_1 + Q_2 + Q_3 \quad . \tag{37}$$

In the special case, where the friction is localized into volume of the "third body" (figure 2) equation (37) develops into:

$$W_f = \Delta U_{e3} + \vec{Q}_3. \tag{38}$$

Here $\Delta U_{e3} = V_3 \Delta u_{e3}$.

3.2. Energy interpretation of Leonardo da Vinci (Amonton's) friction coefficient

According to thermodynamic theory of strength [2], the structure parameter should be related to the portion of the accumulated plastic deformation that is responsible for strain hardening. This portion is uniquely and integrally defined by the density of the potential component of internal energy (that is, the latent energy density Δu_e) of various defects and damages that accumulate in a plastically strained material. With this in mind, if we neglect the heat effect Q of friction, one will infer from the thermodynamic analysis of friction of equations (27)-(28) that the Amonton (Leonardo da Vinci) friction coefficient is

$$\mu = \frac{\Delta U_e}{\mu^* Nl} = \frac{F}{N}; \quad F = \frac{\Delta U_e}{l}; \qquad Q \cong 0, \quad \mu^* = 1. \tag{39}$$

Consequently, the coefficient of friction has a very deep physical sense. On the one hand, it is the parameter which generally characterizes the resistance of relative displacement (movement) of surfaces, for it reflects the portion of energy, which «is done by friction away» as accumulated latent energy ΔU_e, by relation to parameter of external forces work $\mu^* Nl$ (energy of external relative movement). On the other hand, it is the generalized characteristic of damage, for it is defined of the latent energy density Δu_e as integral characteristic of the structure defectiveness measure, because this energy is the generalized parameter of damage. Here too, coefficient of friction generally reflects the structural order (disorder) of deforming contact volume, since the parameter $\Delta U_e = \Delta u_e V_f$ is defined of the energy of defects and damages of different types, that are accumulated into contact volumes V_f solids.

Therefore, coefficient of friction is a true and generalized parameter of tribosystem state. From this conclusion we can say that the analysis of the evolution of the states of a tribosystem is primarily an analysis of the latent deformation energy accumulated within the contact friction volumes.

3.3. Generalized experimental friction curves

The dependences obtained for the friction coefficient μ are in agreement with experimental curves $\mu = \mu(N, v)$ (figure 3-5).

A subsequent analyses of modern experimental data using equations (23)-(34) has shown that the experimental friction curves (figure 3-5) of type $\mu = \mu(N, v)$ are generalized friction curves that reflect the evolution (the change in the friction coefficient) of tribosystem.

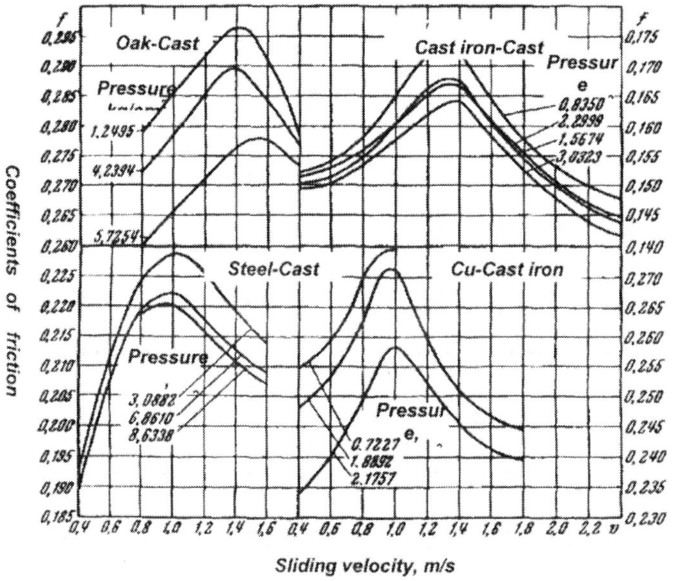

Figure 3. P.Conti's experimental results [9].

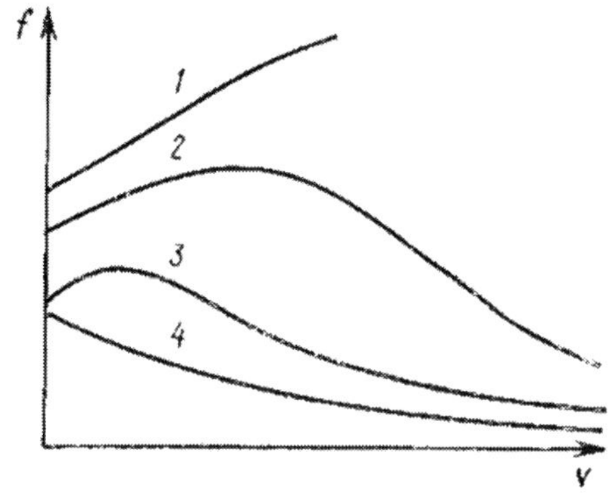

Figure 4. Generalized friction experiments in Kragelsky's interpretation [7]: v - sliding velocity (load: 1-small; 2 and 3 – medium; 4 - considerable).

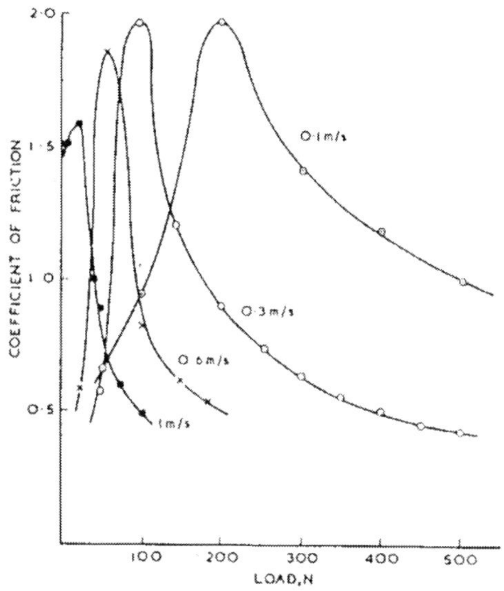

Figure 5. Watanabe's [10] experimental results of Nylon 6 – Steel pair of friction.

3.4. Structural-energy regularities of rubbing surfaces evolution

We propose an energetic interpretation of the experimental friction curves $\mu = \mu(N, v)$ (figure 6). According to our concept [1, 11], the ascending portion of the friction coefficient curve μ is mainly controlled by processes associated with the accumulation of latent energy ΔU_e in various structural defects and damages. Here the increase in μ is due to the increasing density of latent (potential) energy Δu_e and the increasing adaptive friction volume V_f. The descending portion of the friction curve is mainly controlled by processes associated with the release and dissipation of energy $Q = \Delta U_T + \vec{Q}$. Here the decrease in μ is due to the decrease in latent energy density within the friction volume V_f or (which is virtually the same) to the decrease of the adaptive friction volume V_{adapt} $(u_e = u_e^*)$ and to the increase of the dissipative volume V_{dis} $(\vec{q}^* = u_e^*)$.

Evolution of tribosystem, presented as a diagram view (figure 6), has an adaptive-dissipative character (29)-(34) and reflects the competitive (dialectical) nature of friction. Evolution curve has the row of principal points (1-5) of transitional tribosystem states, which strictly obeys the balance principle of friction; there are more characteristic areas of tribosystem behavior between these points. These areas reflect the common properties of nonlinear dynamic of evolution.

So, in figure 6 it is possible to see the following conventionally designated points and stages: 0-1 – a stage of static friction and deformational strengthening; 1 – a point of limit for deformational strengthening; 1-2 – a stage of pumping of excess energy; 2 – a point of gripping (adhesion) and transition of outer friction into internal (critical non-stability): 2-3 – a stage of forming dissipation structures (formation of heat fluctuation in friction volume); 3 – a point of minimum compatibility (maximum frictionness); 1-2-3 - a stage of selforganization; 3-4 – a stage of compatibility; 4 – a point of wearlessness (anormal-low friction); 5 – a point of thermal adhesion.

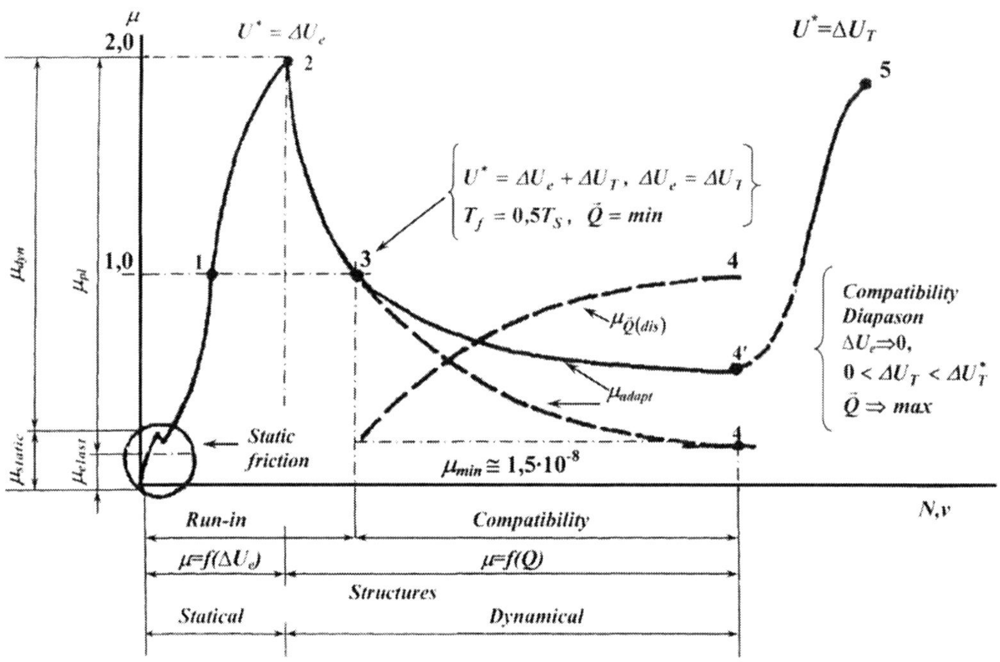

Figure 6. Structural-energy diagram for evolution of rubbing surfaces [1, 14].

An ideal evolution of tribosystem is symmetrical. The process starts and finishes within areas of elastic behavior. A plastic maximum (a superactivated condition) exists between them as a condition of selforganisation and adaptation.

In the most general case evolution (adaptation) regularities of tribosystems may be presented as a 2-stage (figure 6). At the first stage (0-2) of adaptation the evolution of friction contact rushes to form some critical volume of friction V_f^* (point 2). It is elementary tribosystem that is the elementary and self-sufficient energy transformer. The first stage - latent energy density growth Δu_e to a limited magnitude Δu_e^* within critical friction volume V_f^*.

This friction volume V_f^* is constant at the second stage of evolution, but here it is evolutionary developed owing to structural transformation; by this one may realize wide spectrum of compatibility friction structures (figure 6). The second stage (2-4) – structural transformation of critical friction volume (elementary tribosystem) V_f^* into adaptive V_{adapt} and dissipative V_{dis} volumes (figure 7). The limit (point 4) of this stage is characterized by a full transformation of adaptive critical volume V_{adapt}^* into V_{dis}^* dissipative.

The volumes mentioned above characterize different regularities of transforming energy of outer mechanical movement at friction. Adaptive volume V_{adapt} is connected with non-reversible absorption of deformation energy. It is in this volume where latent deformation energy Δu_e accumulates and where the centres of destruction initially emerge (birth). Dissipative volume V_{dis} is capable of reversible transformation (dissipate) of outer movement energy. It doesn't accumulate latent deformation energy owing to reversible elastic-viscous-plastic deformation.

Suggested theoretical and calculation evaluation [1, 11] showed that dissipative friction volume performs reversible elastic energy transformation of outer mechanical movement with density $\vec{q}^*$ equal to critical density of latent energy u_e^*.

Culmination of tribosystem evolution is its final and limited condition of point 4 – a state of anomalously low friction and wearlessness (maximum efficient).
A schematic evolution of the contact volume of friction in diagram's points 1-5 is presented in figure 7.

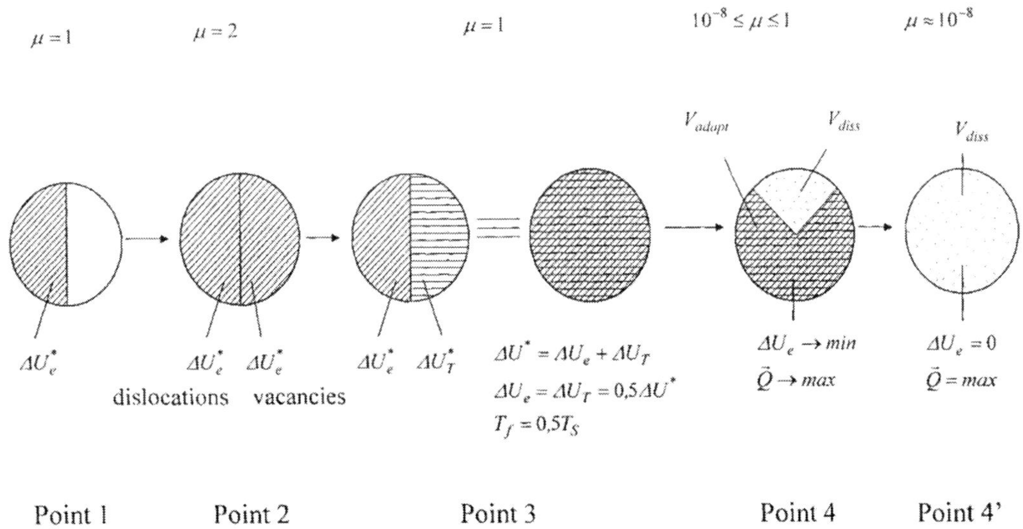

Figure 7. A schematic evolution of the contact volume of friction in diagram's points 1-5 [1].

Calculation shows [1] that at an ideal tribosystem evolution an adaptive (Amontons) friction coefficient μ_{adapt} in a point 2 of a diagram falls abruptly down, reaching in a point 4 the value of elastic friction coefficient μ_{elast}. For point 4 of compatibility area 3-4 an equation of energy balance (33) showed be put in the following way:

$$\mu_{adapt} = \mu^* - \mu_{dis} = 1 - \mu_{dis} = \mu_{plast} = 0 = \mu_{elast}\,;\, \mu^* = 1{,}0\,. \qquad (40)$$

Thus, point 4 stands for an ideal evolution of contact friction volume a condition of ideal elastic-viscous-plastic deformation. Equation (40) shows as a matter of fact exactly it, i.e. Amontons friction coefficient μ_{adapt} being in its essence plastic friction coefficient μ_{plast} has a minimum value equal to zero. It follows then, that plastic friction became elastic with friction coefficient μ_{elast}. It means that plastic deformation of contact volume friction is implemented with the maximum dynamic dissipation ($\vec{Q} = \max$) of accumulated latent energy. That is why the value of accumulated energy in point 4 is equal to zero ($\Delta U_e = 0$). This fact proves an ideal condition at full evolution of contact volume. From the physics point of view this condition may be explained by the full dissipation of accumulated energy ΔU_e^* in point 2 and by newly emerged structures of point 4 in the form of elastic energy of interaction between them (dynamic dissipation energy $\vec{Q}^*$). Here $\mu_{dis} = 1{,}0$. The structural elements themselves are defectlessness - $\mu_{adapt} = 0$, and friction is elastic - $\mu = \mu_{elast}$.

It has been demonstrated [1] that value of minimum adaptive friction volume $V_{adapt}^{\min}$ corresponding to the zero meaning of plastic friction component μ_{adapt} is not equal to zero, but is equal to some minimum structural element of deformed solid body.

3.5. About mechanical (nano) quantum of dissipative friction structures

The result of ideal elementary tribosystem (contact) evolution is forming of unique nanostructure – a mechanical (nano) quantum. Strict notions about mechanical quantum have been obtained [1] considering equation of quasiideal solid body for point 4 of diagram of friction evolution –

$$\vec{Q}^* = \vec{S}_Q T = \mu_{dis}^* N l_f = V_f^* u_e^* = V_f^* \vec{q}_*, \tag{41}$$

which is particular case of solving equation of energy friction balance (29) at $\mu_{adapt} = 0$ and $\mu_{dis} = 1 = \mu_{dis}^*$. Here $\vec{S}_Q$ – inertia entropy of compatible friction volume; T - characteristic temperature of contact friction volume; l_f - linear size of elementary contact.

Correspondingly, in conditions of maximum compatibility (point 4) when tribosystem implements full evolution cycle of adaptation with formation of most perfect dissipative structure, the behaviour of structure is subject to equation of quasiideal solid body condition. So, it is to be presumed that, interaction between elements of this structure, are minimized – a condition of ideal elasticity in dynamics. Equation (31) with taking into account Plank-Boltzmann formula $S = k \ln W$ and real number of atoms oscillators N_f in the volume of elementary tribosystem (contact) V_f^* is brought to the form explaining friction regularities from the point of view of system evolution:

$$\mu_{\text{diss}} = \frac{\vec{S}_Q T}{N l_f} = \frac{k T N_f \ln W}{N l_f}, \tag{42}$$

$$1 \mu_{adapt} = 1 - \mu_{diss} = 1 - \frac{k T N_f \ln W}{N l_f} = 1 - \frac{\vec{S}_Q T}{N l_f} = \frac{S_U T}{N l_f}, \tag{43}$$

Where k - Boltzmann constant; W - condition probability; S_U - configuration entropy of friction, contact volume.

Tribosystem always tends to some optimal condition, characterized, i.e. to a most probable condition $W' = N_f \ln W$ for the given friction conditions.

Analysis and solution of these equations [1] allows to demonstrate the principle of constant probability value (parameter of tribosystem condition (order)) W for the whole range of compatible friction precisely $\ln W = 3$ and $W = e^3 = 20{,}08553696...$

The value of thermodynamic probability W equal to 20,08553696... was interpreted [1, 11-13] as a minimum value of linear, atomic oscillators in one of three directions of minimum adaptive friction volume $V_{adapt}^{\min}$ corresponding to condition of practically absolute elastic friction – anomalously-low friction (safe deformation threshold). Then the number of atomic oscillators in this volume equals $V_Q = \left(e^3\right)^3 = (20{,}08553695...)^3 = 8103{,}083969$ atom's oscillators.

It is the universal size (volume) of mechanical quantum [1, 11-13].

On the other hand, adopting the meaning of Boltzmann entropy S, a universal friction constant $R_f = k N_f$ [1,11-13] is obtained, which characterizes in physical meaning «energetical size» of elementary tribosystem (TS), containing in ideal conditions the same number of atomic oscillators N_f (mechanical quanta N_Q):

$$R_f = k \cdot N_f = k \cdot W^3 \cdot N_Q = R_{MQ} \cdot N_Q, \qquad \frac{J}{grade \cdot TS}; \tag{44}$$

$$R_{MQ} = k \cdot W^3, \qquad \frac{J}{grade \cdot MQ}, \tag{45}$$

where R_{MQ} - universal constant of deformation at friction.

As it follows from calculations [1] the size of minimum adaptive friction volume $V_{adapt}^{\min}$ coincides in its value with the size of submicroscopic area in crevice mouth, which is equal for metals $(4...9)\cdot 10^{-6}$ mm, i.e. of critical volume size responsible to fracture. Thus the size of minimum adaptive friction volume $V_{adapt}^{\min} = V_{elast}$, can be presented as the size of some mechanical quantum.

Table. Parameter W for Metals and Steels [1].

Metals and steels	$E\times 10^{-3}$, MPa	$(u_e^*)\Delta H_S \times 10^{-3}$, MJ/m^3	$E/3E_r$
Cr	235,4	8,5	27,69
Mg	44,4	1,9	23,37
Ag	79,0	3,7	21,35
Au	78,7	4,0	19,67
Co	200,1	10,6	18,88
Fe	211,4	9,9	21,35
Ta	184,4	10,6	17,39
Ti	105,9	6,7	15,8
Nb	104,0	9,2	11,3
Zr	95,6	5,7	16,77
Mo	316,9	12,0	26,4
W	392,4	14,4	27,25
Ni	201,1	9,4	21,39
Iron	210,9	10,1	20,88
20	200,1	9,5	21,06
1Kh13	206,0	8,9	23,14
3Kh13	218,8	9,2	23,78
Kh18N9T	199,1	9,4	21,19
Kh18M9	199,1	9,6	20,74
30Kh	214,1	10,2	20,99
30N3	207,5	10,3	20,11
40	209,4	9,7	21,58
30G2	207,2	10,0	20,72
30KhGN3	208,0	10,2	20,4
G13	204,0	10,0	20,4
50S2G	196,2	10,3	19,05
U8	198,0	10,3	19,22
U12	198,0	10,4	19,04

$\Delta H_S = 3E_r$, $E/3E_r$=20,77 $E/3E_r = 20,77$.

This mechanical quantum constitutes a minimum number of atoms capable to provide such a configurational distribution (structure) which obtains the property of reversibly taking and dissipating (recovering) energy of outer mechanical movement. It also constitutes minimum structure in conditions of plastic deformation and it is formed at tribosystem transition (deformed volume) through an ultimately activated (critical) condition (see figure 6) due to development of selforganisational tribosystem adaptation processes. Mutual rotation-oscillation movement of these mechanical quanta in respect of each other within elementary tribosystem (contact) determines condition of most perfect dissipative friction structure. Properly speaking, such condition is described by equation of quasiideal solid body condition (41), a condition when interaction between structural elements (mechanical

quanta) is minimized – a condition of ideal elasticity of quasiviscous flow. Calculation friction coefficient between quanta equals about 10^{-8} [1, 11, 13].
A conclusion that mechanical quantum constitutes a minimum structural form at plastic deformation (friction) is supported by calculation. If values of elasticity modules E correspond to atomic (true) elastisities E_r then values equal to 60 are obtained, where $60 = 3W$ can be interpreted as a characteristic of volume elasticity of one mechanical quantum – minimum adaptive friction volume V_{adapt}^{min}. Calculation assessment of parameter $W \cong 20 = E / 3E_r$, done for various metals and steels gives an average value 20,77 ((Table)); $\Delta H_S = 3E_r$ - entalpy of melting.
A conclusion is made [1] that the number of atoms (mechanical quantum (MQ)) within volume of one elementary tribosystem (TS) in conditions of ideal tribosystem evolution is a constant value. Thus, it is possible to speak about the quantity of substance equal by mass to one elementary tribosystems and to one mechanic quantum.

3.6. Synergism of tribosystem and optimum states

Mechanical quantum is dynamic oscillator of dissipative friction structure. An ideal quasielastic contact condition at its full evolution constitutes effect of most fully dissipated energy of outer mechanical movement throughout newly formed (by mechanism of selforganization) structural elements – mechanical quantums (dynamic oscillators) which most fully realize their rotationary – oscillatory behavior in relation to each other within elementary tribosystem volume. Their resistance to relative interaction here is minimally elastic and corresponds to elasticity of ideal atomic (thermodynamically balanced) interactions at the level of electron orbits.
Universal constants of mechanical quantum and elementary tribosystem (material point) determine quantum model of surface damping:

$$\mu_{dis} = \frac{3R_{MQ}T_{n_i}}{Nl_f} = \frac{U_{1Q^{n_i}}}{U_{1Q^{n_*}}} = \frac{n_i}{n_*} = 1 - \mu_{adapt}; \tag{46}$$

$$\mu_{adapt} = 1 - \frac{n_i}{n_*} = \frac{n_{dest}}{n_*}, \tag{47}$$

taking into account destruction quantums n_{dest} (non-reversible process component) and damping quantums n_i (reversible, elastic component – fatigue number), and also probability evolution tribosystem model to a most ordered condition:

$$\mu_{adapt} = 1 - \mu_{dis} = 1 - \frac{R_f T \ln W_i}{Nl_f} = 1 - \frac{\ln W_i}{\ln W_*}, \tag{48}$$

where $3R_{MQ}T = U_{1Q}$ - energy of one mechanical quantum; W_i and W_* - current and ultimate probabilities of tribosystems compatibility conditions.
According to a model of quantum surface damping at friction in state of most complete evolution (adaptation) of elementary tribosystem all mechanical quantums with the exeption of one elasticity and reversibly transform energy of outer impact (mechanic movement). One mechanical quantum of radiation (≈ 8103atoms) – is a minimum loss (essence of wearlessness or other wear primary standard).
Linear size of quantum is equal to diameter of spherical ideal crystal with atomic roughness:

$$D_{MQ} = 2 \cdot W \cdot \bar{d}_a \cdot (3/4 \cdot \pi)^{1/3} = 7{,}177 nm. \tag{49}$$

Here $\bar{d}_a$ - mean atomic diameter for metals; $W = e^3$ - parameter of state for mechanical quantum [1].
Mechanical quantum (figure 8) can be examined as the elementary nanostructure of metal's solid body.

Figure 8. Model of elementary nanostructure of friction (8103 atomic cubical cells) [1, 11-14].

Calculations have shown [1] the number N_Q of such mechanical «quanta» (subtribosystems) within the elementary tribosystem's volume $V_f^* = V_{dis}^*$ to be $0{,}63 \cdot 10^8$, which is close to the safe number n_* of fatigue cycles.

According to the quantum damping model of surfaces under friction, when we have the state of more full evolution of elementary tribosystem, the all mechanical quanta to elastic and reversible transform the energy of external mechanical motion. Only one mechanical quantum (8103 atoms) is the minimum loss (the essence of «wearlessness»).

In these terms (point 4) only one mechanical quantum [1,11-14] is the lost – standard wear. The tribosystem (friction contact) has the ideal damping properties – «wearlessness».

The principle of mechanical quantum determines nanoquantum levels of all friction parameters of compatible (optimal) tribosystems and other.

4. Nanoquantum models of tribosystem maximum capacity for work

4.1. Gear wear calculation principle

The all parameters of compatibility (optimal) friction have to be in quanta levels - commensurable with the parameters of the one mechanical quantum – standard of wear.

So, all heavy-loaded tribosystems it is necessary to examine with position of tribosystem ideal evolution. This ideal state of tribocontact is true indicator of tribosystem state for practical examples of tribology. It is the standard of maximum tribosystems efficiency - anomalously low friction and wearlessness.

The state of friction contact under its most full evolution is the characteristic with exploitation of hard loaded Hertzian contact, for example, on the surfaces of gear wheels teeth and systems of wheel-rail and other. We can examine the active surface of gear wheel, which consist of equilibrium spherical form asperities after run-in. During one revolution of gear wheel each asperity of gear wheel teeth is loaded one time too. Under it the loss of one friction contact is equal to one mechanical (nano) quantum. Therefore, the whole contact volume is fatigue failured during about 63 millions cycles. The linear wear h_* of gear wheel is equal to diameter size $Q_{TS} = 2{,}85 \cdot 10^{-6}$ m of an equilibrium friction volume V_f^* (figure 9) [15]. It is the physical criterion of wear. One may understood that the constructive (limiting) criterion of gear teeth is equal to the limit clearance between tooth surfaces. For example, it is about $0{,}3 \div 0{,}4$ modulus of gear wheel tooth wear when we have the loss of bend strength of teeth.

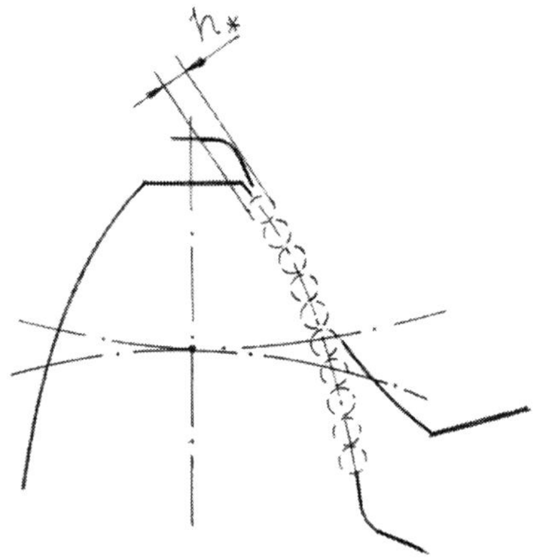

Figure 9. Model of an active surface of gear wheel with equilibrium spherical form asperities.

Thus, an elementary nano-structure of deformed solids may examine as the standard of wear and to apply with optimization the life time of real hard pressed Hertzian contact systems.

4.2 Estimation of bearing capacity for work of internal combustion engines

Take the engine with a frequency of $n = 1500\,\text{min}^{-1}$ shaft rotation. Take the limit wear (linear) of bearing is equal to - $h^* = 0{,}1\,\text{mm}$. We know the line size elementary tribosystem- $D_{TS} = 2{,}85\,\text{mkm} = 2{,}85 \cdot 10^{-6}\,m$. For every revolution shaft one elementary tribosystem (equilibrium, run-in contact) loses one mechanical quantum. The number of turns required to wear one elementary tribosystem equals the number of mechanical quantum in this tribosystem.

So, there are $n_{MQ} = 0{,}63 \cdot 10^8$ revolutions.

Now you can define the time of wear for one elementary tribosystem:

$$t_{TS} = \frac{n_{MQ}}{n} = \frac{0{,}63 \cdot 10^8}{1500} = 42000\,\text{min} = \frac{42000}{60} = 700\,\text{hour} = \frac{700}{24} = 29{,}166\,\text{day} = \frac{29{,}166}{365} = 0{,}0799\,\text{year}$$

Now let's define the number of layers of elementary tribosystems into linear wear - 0,1 mm :

$$a_{h^*} = \frac{h^*}{D_{TS}} = \frac{1 \cdot 10^{-4}}{2{,}85 \cdot 10^{-6}} = 0{,}35 \cdot 10^2 = 35.$$

Now, let's define the time of wear of shaft-bearing system with the ultimate linear given wear - $h^* = 0{,}1\,\text{mm}$, namely:

$t_{motor} = t_{TS} \cdot a_{h^*} = 0{,}0799 \cdot 35 = 2{,}7968\,\text{year}$.

Total we have 2,7968 years of continuous operation on limit load.

If you work 8 hour per day, then we get the following result:

$2{,}7968 \cdot 3 = 8{,}39\,\text{year}$.

This is a very real result for modern cars. And if the ride is not 8 hours per day and less durability increases considerably-really to decades.

For this result we have the wear rate - $i = 4\ nm/h$. It is good correspondence with the data for wear rate - $i = 5\ nm/h$ by Prof. F. Franek [15].

4.3 The principle of critical velocity of rolling wheels

This speed limit is determined by the principle of filling the entire nominal friction sliding system area with elementary tribosystems, damping process. Above this speed happens full unloading tribosystem, detachment of wheels from the surface of the rail as distorts the principle of minimum resistance to movement (the principle of one elementary tribosystem or irreversibility). In this case, all mechanical quantuns of elementary tribosystem will repel the wheel. There will be no quantum which activates a process to maintain the system in an excited state.
The calculation will be performed in the following order. Denote elementary nominal contact area. By definition [1], on elementary, nominal area of tribosystem can accommodate and work $n_*^{TS} = 0{,}63 \cdot 10^8$ elementary tribosystems. Each elementary tribosystem (for model of spherical roughness) has a size of $D_{1TS} = 2{,}85 \cdot 10^{-6}\ m$ and is able to implement rolling wheel path in the elementary act of rolling at length of this tribosystem.
Thus, if per unit of time all elementary tribosystems work on elementary nominal area of friction time, then the path passable wheel per unit of time, equal to
$L_{\Sigma TS} = D_{1TS} \cdot \overset{*}{n}_{TS} = 2{,}85 \cdot 10^{-6} \cdot 0{,}63 \cdot 10^8 = 179{,}55\ m$.
Consequently, the critical speed of wheel rolling is equal
$v_* = L_{\Sigma TS} \cdot 3600 = 646{,}38\ km/h$.
This result is close to modern speed of 574,8 km/h (TGV, France).

5. Conclusions

5.1. Structural-energy analysis of the friction process allows us to examine the friction process as the evolution process;
5.2. From the energy balance equations of friction follows that the evolution of tribosystem (contact) has an adaptive-dissipative character.
5.3. Experimental friction curves of $\mu = \mu(N, v)$ type may be examined as generalized friction experimental curves;
5.4. The fuller evolution of tribosystem has symmetrical view - the friction process is started and finished within elastic area.
5.5. Under fuller evolution of friction contact (elementary tribosystem) the unique nanostructure is formed; the basis of this structure is the mechanical (nano) quantum and the contact (material point of mechanics) consists of about $0{,}63 \cdot 10^8$ such quantums.
5.6. We can examine the mechanical quantum as the least structural form of solid material body and the standard of wear.
5.7. All parameters of compatibility (optimal) friction have to be in quanta levels - commensurable with the parameters of the one mechanical quantum.
5.8. Interaction between nanoquantums is nature the net elasticity. The value of the coefficient of friction between mechanical quantums has order $\mu_{MQ} = 1{,}587 \cdot 10^{-8}$.
5.9. Exploitation of gear wheels and other heavy-loaded tribosystems (Hertzian contact) are subjected to model of nanoquanta damping, when one mechanical quantum is the standard of wear (subtribosystem).

References

1. Fedorov, S.V., The Foundations of Triboergodynamics and Physico-Chemical Prerequisits of Compatibility Theory (in Russian), Kaliningrad State Technical University Press, Russia, 2003.

IOP Conf. Series: Materials Science and Engineering **295** (2018) 012038 doi:10.1088/1757-899X/295/1/012038

2. Fedorov, V.V., Thermodynamic Aspects of Strength and Fracture of Solids" (in Russian), Science, Tashkent, 1979.
3. Fedorov, V.V., Kinetics of Damage and Fracture of Solids" (in Russian), Science, Tashkent, 1985.
4. Fedorov, V.V. Ergodynamic concept of failure, Strength of Materials (Translated from Russian), vol. 23, no. 8, 883-889, 1991.
5. Ivanova, V.S., Fatigue Fracture of Metals (in Russian), Moscow, USSR, 1963.
6. Friedel, J, 1964, Dislocations, Oxford.
7. Kragelskii, I.V., Dobychin, M.N., Kombalov, V.S., Friction and Wear Calculation Methods (in Russian), Mechanical Engineering, USSR, 1977.
8. Bowden, F.P., Tabor, D., The Friction and Lubrication of Solids, Oxford, 1964.
9. Conti, P., Sulla resistenza di Attrito. Royal Akademia dei Lencei, vol. II. 1875.
10. Lancaster, I.K., Basic mechanisms of friction and wear of polimers. Plastics and Polymers, vol. 41, 296-306, 1973.
11. Fedorov, S.V., Energy model of sliding friction coefficient and generalized regularities of tribosystems evolution, 19th International Colloquium "Tribology-Industrial and Automotive Lubrication", TAE, Esslingen, Germany, 2014, www.tae.de info@tae.de
12. Fedorov, S.V., The friction coefficient and its relation to the contact fatigue characteristics of materials. Industrial Laboratory, vol. 61, no.1, 41-49, 1995.
13. Fedorov, S.V., Generalized energy model of sliding friction coefficient and regularities of tribosystem evolution, V WTC2013, Turino, Italy, 2013, ISBN 9788890818509.
14. Fedorov, S.V., The mechanical quantum of dissipative friction structures is the elementary tribonanostructure, IV WTC2009 Transactions, Kyoto, Japan, 926, 2009.
15. Franek, F., Wopelka, T., Jech, M., On-board applicable high-resolution wear measurement technique for internal combustion engines, BALTTRIB'2011, Kaunas, Lituania, 196-201, 2011.
16. Fedorov, S.V., Calculation of the true friction volume. Friction & Lubrication in Machines and Mechanisms, no. 5, 3-7, 2010.

Research About the Corosive Effects of $FeCl_3$ in the Aeration Wastewater Basin

***Panaitescu C*[1], *Petrescu M.G.*.[2]**
[1] *Petroleum-Gas University of Ploiesti, Department of Petroleum Refining and Environmental Protection, 100680, Ploiesti, Romania*
[2]*Petroleum-Gas University of Ploiesti, Department of Mechanical Engineering, 100680, Ploiesti, Romania*

Abstract: Biological aeration of industrial wastewater is a very impressive process in the treatment of wastewater. The involvement of chemical reagents in this process, however, implies the intensification of the corrosion processes due to both pollutants in the wastewater and the chemical reactions that occur when the coagulation / flocculation reagents are added. This paper explores the action of ferric chloride ($FeCl_3$) on metallic parts in the aeration basin. The most affected structures are metal. At the classical basins the aeration systems were made of P295GH materials. The corrosion produced is uneven. The analysis of the high degree of corrosion was done according to the national and international standards. Finally, the paper supports the replacement of the existing aeration system with an anticorrosive material.

Keywords: wastewater, biological aeration system, corrosive effects, $FeCl_3$

1. Introduction

The need for proper operation of the wastewater aeration system is an important desideratum [1,2]. Therefore, stopping the corrosion process must be effective. In this regard, finding suitable materials from which to build this system is a permanent concern of specialists in this field.
The accomplishment of the aerobic treatment process is an important wastewater in the industrial wastewater treatment plant [3,4,5]. The need for the addition of ferric chloride($FeCl_3$) to this stage leads to the corrosion process of the system that carries out this process. Ferric chloride is added to increase sedimentation speed and uniformly develop active mud flocs. Due to the fact that $FeCl_3$ is highly corrosive, in the biological stage the corrosion phenomena lead to additional costs reflected in the cost price of the purified water [6,7,8]. This paper was based on monitoring the corrosion process in the aeration basin and finding a solution for the construction of a new aeration system.

2. Experimental part

The flow rate at the entrance to the treatment plant and at the exit from the biological treatment step was measured with an inductive electromagnetic flowmeter. The corrosion process is decisively influenced by the cathodic reaction. Because the temperature in the aeration basin is 32°C, the initial

oxidation rate is high [9-14]. The technological parameters that influence the corrosion process are: pH, temperature, amount of added $FeCl_3$.
Oxygen was continuously measured with dissolved oxygen traductor and through the automation system the quantity is adjusted by the air introduced into the aeration tank and the period of denitrification.
The aeration basin was divided into four areas for a proper monitoring of physicochemical parameters (Figure 1). The anoxic zones have been alternated with the oxic zones. The monitoring period was 12 months.

Figure 1. Aeration stage basin with four area

3. Results and discussion

The recorded flows at the entrance wastewater treatment plant and for effluent of the biological treatment stage are shown in the figure 2. It can be seen that it varies between the 405 l/s and 460 l/s.

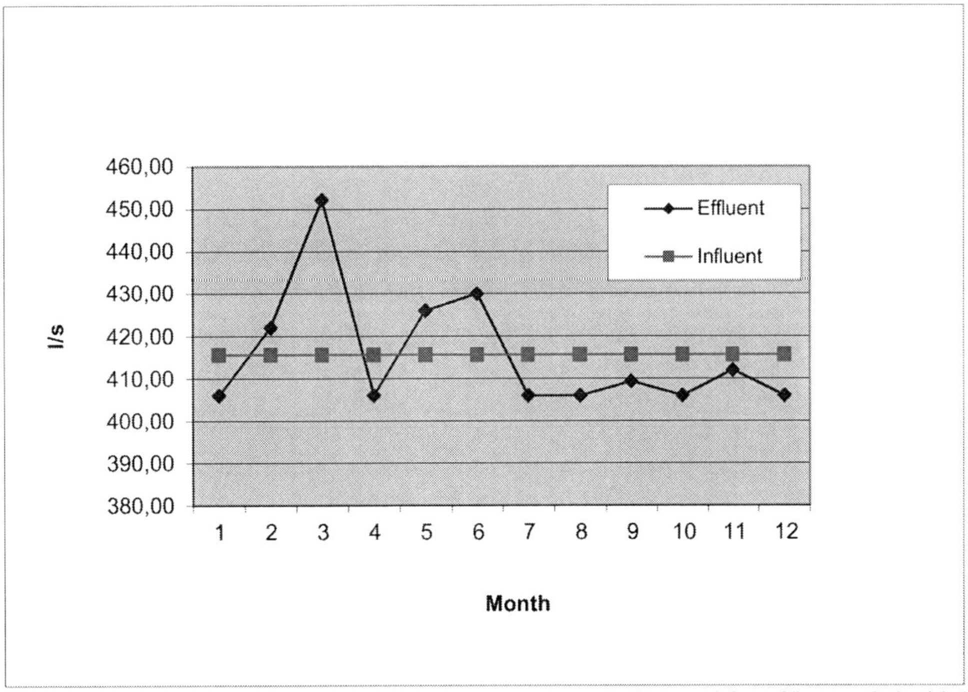

Figure 2. Flows recorded at the entrance wastewater treatment plant and for effluent of the biological treatment stage

The pH value recorded in the aeration tank is shown in the figure 3. The monitoring was performed over a period of 12 months. The pH was measured in all four areas of the aeration basin.-The influence of pH is decisive in the aeration process. Addition of ferric chloride does not greatly influence the pH value, but the corrosion process is directly proportional to it. Value of pH acid can lead to the corrosion process occurring and intensifying. The minimum recorded value of pH in the aeration basin was 7.3 and the maximum value was 7.8.

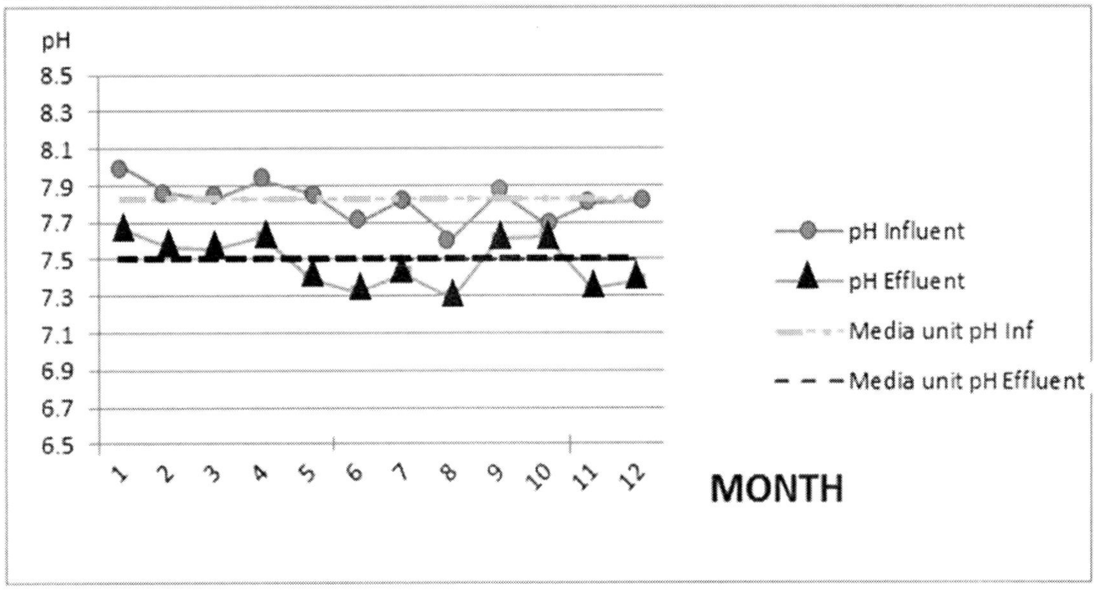

Figure 3. Value of pH monitoring 12 month

The temperature in the aeration basin depends on the temperature of the wastewater coming from the primary stage. Following the monitoring, the temperature was found to be in the range of 27°C-32°C.

The working temperature in the aeration tank depends on the temperature of the process water coming from the primary stage.
On the metallic surface of the aeration systems deposits take place. Thus, the material from which this system is built must resist corrosion phenomena.
At the classical basins the aeration systems were made of P295GH materials. The properties of this materials are presented in the Table 1.

Table 1. The chemical composition of P295GH [1]

Grade	The Element Max (%)						
	C	Si	Mn	P	S	Al(min)	N
P295GH	0.08-0.20	0.40	0.80-1.50	0.025	0.015	0.020	0.012
	Cr	Cu	Mo	Nb	Ni	Ti	V
	0.30	0.30	0.08	0.020	0.30	0.03	0.02

The corrosion produced is uneven. Its aspects of the aeration system are shown in the figure 4. The analysis of the high degree of corrosion was done according to the national and international standards. Finally, the paper supports the replacement of the existing aeration system with an anticorrosive material.

Figure 4. Corrosive effect of $FeCl_3$ about aeration system

The aeration system provides oxygen for the biological treatment process. Damage to the aeration system by swelling and peeling has led to the propagation of cracks along the entire length of the air supply. Crack propagation occurred longitudinally and transversely.
The adhesion of the oxidation layer to the surface of the system causes it to suffer damage and lead to significant oxygen losses
The main parameters used to assess the oxygen transfer performance are the oxygen transfer efficiency, global oxygen transfer coefficient, and the standard oxygen transfer rate in water. The distribution of gas bubble size depending on the surface velocity is very important in the aeration process. These determine the mass transfer coefficients and decrease or increase the efficiency of the aeration process. As a result of the corrosion phenomenon, the aeration becomes uneven and the amount of suspended matter increases. Thus, it is necessary to add additional amounts of ferric chloride which maintain an acceptable value of the sedimentation rate. The average concentrations of chlorides and $FeCl_3$ in the aeration basin, monitored for 12 months, are shown in figure 5.

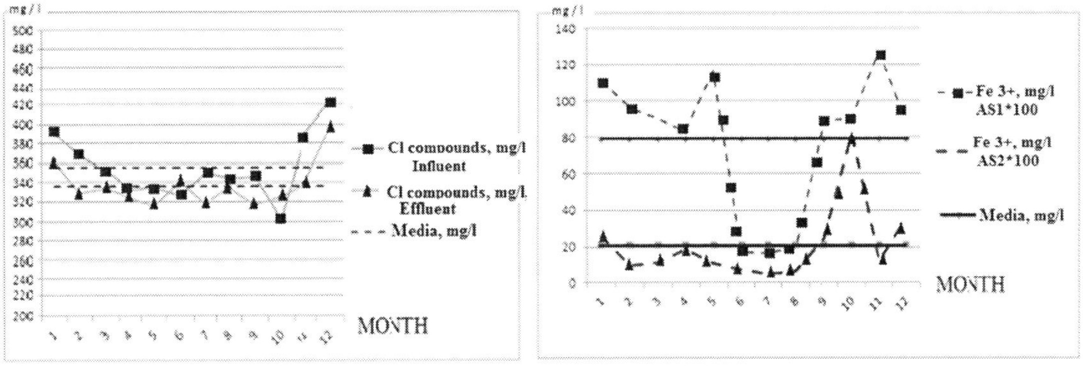

Figure 5. The average concentrations of chlorides and $FeCl_3$ in the aeration basin

At the most advanced stage of the corrosion of the operating system, metal specimens were introduced into the aeration basin for 4000 h. The corrosion rates recorded were the following:

- S 235JR: k'_g=0.047-0.049 g/m^{2}h
- Al CuZn28Sn1: k'_g=0.032-0.044 g/m^{2}h
- AISI Type 304L stainless steel: k'_g=0.0020-0.0024 g/m^{2}h

Corrosion rates were relatively high. On the surface of the specimens appeared local forms of corrosion due to the pearl chloride. There were also in-depth cracks in the Magna flow method. Damage to aeration systems by swelling (blisters) and peeling has led to the propagation of cracks along the entire length of the air supply to the bottom of the aeration tank. The crack propagation occurred transversally and longitudinally. In anaerobic areas of the pool, corrosion is more reduced in relation to aerobic areas where the phenomenon is more intense. In aerobic areas the corrosion rate is higher. Oxygen introduced into the stage also has a uniform mixing role. The dysfunctionalities resulting from the corrosion process cause the temperature to be uneven, resulting in higher temperature gradients. They favour the intensification of the local corrosion process.

4. Conclusions

The paper was based on a study of corrosion processes in the aeration basin of a wastewater treatment plant. The analysis of the high degree of corrosion was done according to the national and international standards. After this study appear the necessity to replacement of the existing aeration system with an anticorrosive material

References

1. Chi-Yuan F A., Field. R., Pisano.W.C., Barsanti. J, Joyc. J.J., Sorenson.H, Sewer and Tank Flushing for Sediment, Corrosion, and Pollution Control, Journal of Water Resources Planning and Management, vol. 127, 3, May/June 2001, 194-201, 2000.
2. Panaitescu, C., Jinescu, C., Mares, A.M., Study on ethylenediamine removal from textile industry wastewater, Revista de Chimie (Bucharest), 67, no.5, p. 925-928, ISSN 0034-7752, 2016.
3. Mishina,F., Nonaka, T., Hikosaka, Y., .Koga, M., Mori, T., Microbial Corrosion of Concrete Pipes, H_2S production from Sediments and Determination of Corrosion Rates, Water Science and Technology WSTED4, 23, 7/9, 1275- 1282, 1991.
4. Ionescu O.N., Ionescu G.C., Minescu M., Research on life prolongation for temperature sensors placed in highly erosive environment, Journal of the Balkan Tribological Association, vol. 19, 2,214-220, 2013.

5. Buzoianu, D.A., Panaitescu, C., Bombos, M., Stoica, M.E., Study on efficiency increasing of biological stage by sequential operating of aeration reactors, Revista de Chimie (Bucharest), 67, no. 5, p. 962-966, ISSN 0034-7752, 2016.
6. Pana I., Minescu M., Assessment procedure of technological consumption in the natural gas distribution systems, Applied Mechanics and Materials, vol. 371, 2013, 586-590, 17th International Conference on Innovative Manufacturing Engineering, IManE 2013; Iasi; Romania; 23- 24 May 2013.
7. Bucuroiu, R., Petrache, M., Vlasceanu, V., Petrescu, M.G., Study on oil wastewater treatment with polymeric reagents, 2016, Scientific Study and Research: Chemistry and Chemical Engineering, Biotechnology, Food Industry, 17, no. 1, p. 55-62, ISSN 1582-540X, 2016.
8. Joyce, James, Odor and Corrosion Prediction and Control in Collection Systems and Wastewater Treatment Plants. Water Environment Federation Workshop Document, 11-12, 2001.
9. Accounting Section of California, Los Angeles County Sanitation Districts Report Pomeroy and Bowlus, 1946.
10. NFPA, National Fire Protection Association, NFPA 820: Standard for Fire Protection in Wastewater Treatment and Collection Facilities, 2008 Edition
11. Minoda T., Yoshida H., Effect of Grain Boundary Characteristics on Intergranular Corrosion Resistance of 6061 Aluminum Alloy Extrusion, Metall. Mater. Trans. A. 33 (2002) 2891–2898. doi:10.1007/s11661-002- 0274-3
12. Metcalf L., Eddy H.P., eds., Wastwater Engineering Treatment and Resourcce Recovery, 5th ed., McGraw-Hill Education, New York, NY, 2014.
13. Stansbury E.E., Buchanan R.A., Fundamentals of Electrochemical Corrosion, ASM International, Materials Park, OH, 2000.
14. Pavelich J.S., Nychka J.A., Additional Inspection Report: Additional Visual Inspection and Corrosion Assessment of the EPCOR Borealis Wastewater Treatment Plant, Edmonton, AB, 2014.

Biotribology of Cartilage Wear in Knee and Hip Joints Review of Recent Developments

Akdogan Gulsen [1,2], *Goncu Merve* [2], *Parlak Meltem* [2]
[1] *Corresponding Author*
[2] *Erciyes University, Department of Biomedical Engineering, 38039, Kayseri, Turkey*
akdogang@erciyes.edu.tr [1,2], *mervegnc4@gmail.com* [2], *meltemprlk@gmail.com* [2]

Abstract: Nowadays, the problem of wear in the knee and hip joints is an important issue that concerns many people and still requires new solutions. In recent years, researchers dealing with knee and hip articular cartilage erosion continue to investigate the subject in terms of biotribology. In this study, recent developments and studies in this relevant area are been examined. By using the basic principles of tribology, useful new methods that can be used in the field of biotribology can be produced. Artificial joints designed using various materials such as metals, ceramics, polymers and composites are still being studied. New studies in this area will affect the development of implant technology. Different alloys or composites are currently being tested for new implant designs. Moving implants with a risk of wear are tested in laboratory conditions in simulator devices before they are used in the human body. Major topics such as nanotechnology, tissue engineering, orthopedics, tribology, biotribology, lubrication, organ transplantation and artificial organs, which are still important today, will be useful in the search for finding suitable solutions in the future in biotribological studies. This review article aims to provide an overview of in-vitro studies at the theoretical and laboratory conditions that must be performed prior to clinical investigation.

1. Introduction

Nowadays, cartilage wear problems are also increasing due to longer life, an increasingly comfortable life style, excessive weight gain, traffic and sports accidents. It is known that millions cases of damaged human cartilage are in need of repair. The need for this repair is found in the cartilaginous tissue, usually in joints with large loads, such as the hip and knee joints. Lubrication theories, such as hydrodynamic lubrication and boundary lubrication, which are well known and have been used for over fifty years in engineering science, are also considered ideal for use in biotribology. Biotribological knowledge also exploits the relevant lubrication theories to describe and explain friction and wear, even under the dynamic loading and sometimes compressed loading conditions encountered in the synovial joints. [1]

Biotribology is concerned with all aspects of tribology related to biological systems. It is one of the most exciting rapidly growing areas of tribological studies. As in many biological systems, for the solution of the problems in the operation of common systems, it is considered as one of the most important issues in terms of how our natural systems work, how diseases develop and how medical interventions are carried out. In this context, biotribology is an interdisciplinary science and technology that examines the surfaces of biological materials such as friction, wear, abrasion and lubrication. [2] [3]

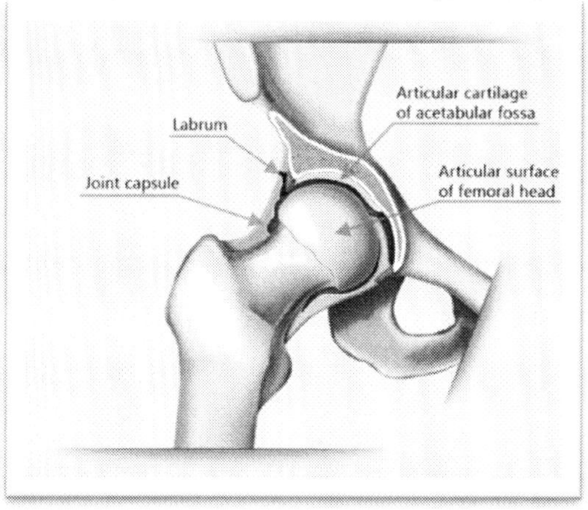

Figure 1. Hip joint and cartilage [4]

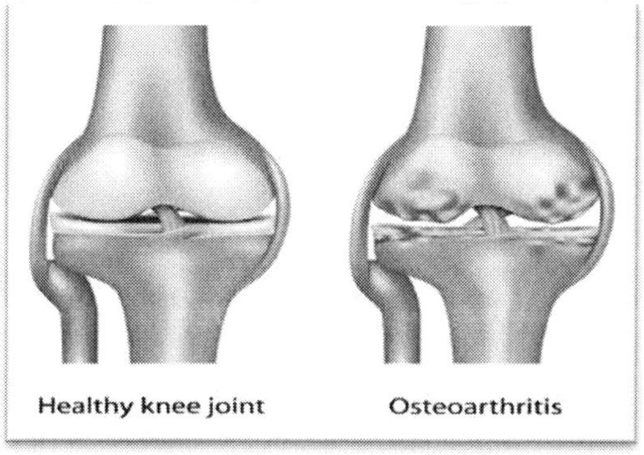

Figure 2. Knee joint and cartilage [5]

For the medical solution of the problem arising from cartilage erosion which is called arthrosis, it is necessary to use the theoretical and practical solutions offered by the biotribology science.
When looking for medical solutions, it has been recognized that the basic principles of tribology and biotribology are valid. They include basic topics such as friction, wear, wear debris, lubrication system, size change and losses. [6]

The knee joint is the most important joint in knee movements. The side and cross-links support the knee joint to ensure it remains in place. The femur and tibia bones in the knee joint are cross-linked to each other. One of the most important problems in the load bearing knee joint is the wear of the cartilage in this region. This wear directly affects the movement of the body. Associated joint diseases arise as arthritis and arthrosis. Arthritis is an inflammatory disease of the joints. Arthrosis refers to wear on joints. Arthrosis, which is usually seen in older people, is only rarely seen in younger people. This problem causes serious joint pain. Among the causes of arthrosis, in addition to factors such as excessive weight, lack of movement, body weight, and unbalanced weight loading, hereditary genetic predisposition is also a factor. [7]

2. How does knee cartilage wear (arthrosis) occur?

The knee joint has a cartilage layer that prevents friction in the bone heads. The cartilage, which is smooth and slippery, covers the moving parts of the bones. Normally, these ends have a soft and elastic cartilage sheath. Continuous load-bearing cartilage gradually disintegrates and loses its flexibility. Finally, the cartilage becomes and fragile. [8]

Cartilage injuries are most common in sudden impact injuries situations. Such injuries can occur while playing a sport or from the impact of a crash in a traffic accident. They are most commonly seen in the knee joint. After cartilage injuries to the knee, meniscus and cross ligament injuries may develop or symptoms may appear. [9] [10]

According to the information provided by the "Institute of Advanced Orthotics and Sports Institute of the United States", the number of "cartilaginous surgery" cases consisting of people of all ages in the United States alone is around 2 million a year.

2.1. Wear of "Cartilage to cartilage

Mechanical-biochemical erosion involves the tribological interaction between the biological and mechanical system responsible for cartilage wear and disintegration. To advance forward in research, the cell response to mechanical stimuli and the related mechanical-biochemical erosion mechanisms must be further investigated for a complete picture of tissue degradation. [11] [12]

If the cartilage is completely eroded, the bone starts to rub against the bone. This situation causes the person pain and a reluctance to move. In some occupational diseases, in those who do not pay attention to proper shoe selection, cartilage wear at an early age may be seen in unilateral weight bearing, when using the joints badly. Apart from the knee joints, arthrosis can also be seen in the waist, hip and other joints. Although age progression is a risk, sometimes the disease can be seen even in young people. If the cartilage wears out once, regional healing is seen in young people, but it is not possible in today's conditions to treat it in elderly people. [13]

When the materials of the body are classified, one of the various materials is cartilage. There are several types of tasks which are performed by various types of cartilage. Depending on its composition, cartilage is classified as cartilage (hyaline), fibrocartilage, or elastic cartilage. [12]

2.2. Wear of "Cartilage to artificial biomaterials"

One of the main reasons for disability worldwide is articular cartilage injury, that is arthritis resulting from degeneration. Although orthopedic surgeons and researchers continue their work on eliminating articular and cartilaginous defects, it is not possible to imitate the biological and biochemical properties of joint cartilage using existing research and development approaches. [14]

It is not possible to produce artificial cartilage with the same properties as the original. In this case, it is necessary to select suitable pairs of materials that are compatible with cartilage and which will move together in the same environment.

There is presently much experimental work in this area to develop the different kinds of artificial material that will come into contact with the cartilage or hold the cartilage and be placed in the body. For example, these materials alumina (Al_2O_3), cobalt-chromium (CoCr), titanium alloys (Ti6Al4V), stainless steel (SS), and cross-linked ultra-high-molecular-weight polyethylene (UHMWPE)'s impact resistance is high, the friction and abrasion resistance is very suitable for cartilage. [15] [16] [17] [18]
There have been many studies involving the friction and abrasion tests of materials that will come into contact with or replace cartilage. The abrasion residues resulting from the friction and abrasion of these artificial materials with articular cartilages play an important role in the choice of material to be used. In various studies, nano-scale rubbing, roughness measurements and protein attachments of different artificial materials used against "bovine articular cartilages" have been examined histologically.

Because of the complex physiology, tissue regeneration and low recovery capacity of cartilaginous tissue, it is very difficult to heal the tissue. Unrepaired or improperly repaired lesions make the cartilage susceptible to osteoarthritis by creating malfunction tissues in terms of biochemical and mechanical strength. Applications such as FDA-approved cell-based treatments such as washing of the joint area, post-abrasive cleaning, lower cartilage piercing and micro-fracture are now being used to heal and restore focal cartilage defects in younger patients. [19] [20]

Despite the fact that cartilage in the tissue engineering that has developed over the last years for repairing damaged or damaged tissue is still cartilaginous tissue similar to natural articular cartilage, due to the complex hierarchical structure of natural cartilage, it is not produced. [21]

Artificial cartilaginous tissue, which tries attempted to mimic natural articular cartilage due to the complex hierarchical structure of true natural cartilage, has not yet been produced in sufficient detail, although it has worked in many of the tissue engineering methods developed in recent years to repair cartilage damage. [22] [23]

3. Treatment methods for cartilage

In adults, joint cartilage healing is insufficient. Unlike other tissues in the body, joint cartilage cannot be renewed after injury. However, due to injuries during sports, they may see regional damage in the joint cartilage of young individuals. In such cases, cartilage regenerative treatments can be performed because only a portion of the joint cartilage is damaged and the rest is intact. Treatment of knee cartilage erosion involves exercises without overloading to strengthen and support the muscles around the joint. Cartilaginous tissue cannot rebuild itself. There is no reversal for this common problem and medication is used before surgical treatment is considered. If there is no solution after all treatment methods have been tried to relieve joint damage after cartilage injuries, the joint is *replaced by on* the artificial joint. For patient comfort, the joints are shaved and replaced with metallic implants by *means of* surgical intervention. [24]

3.1. Classical treatment methods for cartilage

Especially in the elderly, the ability of the joint cartilage to heal is inadequate. Unlike other tissues in the body, joint cartilage cannot renew itself after injury. For this reason, improvement in cartilage is not expected, and fluid which will make the joint less inflamed is injected, in order to reduce the friction by increasing the amount of ambient fluid. This method, which can be likened to the lubrication of hinges mechanically, is not effective in the long term because it is not possible to keep the fluid permanently in the liquid hinge region. [25]

In the case of drug supplements taken as oral tablets, a substance called "glucosamine" is used, this is already found in the body and is thought to be beneficial in the repair and shaping of joint cartilage. However, some people have also shown a reaction to this. It forces the muscles around the cartilage to work to reduce cartilage-cartilage friction and also to reduce excessive weight. [26] [27]

While there are many new techniques in the treatment of cartilage injuries, there is no clarification as to whether some lesions are in need of treatment or which treatment model is more appropriate for a given lesion.

New applications are increasing, but they are expensive. More objective data on cartilage injuries and newer efficient methods are needed to evaluate post-surgical outcomes. There are several published studies about the results of total joint replacement after post-surgical image evaluation and after surgery. In the future, more research will need to be done on cartilage surface reconstruction clinically.

In this regard, the development of common evaluation tools and standards of different engineering branches and orthopedic surgeons will achieve significant results. [28]

The International Cartilage Repair Association (ICRS), which was established in 1997, strives to establish a standardization system for the assessment of cartilage damage and repair. [29]
Engineers from different disciplines and orthopedic surgeons continue to work on the search for new solutions in the treatment of cartilage damage. Engineering studies on non-destructive cartilage imaging are another engineering subject that has been studied. [30]
The use of drugs that affect the gene system is being discussed in the treatment of rheumatic disorders in various articular cartilages. Innovations in cartilage imaging technology, interdisciplinary work such as stem cell technology, tissue engineering, bioengineering, and scientific discussions are ongoing. [31] [32]

3.2. Current new treatment methods for cartilage damage

After mechanical impact is applied to the joint region, progressive deterioration and damage to the cartilage occurs. Later, pain, swelling and loss of joint functions occur. After the treatment of the cartilage system, the advantages and disadvantages of the procedure can also be discussed. Many efforts have been made to develop tissue-engineered grafts or patches to repair focal cartilage and cartilage defects. [33] [34]

3.2.1 Micro-fracture method. In this method, small holes are created in damaged cartilage areas. Then, the stem cells in the bone marrow are removed and transported to the damaged small hole areas, creating a cartilage patch. Cylindrical solid cartilage taken from the surfaces of "non-load bearing" areas can be transferred as patches to damaged areas of need. The advantage of this method is that it is completely arthroscopic, no complicated equipment is required, and it is effective and cheap. The disadvantage is that the newly formed cartilage is structurally different and weaker than the normal joint cartilage. The effect is reduced in large areas of 3 cm^2. [35][36]

3.2.2 Cartilage transplant method. Scientists are continuing their new work on in-vitro models to develop new treatment technologies that can be applied now and in the future. After the patient's own solid cartilage cells are produced in the laboratory environment, they are then transplanted in the damaged region. Autogenous cartilage transplantation, whereby cartilage can be taken from one part of the body and transported to another, is a generally accepted method in reconstructive surgery. In-vitro engineering studies for cartilage tissue are promising. The host's response to the implant materials determines whether the size of the implant materials to be transported will be small/large. [1] [37] [38]

3.2.3. Cartilage and meniscus allografts method. The meniscus was known as a useless surplus in the past, but today it is known to have very important functions. The functions of the meniscuses are as follows load transportation, load transfer, and joint lubrication and helping with the screw home mechanism. In the absence of anterior cruciate ligaments, it aids secondary anterior-posterior stabilization. Nowadays, no implant can replace total meniscus loss. It is indicated only in segmental defects.[39]

This method is a good alternative to knee prosthesis in young patients with large bone and cartilage damage. [40]

3.2.4 Mesenchymal stem cell repair (MSC) method. Stem cells, according to their period of development, are divided into the following groups embryonic, stimulated pluripotent and adult cells. Among them are embryonic stem cells, cartilage, bone, and muscle, but are not yet differentiated. Joint cartilage injuries (arthritis) cause permanent damage to the joint. In this case, there are many people in this world who live as disabled people.

The work done by "mesenchymal stem cell repair" is one of the most important current areas of research for disability. Muscles, bones, cartilages, etc., have the potential to transform into different tissues in the body. MCS is found in the gel matrix of human bone marrow. If it is placed in the damaged cartilage region, new cartilage tissue develops. [20] [23]

4. Biotribological research support for cartilage treatments.

4.1. Simulation programs

By using biomimetic science in real joint systems, problems are solved with the support of engineering subjects such as biomechanics, biotribology, contact mechanics, and lubrication systems. When it is defined by the classical macroscopic approach in tribology, properties such as roughness and friction coefficient of surfaces, properties of lubrication and properties of lubricant and relative speeds of moving surfaces are within the scope of this heading.

Simulation applications can be made for different sample applications by changing all the contribution properties in this regard. There are also different studies based on the assumption that joint movement is an important part of mechanical conduction in the synovial joints. Software and programs that simulate different applications where sliding, rotating, or different versions of these exist together have also been developed. [41] [42]

4.2. Mechanical fatigue tests

Mechanical fatigue testing machines have also been used in studies to observe the in-vitro results of applications such as wear, friction, and lubrication.

There is still a lack of information about the dynamic process of cartilage damage, which is the result of the negative consequences of social effects. Experimental system studies are carried out to perform a series of cycles designed to produce mechanical fatigue in in-vitro environments that fulfill in-vivo conditions. Different results are obtained by changing the specific cycle loading numbers to obtain the functional stress, strain, wear, and texture response results. The surface analyses obtained are examined histologically and microbiologically. Fatigue tests performed under pressure or without pressure are also used. In mechanical fatigue tests, the size of the load, the applied frequency, the amplitude of the alternative load, the thickness of the sample, etc., affect how the result will be formed. [43]

4.3. Wear and friction tests

In different studies, a variety of materials were applied to see the results of friction and abrasion of biomaterials on cartilage. It is the UHMWPE that causes the least wear and tear on these materials and the most damaging is CoCr alloy.

Scientists are currently working to develop new treatment methods for degenerative joint diseases by performing different studies to understand the tribological properties of joint cartilage. The results obtained by the sliding and rotation mechanisms in the experiment on the friction and wear behavior of bovine cartilage applied in dry and wet environment using liquid film formation and boundary lubrication technique were evaluated. The experiments were repeated with a very high number of cycles. [19] [44] [45]

4.4. Lubrication tests

Cartilage of different shapes and composition performs many functions in the body. There are three types of cartilage: joint cartilage, fat and elastic cartilage. Cartilage functions as a bonding material or as lubrication material on the loading surface. Thanks to cartilage support, organs such as the ears, nose and food pipe remain intact. Lubrication of the joints is performed by means of cartilage which is a natural lubrication mechanism using bio-fluid synovial and bio-fat. [12] [46]

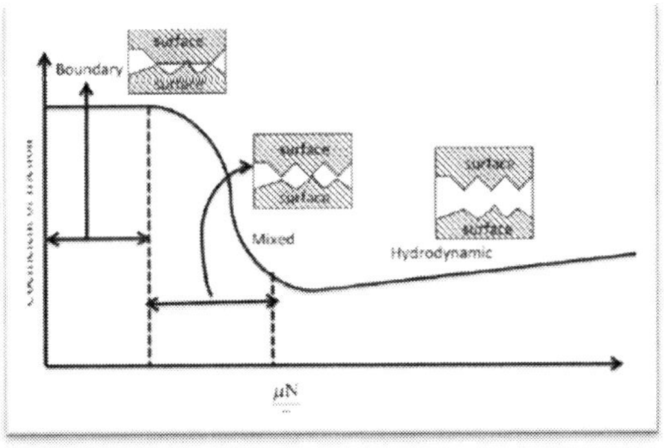

Figure 3. Stribeck curve diagram [47]

1.1. Knee and hip joints are living tribological systems with a very low coefficient of friction and no similarities. It is not possible to produce a material or mechanism with these properties. The reason for the low friction between the cartilaginous surfaces is believed to be due to the sliding molecular layers between the cartilaginous surfaces.

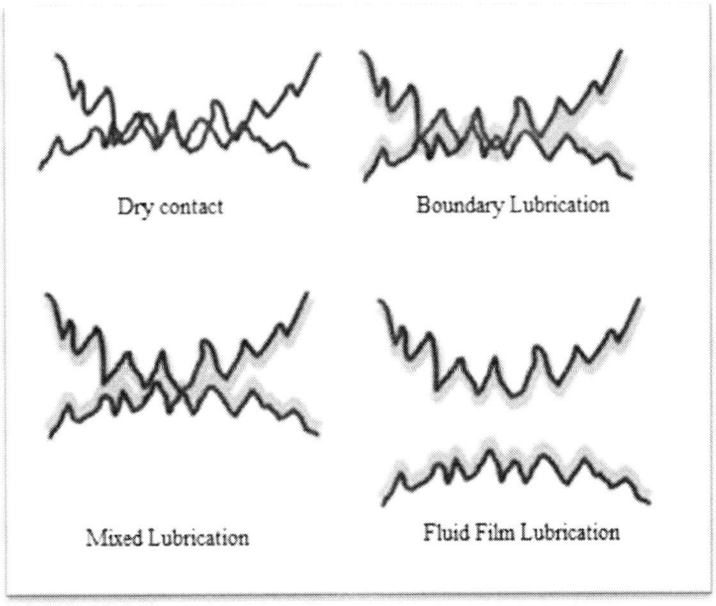

Figure 4. Diagrammatic representation of dry contact and lubrication regimes [48]

Thanks to the excellent tribological mechanism of cartilage, there is a unique lubrication system consisting of both the presence of a pressurized liquid medium and a boundary lubrication mechanism supported by large fat molecules attached to the surface. [49] [50]

4.5. Biocorrosion tests

Rapid degradation of the biomaterial is caused due to the microbial film layer on the surface of the biomaterials. The mechanism of formation of biocorrosion is still not fully resolved. Very rapid biocorrosion occurs in the environment of hydrogen and oxygen in the metals. Polymers also develop as corrosion/degradable. When in-vitro and in-vivo experiments are performed to produce different implants suitable for the body, the results give a certain amount of results. However, no any artificial materials give results equivalent to the natural body materials.

In recent years, polymer biomaterials have to have properties that vary from place to place. The biodegradable property is desirable in fracture fixation plates, but not in implants that involve the hip and articular cartilage. Meanwhile, the hope that the interstitial spaces of artificial polymers, which have started to be used as scaffolding structures in tissue engineering, can be filled with the structure to be developed by the patient's one's own stem cells, opens new horizons. Artificial biomaterials are being researched natural or artificial polymer types for different applications. [51]
Biodegradable polymeric biomaterials are used as fracture fixation plates. They are also used as 3D scaffolds in tissue engineering thanks to their porous construction. This property of biodegradable polymers also makes it the preferred material for the development of controlled drug release devices. The development of these properties is hoped to lead to the introduction of polymers to be used in cartilage structure. [52] [53] [54] [55]

4.6. Abrasively Bio-Corrosion Tests

It is inevitable that abrasive particles will form after friction. Normally, the particles are eliminated and bleeding can be stopped. If the resulting particle load exceeds the elimination capacity, failure occurs. [56]
If residues are trapped in cartilaginous interfaces, increased wear due to mechanical friction occurs and wear increases the friction of the intermediate zone and causes more wear and tear. The debris particle will accelerate the onset of wear on the environment. Preventing corrosion and mechanical abrasion will also slow the formation of debris. [57] [58] [59]

5. Conclusion

The latest developments in classification and different approaches to treatment are constantly ongoing. In small blood vessels in small damaged chambers that are intentionally created in cartilage, there is a possibility that the stem cells may turn into cells that resemble cartilage if the surrounding conditions are favorable. Recently, mold implants called "matrix" have been developed so that the said root cell can be held more tightly in damaged mini chamber regions. Any work that can be done in order to reduce friction between cartilages will be beneficial. However, if the damaged cartilage layer is to be removed from the medium, the artificial biomaterials that will retain it should be selected to form the appropriate pair. Prop*erly selected pairs exhibit no* wear behavior at first, but wear can occur over time.

No artificially produced cartilage will ever be as good as natural original cartilage. However, more research on cartilage will continue to be carried out in order to allow people to lead a more productive and comfortable life with a longer life expectancy.

References

1. J. Katta, Z. Jin, E. Ingham, and J. Fisher, "Biotribology of articular cartilage-A review of the recent advances," Med. Eng. Phys., vol. 30, no. 10, pp. 1349–1363, 2008.
2. Z. R. Zhou and Z. M. Jin, "Biotribology: Recent Progresses and Future Perspectives," Biosurface and Biotribology, vol. 1, no. 1, pp. 3–24, 2015.
3. M. J. Furey and B. M. Burkhardt, "Biotribology: Friction, wear, and lubrication of natural synovial joints," Lubr. Sci., vol. 9, no. 3, pp. 255–271, 1997.
4. New York Dynamic Neuromuscular Rehabilitation &Physical Therapy, New York, 2017, https://nydnrehab.com/what-we-treat/hip-pain/hip-groin-pain/
5. Health Navigator, New Zealand, ttps://www.healthnavigator.org.nz/medicines/g/glucosamine/.
6. M. O. Brien and F. Burke, "European Surgical Orthopaedics and Traumatology," Eur. Surg. Orthop. Traumatol., pp. 2717–2727, 2014.
7. F. M. Maclean, "Arthritis and other proliferative joint diseases, " Diagnostic Histopathology, vol. 22, no. 10. pp. 369–377, 2016.
8. M. Hudelmaier et al., "Age-related changes in the morphology and deformational behavior of knee joint cartilage," Arthritis Rheum., vol. 44, no. 11, pp. 2556–2561, 2001.
9. W. W. Curl, J. Krome, E. S. Gordon, J. Rushing, B. P. Smith, and G. G. Poehling, "Cartilage injuries: A review of 31,516 knee arthroscopies," Arthroscopy, vol. 13, no. 4. pp. 456–460, 1997.
10. J. a. Buckwalter, "Mechanical Injuries of Articular Cartilage," Iowa Orthop J, vol. 12, pp. 50–57, 1992.
11. R. L. Trevino, J. Stoia, M. P. Laurent, C. A. Pacione, S. Chubinskaya, and M. A. Wimmer, "Establishing a live cartilage-on-cartilage interface for tribological testing," Biotribology, vol. 9, pp. 1–11, 2017.
12. J. M. Mansour, "Biomechanics of Cartilage," Kinesiol. Mech. pathomechanics Hum. Mov., pp. 66–79, 2009.
13. J. Sanchez-Adams, H. A. Leddy, A. L. McNulty, C. J. O'Conor, and F. Guilak, "The mechanobiology of articular cartilage: bearing the burden of osteoarthritis," Current rheumatology reports, vol. 16, no. 10. p. 451, 2014.
14. D. F. Duarte Campos, W. Drescher, B. Rath, M. Tingart, and H. Fischer, "Supporting Biomaterials for Articular Cartilage Repair.," Cartilage, vol. 3, no. 3, pp. 205–21, 2012.
15. V. Saikko, T. Ahlroos, O. Calonius, and J. Keränen, "Wear simulation of total hip prostheses with polyethylene against CoCr, alumina and diamond-like carbon," Biomaterials, vol. 22, no. 12, pp. 1507–1514, 2001.
16. a Wang, a Essner, V. . Polineni, C. Stark, and J. . Dumbleton, "Lubrication and wear of ultra-high molecular weight polyethylene in total joint replacements," Tribol. Int., vol. 31, no. 1–3, pp. 17–33, 1998.
17. S. M. Kurtz, "A Primer on UHMWPE," in UHMWPE Biomaterials Handbook: Ultra High Molecular Weight Polyethylene in Total Joint Replacement and Medical Devices: Third Edition, 2015, pp. 1–6.
18. F. S. Senatov, A. N. Kopylov, N. Y. Anisimova, M. V. Kiselevsky, and A. V. Maksimkin, "UHMWPE-based nanocomposite as a material for damaged cartilage replacement," Mater. Sci. Eng. C, vol. 48, pp. 566–571, 2015.

19. S. M. T. Chan, C. P. Neu, K. Komvopoulos, a. H. Reddi, and P. E. Di Cesare, "Friction and Wear of Hemiarthroplasty Biomaterials in Reciprocating Sliding Contact With Articular Cartilage," J. Tribol., vol. 133, no. 4, p. 41201, 2011.
20. Z. Ge, C. Li, B. C. Heng, G. Cao, and Z. Yang, "Functional biomaterials for cartilage regeneration," Journal of Biomedical Materials Research - Part A, vol. 100 A, no. 9. pp. 2526–2536, 2012.
21. J. Y. Rho, L. Kuhn-Spearing, and P. Zioupos, "Mechanical properties and the hierarchical structure of bone," Med. Eng. Phys., vol. 20, no. 2, pp. 92–102, 1998.
22. M. Ochi, N. Adachi, H. Nobuto, S. Yanada, Y. Ito, and M. Agung, "Articular Cartilage Repair Using Tissue Engineering Technique—Novel Approach with Minimally Invasive Procedure," Artif. Organs, vol. 28, no. 1, pp. 28–32, 2004.
23. R. L. Mauck and J. A. Burdick, "Engineering cartilage tissue," in Tissue Engineering: From Lab to Clinic, 2011, pp. 493–520.
24. R. Shenoy, P. S. Pastides, and D. Nathwani, "(iii) Biomechanics of the knee and TKR," Orthop. Trauma, vol. 27, no. 6, pp. 364–371, 2013.
25. D. Eyre, "Collagen of articular cartilage.," Arthritis Res., vol. 4, no. 1, pp. 30–5, 2002.
26. H. Forster and J. Fisher, "The influence of loading time and lubricant on the friction of articular cartilage," Arch. Proc. Inst. Mech. Eng. Part H J. Eng. Med. 1989-1996 (vols 203-210), vol. 210, no. 28, pp. 109–119, 1996.
27. J. Desrochers, M. W. Amrein, and J. R. Matyas, "Microscale surface friction of articular cartilage in early osteoarthritis," J. Mech. Behav. Biomed. Mater., vol. 25, pp. 11–22, 2013.
28. M. Brittberg, "Evaluation of cartilage injuries and cartilage repair," Osteologie, vol. 9, no. 1, pp. 17–25, 2000.
29. R. J. Williams, L. Peterson, and B. J. Cole, Cartilage repair strategies. 2007.
30. M. Cucchiarini, C. Henrionnet, D. Mainard, A. Pinzano, and H. Madry, "New trends in articular cartilage repair.," J. Exp. Orthop., vol. 2, no. 1, p. 8, 2015.
31. T. Minas, "A primer in cartilage repair," J. Bone Jt. Surgery, Br. Vol., vol. 94–B, no. 11 Supple A, pp. 141–146, 2012.
32. R. Mardones, C. M. Jofre, and J. J. Minguell, "Cell therapy and tissue engineering approaches for cartilage repair and/or regeneration," International Journal of Stem Cells, vol. 8, no. 1. pp. 48–53, 2015.
33. G. Musumeci et al., "New perspectives for articular cartilage repair treatment through tissue engineering: A contemporary review.," World J. Orthop., vol. 5, no. 2, pp. 80–8, 2014.
34. R. M. Thiede, Y. Lu, and M. D. Markel, "A review of the treatment methods for cartilage defects.," Vet. Comp. Orthop. Traumatol., vol. 25, no. 4, pp. 263–272, 2012.
35. H. Chen et al., "Drilling and microfracture lead to different bone structure and necrosis during bone-marrow stimulation for cartilage repair," J. Orthop. Res., vol. 27, no. 11, pp. 1432–1438, 2009.
36. Q. Wang et al., "Effect of low-energy shock waves in microfracture holes in the repair of articular cartilage defects in a rabbit model," Chin. Med. J. (Engl)., vol. 124, no. 9, pp. 1386–1394, 2011.
37. A. Haisch et al., "Macroencapsulation of human cartilage implants: Pilot study with polyelectrolyte complex membrane encapsulation," Biomaterials, vol. 21, no. 15, pp. 1561–1566, 2000.

38. S. Marlovits, P. Zeller, P. Singer, C. Resinger, and V. Vécsei, "Cartilage repair: Generations of autologous chondrocyte transplantation," Eur. J. Radiol., vol. 57, no. 1, pp. 24–31, 2006.

39. C. R. Allen, E. K. Wong, G. A. Livesay, M. Sakane, F. H. Fu, and S. L. Y. Woo, "Importance of the medial meniscus in the anterior cruciate ligament-deficient knee," J. Orthop. Res., vol. 18, no. 1, pp. 109–115, 2000.

40. E. K. Danso, J. T. J. Honkanen, S. Saarakkala, and R. K. Korhonen, "Comparison of nonlinear mechanical properties of bovine articular cartilage and meniscus," J. Biomech., vol. 47, no. 1, pp. 200–206, 2014.

41. V. L. Popov and S. G. Psakhie, "Numerical simulation methods in tribology," Tribol. Int., vol. 40, no. 6, pp. 916–923, 2007.

42. M. a Wimmer et al., "Tribology approach to the engineering and study of articular cartilage.," Tissue Eng., vol. 10, no. 9–10, pp. 1436–1445, 2004.

43. J. T. Kaplan, C. P. Neu, H. Drissi, N. C. Emery, and D. M. Pierce, "Cyclic loading of human articular cartilage: The transition from compaction to fatigue," J. Mech. Behav. Biomed. Mater., vol. 65, pp. 734–742, 2017.

44. K. G. Budinski, Guide to Friction, Wear, and Erosion Testing. 2007.

45. A. C. Cilingir, "Effect of Rotational and Sliding Motions on Friction and Degeneration of Articular Cartilage under Dry and Wet Friction," J. Bionic Eng., vol. 12, no. 3, pp. 464–472, 2015.

46. J. P. Gleghorn and L. J. Bonassar, "Lubrication mode analysis of articular cartilage using Stribeck surfaces," J. Biomech., vol. 41, no. 9, pp. 1910–1918, 2008.

47. N. W. M. Zulkifli, M. A. Kalam, H. H. Masjuki, M. Shahabuddin, and R. Yunus, "Wear prevention characteristics of a palm oil-based TMP (trimethylolpropane) ester as an engine lubricant," Energy, vol. 54, pp. 167–173, 2013.

48. D. Dowson and Z.-M. Jin, "Metal-on-metal hip joint tribology," Proc. Inst. Mech. Eng. Part H J. Eng. Med., vol. 220, no. 2, pp. 107–118, 2006.

49. S. Jahn, J. Seror, and J. Klein, "Lubrication of Articular Cartilage," Annu. Rev. Biomed. Eng, vol. 18, pp. 235–58, 2016.

50. G. W. Greene, A. Olszewska, M. Osterberg, H. Zhu, and R. Horn, "A cartilage-inspired lubrication system," Soft Matter, vol. 10, no. 2, pp. 374–382, 2014.

51. J. Park and R. S. Lakes, Biomaterials. 2007.

52. L. S. Nair and C. T. Laurencin, "Biodegradable polymers as biomaterials," Progress in Polymer Science (Oxford), vol. 32, no. 8–9. pp. 762–798, 2007.

53. X. Miao and D. Sun, "Graded/gradient porous biomaterials," Materials (Basel)., vol. 3, no. 1, pp. 26–47, 2010.

54. F. Mahyudin, L. Widhiyanto, and H. Hermawan, "Biomaterials in orthopaedics," in Advanced Structured Materials, vol. 58, 2016, pp. 161–181.

55. I. B. Beech and J. Sunner, "Biocorrosion: Towards understanding interactions between biofilms and metals," Current Opinion in Biotechnology, vol. 15, no. 3. pp. 181–186, 2004.

56. C. Valero Vidal and A. Igual Muñoz, Bio-Tribocorrosion in Biomaterials and Medical Implants. 2013.

57. Q. Chen and G. A. Thouas, "Metallic implant biomaterials," Materials Science and Engineering R: Reports, vol. 87. pp. 1–57, 2015.

58. J. L. Gilbert, "Mechanically assisted corrosion of metallic biomaterials," in ASM Handbook, Volume 23: Materials for Medical Devices, vol. 13, 2012, pp. 79–89.

59. R. J. Narayan, M. Marek, and C. Jin, Biomaterials, Corrosion and Wear of. 2006.

Effects of Weldment Layer on the Tungsten Carbide Hardfacing Microstructure

***Mohd Tobi A L*[1], *Nagentrau M*[2], *Kamdi Z*[2], *Ismail M I*[2], *Omar A S*[3]**
[1]*Faculty of Engineering Technology, Universiti Tun Hussein Onn Malaysia, Batu Pahat, Malaysia,*
[2]*Faculty of Mechanical and Manufacturing Engineering, Universiti Tun Hussein Onn Malaysia, Batu Pahat, Malaysia,*
[3]*Centre for Diploma Studies, Universiti Tun Hussein Onn Malaysia, Batu Pahat, Malaysia.*
abdlatif@uthm.edu.my

Abstract: Hardfaced carbon steel blade is used to mix ilmenite ore with sulphuric acid as part of a production process inside a digester tank. The carbon steel blade wear life is improved by applying tungsten carbide (WC) hardfacing using shielded metal arc welding (SMAW). This study addresses the effect of welding layers on the microstructure and distribution of WC hardfaced. The influence of different number of welding layer (1 layer, 2 layers and 3 layers) on WC hardfacing microstructure and hardness value were analysed using scanning electron microscope (SEM) and micro-Vickers hardness tester respectively. It is revealed that the growth of large-sized carbide and distribution uniformity of small-sized carbide in hardfacing region is increasing with number of welding layers applied during the hardfacing. The volume fraction analysis of the WC in the hardfacing region shows linear increase with the increase of number of layers. Small-sized carbide grows larger with increasing number of welding layer applied in the matrix region. The continuous supply of heat energy from multiple layers of welding assists on the growth of WC in the hardfaced region increasing its hardness and subsequently increasing the wear resistance.

Keywords: Hardfacing, tungsten carbide, layer, carbon steel blade

1. Introduction

Surface coating has been considered as an effective approach in increasing the lifetime of engineering components [1-3]. In addition, surface treatment technology is widely used in the industry for protecting working tools from wear and resisting corrosion [4]. Hardfacing welding technique is one of the surface treatments commonly practiced in the industry [5]. The service life of components subjected to abrasive wear can be extended by hardfacing deposit on the parts exposed to the wear process [6]. Components operating under severe condition and exposed to wear and corrosive environments can be hardfaced with coating material in order to provide wear protection [7]. Tungsten carbide (WC) is widely used as hardfacing material because of its high hardness, high thermal conductivity and good wettability [8, 9]. Katsich and Badisch reported that WC coatings are broadly utilized in enhancing the lifetime of engineering components that operate in severe environment [10-11].

Numerous studies have attempted to explain the effect of number of welding layers influences on the performance of hardfacing coating. Coronado et al. highlighted that abrasive wear resistance will be higher if three layers of welding is applied [12]. Buchely et al. states that higher number of hardfacing layer will give a better wear resistance of the coating [13]. Microstructural characteristics, especially carbide distribution influences the wear resistance of hardfacing coating. Desai et al. [14] and Shetty et al. [15] investigated the dependency of abrasive wear mechanism on the microstructure of hardfacing coating. They discovered that the amount, size and distribution of carbides have great influence on the wear rate. Leech et al. shows that high uniformity of WC distribution in a WC-Ni based metal matrix composite is able to provide better resistance against wear in the condition of pure abrasion [16]. It is found that closely spaced uniformly distributed carbide ensured abrasives cannot penetrate effectively into the coating matrix, resulting in higher wear resistance [17].

The presented research work has the main goal to evaluate the influence of WC welding layers on the overall carbide distribution in the coating. Elemental composition, micrograph and volume fraction analyses were conducted. The number of welding layers influence on carbide distribution in the coating are studied along with its hardness values in order to improve the wear resistance of the hardfaced coating.

2. Sample preparation

2.1. Blade and electrode

Figure 1 illustrates the BS3100 carbon steel Grade A3 blade that acts as the base metal in the hardfacing process. Carbon percentage in the blade determines the grade of the blade. Carbon (C) and Manganese (Mn) are the major elements in carbon steel. Meanwhile, other minor elements are also present as provided in Table 1. WC electrode enclosed in tubular mild steel is used as the hardfacing electrode. WC is hard material with excellent corrosion resistance which is suitable for hard surfacing of speed mullers, pump impellers cutter knives, and construction machineries. The electrode is approximately 6 mm diameter and 350 mm long. Table 2 presents the electrode chemical composition.

Figure 1. Hardfaced carbon steel blade.

Table 1. Elemental composition of BS3100 carbon steel.

Element	C	Mn	Si	S	Mo	P	Ni	Cr	Fe
Composition (%)	0.32	0.74	0.25	0.01	0.02	0.02	0.17	0.29	balance

Table 2. Elemental composition of tungsten carbide (WC) electrode.

Element	C	Si	Mn	W	Fe
Composition (%)	3.1	0.4	1.5	60.2	balance

2.2. Hardfacing method and welding condition

Shielded Metal Arc Welding, better known as SMAW, was assigned for this particular welding condition due to its extensive usage in the industries. The portability of its welding equipment and relative simplicity of SMAW welding makes it more economical without requiring any complex equipment compared to other welding methods. Besides that, SMAW welding is well-known for its versatility as it is readily applied to different applications with wide electrode choices. The low cost SMAW welding is also adaptable to remote locations and confined spaces. Horizontal welding direction in a flat position approach is practiced during hardfacing deposition of WC on carbon steel blade as shown schematically Figure 2. The position of weld deposit and specimen cut location is presented in Figure 2. Table 3 shows the hardfacing condition using SMAW welding. 1, 2 and 3 layers of hardfacing are employed to examine the influence of number of layer on the microstructure of WC hardfacing coating. Welding current, blade pre-heat and electrode drying are maintained to investigate welding layer effect on the hardfacing. Table 4 exhibits the welding parameters employed.

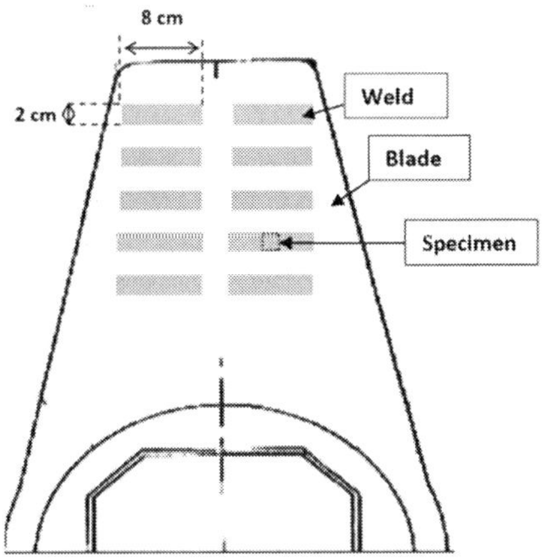

Figure 2. Schematic diagram of hardfacing deposit on carbon steel blade.

Table 3. Welding conditions using SMAW welding method.

Welding size	Welding speed	Polarity	Electrode feed rate	Welding length	Electrode length
2.0 cm	0.0025 m/s	DC	0.0035 m/s	8 cm	35 cm

Table 4. Specimens with different number of welding layer.

Specimen	Layer	Current (A)	Blade Preheat	Electrode drying
1	1	150	Yes	Yes
2	2	150	Yes	Yes
3	3	150	Yes	Yes

2.3. Sample cross section

Three specimens at various locations from hardfaced carbon steel blades are sectioned as illustrated in Figure 3. The specimens are sectioned using EDM (Electrical discharge machining) wire cutting with a dimension of 10 mm x 10 mm x 25 mm as shown in Figure 3a. The rough surface of sectioned specimens are flattened using grit paper starting from 100, 240, 500, 800, 1000, 1200 and 2000, which are of particle size of 162.0 μm, 58.5 μm, 21.8 μm, 18.3 μm and 10.3 μm respectively and followed by polishing process. Three positions A, B, and C are focused on each specimen for microstructure analysis as illustrated in Figure 3b.

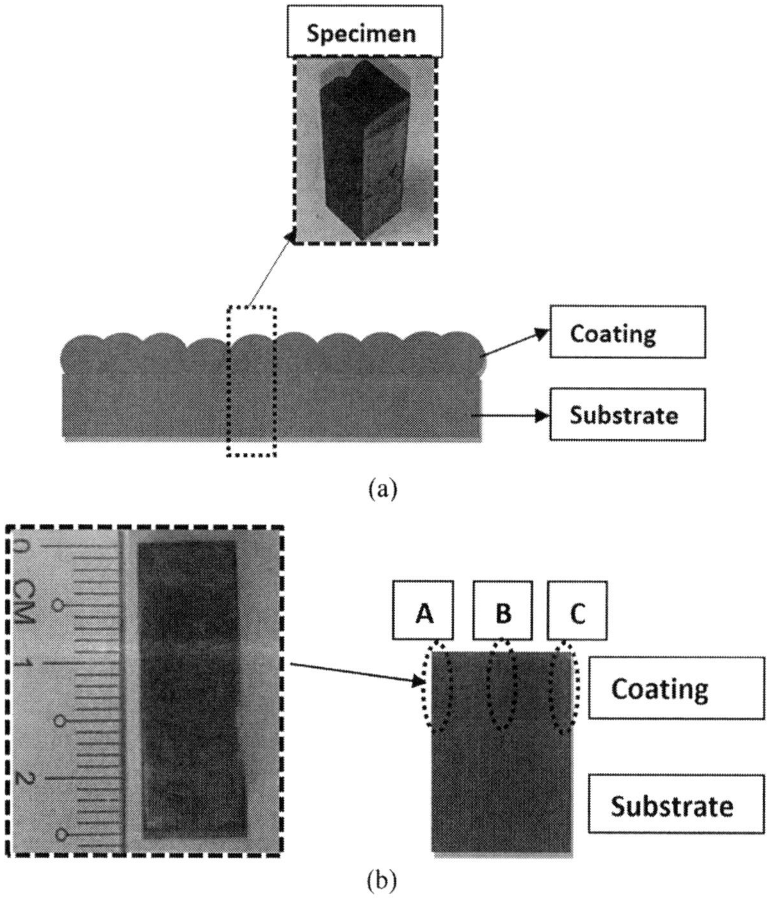

Figure 3. (a) Schematic view of the sectioned specimen from carbon steel blade and (b) specimen cross-section schematic position for coating thickness and SEM analyses.

3. Results and discussion

3.1. Micrograph of the WC hardfacing

SEM is used to investigate the microstructure of the specimens. The carbide distribution pattern is observed from the SEM images. Figure 4 shows general WC hardfacing cross-sectional view. It is noted that hardfacing specimen is made up of carbide region, non-carbide (matrix) and substrate regions. The substrate region is distinguished as being darker compared to the coating region due to the absence of tungsten element. Large-sized carbide (of about 1 mm in diameter) tends to concentrate at the lower portion of coating region (near the coating-substrate interface).

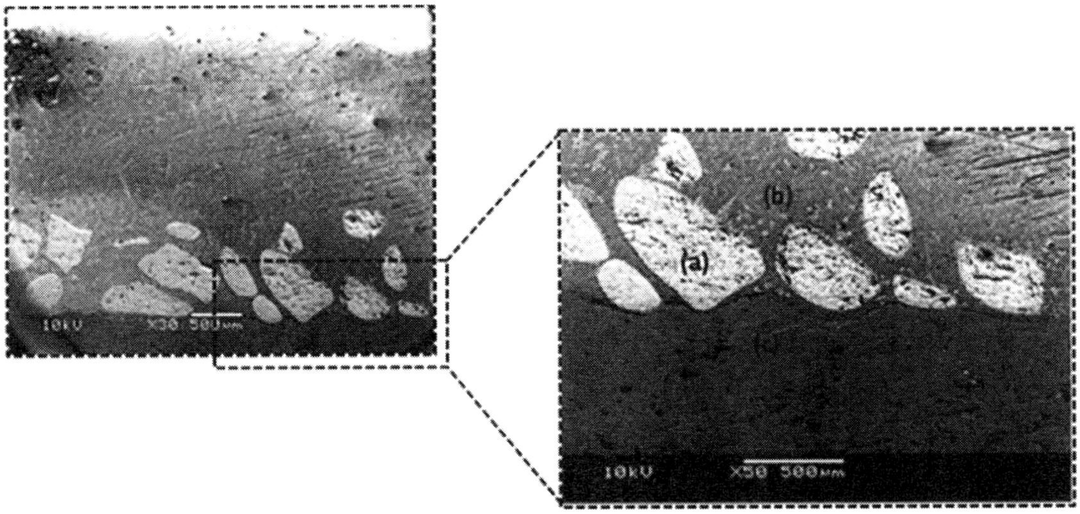

Figure 4. General WC hardfacing cross-sectional SEM image view; a) large-sized carbide region, b) non-carbide (matrix) region, and c) substrate.

EDS analysis is conducted to determine the elemental composition of the substrate and the coating. EDS analysis results are presented in Figure 5a, 5b and 5c for the carbide region, non-carbide (matrix) region and substrate region, respectively (as annotated in Figure 4). It is apparent from the figures that tungsten (W), carbon (C), oxygen (O) and iron (Fe) are identified in the EDS analysis. As expected, a high percentage of W is found in the carbide region. The non-carbide (matrix) region is rich in both W and Fe indicative of carbide and binder in close proximity. The substrate region mainly consisted of Fe.

3.2. Effect of welding layer on carbide growth

Figure 6 exhibit the effect of hardfacing layer on the microstructure of the coating. The present study indicates higher number of weld layers (three layers) caused more large-sized carbide presence in in the coating region compared to lower number of layers. It can be seen progressively the large-sized carbide presence in the coating increasing with increasing number of layers applied. Figure 7 illustrate the weldment layer effects on the WC hardfacing coating microstructure in non-carbide (matrix) region. The results of the present study show that as the number of layers increased, the nucleation and growth of the WC particle is increasing in size. The SEM images shown in Figure 7 highlight the nucleation process of the WC in dendritic shape for 1 layer case. As the number of layers increases to 2 and 3 layers, the WC particles become larger. Continuous deposition of multiple weld layers (three layers) can supply continuous heat energy for the carbide to nucleate and growth.

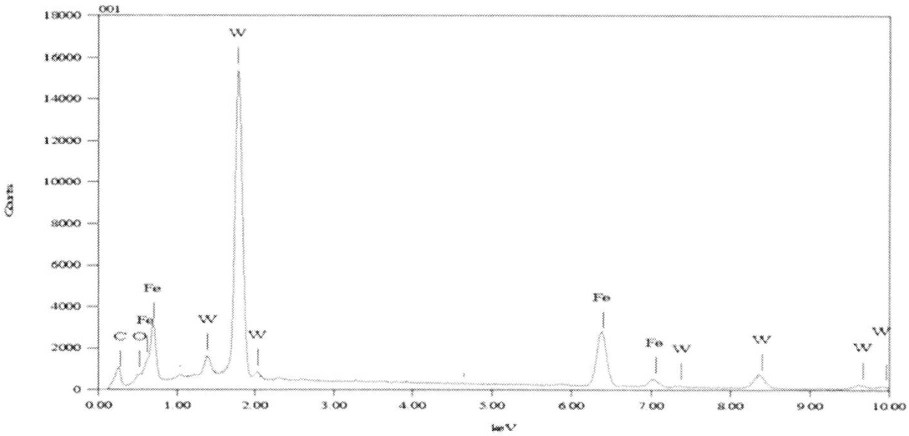

(a)

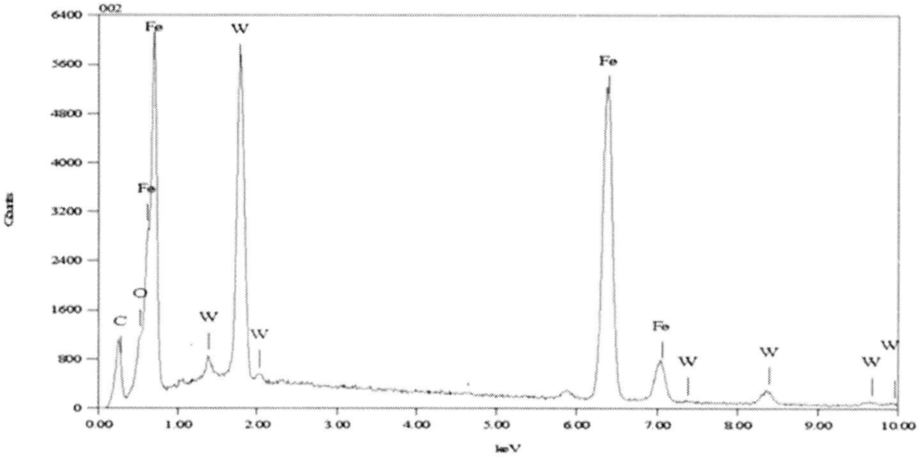

(b)

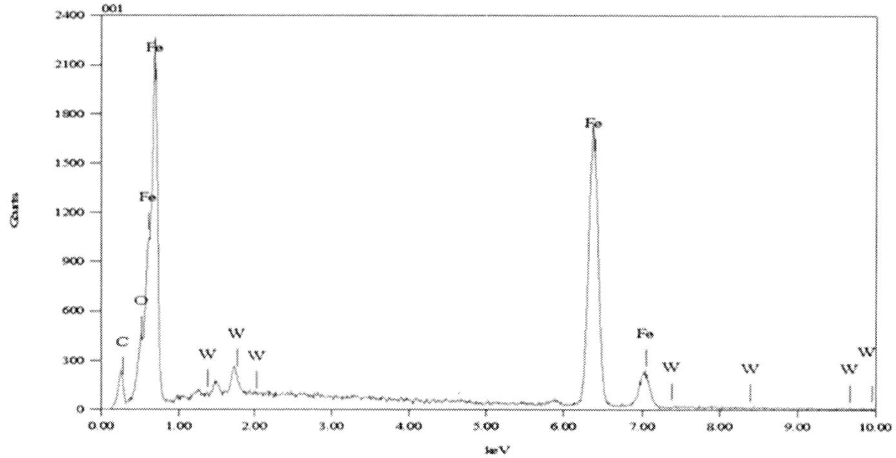

(c)

Figure 5. EDS elemental composition analysis of (a) carbide region, (b) non-carbide (matrix) region and (c) substrate region.

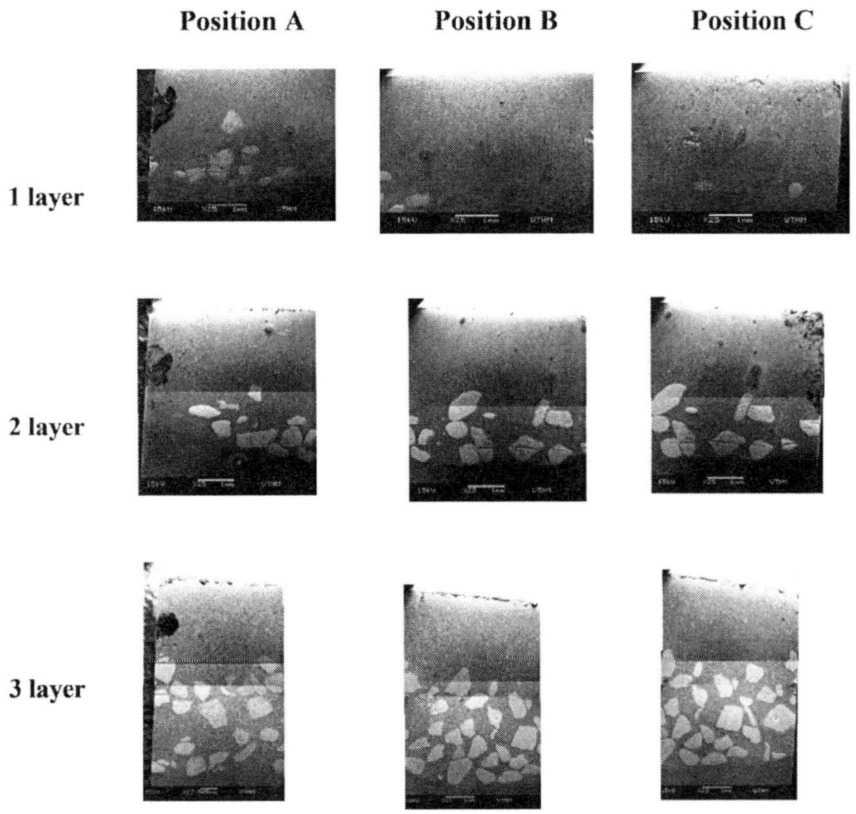

Figure 6. Microstructure of overall coating region for different number of welding layer.

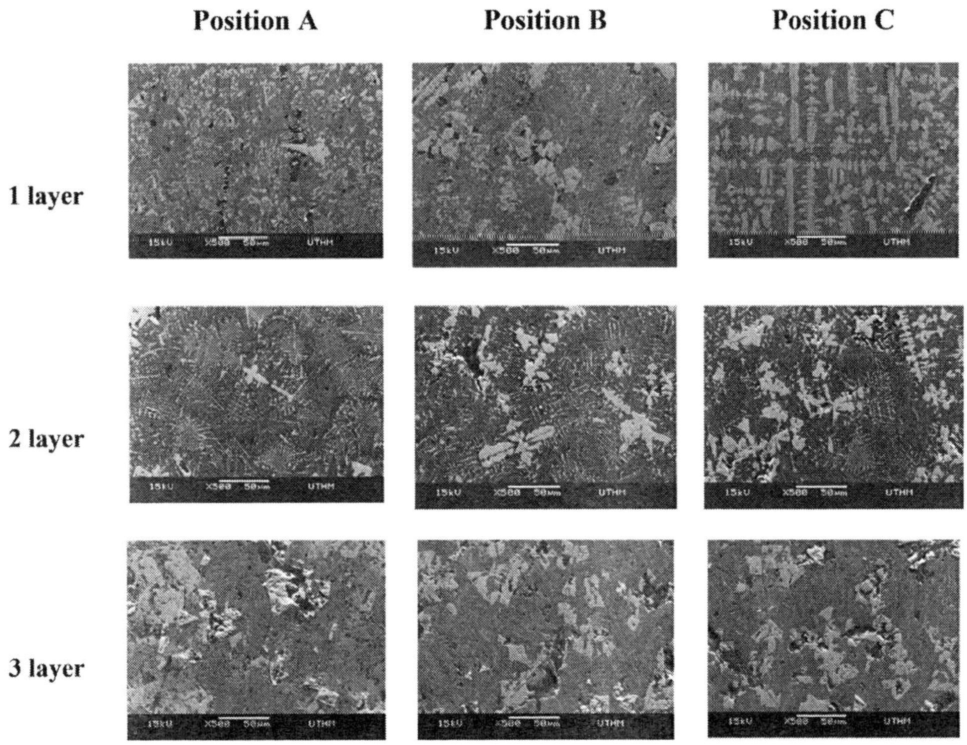

Figure 7. Microstructure of non-carbide (matrix) region for different number of welding layer.

3.3. Volume fraction analysis

WC volume fraction analysis on WC hardfaced microstructure was conducted. The volume fraction analysis is performed to calculate WC particle percentage in overall coating region. The higher percentage of WC volume fraction indicates the abundance of WC particles in the coating region. Table 5 presents the volume fraction of WC (%) with different number of welding layer. Higher WC volume fraction is registered for high number of weld layers (3 layers). Heat energy supplied due to continuous deposition of multiple weld layers (three layers) enable carbide growth, hence increasing the carbide content in overall coating region in almost linear rate of increase.

Table 5. Volume fraction of WC (%) with different number of welding layer.

Specimen	Layer	Volume fraction of WC (%)
1	1	3.4
2	2	26.64
3	3	50.12

3.4. Microhardness analysis

Hardness tests are performed using micro-Vickers hardness tester of 0.5 HV load on specimens for different position on substrate, carbide and non-carbide (matrix) regions respectively. Figure 8 shows the hardness values of the substrate, carbide and non-carbide regions. As expected, the carbide region had higher hardness compared to the non-carbide region. The hardness value registered for the carbide regions is higher (1795 HV) compared to the non-carbide region (814 HV). The hardness value will be higher if the indentation is biased toward on carbide region.

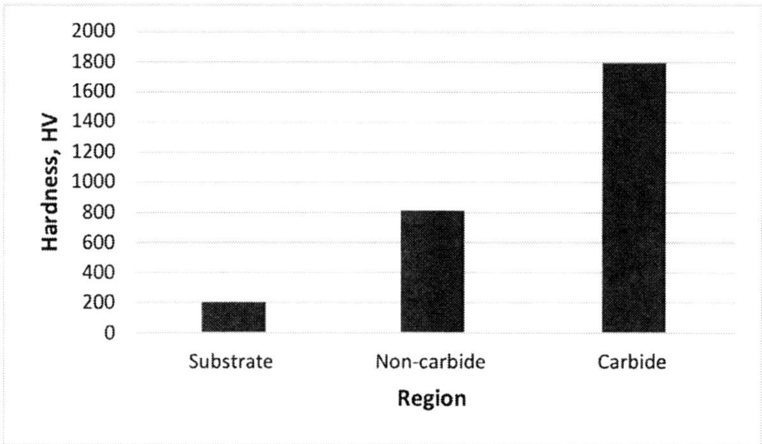

Figure 8. Hardness value for different regions of the hardfaced carbon steel blade specimen.

The hardness value of non-carbide region on the hardfaced blade with different number of weld layers is illustrated in Figure 9. Higher hardness value is registered for high number of weld layers (3 layers) compared to lower number of layers (1 layer). This supports the evidence of growth of carbide in non-carbide (matrix) region (Figure 7). Hardness is frequently used as an indicator of wear resistance of a material. Higher hardness will lead to greater wear resistance for the hardfaced blade [18]. Thus, high number of weld layers can increase the hardness of the hardfacing and subsequently enhancing the wear resistance.

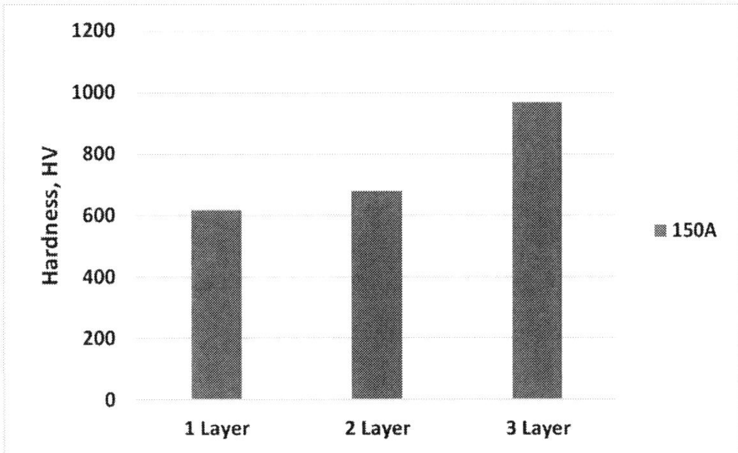

Figure 9. Hardness value of non-carbide (matrix) region for different number of welding layers

4. Conclusion

A microstructural analysis of tungsten carbide (WC) hardfacing on a carbon steel blade is examined. The influence of number of weld layers on coating microstructure, elemental presence, volume fraction and hardness are studied in details. The presented results allow the following conclusions to be made:

- Large-size carbide is presence in the coating near the substrate interface and the volume is increasing linearly with increasing number of weldment layers.

- Continuous welding of multiples hardfacing layers supplies continuous heat energy to the weldment causing WC to nucleate and growth in the non-carbide (matrix) region. This subsequently increases the overall coating hardness.

- Increasing number of layers will increase the hardness of the non-carbide (matrix) region thus increasing the overall coating hardness. This will provide better wear resistance of the hardfacing.

Acknowledgement

The authors acknowledge the financial support by the Ministry of Education Malaysia and Universiti Tun Hussein Onn Malaysia. This research is supported by the Short Term Grant (STG), Vot. No. U643 and Knowledge Transfer Programme (KTP), Vot. No. 1483.

References

1. Mohd Tobi, A. L., Harimon, M. A., Ismail, A. E., Saad, A. A., & Azalan, A. A. (2015). Investigation on the Fretting Wear of a Coated Substrate: Interlayer Stress Behaviour. In Applied Mechanics and Materials (Vol. 699, pp. 311-317). Trans Tech Publications.
2. Mohd Tobi, A. L., Harimon, M. A., Saad, A. A., & Karim, R. M. (2013). Investigation on the Fretting Wear of A coated Substrate with Interlayer. In Applied Mechanics and Materials (Vol. 315, pp. 909-913). Trans Tech Publications.
3. Mohsin, M. L., Tobi, A. L. M., Siswanto, W. A., & Tamin, M. N. (2014, November). Finite element analysis of stress intensity factor of pre-cracked coated substrate under contact sliding. In Electronics Manufacturing Technology Conference (IEMT), 2014 IEEE 36th International (pp. 1-4). IEEE.

4. Mohd Tobi, A. L., Harimon, M. A., Saad, A. A., & Azalan, A. A. (2013). Fretting Wear of Coated Substrate with Interlayer: Substrate Stress Behaviour. In Applied Mechanics and Materials (Vol. 372, pp. 516-521). Trans Tech Publications.
5. Nagentrau, M., Tobi, A. M., Kamdi, Z., Ismail, & Sambu, M. (2017, June). Microstructure Analysis of Tungsten Carbide Hardfacing on Carbon Steel Blade. In IOP Conference Series: Materials Science and Engineering (Vol. 203, No. 1, p.012014). IOP Publishing.
6. Marimuthu, K., & Murugan, N. (2013). Prediction and optimisation of weld bead geometry of plasma transferred arc hardfaced valve seat rings. Surface Engineering, 19(2), 143-149.
7. Bharath, R. R., Ramanathan, R., Sundararajan, B., & Srinivasan, P. B. (2008). Optimization of process parameters for deposition of Stellite on X45CrSi93 steel by plasma transferred arc technique. Materials & Design, 29(9), 1725-1731.
8. Upadhyaya, G. S. (1998). Cemented tungsten carbides: production, properties and testing. William Andrew.
9. Jankauskas, V., Antonov, M., Varnauskas, V., Skirkus, R., & Goljandin, D. (2015). Effect of WC grain size and content on low stress abrasive wear of manual arc welded hardfacings with low-carbon or stainless steel matrix. Wear, 328, 378-390.
10. Katsich, C., & Badisch, E. (2011). Effect of carbide degradation in a Ni-based hardfacing under abrasive and combined impact/abrasive conditions. Surface and Coatings Technology, 206(6), 1062-1068.
11. Mohd Tobi, A. L., Kamdi, Z., Ismail, M. I., Nagentrau, M., Roslan, L. N. H., Mohamad, Z. & Latif, N. A. (2017, January). Abrasive Wear Failure Analysis of Tungsten Carbide Hard facing on Carbon Steel Blade. In IOP Conference Series: Materials Science and Engineering (Vol. 165, No. 1, p. 012020). IOP Publishing
12. Coronado, J. J., Caicedo, H. F., & Gómez, A. L. (2009). The effects of welding processes on abrasive wear resistance for hardfacing deposits. Tribology International, 42(5), 745-749
13. Buchely, M. F., Gutierrez, J. C., Leon, L. M., & Toro, A. (2005). The effect of microstructure on abrasive wear of hardfacing alloys. Wear, 259(1), 52-61.
14. Desai, V. M., Rao, C. M., Kosel, T. H., & Fiore, N. F. (1984). Effect of carbide size on the abrasion of cobalt-base powder metallurgy alloys. Wear, 94(1), 89-101.
15. Shetty, H. R., Kosel, T. H., & Fiore, N. F. (1982). A study of abrasive wear mechanisms using diamond and alumina scratch tests. Wear, 80(3), 347-376.
16. Leech, P. W., Li, X. S., & Alam, N. (2012). Comparison of abrasive wear of a complex high alloy hardfacing deposit and WC–Ni based metal matrix composite. Wear, 294, 380-386.
17. Scandella, F., & Scandella, R. (2004). Development of hardfacing material in Fe-Cr-Nb-C system for use under highly abrasive conditions. Materials science and technology, 20(1), 93-105.
18. Chotěborský, R., Hrabě, P., Müller, M., Válek, R., Savková, J., & Jirka, M. (2009). Effect of carbide size in hardfacing on abrasive wear. Research in Agricultural Engineering, 55(4), 149-158.

Catalytic aided electrical discharge machining of polycrystalline diamond - parameter analysis of finishing condition

M.A. Haikal Ahmad[1], ***M. Zulafif Rahim***[2,*], ***M. F. Mohd Fauzi***[3], ***Aslam Abdullah***[4], ***Z Omar***[5], ***Songlin Ding***[6], ***A.E Ismail***[7], ***and M. Rasidi Ibrahim***[8]

[1,2,3,4,5,7,8]*Faculty of Mechanical and Manufacturing Engineering, University Tun Hussein Onn Malaysia (UTHM) P.O. Box 101, Parit Raja, Batu Pahat 86400, Johor, Malaysia.*

[6]*School of Aerospace, Mechanical and Manufacturing Engineering, Royal Melbourne Institute of Technology (RMIT) University, Melbourne, Australia;*

hd160081@siswa.uthm.edu.my[1], zulafif@uthm.edu.my[2,*], ffaizfauzi@gmail.com[3], aslam@uthm.edu.my[4], zamri@uthm.edu.my[5], songlin.ding@rmit.edu.au[6], emran@uthm.edu.my[7], rasidi@uthm.edu.my[8],

Abstract. Polycrystalline diamond (PCD) is regarded as among the hardest material in the world. Electrical Discharge Machining (EDM) typically used to machine this material because of its non-contact process nature. This investigation was purposely done to compare the EDM performances of PCD when using normal electrode of copper (Cu) and newly proposed graphitization catalyst electrode of copper nickel (CuNi). Two level full factorial design of experiment with 4 center points technique was used to study the influence of main and interaction effects of the machining parameter namely; pulse-on, pulse-off, sparking current, and electrode materials (categorical factor). The paper shows interesting discovery in which the newly proposed electrode presented positive impact to the machining performance. With the same machining parameters of finishing, CuNi delivered more than 100% better in Ra and MRR than ordinary Cu electrode.

Keywords: EDM, PCD, Graphitization, Erosion

1. Introduction

PCD is formed by a sintering process that composed diamond particles and metallic binder of cobalt under high pressure and high temperature condition. PCD is typically used as the cutting tools because of its exceptional wear resistance that associated with its extremely hardness and high transverse rupture strength. Nevertheless, this extremely hard material is difficult to be fabricated by conventional method while EDM was found as a suitable method to machine PCD with the ability to produce great surface roughness [1, 2].

In the research relating to the PCD EDM, most researchers focus on parameter optimization to increase the machining performance of material removal rate (MRR), electrode wear rate (EWR) and surface roughness (SR). Research on the chemical reaction of the PCD, while using certain electrode to improve the machining performance is very limited.

It was theorized that the nickel composition in the electrode will act as the graphitization catalyst which speed up the EDM erosion process of diamond type workpiece, in this case PCD. As mentioned by Erden et al. [3], electrode material influences the removal mechanism of EDM as the suspended material will naturally produce by the re-solidified electrode material and finally influence the removal characteristic.

As reported by J.Yan et al., more than 500% improvement was achieved by the catalyst electrode Copper Nickel, CuNi in comparison with ordinary electrode of copper, Cu. However the reported investigation is limited to only in Micro EDM application [4]. As different significantly in erosion parameter, the influence of this catalyst electrode in normal EDM is remained uncertain thus require further investigation.

2. Methodology

In EDM, the performance of machining are generally influenced by the parameters selection and types of electrode used. The PCD used in this investigation is limited to CTX002 grade, with the dimension 0.7mm x 0.5 mm for width and thickness respectively (Figure 1). Table 1 shows the properties of the PCD used.

Table 1. The properties of CTX002 PCD [5, 6]

PCD type	Grain size (μm)	Total of diamond fraction (volume %)	Total of cobalt fraction (volume %)
CTX002	2	84.8	15.2

Figure 1. Polycrystalline diamond

Suitable electrode and parameter selection plays an important part to improve the EDM performance. In this study, the performance is defined by the MRR and SR. This study was not only focused on the influence of numerical machining parameter but also the electrode material of Cu and CuNi which was regarded as a categorical factor in the DOE analysis. Three numerical parameters that were investigated in this research are; pulse-On, pulse-Off and sparking current.

The process was considered as high voltage machining as high peak voltage was fixed at 120V while the depth of cut is fixed to 0.1 mm for every experimental run. Negative polarity was used as found the best for finishing process [7]. The DOE technique of 2-level full factorial design with combination with 4 center points was applied to analyze the process as well as to develop new mathematical model for the respond prediction. The center point analysis is required for the curvature test of the developed model. Table 2 shows the machining parameter used in this investigation. The parameters were chosen as refer to the published article [8] which was mentioned as the parameter that is normally used by the industries.

Table 2. Parameter of EDM process according DOE layout

Factor	Name	Unit	Low(-)	Middle	High(+)
A	Pulse-On	μs	1	3	5
B	Pulse – Off	μs	1	3	5
C	Current	Ampere	1	3	5
D	Electrode	-	Cu	-	CuNi

The experiment is conducted by using a 4-axis SODICK EDM die-sinking machine (Figure 2) with the kerosene as a dielectric fluid. An Ultrasonic cleaning machine was used to clean the electrode and workpiece before and after the EDM process. This cleaning process of using Acetone liquid as the cleaning agent is vital to remove any deposited debris. The surface roughness of eroded surface was measured by Surface Roughness Tester SJ-400 Mitutoyo according to EN ISO 4288. The Material Removal Rate (MRR) was calculated by using the formula (1).

Figure 2. 4-axis SODICK EDM die-sinking

$$MRR = \frac{Volume\ of\ material\ removed, V\ (\mu m^3)}{erosion\ time, t\ (\mathrm{sec})} \quad (1)$$

IOP Conf. Series: Materials Science and Engineering **295** (2018) 012042 doi:10.1088/1757-899X/295/1/012042

3. Result and Discussion

All the significant factors were determined through Analysis of Variance (ANOVA) with 24 experiments for each response. Augment design were implemented as the model curvature was found significant as stated by the initial ANOVA analysis. Figure 3 and Figure 4 show the post augmented ANOVA for both responses. As indicated by the p-value of lower than 0.05, Figure 3 shows the significant factors for MRR which are single factor B (pulse-on), single factor C (pulse-off), single factor D (electrode material) and several interaction factors

ANOVA for Response Surface Reduced Cubic model

Analysis of variance table [Partial sum of squares - Type III]

Source	Sum of Squares	df	Mean Square	F Value	p-value Prob > F	
Block	5.65	1	5.65			
Model	57.60	11	5.24	45.53	< 0.0001	significant
A-current	*0.075*	*1*	*0.075*	*0.65*	*0.4240*	
B-pulse-on	*1.71*	*1*	*1.71*	*14.88*	*0.0004*	
C-pulse-off	*4.53*	*1*	*4.53*	*39.41*	*< 0.0001*	
D-electrode	*0.57*	*1*	*0.57*	*4.99*	*0.0305*	
AB	*0.80*	*1*	*0.80*	*7.00*	*0.0112*	
AC	*1.39*	*1*	*1.39*	*12.11*	*0.0011*	
AD	*0.37*	*1*	*0.37*	*3.25*	*0.0781*	
BC	*0.83*	*1*	*0.83*	*7.24*	*0.0100*	
C^2	*2.28*	*1*	*2.28*	*19.79*	*< 0.0001*	
ABC	*0.82*	*1*	*0.82*	*7.12*	*0.0106*	
AC^2	*1.55*	*1*	*1.55*	*13.47*	*0.0006*	
Residual	5.18	45	0.12			
Lack of Fit	*1.91*	*13*	*0.15*	*1.44*	*0.1966*	*not significant*
Pure Error	*3.27*	*32*	*0.10*			
Cor Total	68.43	57				

Figure 3. ANOVA for material removal rate, MRR

As shown in Figure 4, all single factors considered in this research was significantly influenced the Ra. Although the single factor of B (pulse-on) indicates the p-value of more than 0.05, it should be included into the developed model in order to maintain the hierarchy of some interaction factors involving factor B.

Analysis of variance table [Partial sum of squares - Type III]

Source	Sum of Squares	df	Mean Square	F Value	p-value Prob > F	
Block	0.018	1	0.018			
Model	0.95	19	0.050	17.40	< 0.0001	significant
A-current	*0.049*	*1*	*0.049*	*17.04*	*0.0002*	
B-pulse-on	*1.015E-003*	*1*	*1.015E-003*	*0.35*	*0.5556*	
C-pulse-off	*0.017*	*1*	*0.017*	*5.91*	*0.0200*	
D-electrode	*0.019*	*1*	*0.019*	*6.70*	*0.0137*	
AB	*6.903E-003*	*1*	*6.903E-003*	*2.41*	*0.1294*	
AC	*9.753E-004*	*1*	*9.753E-004*	*0.34*	*0.5634*	
AD	*0.12*	*1*	*0.12*	*40.72*	*< 0.0001*	
BC	*5.253E-003*	*1*	*5.253E-003*	*1.83*	*0.1842*	
BD	*0.049*	*1*	*0.049*	*17.19*	*0.0002*	
CD	*0.014*	*1*	*0.014*	*4.79*	*0.0350*	
A^2	*0.023*	*1*	*0.023*	*7.90*	*0.0079*	
B^2	*0.026*	*1*	*0.026*	*9.02*	*0.0048*	
C^2	*0.017*	*1*	*0.017*	*5.75*	*0.0216*	
ABC	*0.10*	*1*	*0.10*	*35.68*	*< 0.0001*	
ABD	*0.16*	*1*	*0.16*	*55.79*	*< 0.0001*	
A^2B	*6.246E-003*	*1*	*6.246E-003*	*2.18*	*0.1486*	
A^2D	*0.016*	*1*	*0.016*	*5.42*	*0.0254*	
AB^2	*0.066*	*1*	*0.066*	*22.83*	*< 0.0001*	
A^2B^2	*0.048*	*1*	*0.048*	*16.62*	*0.0002*	
Residual	0.11	37	2.869E-003			
Lack of Fit	*0.019*	*5*	*3.742E-003*	*1.37*	*0.2618*	*not significant*

Figure 4. ANOVA for surface roughness, Ra

Based on Figure 5 and Figure 6, the R-squared value for both responses were found between 0.8994 and 0.9176 respectively. Close the value to 1 indicates that all significant factors were included into the developed model [9].

Std. Dev.	0.34	R-Squared	0.9176
Mean	1.52	Adj R-Squared	0.8974
C.V. %	22.32	Pred R-Squared	0.8464
PRESS	9.64	Adeq Precision	22.852

Figure 5. Source of variation analysis for material removal rate, MRR

Std. Dev.	0.054	R-Squared	0.8994
Mean	0.68	Adj R-Squared	0.8477
C.V. %	7.83	Pred R-Squared	N/A
PRESS	N/A	Adeq Precision	16.525

Figure 6. Source of variation analysis for surface roughness, Ra

Figure 7 shows the 3D surface response graph of MRR for both electrodes with the used of optimized pulse-off parameter of 1μs for both electrodes. Although no different was observed with 5A current used, copper nickel electrode recorded higher MRR of 0.740μm3/s as compared to the copper electrode of 0.324μm3/s at lowest current used. For Ra, the graph as shown in Figure 8 was obtained by the optimize pulse-off parameter which are 2.6μs and 3.3μs for Cu and CuNi respectively. With the range of the investigated parameter, 0.403μm was found as the minimum Ra obtained by Cu electrode while CuNi was obtained more than 100% improvement with the Ra of 0.164μm.

It is known that electrode material will migrate to workpiece in the small discharge gap. As the high frequency sparks occur in a microns size gap, the temperature in the gap is extremely high [4]. Diamond is a metastable phase of carbon, and could convert to graphite at temperature approximately 2273K [8]. With intermediate reactivity of Ni from CuNi electrode, conversion of diamond to graphite can be accelerated. Based on this basic principle, graphite to diamond conversion will happen at lower temperature, with assistance of naturally produced suspended Ni particles which also migrated to workpiece which causes to the graphitization and diffusion-based chemical reaction [4]. These phenomena will then greatly reduce the energy required to erode the surface.

Remarkably, the voltage feedback signal collected by the oscilloscope (Figure 9) from the process in contradiction of the erosion norm in which the higher short circuit (non-energized pulse) tyoically decreases the MRR. In this research, CuNi was offered better performance, but yet short circuiting pulse was frequently appeared. Conversely for Cu electrode, although higher in sparking energy (by means of less short circuit pulse), it does not offer better in the performances. This proved that the diffusion phenomenon that happened while the electrode touching the workpiece (during short circuiting) greatly influenced the erosion performance of Ra and MRR.

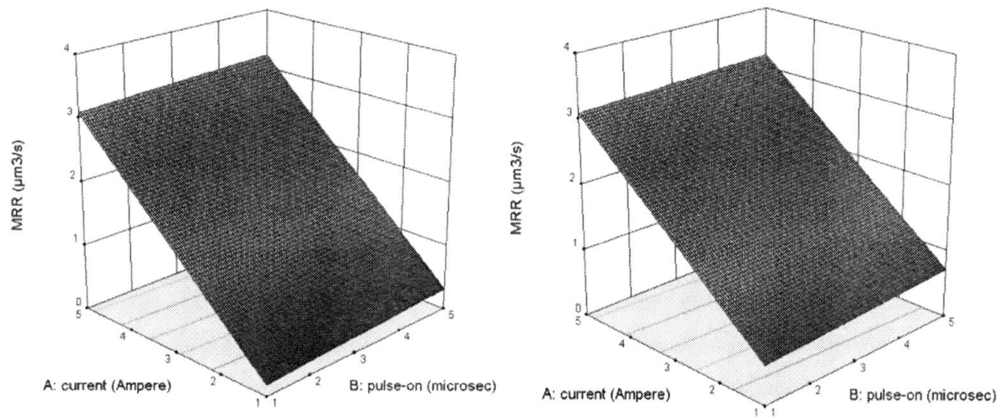

Figure 7. Response for MRR obtained by different electrodes a. Cu electrode b. CuNi electrode

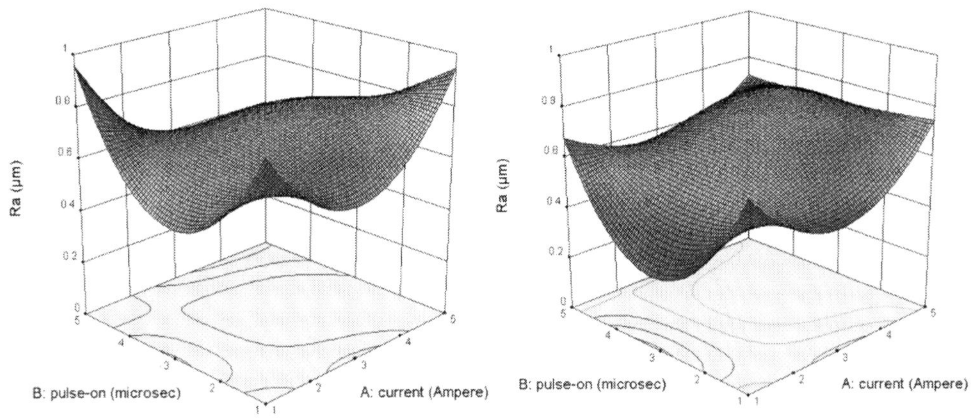

Figure 8. Response for MRR obtained by different electrodes a. Cu electrode b. CuNi electrode

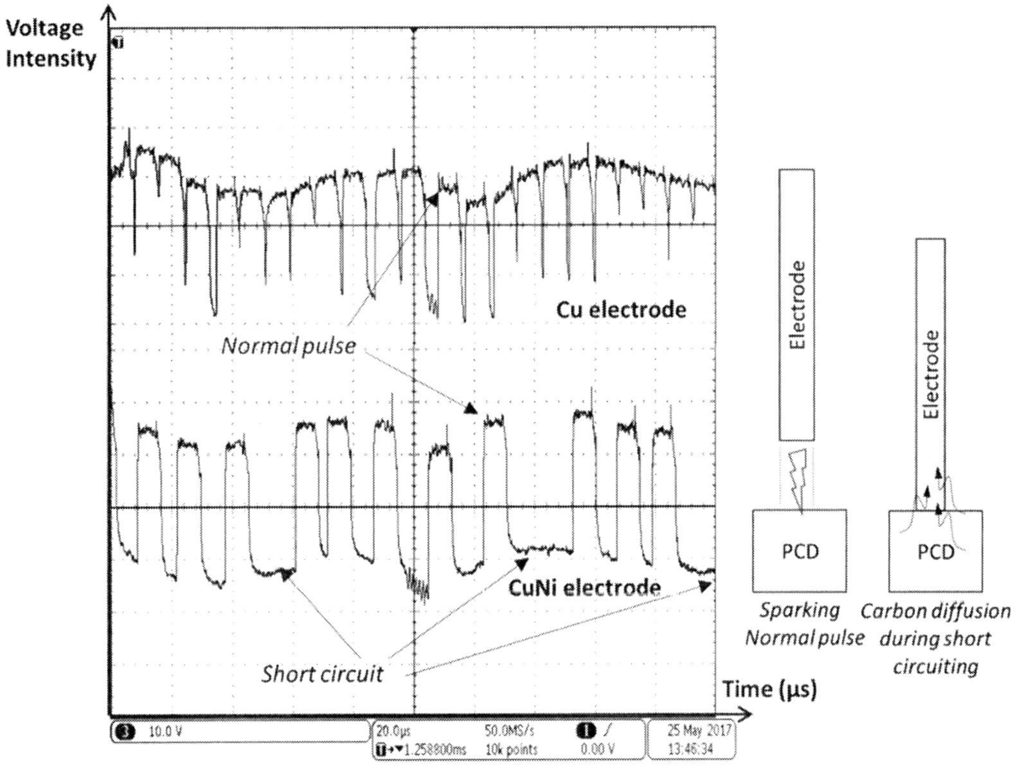

Figure 9. Voltage feedback signal obtained while using: Pulse-on of 5μs, Pulse-off of 2μs, and Current of 1A

For each response, mathematical model or equation was developed with 95% interval of confident. The mathematical model for both response were given by:

$$MRR = 1.04 + 0.20(A) + 0.22(B) - 0.36(C) + 0.10(D) + 0.1(AB) - 0.2(AC) - 0.11(AD)C^2) + 0.16(ABC) + 0.95(AC^2) \quad (2)$$

$$Ra = 0.68 + 0.17(A) + 0.03(B) + 0.02(C) - 0.03(D) - 0.02(AB) + 5.52E - 3(AC) + 0.06(AD) + 0.01(BC) + 0.04(BD) - 0.02(CD) - 0.23(A^2) - 0.24(B^2) + 0.12(C^2) + 0.06(ABC) + 0.07(ABD) - 0.06(A^2B) - 0.04(A^2D) - 0.2(AB^2) + 0.46(A^2B^2) \quad (3)$$

Confirmation run were conducted to prove the validity of the developed model. Three parameters combinations of experiments were used for both responses as shown in Table 3. As shown in Table 4, the results demonstrated that the model is reasonably accurate for the response prediction as the percentages are below than 15%.

Table 3. Parameters of the confirmation run

Run	Current (A)	Pulse-on (μs)	Pulse-off (μs)
1	5	5	5
2	5	5	1
3	1	5	1

Table 4. Predicted and actual responses values for copper electrode

Run	Electrode	Expected Value		Experimental Value		Percentage Error	
		MRR (μm3/s)	Ra (μm)	MRR ($\mu m^3/s$)	Ra (μm)	MRR (%)	Ra (%)
1	Cu	2.761	0.722	2.811	0.713	1.81	1.25
2		3.247	0.485	3.262	0.483	0.46	0.41
3		0.324	0.960	0.333	0.980	2.78	2.08
1	CuNi	2.757	0.884	2.694	0.873	2.29	1.24
2		3.242	0.729	3.251	0.713	0.28	2.19
3		0.740	0.686	0.737	0.667	0.41	2.77

4. Conclusion

The influence of copper and copper nickel electrode for MRR and Ra were determined experimentally. In the case of MRR, there is no big difference value recorded by both electrodes at highest peak current used. However, at lower peak current of 1A, better MRR was obtained by the CuNi electrode. Similarly for Ra, the influence of electrode material can be clearly seen at low current used where more than 100% improvement was obtained with the used of CuNi electrode. It can be concluded from this experiment that the electrode that is containing graphitization catalyst material of nickel was contributed to better machining performance (better surface roughness and material removal rate). However, further investigation is required to elucidate the removal mechanism involved which was caused to this improved performance.

Acknowledgement

The authors would like to acknowledge the Ministry of Higher Education (MOHE) Malaysia under Fundamental Research Grant Scheme (FRGS) grant, Vot 1586 and Universiti Tun Hussein Onn Malaysia (UTHM) under Postgraduate Research Grant (GPPS), Vot U805 for financial support.

References

1. Rahim, M.Z., S. Ding, and J. Mo, *Electrical discharge grinding of polycrystalline diamond–Effect of wheel rotation.* Machining science and technology, 2016. **20**(1): p. 62-78.
2. Li, G., et al., *Experimental study on quality of PCD tools machined by different electric discharge grinding processes.* Cogent Engineering, 2016. **3**(1): p. 1228234.
3. Erden, A., *Effect of Materials on the Mechanism of Electric Discharge Machining (E.D.M.).* Journal of Engineering Materials and Technology, 1983. **105**(2): p. 132-138.
4. Yan, J., K. Watanabe, and T. Aoyama, *Micro-electrical discharge machining of polycrystalline diamond using rotary cupronickel electrode.* CIRP Annals - Manufacturing Technology, 2014. **63**(1): p. 209-212.
5. Rahim, M.Z., et al., *ELECTRICAL DISCHARGE GRINDING VERSUS ABRASIVE GRINDING IN POLYCRYSTALLINE DIAMOND MACHINING.* Jurnal Teknologi, 2015. **74**(10).
6. Rahim, M., et al., *Electrical discharge grinding versus abrasive grinding in polycrystalline diamond machining—tool quality and performance analysis.* The International Journal of Advanced Manufacturing Technology, 2015: p. 1-15.
7. Rahim, M.Z., S.L. Ding, and J. Mo. *Electrical Discharge Grinding (EDG) of Polycrystalline Diamond-Effect of Machining Polarity.* in *Advanced Materials Research.* 2014: Trans Tech Publ.
8. Zulafif Rahim, M., S. Ding, and J. Mo, *Electrical Discharge Grinding of Polycrystalline Diamond—Effect of Machining Parameters and Finishing In-Feed.* Journal of Manufacturing Science and Engineering, 2015. **137**(2): p. 021017-021017.

Mechanical and Micro structural Characterization of Al-Al/B4C Laminated Composites Depending on Manufacturing Parameters

Göçer A[1], Nair F,[2] Karamış M B[3]
[1]Erciyes University, Department of Mechanical Engineering, 38039, Kayseri, Turkey
[2]Erciyes University, Department of Mechanical Engineering, 38039, Kayseri, Turkey
[3]Erciyes University, Department of Mechanical Engineering, 38039, Kayseri, Turkey
abdullahgocer@erciyes.edu.tr[1], fnair@erciyes.edu.tr[2], karamisb@erciyes.edu.tr[3]

Abstract: In this study, it is aimed to produce a layered composite structure with Al sheath and Al / B4C cored by extrusion technique and to characterize the mechanical and micro structure of the products. In the production of composite billets, powder in tube (PIT) method was used. In this method, Al7075 (ø 30 mm) tube is filled with Al2124 / B4C powder mixtures. The semi-finished composite billets were sintered at 500 ° C for 1 hour and extruded at the extrusion ratios of R = 14 R = 9 and R = 6.25 at the same temperature to obtain circular cross-section rods. The density changes, hardness distributions and sheath-core section ratios of these bars were determined. The changes of these values depending on the extrusion rate and B4C ratio were investigated. It is also revealed in the micro structure changes depending on these ratios.

Keywords: Bimetal Composite, Laminated Metal Composite, Extrusion, Density, Micro structure, Hardness

1. Introduction

Machine elements in the form of hollow, hollow or hollow tubes with circular or special sections are widely used in the engineering field. These elements are now produced according to the needs of different materials, both inside and outside. This requirement can be in the form of requiring different features inside and outside of the machine element, or requiring a new feature. Layered metal composite materials, a type of material developed for such situations, are a different type of typical composite materials formed by joining the interfaces of two or more materials.

Aluminum has a heavy use in machinery, automotive and aeronautics due to reasons such as high strength / density feature, ease of workability. Along with this, the work of composite materials to improve the properties of aluminum or to use the superior properties of aluminum in different areas is also increasing. When we examine layered metal composites, it is seen that aluminum is used in the production of lighter and more economical conductive materials [1, 2, 3] together with copper. Or we can see that aluminum is used together with materials like steel to obtain structures with high strength-corrosion resistance [4]. Likewise, studies have been carried out in order to combine the superior properties of two different aluminum alloys [5].

2. Material and Method

In layered metal composites, the outer part of the material is called "sheath" and the inner part is called "core". For the layered metal composites produced in this study, 7xxx series aluminum alloy was used as sheath, 2xxx series aluminum powder with average size of 30 μm and B4C powder with average size of 20 μm were used (Figure 1). The production was carried out by extruding the powder compacted tube into it.

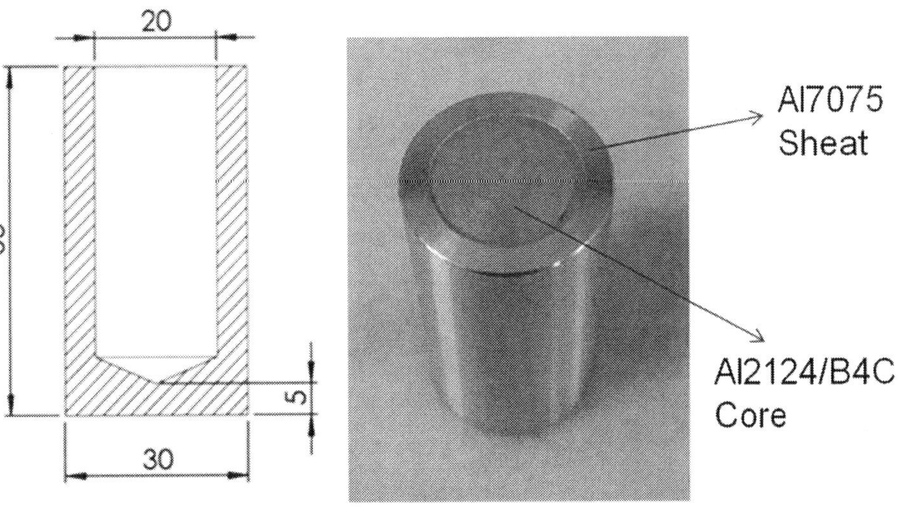

Figure 1. Sheath dimensions (mm) and prepared billets for extrusion.

Al2124 / B4C powder mixtures containing 5%, 10% and 15% B4C in volumetric proportions were used in the sheaths which the oxides removed with the grinding operation. This powder mixture is compressed into sheats at a pressure range of 400-500 kPa. The billets obtained in this way were sintered at 500 ° C for 1 hour and then immediately subjected to extrusion. The billets were extruded at the extrusion ratios of R = 14, R = 9 and R = 6.25 by decreasing the diameters of 8, 10 and 12 mm. Table 1 shows the technical data of 12 different billets with extrusion process.

Table 1. Technical data of the tickets entering the extrusion process

	Sheat	Core	D_{out}	D_{in}	Extrusion Ratio
No 1 Billet	Al7075	% 5 B_4C-Al2124	30 mm	20 mm	14
No 2 Billet	Al7075	% 5 B_4C-Al2124	30 mm	20 mm	9
No 3 Billet	Al7075	% 5 B_4C-Al2124	30 mm	20 mm	6,25
No 4 Billet	Al7075		30 mm	20 mm	14
No 5 Billet	Al7075		30 mm	20 mm	9
No 6 Billet	Al7075		30 mm	20 mm	6,25
No 7 Billet	Al7075	% 10 B_4C-Al2124	30 mm	20 mm	14
No 8 Billet	Al7075	% 10 B_4C-Al2124	30 mm	20 mm	9
No 9 Billet	Al7075	% 10 B_4C-Al2124	30 mm	20 mm	6,25
No 10 Billet	Al7075	% 15 B_4C-Al2124	30 mm	20 mm	14
No 11 Billet	Al7075	% 15 B_4C-Al2124	30 mm	20 mm	9
No 12 Billet	Al7075	% 15 B_4C-Al2124	30 mm	20 mm	6,25

Table 2 Some properties of materials used.

Material	Density(gr/cm^3)	Powder size
Al 7075	2,81	-
Al 2024	2,767	30 μm
B_4C	2,52	20 μm

After 12 different billets were extruded under the same conditions, samples were taken from the obtained circular materials (Fig. 2) in suitable sizes for hardness and density examination. Hardness values were measured as micro vickers hardness measurements. The density measurements of the materials were made with Archimedes principle. The productions and tests were carried out at the Composite Laboratories of the Mechanical Engineering Department of Erciyes University.

Figure 2. a) Extrusion mold set and press b) Extrusion billet and product.

3. Results

3.1. Hardness Dispersion

The hardness values at the points shown in Fig. 3 were taken from the sections of the rods obtained from the extrusion end. The hardness test made on the cross section will show the behavior of compacting the material, especially the powdered core, after the extrusion process. For the extrusion process is carried out under warm conditions, it is not expected that the deformation hardens at high rates. When we look at the hardnesses of the Al 7075 bars (Figure 4), the hardness values are increased by at most 7-8% with increasing extrusion rate. However, in composite structures, the situation is somewhat different. The hardness value increases when going to the boundary of the sheath, it falls when it goes to the core region. It is believed that the rise in the sheath zone is due to the fact that the Al 7075 region is more deformed at the boundary, due to the different flow behaviors of the two separate material structures. The hardness in the core region is relatively lower than that of the powder structure.

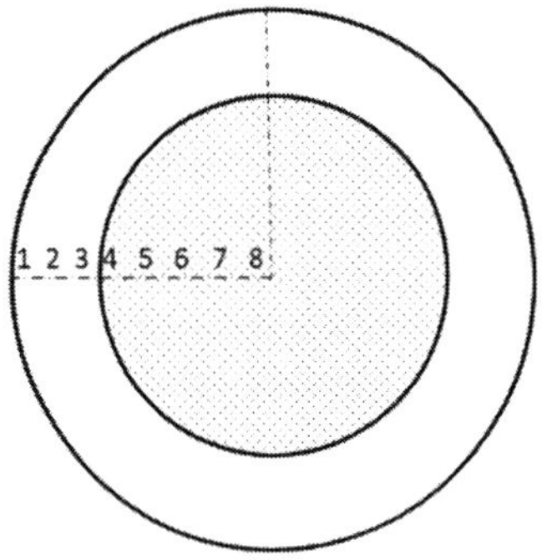

Figure 3. The points taken hardness values

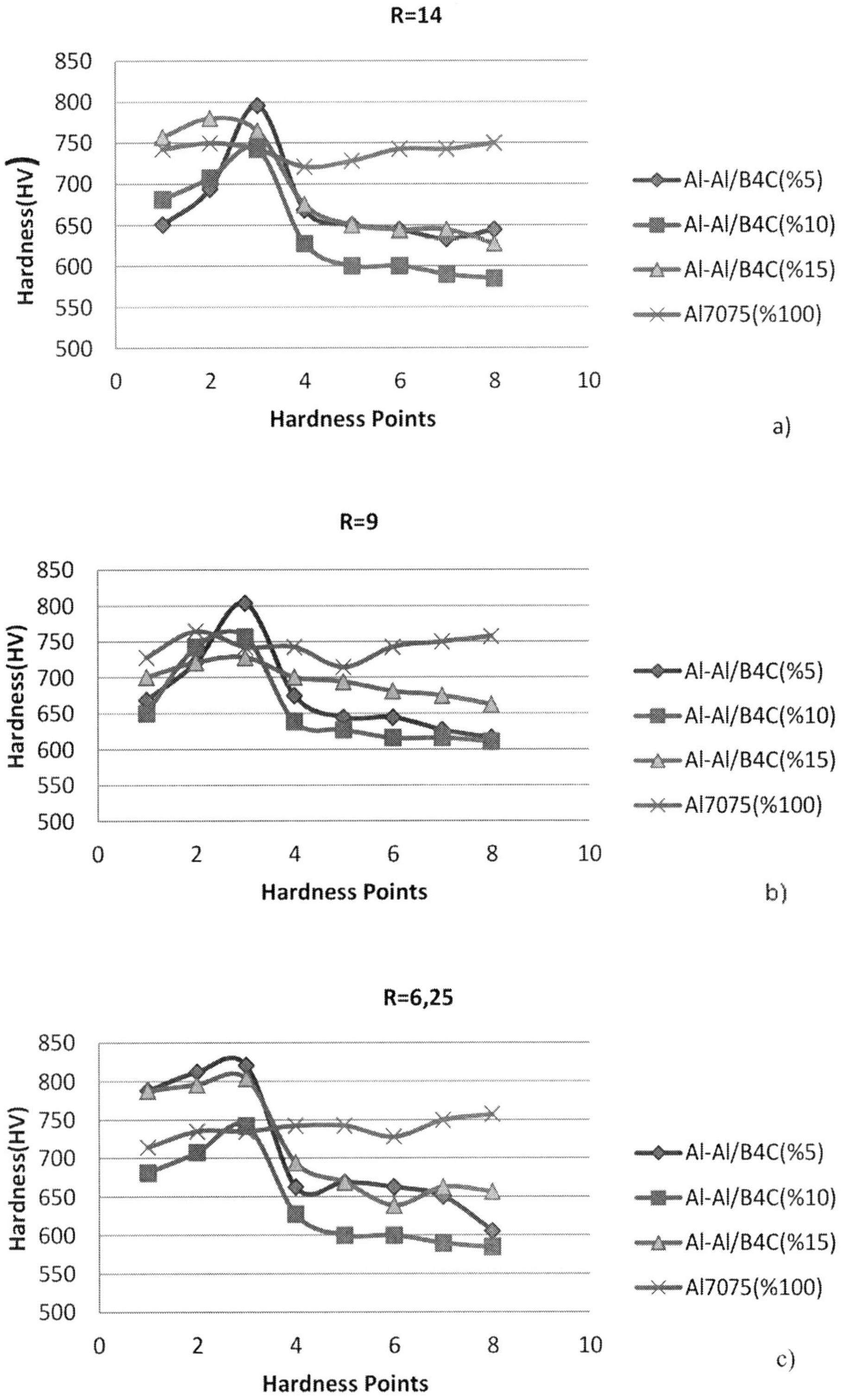

Figure 4. Cross-section hardness distributions depending on extrusion ratios a) R=14 b) R=9 c) R=6.25

When we examine the effect of varying the amount of B4C in the same extrusion ratios, the material has 5% B4C ratio is the highest intersection hardness. As the B4C ratio increases, the hardness distribution in the cross section becomes more stable. The hardness distribution in the core is increased by B4C but not by the extrusion rate (Fig. 5).

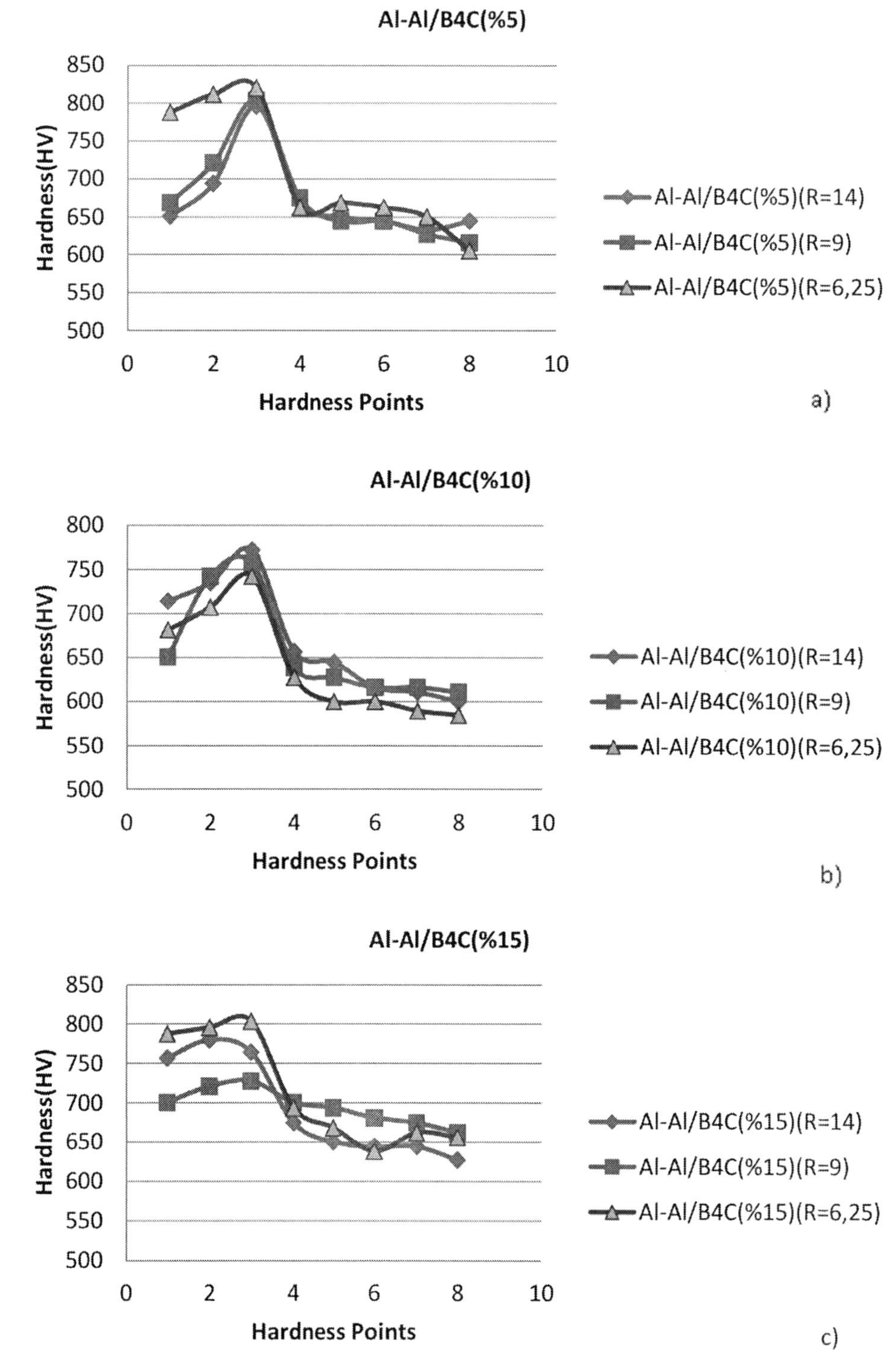

Figure 5. Hardness distributions on the cross section with the B4C ratios

3.2. Density and Cross Section Change

The sheath radius and the core radius of the composite material are proportioned and are defined as "β". And we looked at the change of β ratio by B4C amount and extrusion rate. It is important for the balancing of the flow characterization that the β ratio is as well as the billet. This ratio is the average β = 1.45 of the billet. β ratio does not change significantly with the extrusion rate. However, with the increase of B4C ratio, β ratio is seen to increase a little. This indicates that the B4C ratio reduces the plastic deformation capability of the core (Fig. 6).

The density of the composite material increases with the increase of B4C ratio as expected. However, there is no significant increase in the density with the increase of the extrusion rate (Fig. 7). It is believed that the reasons for this are the compression of the powder into the sheath with the minimum porosity and the hot extrusion process.

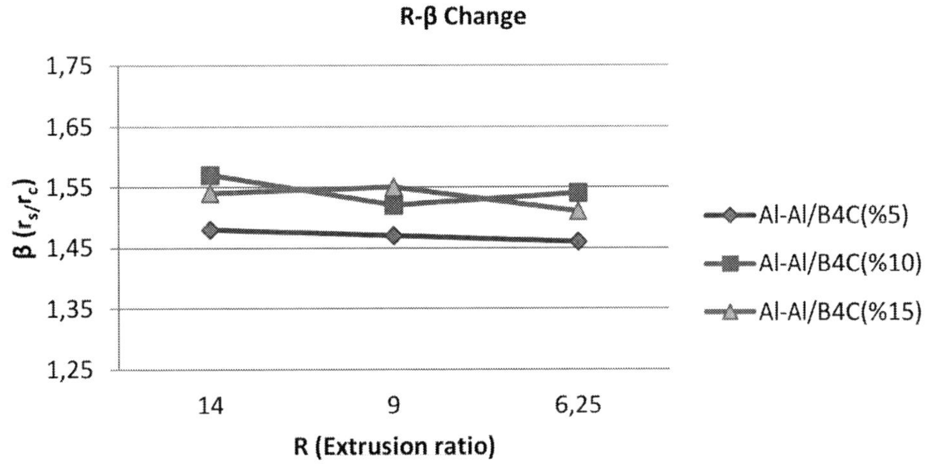

Figure 6. Change of β ratio by B4C amount and extrusion ratio.

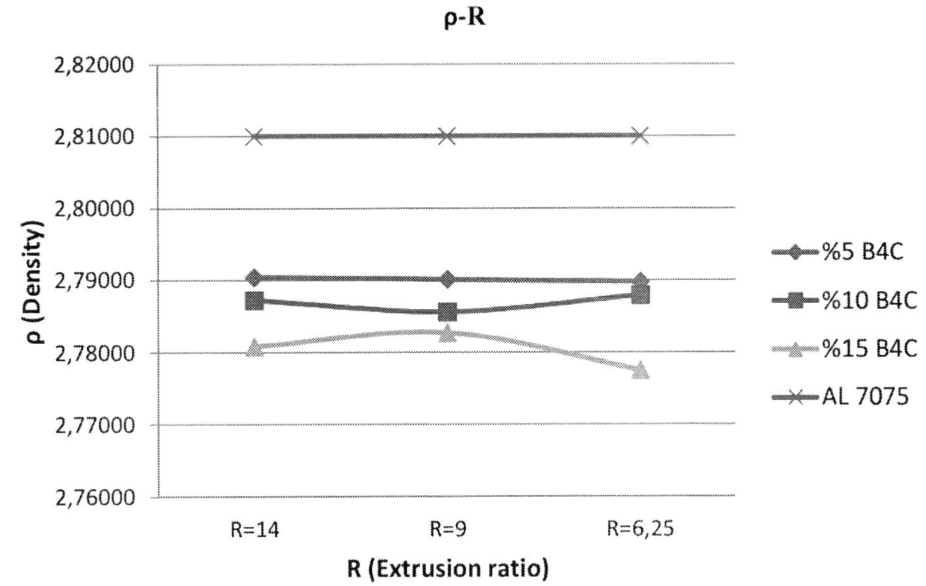

Figure 7. Change of density by extrusion rate.

3.3 Micro structure

When we look at the microstructures of layered metal composites, it appears that the grain sizes of Al7075 are slightly larger than the Al2124 grain sizes (Figure 8). This difference is thought to be due to the fact that Al2124 is produced by sintering in the form of powder, even though the both materials pass the same process conditions. When the sheath-core regions of layered metal composites produced in the form of Al7075-Al2124 and Al7075-Al2124 / B4C are examined, it is seen that the joining is very well performed.

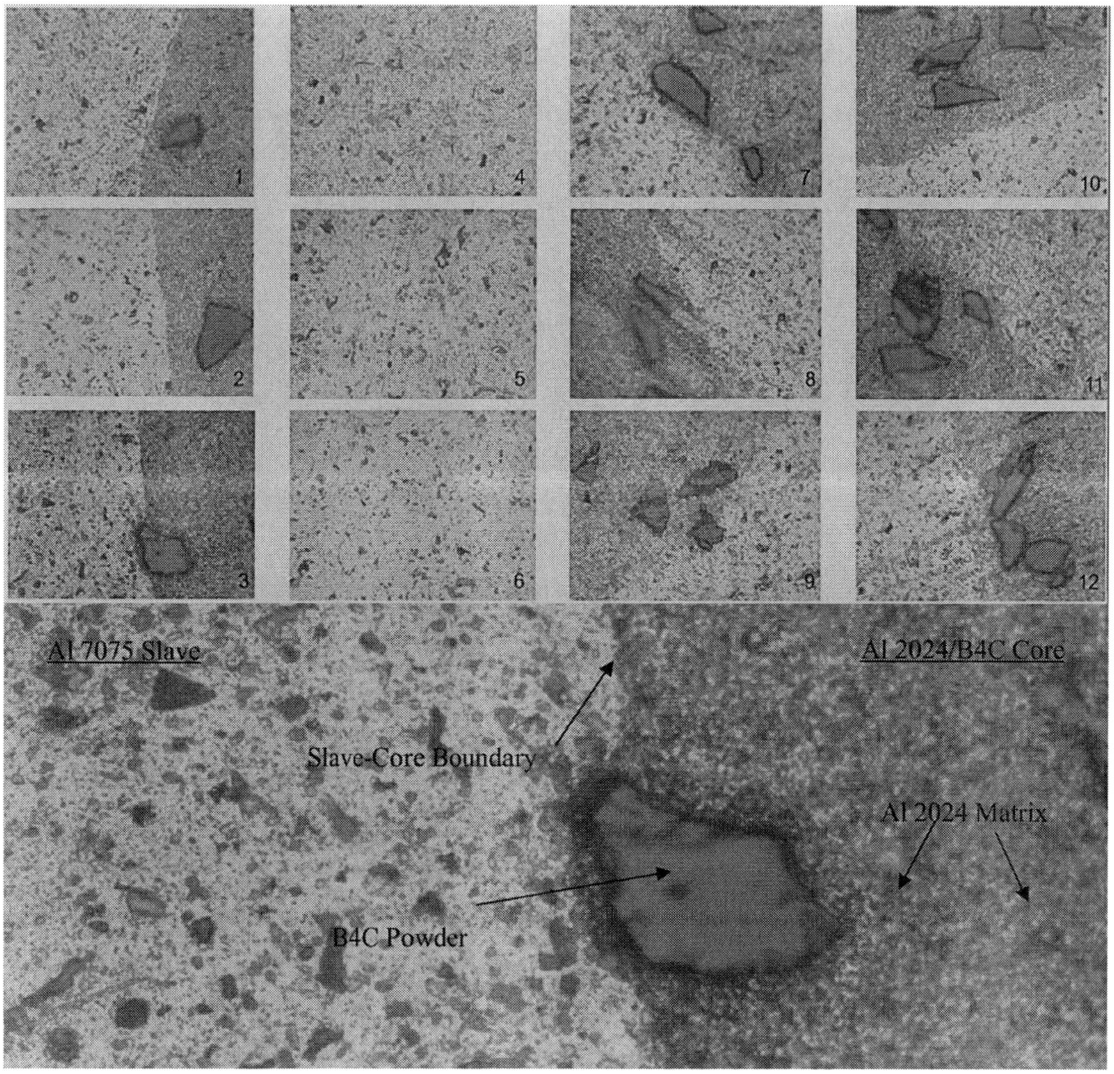

Figure 8. Microstructural images taken from materials.

4. Conclusion

The mechanical and microstructural changes of the produced composite bars according to the production parameters were investigated. The results obtained can be summarized as follows:

1. Hardness values rises in the sheath areas close to the sheath-core boundary. This can be caused by the high rate of plastic deformation in these areas.
2. The hardness decreases from the core to the center. This is due to the flow characteristic of the extrusion
3. The hardness does not increase with the increase of B4C ratio. The increased porosity caused by the increased B4C causes the hardness not to rise.
4. The hardness is increased by the increase of the extrusion rate. Because of this reason the porosity is decreasing.
5. Homogeneous distribution is observed in the core region.
6. As the extrusion ratio increases, the solid solutions precipitated on the grain boundaries become smaller and their dispersions become homogenous.
7. β ratio shows a slight change. The reason for this is thought to be that oiling is not good and billets are not produced as standard.

Acknowledgement

This project was supported by the Scientific Research Projects Unit of Erciyes University (Project number: FDK-2015-6265).

References

1. Rhee, K. Y., Han, W. Y., Park, H.J., Kim, S. S., (2004) "Fabrication of aluminum copper clad composite using hot hydrostatic extrusion process and its material characteristics", Materials Science and Engineering A, vol. 384, pp. 70-76.
2. Park, H. J., Na, K. H., Cho, N. S., Lee, Y. S., Kim, S. W., (1997) "A study of the hydrostatic extrusion of copper-clad aluminium tube", Journal of Materials Processing Technology, vol. 67, pp. 24-28.
3. Kang, C. G., Kwon, H. C., (2002) "Finite element analysis considering fracture strain of sheath material and die lubricant in extrusion process of Al/Cu clad composites and its experimental investigation", International Journal of Mechanical Sciences, vol. 44, pp. 247-267.
4. Xunzhong, G., Jie, T., Wentao, W., Huaguan, L., Chen, W., (2013) "Effects of the inner mould material on the aluminium–316L stainless steel explosive clad pipe", Materials and Design, vol.
5. Son, I. S., Lee, S. P., Kim, W. C., Moon, J. S., Lee, S., Lee, J. S., Kim, Y. B., Lee, G. A., Bae, D. S., (2014) "Effect of hydro co-extrusion on microstructure of duo-cast Al 3003/Al 4004 clad materials", Transactions of Nonferrous Metals Society of China, vol. 24, pp. 75-80.

IOP Conf. Series: Materials Science and Engineering **295** (2018) 012044 doi:10.1088/1757-899X/295/1/012044

The Effect of Coatings on the Wear Behavior of Ti6Al4V Alloy Used in Biomedical Applications

Danışman Ş.[1]***, Odabas D.***[2]***, Teber M.***[3]
[1-2]*Erciyes University, Department of Mechanical Engineering, 38039, Kayseri, Turkey*
[3]*HES Hacılar Electricity Industry and Trade Inc., Kayseri, Turkey;*
sdanisman@erciyes.edu.tr[1]*, odabasd@erciyes.edu.tr*[2]*, muharremteber@hotmail.com*[3]

Abstract: The properties expected from implant materials are biocompatibility, long service life and wear resistance. The wear resistance of the implant materials varies according to the type of implant, usage area and the movement. The ability of implant material to be more compatible with biological tissues and to increase the useful life depends on the surface properties. Today many different kind of surface modification techniques are applying on medical and dental implant surfaces to improve surface specifications and wear resistance.

In this study TiN, TiAlN, TiCN coatings were applied on Ti6Al4V alloy used as implant material by reactive magnetron sputtering method. The wear resistances of uncoated Ti6Al4V alloy and TiN, TiAlN, TiCN coatings were investigated at room temperature under dry conditions. The wear resistance at different load and different sliding rates were compared using an wear tester. The wear losses, wear track widths and friction coefficients of coated and uncoated Ti6Al4V alloys are taken into account for comparison. The results show that a significant improvement in wear resistance of the alloy with the coating is achieved and TiN-coated Ti6Al4V alloy has the highest wear resistance.

Keywords: Biomedical, Ti6Al4V alloy, TiN, TiAlN, TiCN coatings, wear.

1. Introduction

Although it is a new field in the scientific sense, the use of biomaterials extends to ancient times in history. The use of gold in dentistry extends up to 2000 years ago. Since the middle of the 19th century serious progress has been made in the use of different materials in the body. During the last 30 years important information has been obtained about understanding the interactions of biomaterials in the body[1-3]. Metallic biomaterials are one of the best materials that best match the mechanical conditions. This is because metallic biomaterials can withstand long-term, variable and sudden loadings without losing their properties[4,5]. The body has a dynamic environment that will create high wear conditions for metals[4-6]. The use of materials with as high a wear resistance as possible provides great convenience in medical applications. The main current problem with implants is the nature of the implant-tissue interface[7-10]. In recent years there has been a serious increase in the use of titanium and titanium alloys. Because titanium alloys have these advantages such as strength, rigidity, low density, corrosion resistance. However, the tribological properties of the most commonly used Ti6Al4V alloy in the biomedical field are not sufficient[11-13]. Improving the surface characteristics of the implant materials is very important. In order to reduce wear, researchers follow two paths: (i) use new, wear resistant materials or (ii) improve the wear resistance of materials by adding alloying elements or performing surface treatments. Thin film hard nitride coatings are seen as a viable way to enhance the wear resistance of metallic materials, thus extending the lifespan of products[14].

Today, there are numerous methods developed to coat the implant material. These include chemical vapor deposition (CVD), physical vapor deposition (PVD), ion beam, laser, plasma spray, vacuum plasma, HVOF (high velocity oxygen fuel thermal spray process), sol gel immersion and EPD[15-21]. Physical vapor deposition (PVD) techniques are used extensively to deposit a wide range of materials on a number of substrates. The general classification of PVD techniques is made according to the methods used to evaporate the target (material to be deposited)[14]. Titanium nitride-based coatings are mostly deposited by magnetron sputtering, cathodic-arc and pulsed laser deposition techniques[14]. DC reactive magnetron sputtering from an elemental target is a popular technique that can produce films with controllable stoichiometry and composition at high deposition rates and on an industrial scale[14].

In this study, wear behaviors and surface properties of Grade 5(Ti6Al4V) alloys, which are getting more and more important in the production of biomedical materials, have been investigated in dry conditions as a function of load and sliding rate parameters with use of wear test instrument . Then TiN, TiAlN, TiCN coated alloys by reactive magnetron sputtering were subjected to wear tests under the same conditions. The wear losses and wear trace widths of coated and uncoated Ti6Al4V alloys were determined after wear test. The results demonstrated that a significant improvement in wear resistance of the alloy with the coating is provided and that the TiN coated Ti6Al4V alloy is more resistant to wear than the uncoated Ti6Al4V and other coating types.

2. Materials and Methods

2.1. Sample preparation and Coating Method

The samples were prepared from the base material Ti6Al4V (grade 5). The chemical composition of the material is given in Table 1. Ti6Al4V alloy developed in 1954 is consumed 75-85% in all titanium alloys due to its heat resistance, strength, flexibility, toughness, workability, weldability, corrosion resistance and biocompatibility[11-13]. Test specimens were cut out of Ti6Al4V plate and prepared in sizes of 20x65x3 mm. Prior to coating, the substrates were ultrasonically cleaned in alcohol and in acetone for 15 min each, rinsed in distilled water and dried.

Table 1. Chemical compositions of Ti6Al4V Alloy (wt. %)

Ti	N	Fe	Al	V	C
Bal.	0.007	0.06	6.11	4	0.01

In this study, TiN, TiAlN, TiCN coatings were made with reactive magnetron sputtering method. The system has a stainless steel vacuum chamber (600 mm in diameter and 800 mm in high) and two unbalanced planar magnetrons powered by two independent 12-kWDC generator. Sputtering is basically the removal of atomized material from a solid by energetic bombardment of its surface layer which is dependent on the exchange of momentum. The high energy particles used are usually those of a heavy inert or reactive gas (argon, the most commonly used inert gas) or the coating material forms a positive ion[22]. The sputtering material is thrown in an atomic state from the coating material which is called the target. The base material is placed in front of the target to interrupt the flow of sputtered atoms. Magnetron sputtering is a coating technique that is used to produce thin films with magnets placed behind the target in order to limit the plasma. The film layers having different properties are obtained by changing the deposition parameter and the target material[22,23]. Figure 1 shows the principle of the magnetron sputtering method[22,23].

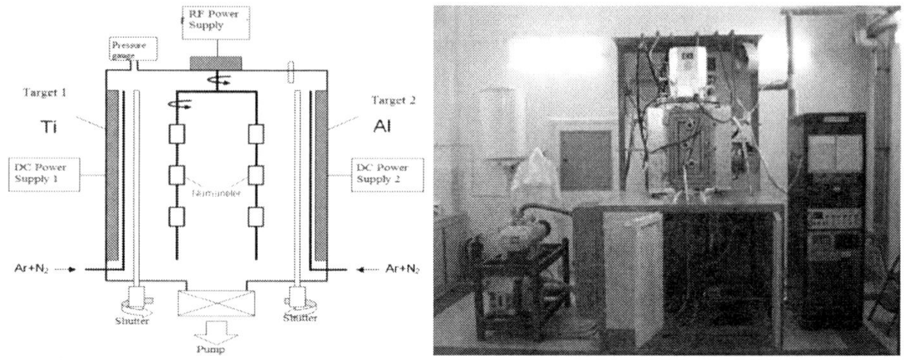

Figure 1.a. Schematic view of the magnetron sputtering method **.b.** PVD magnetron-sputtering system

The test specimens were divided into four groups with uncoated and coated. Then the substrates to be coated are suspended in the middle holders and rotated during coating as shown in Figure 1.a.

As the coating process parameters, 5000 W target power, 80 V bias voltage was used in the experiment. The coating process applied to Ti6Al4V and the silicon wafers was conducted as follows: A pressure of 5 mtorr Ar(~25 sccm) and target power of 4500 W were used for about 5 minutes to clean the target. Then a pressure of 5 mtorr, a target voltage of 150 W and DC voltage of 750 V were used for about 25 minutes to clean the substrate. In order to improve adhesion to the substrate, first of all, the Ti coating was applied at a target power of 4000 W and 15 sccm argon flows for 10 minutes and then the other coating was applied using a target power of 5000 W, -80 V bias voltage and argon + N_2 atmosphere in the coating process. TiN, TiAlN and TiCN coatings were applied for 3 hours during the coating experiment.

2.2. Wear Tests

The wear tests of the uncoated Ti6Al4V alloys and coated samples were performed on a wear device as shown Figure 2.a. In the wear tests, Plint's TE 53 SLIM multi-purpose friction and wear test device was used.

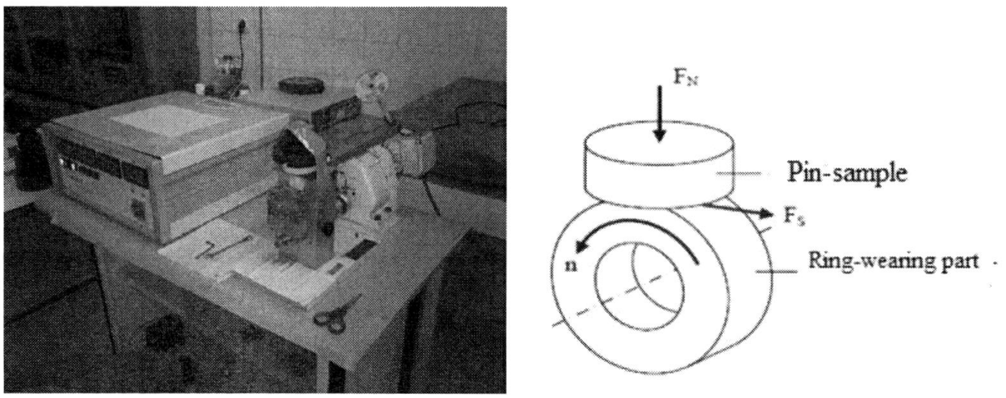

Figure 2.a. Plint's TE 53 SLIM multi-purpose friction and wear test device **b.** Contact geometry and forces for pin-on-ring system.

The wear behaviours of TiN, TiAlN, TiCN coated by PVD-reactive magnetron sputtering and uncoated Ti6Al4V samples were investigated in a wear tester adapted to pin-on-ring system (ASTM G-77) (Fig. 2.a and Fig. 2.b). DIN 100Cr6 material with 26 mm diameter and 5 mm height(bearing outer ring) was used as wearing part in the wear tests. İts hardness is 61 RC and surface roughness R_a

0.12 μm. Counter wearing part material composition is as follows: C 0.99%, Mn 0.38%, Cr 1.42%, Mo 0.02% and Fe balance. The experiments were carried out at room temperature under dry friction conditions using constant sliding speed (0,5 m / s and 0.9 m / s) and normal load (5, 15 and 30 N) were used as variable parameters. The test time was stable at 15 min.

3. Experimental Results and Discussion

3.1. Coating Characterization

In this study, TiN, TiAlN, TiCN coatings was made with reactive magnetron sputtering method. Before coating, strength and hardness of all samples were increased by plasma nitriding at 25 ° C in an atmosphere of 25% Ar + 75% N2 for 1 hour at 650 ° C. The hardness of the Ti6Al4V samples were measured under 100 g load by Struers microhardness device before and after plasma nitriding. The hardness of the untreated Ti6Al4V alloy is 384 HV (Vickers hardness) and has an average hardness value of 592 HV after plasma nitriding. After coating, the hardness of TiN, TiAlN, TiCN were measured by CSEM nanohardness tester using a square-based diamond Vickers tip. The hardness values of the coated samples with thicknesses of average 2 micrometers were found to be in the range of 3500-1050 Hv. hardness values were measured approximately 3500 HV for TiN, 1960 HV for TiAlN and 1050 HV for TiCN, respectively.

The surface roughness values were measured in the test specimens by Mitutoyo Surftest roughness device. The measured average surface roughness values of samples before and after coating are shown in Table-2.

Table 2. Average surface roughness Ra (μm) values for Ti6Al4V alloy and coatings

Step no	Ti6Al4V	TiN	TiCN	TiAlN
1	0.31	0.19	0.18	0.55
2	0.54	0.66	0.26	0.51
3	0.38	0.56	0.21	0.24
4	0.52	0.47	0.17	0.30
Average values	0.44	0.47	0.21	0.4

As shown in figure 3 , AFM analyzes to measure surface roughness were also performed and the results were in agreement. The AFM images also show that TiCN has the lowest surface roughness.

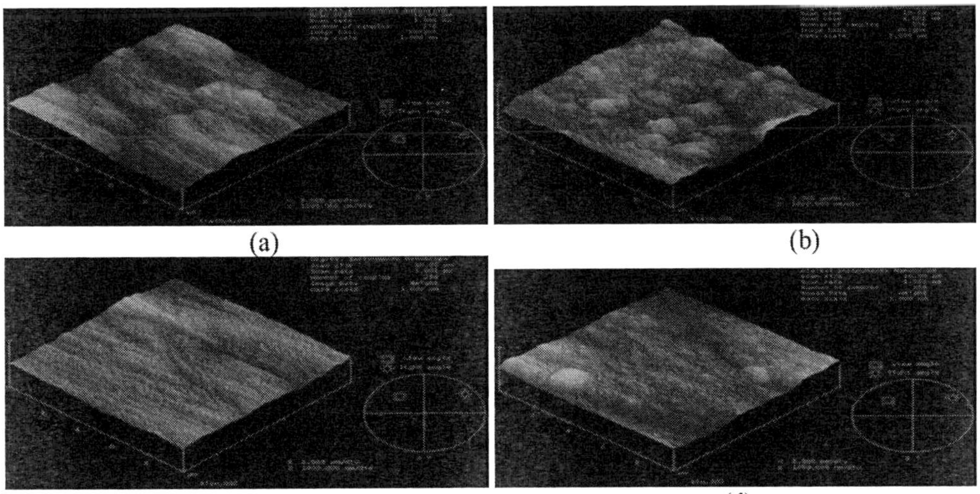

(a) (b) (c) (d)

Figure 3. AFM images a)Ti6Al4V, b)TiAlN, c) TiCN, d) TiN

The coating thicknesses were measured from silicon wafer cross-sections by SEM analysis. As shown in figure 4, the thicknesses of the coatings on the silicon wafer are very stable. The thicknesses were about average 2 μm. The coating exhibited columnar crystal structure[24].

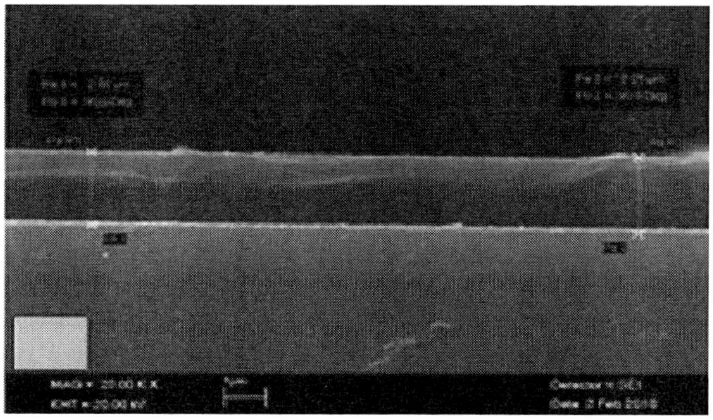

Figure 4. Cross sectional SEM image for the TiCN coating thickness

3.2.Results of Wear Test

Before and after the wear test, each specimen and wearing part were cleaned with alcohol, the weight loss of the specimen and wearing part after the test and the wear track widths on the samples and the average coefficients of friction in steady-state sliding regime were determined. In the pin-on-ring wear tests, experiments were carried out using 24 samples

A sliding distance depend on velocity and time. Friction forces (F_s) depending on sliding distance were recorded using Picolog (Data Loger) in the wear tests. The friction force was found from the steady state friction zone as shown in the figure 5 for every experiment. In the literature, similarly, TiAlN showed a long-running in process and a high steady state friction coefficient [14].

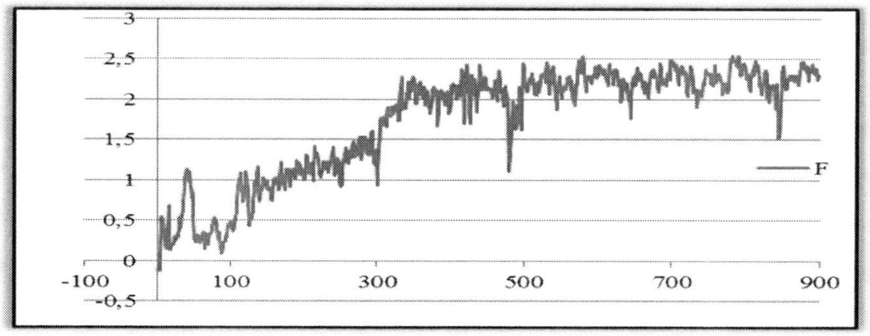

Figure 5. The friction force from the the steady state friction zone for TiAlN.

To calculate the friction coefficient in the wear test, the equation 1 was used. In the literature the change of friction coefficient with time during sliding abrasion was obtained [24]. The friction coefficient can be divided into an adhesion component and a deformation component, which result from the adhesion force and plastic deformation force, respectively[24].

$$\mu = Ft/Fn \quad (1)$$

Where; μ is friction coefficent, Fn is normal load, Ft is friction force (the graphical value).

According to the results of the wear test, the weight loss, wear track widths, coefficient of friction of the samples depending on the test parameters are given in Table 3.

Table 3. The influence of the sliding rate and load on wear behaviour.

Materials	Sample no	Speed (ms^{-1})	Load [N]	Time (min)	Weight loss (mg)	Wear track (mm)	Frictional coefficient μ
	1	0.5	5	15	2.1	3.360	0.399
	2	0.9	5	15	3	3.703	0.727
Ti6Al4V	3	0.5	15	15	3.7	3.828	0.216
	4	0.9	15	15	4.6	4.095	0.177
	5	0.5	30	15	4.9	4.240	0.133
	6	0.9	30	15	6.8	4.525	0.073
	1	0.5	5	15	0.2	0.610	0.502
	2	0.9	5	15	0.2	0.680	0.749
TiN	3	0.5	15	15	0.3	0.760	0.463
	4	0.9	15	15	0.4	0.897	0.389
	5	0.5	30	15	0.5	0.800	0.336
	6	0.9	30	15	0.7	1.508	0.314
	1	0.5	5	15	0.3	0.679	0.702
	2	0.9	5	15	0.6	1.025	0.591
TiCN	3	0.5	15	15	0.4	2.348	0.363
	4	0.9	15	15	0.7	2.815	0.346
	5	0.5	30	15	2.2	3.683	0.225
	6	0.9	30	15	8.5	4.468	0.160
	1	0.5	5	15	0.2	0.856	0.690
	2	0.9	5	15	0.3	0.970	0.556
TiAlN	3	0.5	15	15	0.5	0.707	0.331
	4	0.9	15	15	1.9	3.710	0.235
	5	0.5	30	15	3.1	3.725	0.212
	6	0.9	30	15	7.1	4.423	0.147

In order to understand the variation in frictional coefficient with load, tests were carried out at a sliding rate, a speed of 0.5 m/s, and loads in the range 5-30 N under dry friction conditions. During the tests, the friction forces were continuously recorded throughout the test by a multichannel recorder . Steady state sliding regime average friction coefficients were determined. When friction coefficients are examined in Table 3, it is seen that the friction coefficient values decrease as the sliding rate and load increases. As shown in Figure 6,7 normal load - speed, normal load - friction coefficient graphs were drawn.

It has been studied the wear of coated Ti6Al4V and found that oxidative wear was prevalent under low normal loads. Because of the increase in temperature with friction, it causes oxidation. Since the Ti alloy has the ability to passivate, the situation is similar for uncoated alloys. Once the normal load reached a critical value, the coefficient of friction reduced [25,26].

The high hardness and adhesion strength of the coatings improve the wear resistance[24]. So, the contact between pin and ring continues as metal-coating. The general contact situation changes with time.

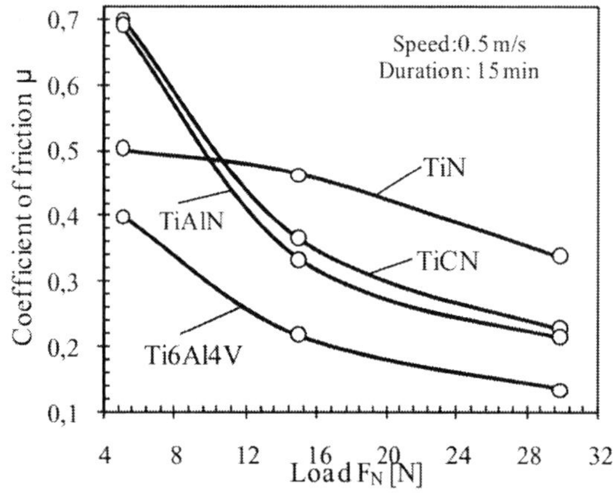

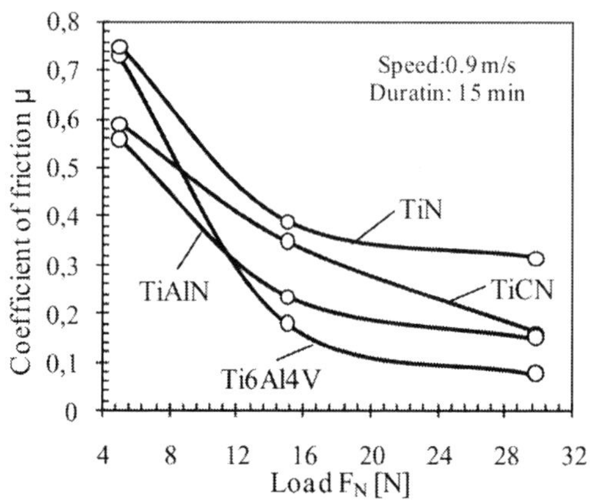

Figure 6. The variation in coefficient of friction with load.(0.5 m/s sliding rate)

Figure 7. The variation in coefficient of friction with load. (0.9 m/s sliding rate)

TiCN and TiAlN have a lower coefficient of friction. The coefficient of friction is proportional to the surface roughness. As shown in Figure 8-9, the surface roughness values of TiAlN and TiCN coatings are low. Although the roughness of the uncoated alloy is higher, the coefficient of friction is lower. The reason for this is based on the oxide layer developing on the surface due to passivation. Figure 9 shows that as the sliding rate increases, the roughness becomes even smaller. The sliding rate promotes oxidation as it increases the temperature. When examined in the literature, it is seen that this is due to the increase of the oxidation together with the increase of the friction heat, and the oxide layer is caused by a kind of lubricant effect [14]. The friction coefficient of TiN was higher than that of TiCN while its wear rate was lower. This indicates that the latter coating can perform as a wear resistant coating at high temperatures [14,25]

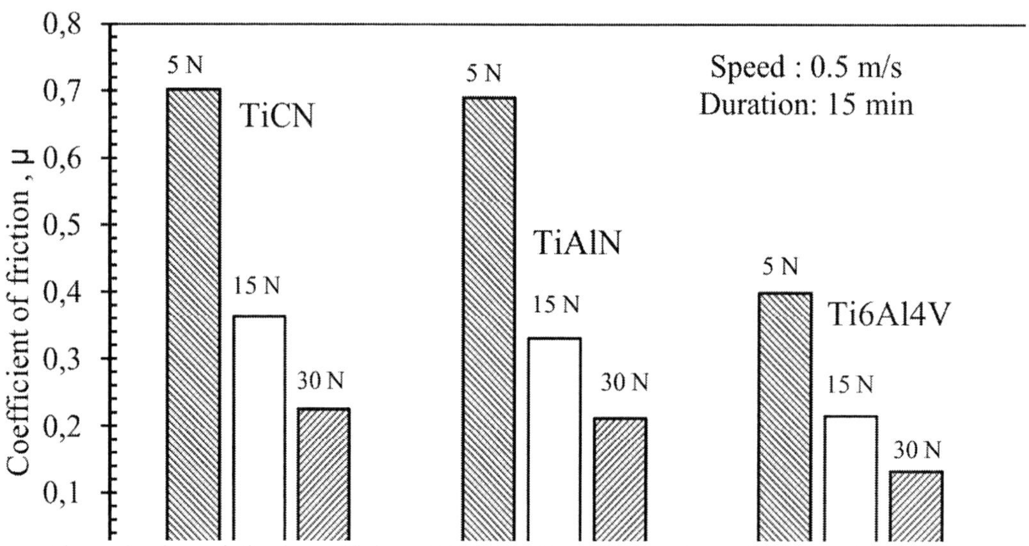

Figure 8. The variation in coefficient of friction with surface roughness (0.5 m/s sliding rate)

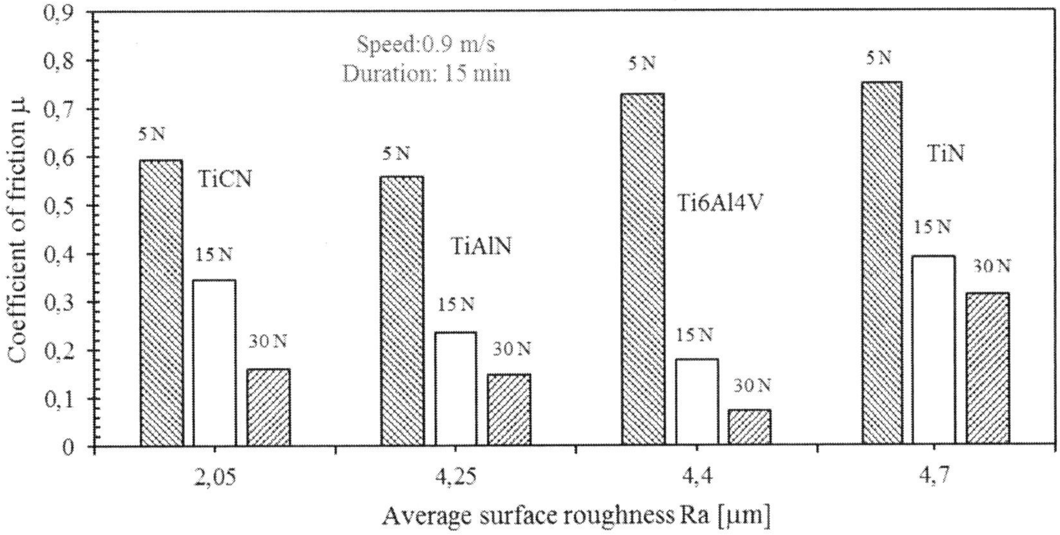

Figure 9. The variation in coefficient of friction with surface roughness .(0.9 m/s sliding rate)

All of the experiments were carried out on the same conditions, with the use of a separate wearing part for each experiment. Weight loss and wear track widths measurements were used to as a measure of wear. Weight loss was measured with sensitive scale (1/10000) .Wear track widths were determined by SEM microscopy for each experiment as shown in Figure 10. Measurements in x and y directions were taken and the average was used as the width of wear track.

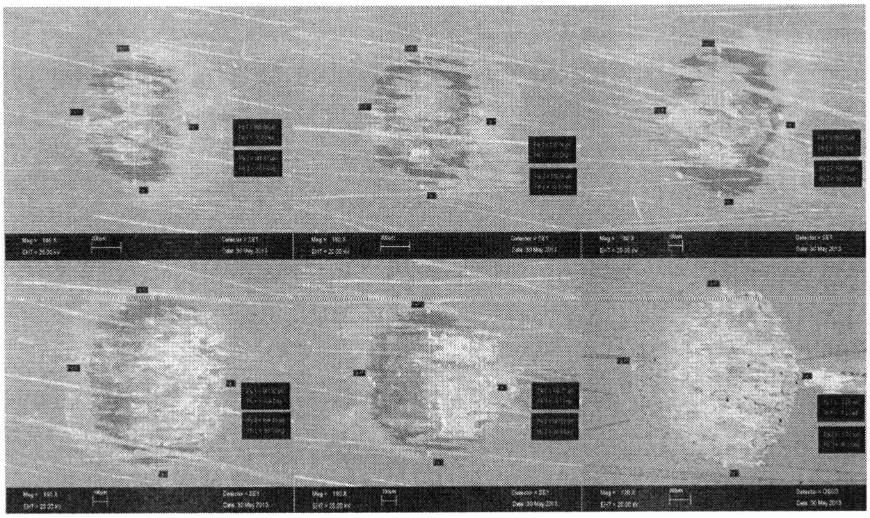

Figure 10. Wear track widths determined by SEM microscopy for Ti6Al4V

When the weight loss is examined according to the test parameters, weight loss increases with increasing load and sliding rate. However, the weight loss of the uncoated samples is higher than the weight loss of the coated samples. The lowest wear loss was observed in TiN. The lowest value among the wear track widths given in Table 3 and Figure 11-12 was seen on the TiN coating. After the abrasion test, surface scans and EDX analysis were performed with Scanning Electron Microscope (SEM). The EDX analyzes obtained from the coated specimens show that the oxidation of the coating is due to friction in the wear zone. This result is supported by the optical microscope images of the wear tracks. In the literature wear analyzes show that wear on Ti-based coatings results from the formation of titanium oxide. Same time the small abrasive particles separated from the coating by the influence of the friction can reveal the sliding wear mechanism [14,26,27].

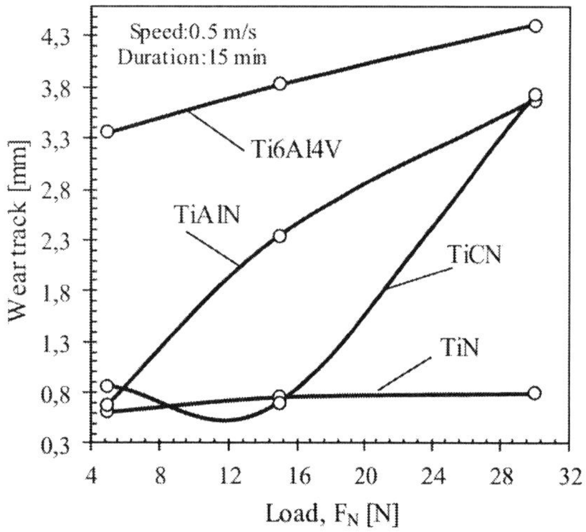

Figure 11 . The variation in wear track with load. (0.5 m/s sliding rate)

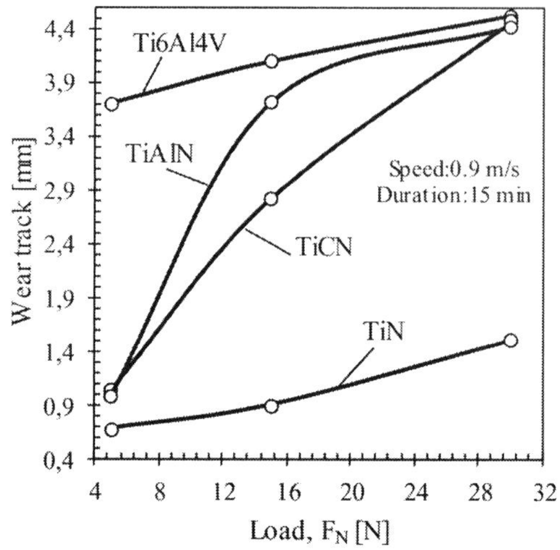

Figure 12 . The variation in wear track with load. (0.9 m/s sliding rate)

It is seen that when the values are compared, there is an increase in the track widths depending on the load and sliding rate. Although it is determined that there is a large difference between uncoated specimens and coated specimens at low loads, this difference is reduced as the load increases. This is also related to the friction coefficient. At higher loads, the friction coefficient decreases. The lowest wear rate achieved in the TiN coating. This result was also proportional to the high hardness of the TiN coatings. TiCN and TiAlN showed low wear trace widths, which is consistent with the friction coefficient values. Furthermore, the coating hardness of the samples directly affects the wear resistance[14,26,27]. The samples with a higher coating hardness showed better wear resistance in this study.

When SEM photo images are examined, it is seen that the depth of the tracks increases in the wear area as the load and sliding rate increases . Wear track images obtained from TiAlN coated samples and wear test parameters are seen on the SEM image in Figure 13.

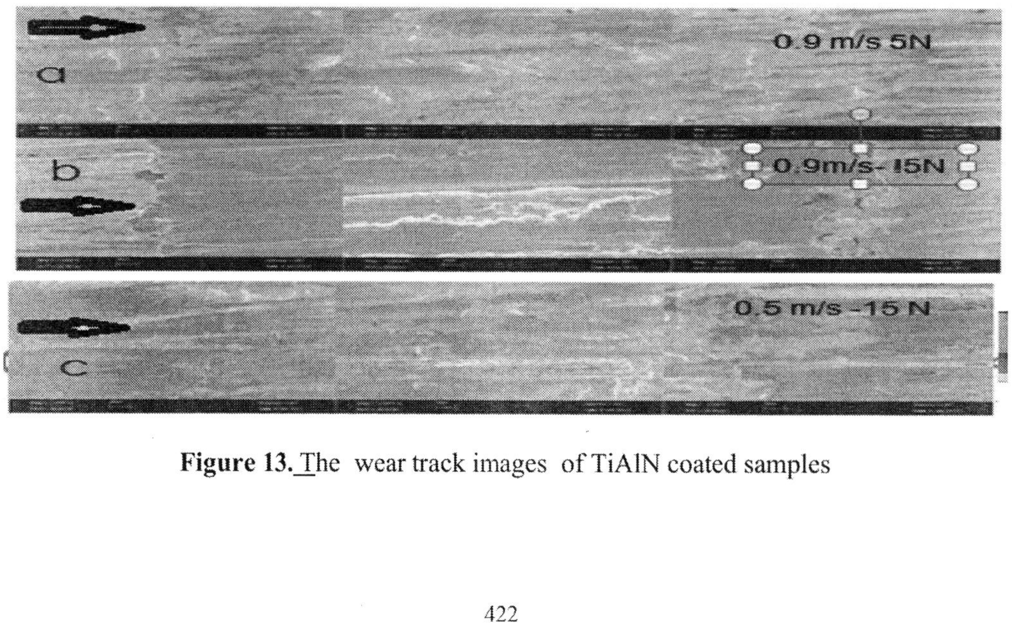

Figure 13. The wear track images of TiAlN coated samples

Depending on the direction of wear, abrasive and adhesive wear mechanisms were observed as shown in the Figure 13 a-b-c. Abrasive wear is more effective . Figure 14 shows wear losses depending on load and speed parameters. The wear loss increases as the load and speed increase, while the lowest wear loss is detected in the TiN coating. It is clear that the coating is effective for increasing the wear resistance of the alloy. Moreover, TiN was the best choice for applications where the predominant wear mechanism is abrasive wear[14,24,28].

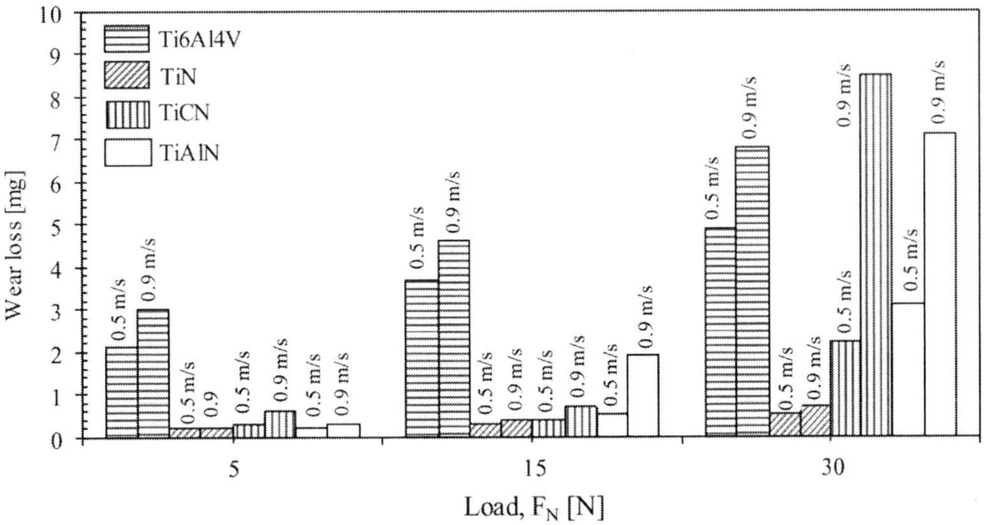

Figure 14. The variation in wear loss with load and speed

4. Conclusions

TiAlN, TiN, TiCN coating was deposited on Ti6Al4V alloy. Wear tests were performed at room temperature and in dry conditions. When samples are compared in terms of wear losses, it is shown that the losses in coated specimens are lower than those of uncoated specimens.

The least surface roughness between the coatings is TiCN coating. The main cause of this result is the TiC(the lubricating effect) formed at the surface of the material by changing the gas flows during the coating. When the values obtained are compared, the order of average surface roughness values of the samples is in the order of TiN> Ti6Al4V> TiAlN> TiCN.

It was determined that the wear widths on the samples were homogeneous and almost parallel to each other. Therefore; the widths of the wear track are taken as a measure of the amount of wear in the specimens. 30N and 0.9 m / s test parameters are used in practice; track widths respectively; Ti6Al4V - 4,53 mm> TiCN - 4,47 mm> TiAlN - 4,42 mm> TiN - 1,51 mm.

Wear track widths of the samples are increasing with increasing load. In uncoated samples at the same load values, the track width values were larger than those of the coated ones. This suggests that the coated samples have a higher wear resistance value. when the trace width and weight loss are taken into consideration, the best abrasion performance is TiN coating and then TiAlN and TiCN in order. The values of the friction coefficient; TiN, TiAlN and TiCN coated specimens were found to be lower in the uncoated samples at a normal load of 30N and a speed of 0.9 m / s.When all samples were examined, it was determined that the average friction coefficient values at 30 N normal load and 0.9 m/ s speed were TiN> TiCN> TiAlN> Ti6Al4V.

Acknowledgements

The authors would like to thank the Scientific Research Units at Erciyes University for their contribution and support (Project No: FBY 09-852).

References

1. Jackson M.J., Ahmed W., Titanium and Titanium Alloy Applications in Medicine, Springer US.,533-576, 2007.
2. He L., Maı Y., Chen Z., Fabrication and Characterization of Nanometer CaP (Aggregate)/Al_2O_3 Composite Coatings on Titanium, Materials Science and Engineering, A 367, 51-56, 2004.
3. Yang D., Lıu C., Lıu X., Qı M., Lın G., EIS Diagnosis on The Corrosion Behavior of TiN Coated NiTi Surgical Alloy, Current Applied Physics, vol.5, 417–421, 2005.
4. Galante J.O., Lemons J., Spector M., Wilson P.D., Wright T.M., "The Biologic Effect of Implant Materials" Journal of Orthopaedic Research, 9, s 760-775, 1991.
5. Paschoal A.L.,Vanâncio E.C., Canale L.C.F., Sılva, O.L., Huerta-Vılca D., Motheo A.J., "Metallic Biomaterials TiN-Coated: Corrosion Analysis and Biocompatibility." Artificial Organs, 27 (5), s 461–464, 2003.
6. Peters, M., Hemptenmacher, J., Kumpfert, J., Leyens, C., Structure and Properties of Titanium and Titanium Alloys, DLR – German Aerospace Center, Cologne, 2003.
7. Leyens, C. and Peters, M., Titanium and Titanium Alloys: Fundamentals and Applications, Wiley-VCH, Weinhem, 2003.
8. Danışman Ş., Savaş S., The Effect of Ceramic Coatings on Corrosion and Wear Behaviour, Journal of the Balkan Tribological Association, Vol.12, No.1, pp 104-113, 2006.
9. Danışman Ş., Savaş S., Işık G., Bendeş O., Özbekler A., Wear Resistant Hard Ceramic Coatings Used in Biomedical Applications, 4 th National Biomechanics Congress, Erzurum, 2008.
10. Probst J., Gbureck U., Thull R., "Binary Nitride and Oxynitride PVD Coatings on Titanium for biomedical Applications." Surface and Coatings Technology, 148, s 226-233, 2001
11. Harman M.K., Banks S.A., Hodge W.A., "Wear Analysis of a Retrieved Hip Implant with Titanium Nitride Coating." The Journal of Arthroplasty, Vol. 12 No. 8, s 938-945, 1997.
12. Vadıraj A., Kamaraj M., "Effect of Surface Treatments on Fretting Fatigue Damage of Biomedical Titanium Alloys." Tribology International, 40, s 82-88, 2007.
13. Vadıraj A., Kamaraj M., "Characterization of Fretting Fatigue Damage of PVD TiN Coated Biomedical Titanium Alloys." Surface and Coatings Technology, 200, s 4538-4542, 2006.
14. Santecchia E., Hamouda A.M.S., Zalnezhad E., Musharavati F., Cabibboc M., Spigarelli S.,Wear resistance investigation of titanium nitride-based coatings, Ceramics International vol.41, 10349–10379, 2015. REVİEW
15. Ceschini L., Lanzoni E., Martini C., Prandstraller D. and Sambogna G., "Comparison of Dry Sliding Friction and Wear of Ti6Al4V Alloy Treated by Plasma Electrolytic Oxidation and PVD Coating", Wear, Vol. 264, pp. 86-95, (2008).
16. Maurer A.M., Brown S.A., Payer J.H., Merrıtt K., Kawalej J.S., "Reduction of Fretting Corrosion of Ti-6Al-4V by Various Surface Treatments." Journal of Orthopaedic Research,11, s 865-873, 1993.

17. Holland L., "Substrate Treatment and Film Deposition in Ionized and Activated Gas." Thin Solid Films, 27, s 185-203, 1975.
18. Wılson R.W, Terry, L.E., "Application of High-rate ExB or Magnetron Sputtering in the Metallization of Semiconductor Devices." J. Vac. Sci. Technol., 13(1), s 157-164, 1976.
19. Schıller S., Heısıg, U., "The Role of Plasmatron-Magnetron Systems in Physical Vapour Deposition Technigues." Thin Solid Films, 54, s 33-47, 1978.
20. Grzesik, W., Zalisz, Z., Nieslony, P., Friction and Wear Testing of Multilayer Coatings on Carbide Substrates for Dry Machining Applications, Surface and Coatings Technology, 155, 37-45, 2002.
21. BarshiliaH.C.,Yogesh K.,Rajam K.S. "Deposition of TiAlN coatings using reactive bipolar-pulsed direct current unbalanced magnetron sputtering", Vacuum, 83,427-434, (2009).
22. Danisman K., Danışman Ş., Savaş S.,, Dalkiran I., "Modelling of the Hysteresis Effect of Target Voltage in Reactive Magnetron Sputtering Process by Using Neural Networks", Surface and Coatings Technology, vol.204, pp.604-614, December 2009
23. Savaş S., Danışman Ş., "Multipass Sliding Wear Behavior of TiAlN Coatings Using a Spherical Indenter: Effect of Coating Parameters and Duplex Treatment", Tribology Transactions , vol.57,pp.242-255 , March 2014
24. W.Cui, G.Qin, J.Duan, H.Wang,"A graded nano-TiN coating on biomedical Ti alloy: Low friction coefficient, good bonding and biocompatibility", Materials Science and Engineering vol.C71, pp 520-528, 2017.
25. Lau K.H., Li K.Y., "Correlation between Adhesion and Wear Behaviour of Commercial Carbon Based Coatings", Tribology International, 39, 115-123, (2006).
26. Wei, M. X. Wang, S. Q., Cui, X. H., and Chen, K. M., Characteristics of extrusive wear and transition of wear mechanisms in elevated-temperature, Tribology Transactions, 53,888–896, 2010.
27. Luo K., Hovsepian P.Eh., Lewis D.B., Münz W.D., Kok Y.N., Kokrem et al., "Tribological properties of unbalanced magnetron sputtered nano-scale multilayer coatings TiAlN/VN and TiAlCrYN deposited on plasma nitrided steels", Surface and Coatings Technology, 193, 39-45 (2005).
28. Mezlini S., et al., "Effect of Indenter Geometry and Relationship Between Abrasive Wear and Hardness in Early Stage of Repetitive Sliding", Wear, 260, 412-421, (2006).

Characterization of Ball Milled Al-Al_2O_3 sub-micron Composites

E. Dagasan [(a)], **E. Gercekcioglu** [(b)], **S. Unalan** [(b)]

[a] *Erciyes University, Engineering Faculty, Department of Materials Science and Engineering, 38039 Kayseri, Turkey*

[b] *Erciyes University, Engineering Faculty, Department of Mechanical Engineering, 38039 Kayseri, Turkey*

Abstract

The purpose of this study is to investigate properties of the composite powders produced by ball milling process. Different weight ratio of high purity Al_2O_3 powders were added to the Al matrix as reinforcing element. Ball milling process was conducted by a planetary type ball mill with WC milling balls and vial at constant parameters like rotating speed, time, Ball-to-Powder ratio and Process Control Agent. Samples that taken from the powder mixtureby various time intervals were analyzed by SEM, XRD and BET surface area and porosity measurement systems.

Keywords: Mechanical Milling, Aluminium Matrix Ceramic Reinforced Composites, Al_2O_3, Ball Milling.

1. Introduction

Mechanical alloying (MA) allows production of homogeneous materials starting from blended elemental powder mixtures [1, 2]. The process is generally achieved by the association of high-energy ball mills.

Mechanical Alloying (MA) is a process that consists of repeated cold welding and fracturing mechanisms of powders (Figure 1.b). And mainly used for creating alloys without melting. Additionally, this technique enable economical and rapid preparation of nanocomposites and ceramics [3]. Although there are different methods to fabricate composite materials, powder metallurgy enables net-shape and cost-effective processing with controlled microstructure and mechanical properties [4].

By the way particulate-reinforced aluminum MMCs have found an extensive commercial application in several sectors namely aerospace, space, automotive and structural [5]. This shows an excellent combination of mechanical and physical properties such as the specific strength and the stiffness, the wear resistance, and the electrical and the thermal conductivities [6, 7, 8].

Many methods are suitable for preparing aluminum matrix composites. In the liquid methods particles are added to the liquid aluminum by stirring before casting [9]. However, the difference in thermal expansion coefficients between ceramic particles and molten Al constituents and the poor wettability between these two become an obstacle to the liquid method for synthesizing Al matrix composite [10]. That makes MA advantageous for the production of Al-MMCs. Many ceramic particles like Al_2O_3

[11], SiC [12], B_4C [13, 14] and SiO_2 [15] are used as reinforcement for Al and its alloys for making properties better.

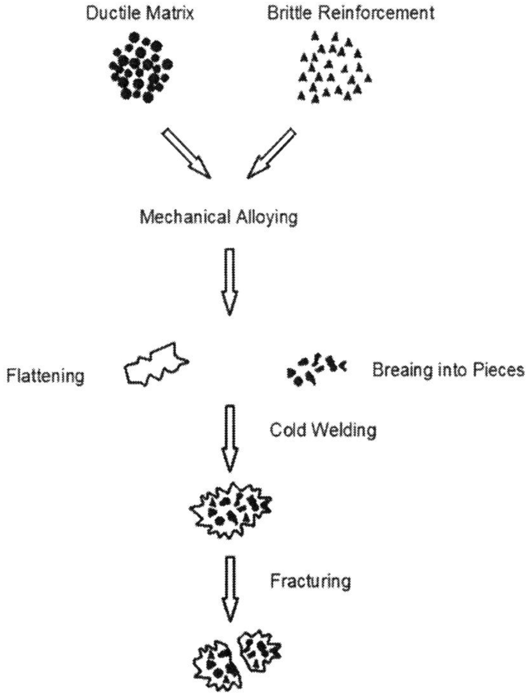

Figure 1. Schematical View of Mechanical Alloying Process.

2. Experimental Details

2.1. Milling Operations

High purity Al matrix was reinforced with high purity Al_2O_3 powders by using a planetary ball mill. Properties of the initial powders are given in Table 1.

Table 1. Properties of the Initial Powders.

	Purity	Particle Size
Al (Alfa Aesar- Al 00010)	99,8 %	-40 +325 mesh
Al_2O_3 (AEE-AL 602)	99,9 %	- 325 mesh

To investigate the effect of Al_2O_3 on the properties of composite powder 5, 10, 15 and 20 wt. % were added to the Al matrix. Ball milling process was conducted by the help of a FRITSCH Pulverisette 6 type planetary ball mill by using tungsten carbide vial and balls. Al-Al_2O_3 mixtures were mixed and loaded into the vial under high purity argon atmosphere by using a MBRAUN GB-2202-PVAC glove box. Parameters that are kept constant for all specimens are given in Table 2.

Table 2. Constant Milling Parameters.

Rotational Speed	250 rpm
Total Ball Weight	336 gr
BPR (Ball-to-Powder Ratio)	10:1
Total Powder Weight	33,6 gr
Vial Capacity	250 ml WC
Milling Atmosphere	High Purity Argon
PCA (Process Control Agent)	% 1.5 Stearic Acid
Total Milling Time	120 hrs

Milling operations were stopped by time intervals of 30 hours to take samples from the powder mixture. Taken samples were then used for SEM, X-RD and BET surface area analyzes.

2.2. SEM and Morphological Studies

To evaluate the morphological changes during ball milling SEM micrograms of specimens were taken by a LEO 440 type scanning electron microscope. Particle size analyzes were then realized on the SEM images by the help of an image processing software named Image ProPlus.

2.3. X-RD Analyses

X-RD analyses were conducted by a Bruker AXS D8 Advance type analyzer with Cu Kα radiation (λ=0,15406 nm). XRD patterns were recorded in the 2θ range 10-90°.

2.4. BET Surface Area and Pore Size Analyses

Surface area and pore size analyses of the composites were measured by a Micromeritics Gemini VII type analyzer. Both single point and multi point results were conducted by using nitrogen as adsorbent.

3. Results

3.1. SEM and Morphological Studies

SEM micrograms of as-received Al and Al_2O_3 are given in Figure 2.

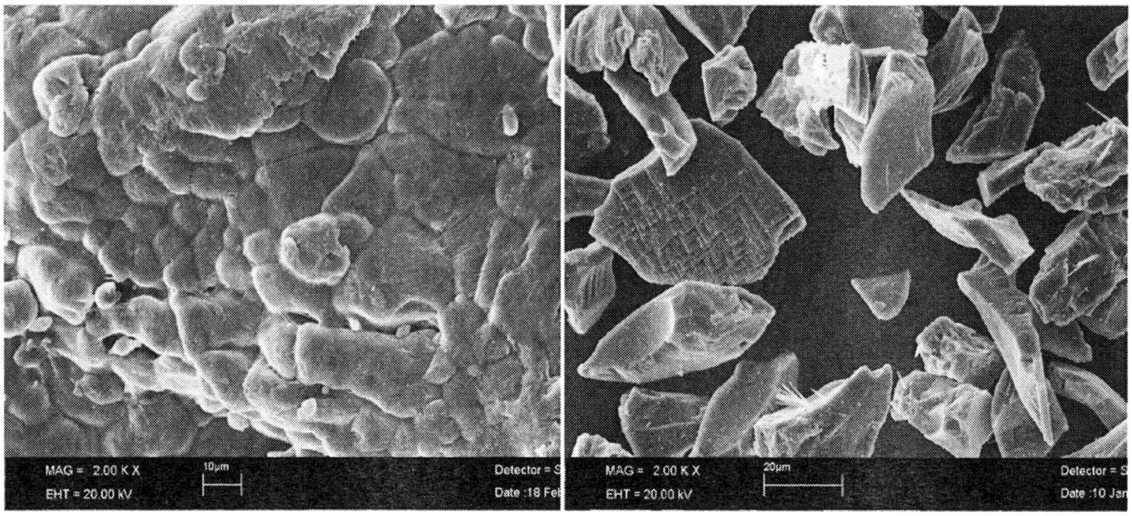

Figure 2. SEM Images of As-Received (a) Al, (b) Al_2O_3.

Images of Al-% 5 Al_2O_3 powder mixture that is ball milled for 30, 60, 90 and 120 hours are given in Figure 3.

Figure 3. SEM Images of the Al- 5 % Al_2O_3 Powder Mixtures Milled for (a) 30 Hours, (b) 60 Hours, (c) 90 Hours and (d) 120 Hours.

It is clear that at initial stages of the milling process (Figure 3.a and 3.b), the powder mixture tend to flatten because of the ductile Al matrix and also the number of the broken particles of brittle reinforcement is less. At the same time it can be also said that particles are sharp cornered at initial stages. But as the milling time goes further particles became round and homogenized by means of particle size.

Figure 4, shows the morphological evaluation of Al-% 10 Al_2O_3 composite powder. As the previous sample flattening tendency goes on. But due to the increase in the weight ratio of Al_2O_3 an increase in the number of broken Al_2O_3 particles is also observed.

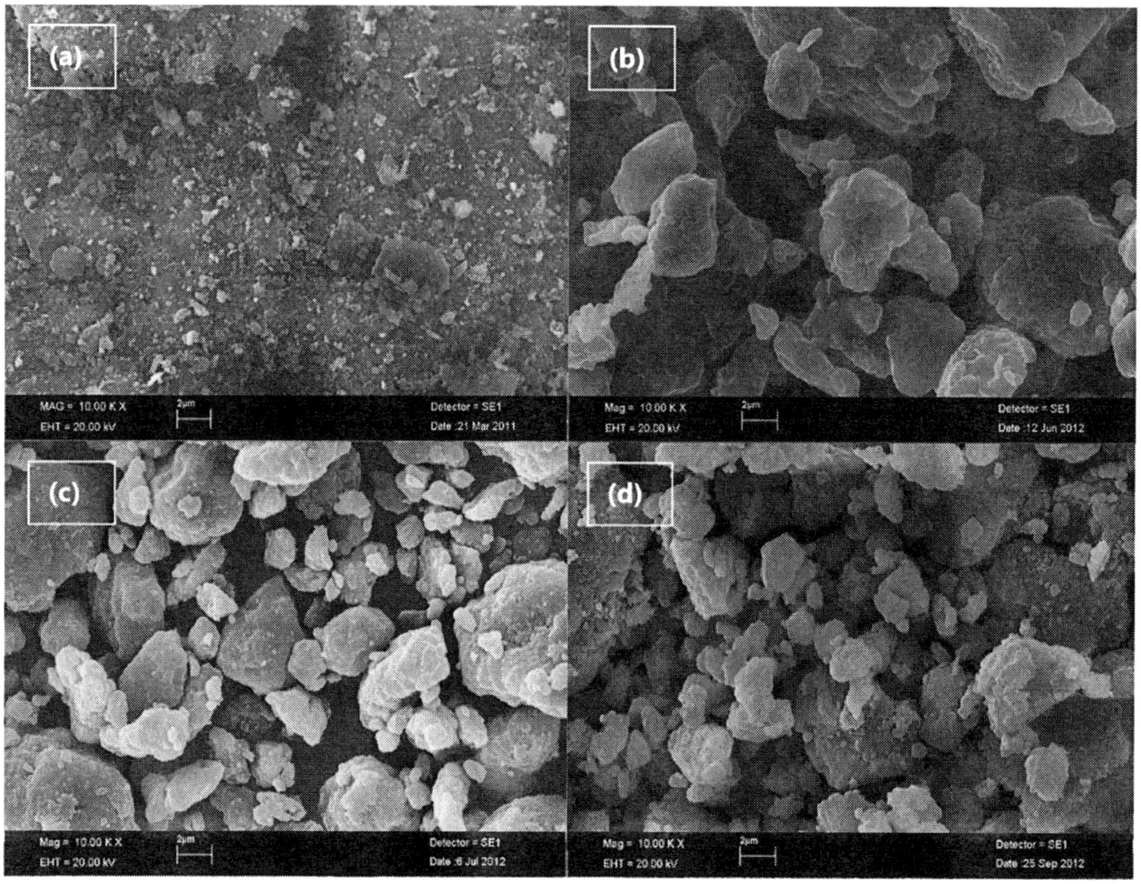

Figure 4. SEM Images of the Al- 10 % Al_2O_3 Powder Mixtures Milled for (a) 30 Hours, (b) 60 Hours, (c) 90 Hours and (d) 120 Hours.

Figure 5 and Figure 6 show microstructures of Al-% 15 Al_2O_3 and Al-% 20 Al_2O_3 powder mixtures milled for different periods, respectively. For the powder mixture of Al-% 15 Al_2O_3 flattening tendency which starts at the initial stages of milling can still be observed after 60 hours of milling. After 90 hours of milling it can be seen that flattening tendency ends and spherical particles can be observed.

When SEM images of Al-% 20 Al_2O_3 specimen is investigated it can be seen that formation of small particles starts earlier because of the increasing Al_2O_3 content.

Figure 5. SEM Images of the Al- 15 % Al_2O_3 Powder Mixtures Milled for (a) 30 Hours, (b) 60 Hours, (c) 90 Hours and (d) 120 Hours.

When the 120 hour milled microstructures of the samples that contain different weight ratio of reinforcing element are compared, it can be easily said that number of spherical particles increase by increasing Al_2O_3 content.

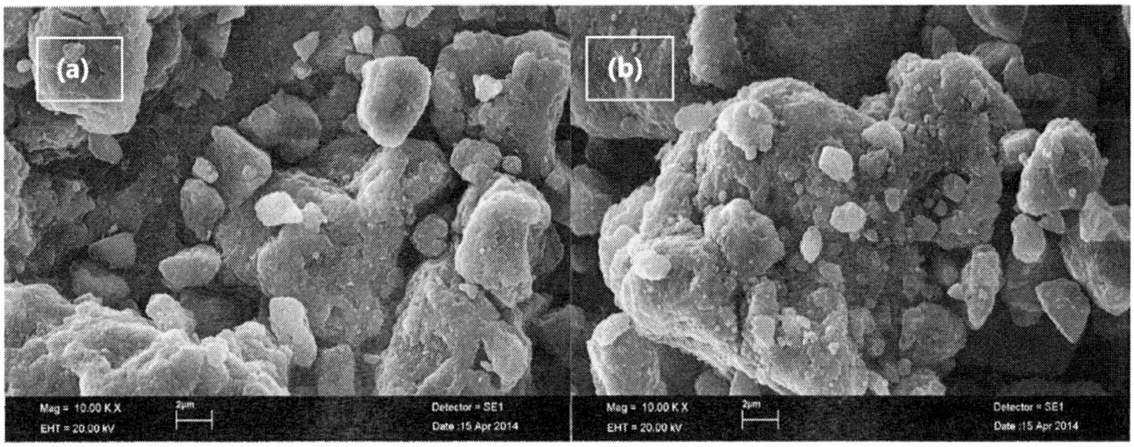

Figure 6. SEM Images of the Al- 20 % Al_2O_3 Powder Mixtures Milled for (a) 30 Hours, (b) 60 Hours, (c) 90 Hours and (d) 120 Hours

While evaluating the particle size change of specimens during milling, 12 images taken at six different magnitudes of 100X, 500X, 1 KX, 2 KX, 5 KX, 10 KX and 20 KX were processed by Image ProPlus software for each specimen. Figure 7 gives an example for the analyze.

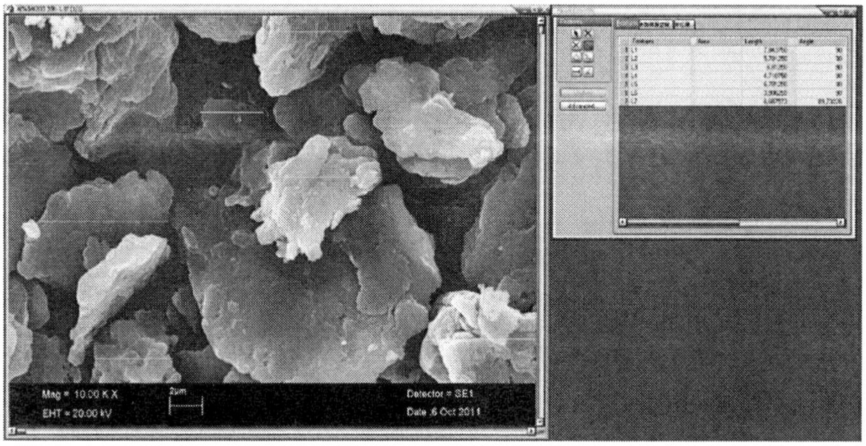

Figure 7. Average Particle Size Measuring Example of 30 Hours Milled Al- 5 % SiC Powder Mixture by Image ProPlus ®.

Figure 8 and Table 3 give the particle size change vs milling time. As expected particle size of each specimen decrease as milling process goes further. After early stages of milling average particle size of the samples changes in a wide range between 6,257 – 120,06 μm. That significant result may be due to the dominant cold welding mechanism during milling. As milling goes further to 60 hours same trend still goes on. But at the progressive stages, because a steady state is reached between cold welding and fracturing mechanisms, particle sizes of the specimens became steady. So final particle sizes of the samples are close to each other.

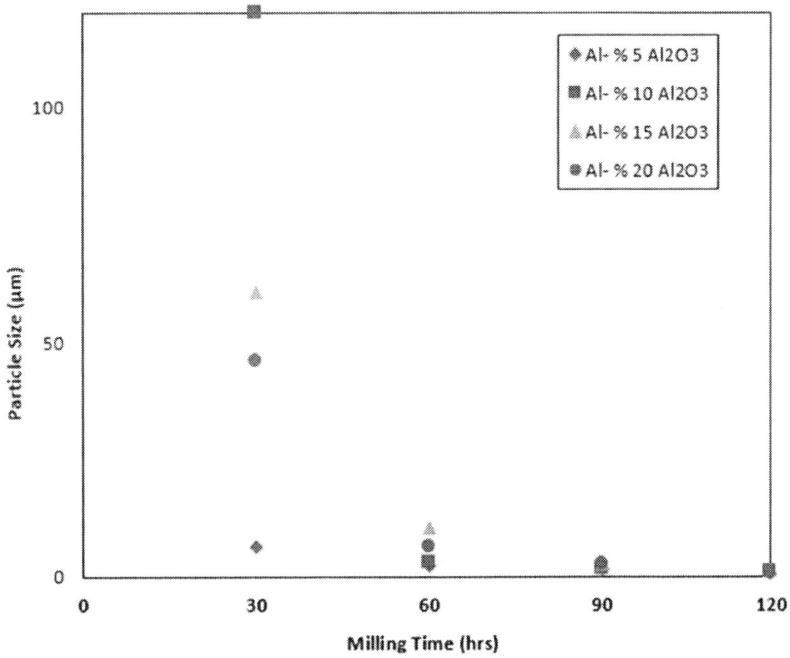

Figure 8. Particle Size Change of Al-Al_2O_3 Composite Powder Mixtures by Milling Time.

Table 3. Particle Size Change against Milling Time.

Milling Time (Hour)	Particle Size of Al-Al_2O_3 Powder Particles After Milling (µm)			
	Al_2O_3 Content (% Wt.)			
	5 %	10 %	15 %	20 %
30	6,257	120,06	60,795	46,035
60	2,280	2,775	10,485	6,347
90	1,544	1,653	2,610	2,701
120	0,696	0,976	1,078	0,512

3.2. XRD Results

Figure 9 gives the XRD pattern of as-received Al and Al_2O_3. In order to investigate if any contamination exists because of the stearic acid added XRD pattern of the stearic acid was also taken.

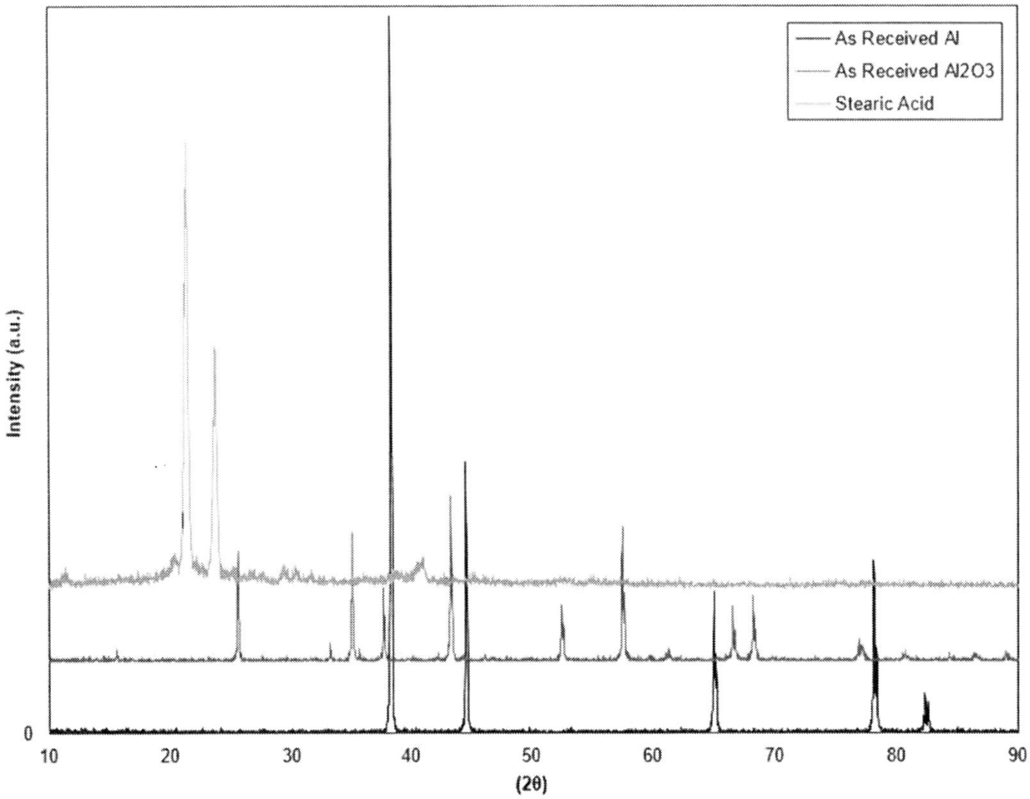

Figure 9. XRD patterns of the as received pure Al, Al_2O_3 and Stearic Acid.

From Figure 10, it is obvious that because of the high Al content of the mixture, all major peaks observed in the pattern match pure Al peaks recorded. But also intensity of those peaks decreases by milling time. For example the intensity of the major Al peak recorded at 2θ=38,38° decreases to 104 after 120 hours of milling. And XRD patterns also show that 2θ degrees increase by increasing milling time. According to the Bragg's Law that result shows the distance between atomic planes decrease by the milling process.

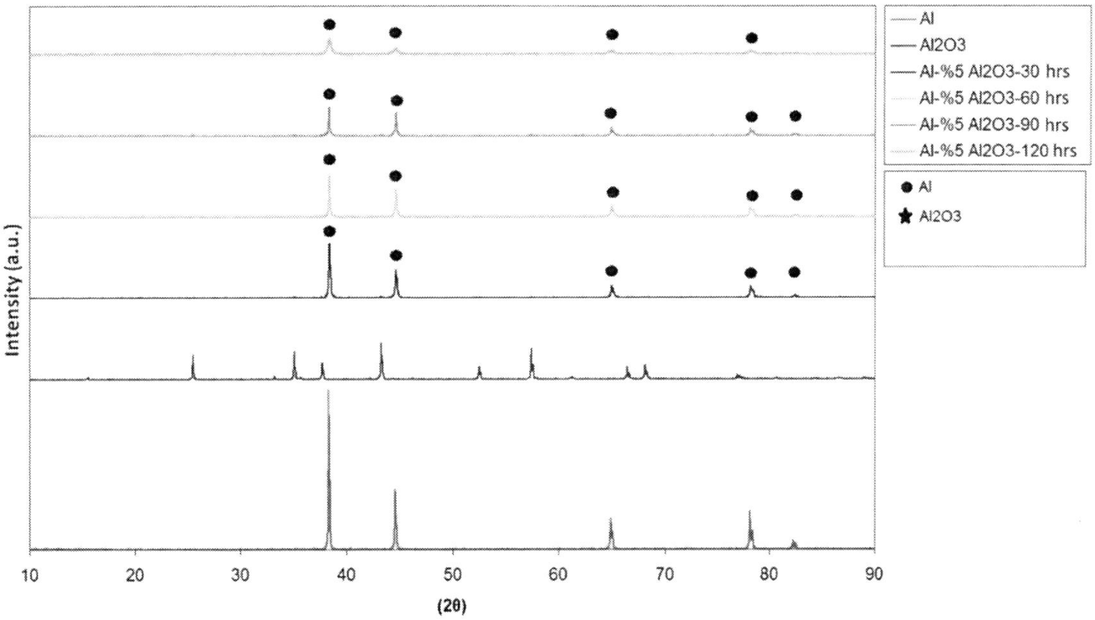

Figure 10. XRD patterns of the Al- 5 % Al_2O_3 Powder Mixtures Milled for 30 Hours, 60 Hours, 90 Hours and 120 Hours.

Figure 11 gives the XRD powder pattern for Al- % 10 Al_2O_3 mixture. Although similar results to the Al- % 5 Al_2O_3 can be said, due to the increasing Al_2O_3 content pure Al_2O_3 peaks like the one at $2\theta =$ 35,14° can be seen initially. But the intensity of that peak decreases and then disappears by increasing milling time.

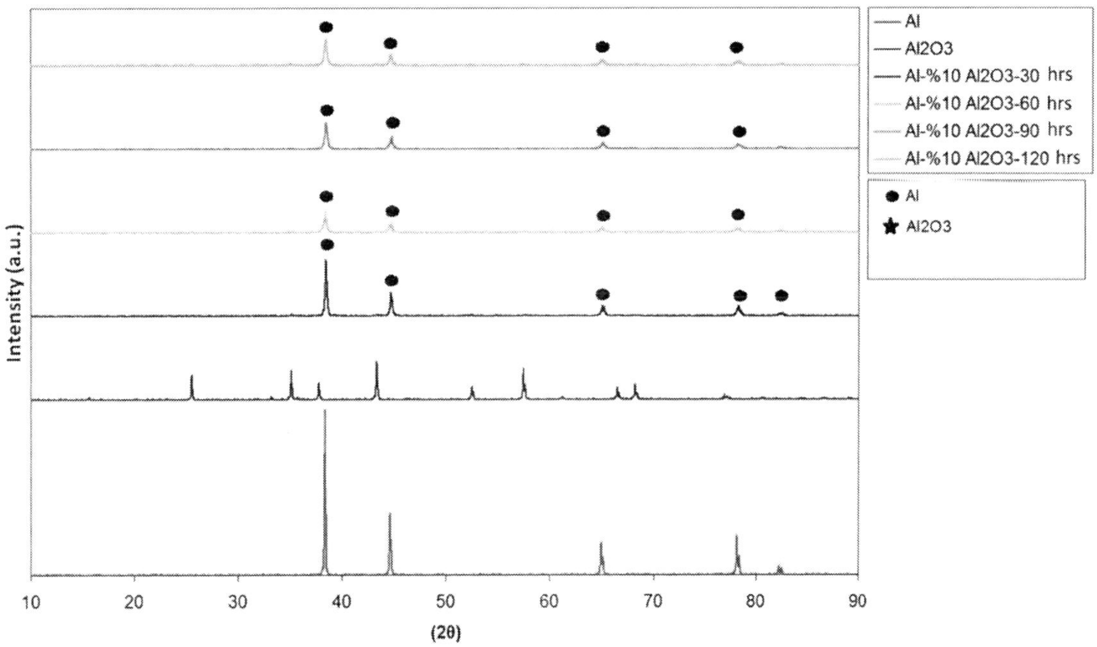

Figure 11. XRD patterns of the Al- 10 % Al_2O_3 Powder Mixtures Milled for 30 Hours, 60 Hours, 90 Hours and 120 Hours.

Al- % 15 Al_2O_3 X-RD pattern can be seen in Figure 12. Significant peaks that appear at 2θ degrees 43,34°, 52,68° and 57,44° can be seen for the first time and belong to pure Al_2O_3. Similar results and peaks can be observed due to Al_2O_3 content at Al- % 20 Al_2O_3 XRD results (Figure 13).

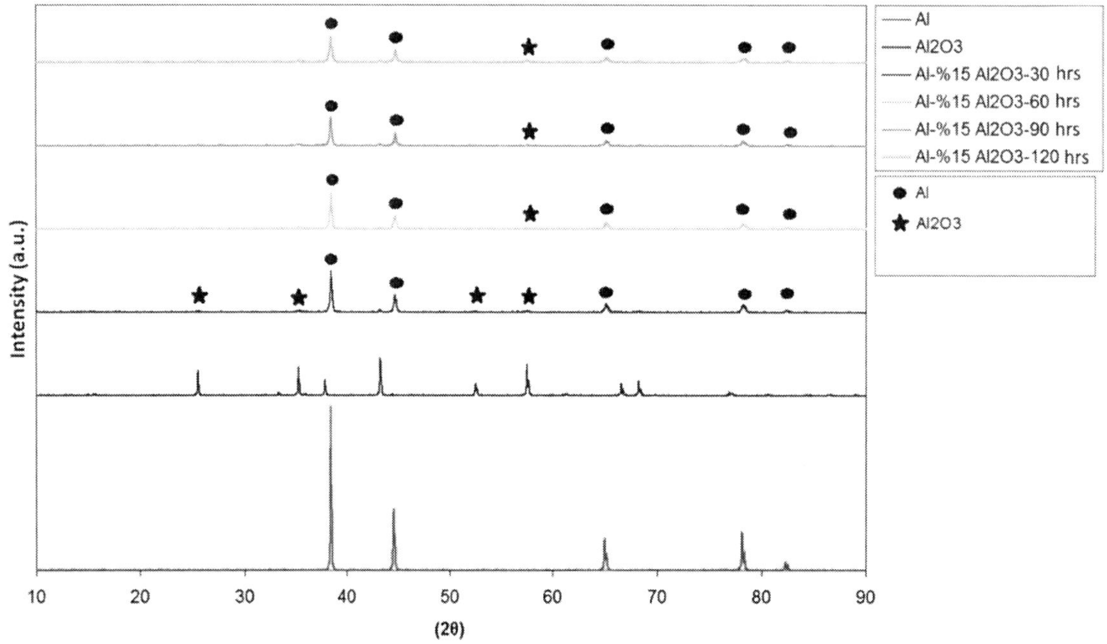

Figure 12. XRD patterns of the Al- 15 % Al_2O_3 Powder Mixtures Milled for 30 Hours, 60 Hours, 90 Hours and 120 Hours.

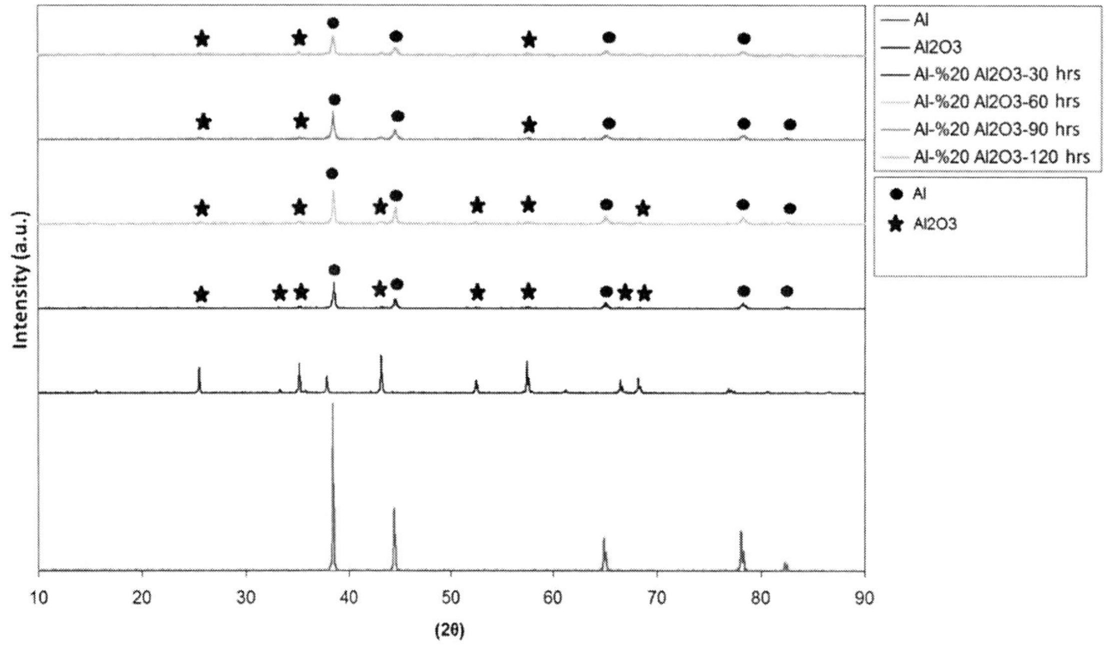

Figure 13. XRD patterns of the Al- 20 % Al_2O_3 Powder Mixtures Milled for 30 Hours, 60 Hours, 90 Hours and 120 Hours.

3.3. BET Surface Area and Pore Size Results

Table 4 gives single point, BET surface area and pore size results for 120 hour milled Al and Al_2O_3 composite powders. Figure 14 summarizes the data graphically.

Table 4. BET, Single Point Surface Area and Pore Size Results.

	Single Point Surface Area (m^2/g)	BET Surface Area (m^2/g)	Pore Size (nm)
Al	0,228	0,092	51,994
Al_2O_3	0,199	0,086	45,441
Al-% 5 Al_2O_3	6,184	4,245	26,888
Al-% 10 Al_2O_3	4,659	1,438	61,720
Al-% 15 Al_2O_3	4,336	2,820	33,827
Al-% 20 Al_2O_3	4,140	3,414	28,113

As can be seen BET and single surface area values are higher than those of as-received Al and Al_2O_3. That significant change in surface area can be associated with decreasing particle size. Generally single point surface area results of the 120 hour milled specimens are close to each other.

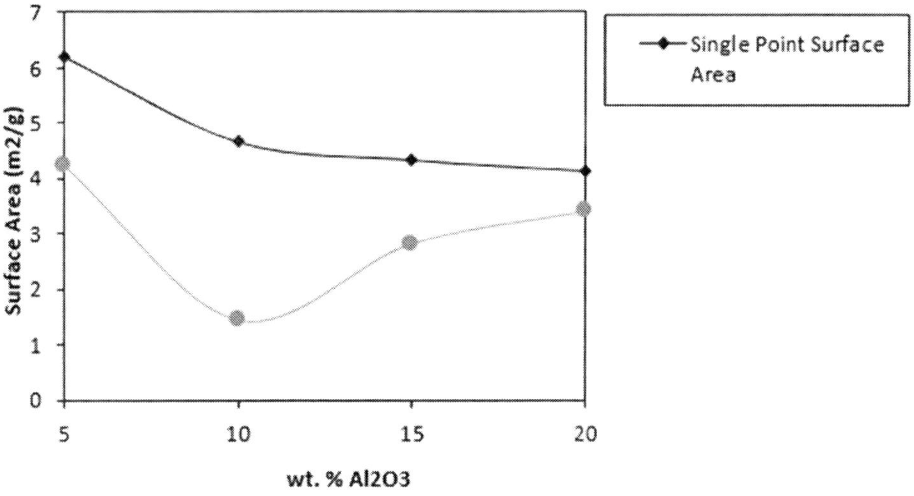

Figure 14. Single and Bet Surface Area Change by Weight Ratio of Al_2O_3.

The highest BET surface area value belongs to Al-% 5 Al_2O_3 composite mixture in accordance with the lowest pore size value (Figure 15).

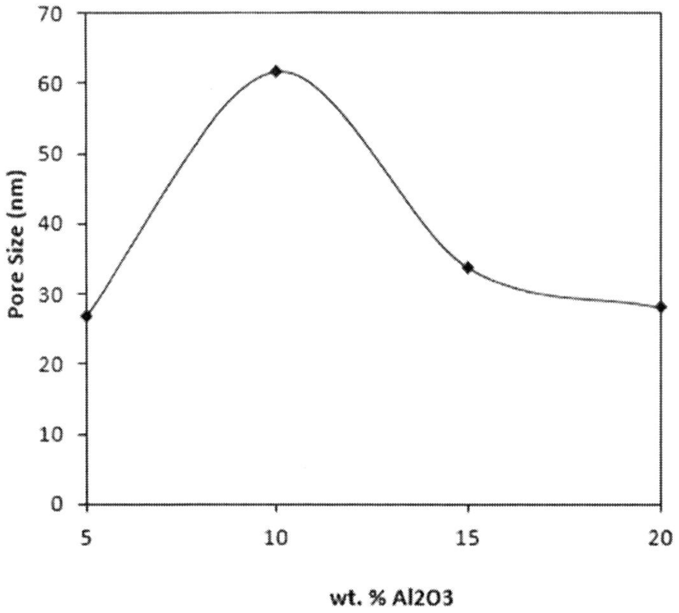

Figure 15. Pore Size Change by Weight Ratio of Al_2O_3.

Acknowledgements

This research was supported by Erciyes University Scientific Research Projects Unit (EU-BAP) with the project number FBA-10-2994.

References

1. Suryanarayana, C. 2001. Mechanical alloying and milling. Progress in Materials Science, 46:1-184.
2. Suryanarayana, C., 2008. Recent developments in mechanical alloying. Reviews on Advanced Materials Science, 18:203-211.
3. Jiang, X., Trunov, M.A., Schoenitz, M., Dave, R.N., Dreizin, E.L., 2009. Mechanical alloying and reactive milling in a high energy planetary mill. Journal of Alloys and Compounds 478:246-251.
4. Zhang, B.Q. Lu, L., ,Lai, M.O. 2003. Physica B, 325:120–129.
5. Sivasankaran, S., Sivasprasad, K., Narayanasamy, R., Kumar Iyer, V., 2010. An investigation on flowability and compressibility of AA 6061100 − x-x wt.% TiO2 micro and nanocomposite powder prepared by blending and mechanical alloying. Powder Technology 201:70-82.
6. Smagorinski, M.E., Tsantrizos, P.G., Grenier, S., Cavasin, A., Brzezinski, T., Kim, G., 1998. The properties and microstructure of Al-based composites reinforced with ceramic particles, Materials Science and Engineering A 244 (1): 86–90.
7. Arik, H., 2008. Effect of mechanical alloying process on mechanical properties of α–Si3N4 reinforced aluminum based composite materials. Materials and Design 29: 1856-1861.
8. Tekmen C., Ozdemir, I., Fritsche, G., Tsunekawa, Y., 2009. Structural evolution of mechanically alloyed Al–12Si/TiB2/h-BN composite powder coating by atmospheric plasma spraying. Surface & Coatings Technology 203:2046-2051.

9. Tavoosi, M., Karimzadeh, F., Enayati, M.H., Heidarpour, A., 2009. Bulk Al–Zn/Al2O3 nanocomposite prepared by reactive milling and hot pressing methods. Journal of Alloys and Compounds 475:198-201.
10. Cambronero, L.E.G., Sánchez, E., Ruiz-Roman, J.M., Ruiz-Prieto, J.M., 2003. Mechanical characterisation of AA7015 aluminium alloy reinforced with ceramics, Journal of Materials Processing Technology 143–144: 378–383.
11. Prabhu B., Suryanarayana C., An L., Vaidyanathan R., 2006. Synthesis and characterization of high volume fraction Al-Al_2O_3 nanocomosite powders by high-energy milling. Materials Science and Engineering A, 425:192-200.
12. Wan-Li, G., 2006. Bulk Al-SiC nanocomposite prepared by ball milling and hot pressing method. Trans. Nonferrous Met. SOC. China 16:398-401.
13. Nie, C., Gu, J., Liu, J., Zhang, D., 2008. Investigation on microstructures and interface character of B4C particles reinforced 2024Al matrix composites fabricated by mechanical alloying. Journal of Alloys and Compounds 454:118-122.
14. Kubota, M., 2010. Solid state reaction in mechanically milled and spark plasma sinteren Al-B_4C composite materials. Journal Of Alloys And Compounds, 504: 319-322.
15. Wang, J., 2008. Mechanical alloying of amorphous Al–SiO_2 powders. Journal of Alloys and Compounds 456:139-142.

Deformations on Hole and Projectile Surfaces Caused By High Velocity Friction During Ballistic Impact

Karamış M. B
Erciyes University, Department of Mechanical Engineering, 38039, Kayseri, Turkey
karamisb@erciyes.edu.tr

Abstract: In this study, the deformations caused by the ballistic impact on the MM composites and on projectile surfaces are examined. The hole section and grain deformation of unreinforced targets are also examined after impact. The relatively high complexity of impact problems is caused by the large number of intervening parameters like relative velocity of projectile and target, shape of colliding objects, relative stiffness and masses, time-dependent surface of contact, geometry and boundary conditions and material characteristics.
The material used in this investigation are 2024 and 7075 aluminum alloys as matrix reinforced with SiC and Al_2O_3 particles. The matrix materials are extensively used in defense applications due to its favorable ballistic properties, moderate strength, high corrosion resistance and super plastic potential. Two different composites were produced; one by casting and the other by lamination.
The ballistic tests of the composite targets were carried out according to NIJ Standard-0101.04, Temperature 21 °C, RH=65% with 7.62 mm projectiles. The bullet weight was 9.6 g and their muzzle velocities were in the range of 770–800 m/s. The projectiles consisted of a steel core, copper jacket and lead material. The composite targets were positioned 15 m from the rifle. The interaction between projectiles and the target hole created after impact were examined by light microscopy and photography. Different damage and failure mechanisms such as petalling, cracking, spalling, dishing, etc., were observed on the target body. On the other hand, dramatic wear and damages on the projectile surface were also observed. The targets were supported with Al-5083 backing blocks having 40 mm thickness.

Keywords: Ballistic, high velocity friction, laminated composite, grain deformation, tribology

1. Introduction

Reinforced aluminium composites are being increasingly considered for structural applications in the defence, aerospace and automotive industries, etc. [1]. MMCs have slowly replaced some of the conventional light weight metallic alloys such as the various grades of aluminium alloys in applications where low weight and energy conservation are important considerations. Some researchers investigated dynamic behaviours of MMCs. The dynamic fracture toughness of SiC whisker reinforced 2124 aluminium alloy is higher than the quasi-static fracture toughness but decreased as the volume fraction of reinforcement increased [2]. in materials with SiC particulate, the ductility increased rapidly with pressure and the mode of damage initiation by particle fracture, and composites with SiC whiskers showed whisker matrix decohesion and strain localization that resulted in shear fracture[3], But very little dependence of the total elongation on the strain rate[4]. Reinforcing a ductile, tough metal with a strong ceramic is one method of achieving such an improvement in properties. Discontinuous reinforcements such as particulates or whiskers tend to produce composites with a greater degree of isotropy in properties as compared with continuous fiber reinforcements [5]. Therefore, the response of the MMC composites to a dynamic loading is a very important property. Although composites have many advantages with regard to mechanical properties; such as higher modules of elasticity, strength, hardness and wear resistance. They have the disadvantage of lower fracture toughness which is affected by many factors such as reinforcement size, volume fraction, distribution, chemical composition and heat treatment of matrix material, etc. The size of the reinforcement and its volume fraction are important variables in determining the properties of a composite [6].

In this study, the deformations caused by the ballistic impact on the MM composites and on projectile surfaces are examined. The hole section and grain deformation of unreinforced targets are also examined after impact.

2. Materials and method

The relatively high complexity of impact problems is caused by the large number of intervening parameters like relative velocity of projectile and target, shape of colliding objects, relative stiffness and masses, time-dependent surface of contact, geometry and boundary conditions and material characteristics.
The material used in this investigation are 2024 and 7075 aluminum alloys as matrix reinforced with SiC and Al_2O_3 particles. The matrix materials are extensively used in defense applications due to its favorable ballistic properties, moderate strength, high corrosion resistance and super plastic potential. Two different composites were produced; one by casting and the other by lamination. The chemical compositions of the matrix material are given in Table 1.

Table 1. Chemical compositions of matrix material used in the manufacture of composites and reinforcement properties

Material	Si	Fe	Cu	Mn	Mg	Cr	Zn	Ti
2024	0.5	0.5	3.8–4.9	0.3–0.9	1.2–1.8	0.1	0.25	0.15
7075	0.4	0.5	1.2-2	0.3	2.1-2.9	0.18-0.28	5.1	0.2

Reinforcement	Particle Size (mm)	Density (gr/cm^3)
SiC	0.25 – 0.5	3.1
Al_2O_3	0.5-1, 1-2, 4-8	3.6

The ballistic tests of the composite targets were carried out according to NIJ Standard-0101.04, Temperature 21 °C, RH=65% with 7.62 mm projectiles. The bullet weight was 9.6 g and their muzzle velocities were in the range of 770–800 m/s. The projectiles consisted of a steel core, copper jacket and lead material. The composite targets were positioned 15 m from the rifle. The interaction between projectiles and the target hole created after impact were examined by light microscopy and photography. Different damage and failure mechanisms such as petalling, cracking, spalling, dishing, etc., were observed on the target body. On the other hand, dramatic wear and demages on the projectile surface were also observed. The targets were supported with Al-5083 backing blocks having 40 mm thickness.

IOP Conf. Series: Materials Science and Engineering **295** (2018) 012046 doi:10.1088/1757-899X/295/1/012046

3. Results and discussion

The deformation behaviour of a composite target is a very important factor regarding the ballistic performance under impact loading conditions. Failure mechanisms are affected by this behaviour and the ballistic performance depends on impact behaviour. The ballistic performance of a composite target depends not only on its hardness but also on its toughness. Generally, brittleness and toughness are contra dictionary properties. If the hardness increases, the brittleness also increases while the toughness decreases. Therefore, the toughness properties are inversely affected by the hardness of the composite. Because of these complex effects, the hardness and toughness properties of a composite target should form an optimum point. This can be achieved by arranging suitable composition and heat treatment of the matrix materials, the type, size, volume fraction, distribution and the shape of the reinforcement.

Fig. 1 shows the hole structure after a projectile is removed from hole. It can be seen that there are many SiC particles in the hole of the MMC. They were first shaken by the high-velocity projectile and then pulled off from the surface and became free particles. The freeing of particles is related to the strength of the bonding between particles and matrix material. When the surface of the hole is examined by macroscopy, it is observed that some galling and yielding surround of the hole can be met. Part of the impact energy is consumed and appear as plastic deformations. When a reinforced composite could not prevent the passing of a projectile, the exit side of the hole may or may not petal, depending on the back side properties of the target. The surrounding area of the hole can be yielded freely and petalling appears after the projectile exits the hole. If the target has lowe reinforcement or without reinforcement, the hole surface or surrounding of hole deformed easily and petalling is occured(fig.2 a). This deformation zone of the hole around is created by projectile puncing effect.

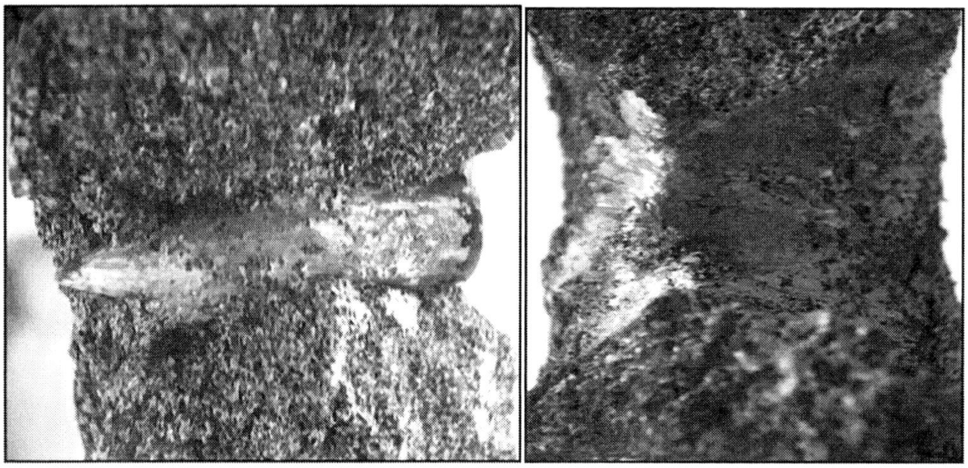

a) b)

Figure 1. The macro illustration of hole created on the composite target a) Hole with projectile b) The hole surface without projectile

When the impact behaviour of the unreinforced target is examined, it can be seen that any cracking or breaking of the target body is not observed other than the petalling flower-like created by the projectile on the back side of the hole (Figure 2(a)).

.

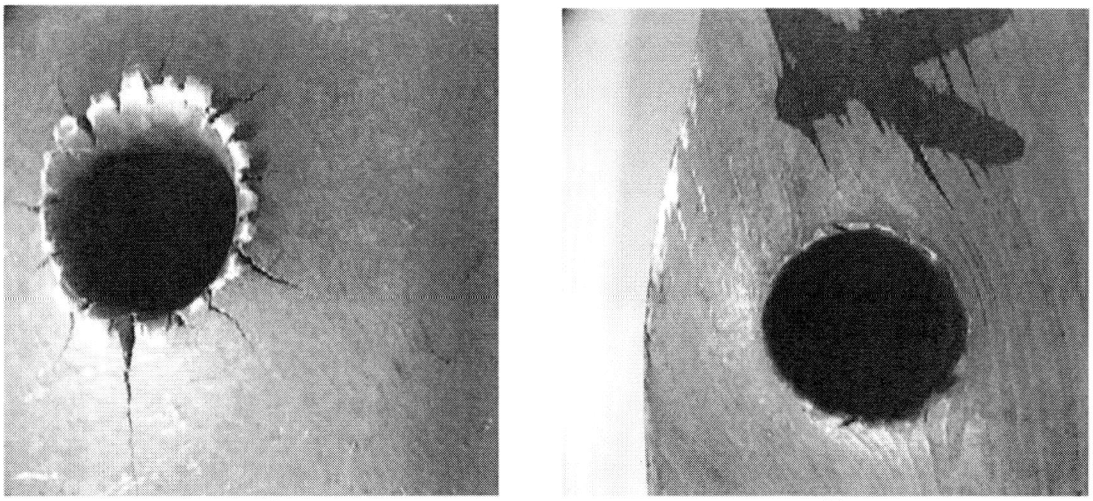

a) b)

Figure 2. Created holes on unreinforced and laminated targets: (a) petalling, (b) plastic deformation at the surrounding of the hole, (c) deformed grains of body material, (d) laminate deformation near the hole section.

The petalling with some soft cracking around the hole is generated from the punching effect of the high velocity impact of projectile. This effect causes plastic deformation along the thickness of the body along the radial direction of the hole (Figure 2(b)..The thickness of the unreinforced target on the hole's radial axis is increased by the punching, however, this increase does not reach the size of the hole diameter. This must be evidence of plastically compressed material structure on both sides of the hole. A certain area around the hole on the unreinforced sample is affected by this plastic deformation and a compression of area took place. At this area, aluminium grains between the projectile and undeformed body material are compressed by the projectile in the radial direction.

Figure 2(c) is evidence of this restricted deformed area. The grains close to the hole surface are also deformed in both the radial and axial directions. Therefore, these grains have a maximum length and a minimum thickness due to extensive deformation It can be seen in Figure 2(d) that the main (macro) deformation direction is axial with the complex one occurring in all directions. Since the laminates have weaker bonding to each other compared with those of unreinforced and particle reinforced target bodies. All deformations occurring in the axial direction on all composite targets used in this study are caused mainly by the frictional forces generated from the friction between the hole and the projectile tip surfaces. When the reinforcement particles close to the hole surface are affected by the punching effect of a projectile, they are compressed as they are forced to move either with the matrix or by translation. They are freely transferred and buried in the matrix and/or projectile surface. Since the particles are harder than the surfaces of the hole and projectiles they can scratch both surfaces easily. The reinforcement particles buried and/or scratched into the surface of the projectile reduce its velocity significantly. This causes a strong friction between the hole and the projectile surface and leads to a deformation of grains surrounding the particles directed along the three axes. The evidence of particle and laminate reinforced composite and buried particle can be seen in Figure 3. The macro deformation of the laminates can be seen in Figure 4. When the projectile punches the laminates and apply a friction to the hole surface of a laminated composite, the laminates are compressed and pushed forward with the projectile tip. Therefore, they are deformed and fractured, and they compress the area around the tip. The laminates close to the back side of the target are affected by the compressing effects of projectile tip like a punch. While the projectile is moving in the hole in the composite body, many mechanical and chemical phenomena [7,8] occur because of the deformation and strong friction mentioned above.

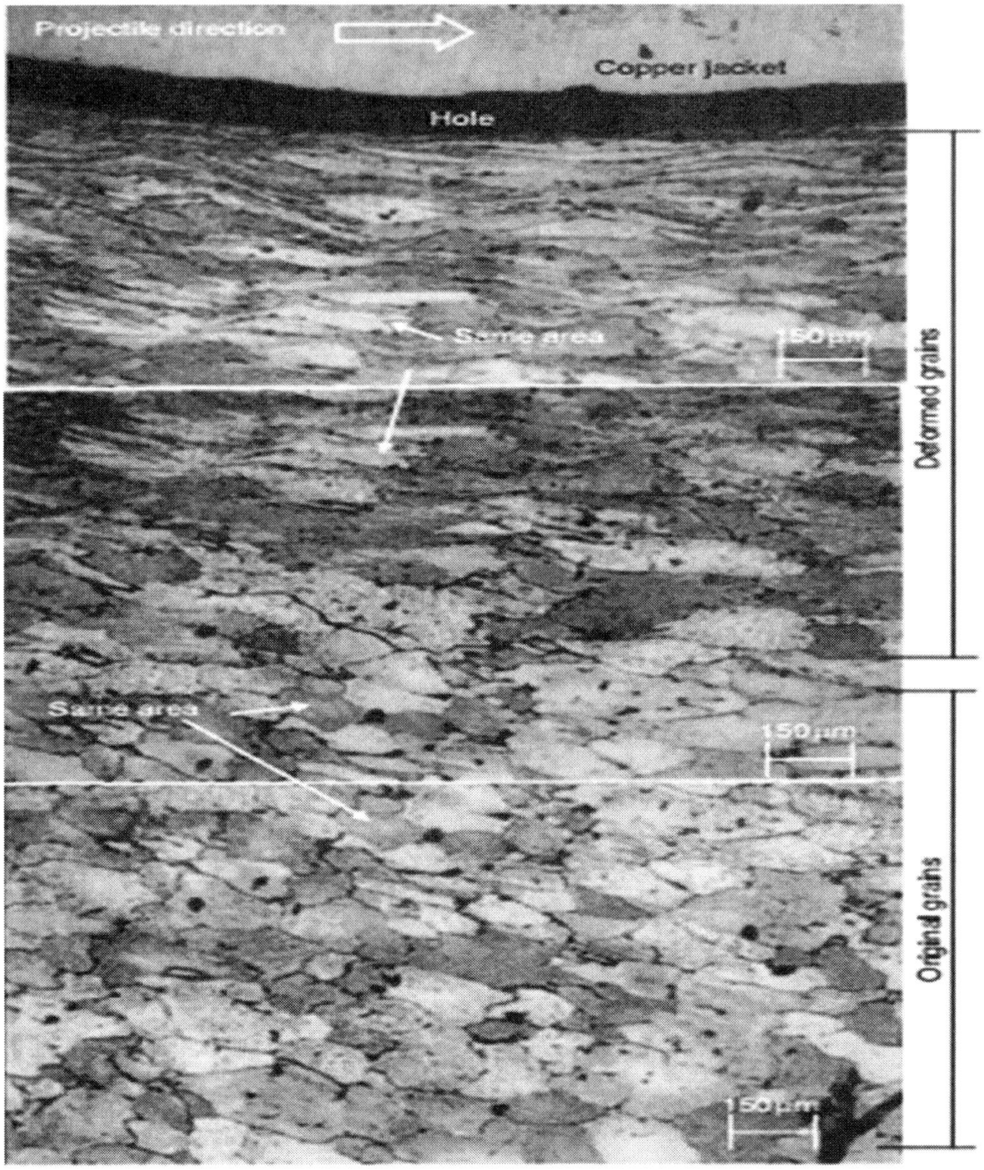
Projectile direction
Copper jacket
Hole
Same area
150 μm
Same area
150 μm
50 μm
Deformed grains
Original grains

2. c)

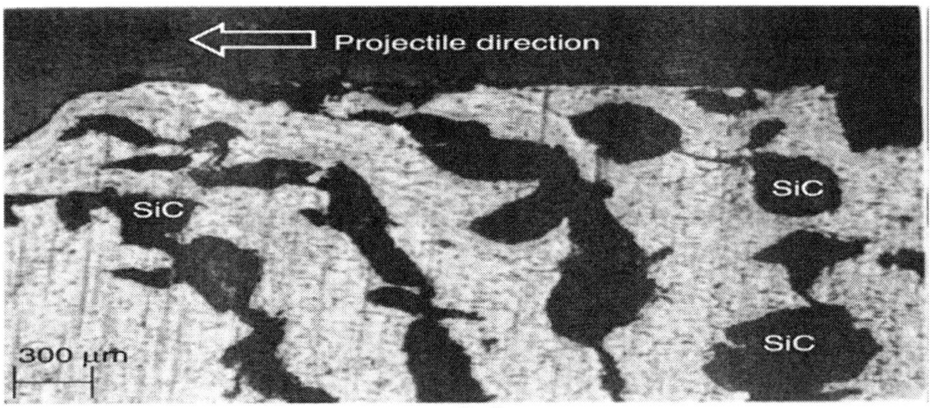
Projectile direction
SiC
SiC
SiC
300 μm

2. d)

IOP Conf. Series: Materials Science and Engineering **295** (2018) 012046 doi:10.1088/1757-899X/295/1/012046

a)

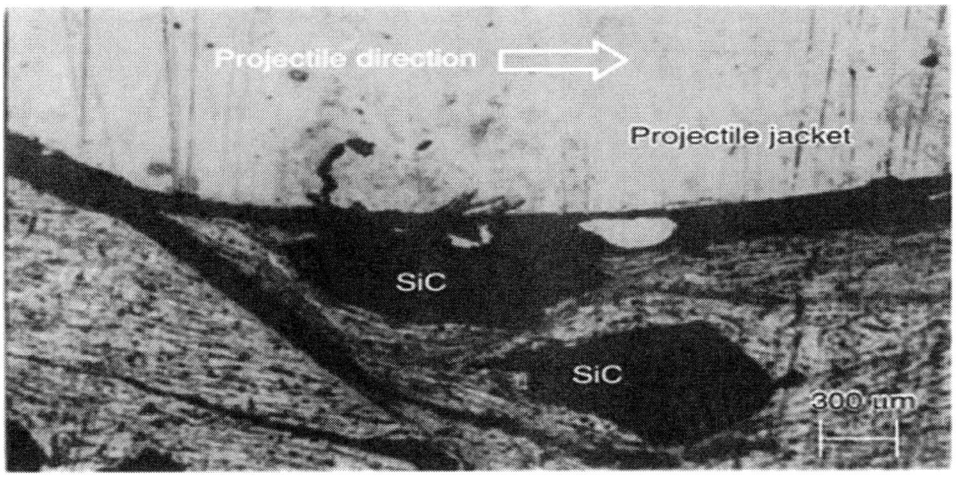

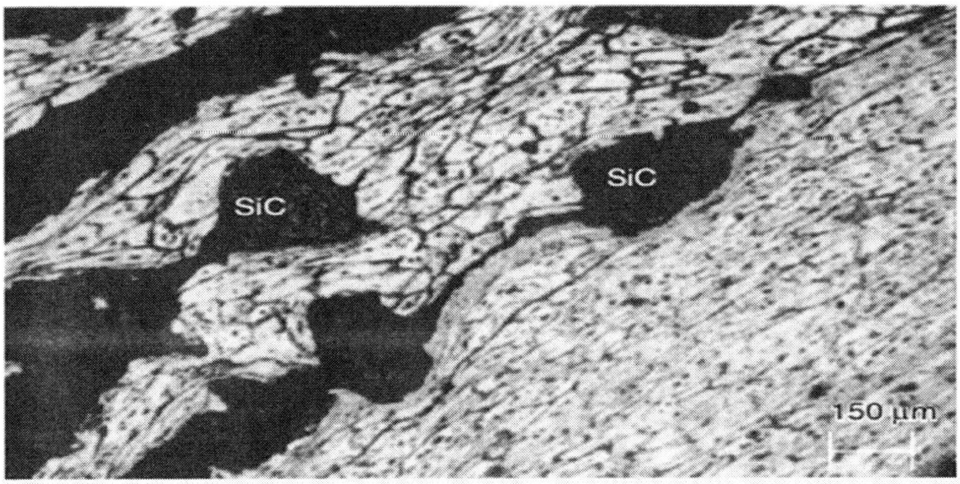

b)

Figure 3. Deformations of the matrix material at the surrounding of reinforcement particles, (a)particle reinforced composite (with SiCs), (b) laminated composite

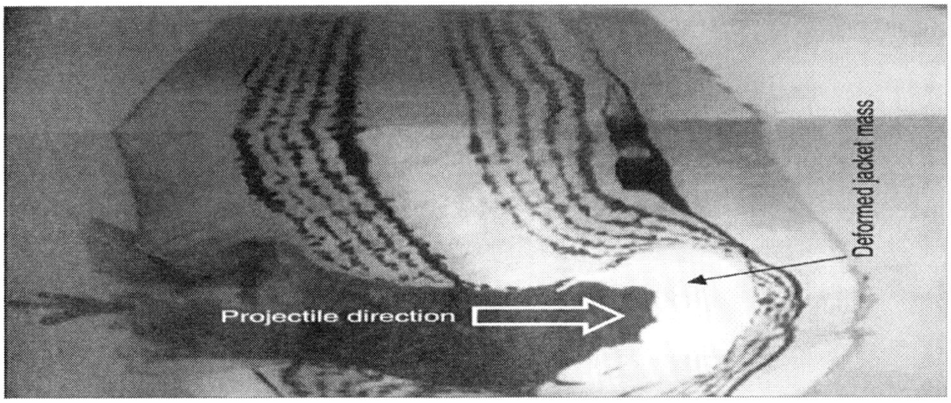

Figure 4. Deformation of laminates at surrounding the hole after friction and punching effects of projectile.

Now, we can examine casted target hole and the projectile surfaces impacted to composite target at high velocity. Fig. 5 shows clearly the frictional traces on the hole surface created by high-velocity projectile. These traces on this frictional area must have been created by the particles pulled out from the matrix area. When a projectile enters into a target, it meets with some particles and the projectile transfers these particles to another zone and/or some other particles adhere to the projectile surface while some are buried into the projectile surface. The particles can scratch both surfaces easily because they are harder than the surfaces of the hole and projectile. As new traces are scratched by the other particles following, some traces are built up over each other. It is also obvious that the path of the projectile in composite can be changed.

Figure 5. Frictional traces on the hole surface

It can be claimed that the stronger the friction, the smaller is the penetration depth. The projectile is first slowed down by the friction of SiC particles and then stopped by other particles embedding into both surfaces. The evidence to this friction mechanism can be seen in Fig. 6. This figure shows that the embedded particles stopped the projectile in the matrix. The hard SiC particles buried themselves into the jacket surface of projectiles and this led to dissipation of the kinetic energy of the projectile and its eventual stop. Fig. 6 b clearly shows that the final deformed shape of the matrix material resembles the particle form which caused the deformation. The plastic yielding is localised around the particles.

The projectile surface is scratched by the SiC particles and the plastic deformation takes place. There are many deep grooves caused by reinforcing particles on the friction surface parallel to the propagation direction.

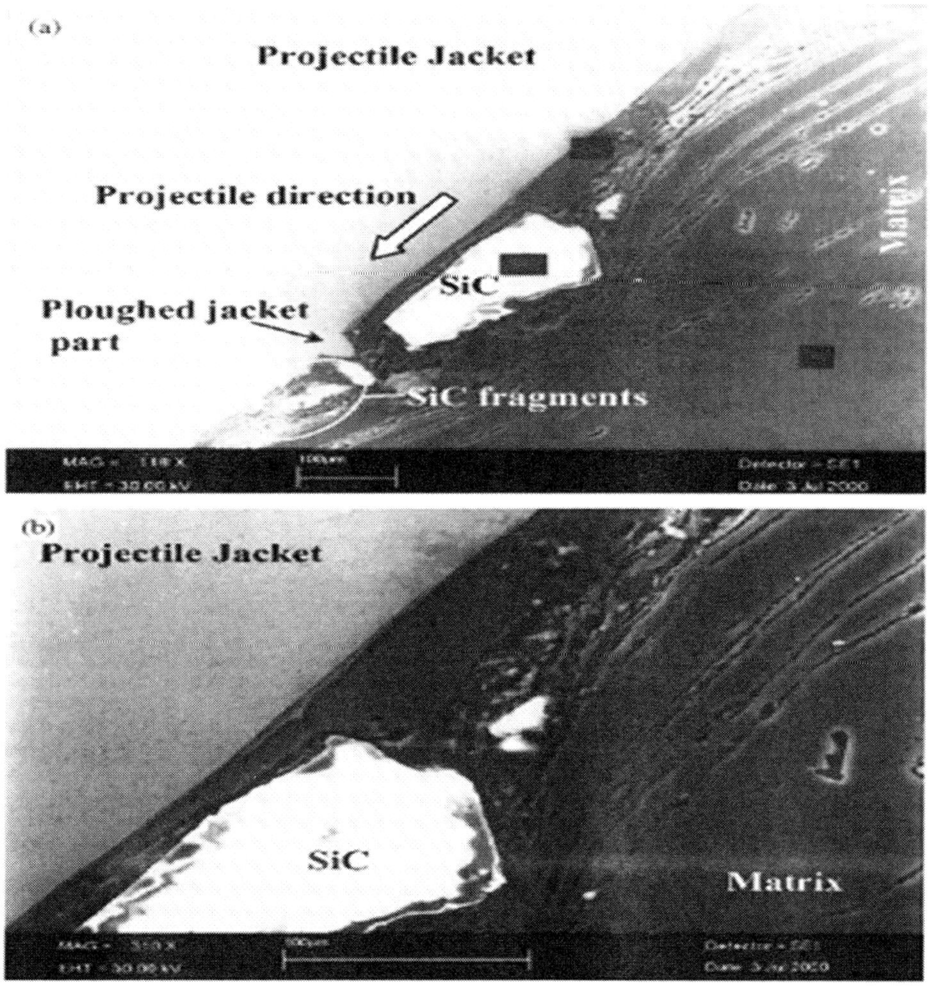

Figure 6. Hard SiC particles buried into the jacket surface; (a) buried SiC particles, and (b) plastically yielding of matrix material at the surrounding SiC particle

Fig. 7 shows the some macro damage on the projectile tip caused by armour plate during impact. It can easily be seen that the projectile tip may have been broken and blunted (Fig. 7a). This breaking often occurs if the projectile impacts a hard zone generated by collecting the hard Al2O3 particles onto the target surface. After the projectile tip enters the armour plate, it has to puss among the hard and coarse Al2O3 particles. Sometimes, this interaction caused to asymmetric forces on the nose of the projectile. These forces deflect projectile line of flight inside the target [9] and may be bend projectile nose (Fig. 7b). This phenomenon may be resembled projectile impact to an inclined plate [10]. It may also detach some parts from projectile nose by breaking, and big craters can be formed (Fig. 7c). Even when the projectile is very sharp, the interception capability of the composite target is very satisfactory. It can be seen from the figure that penetration of the projectile tip to the target is too hard and the projectile lost some parts from its nose. Some deep valleys generated from friction with Al2O3 particles can nose (Fig. 7d) also be observed on the tip.

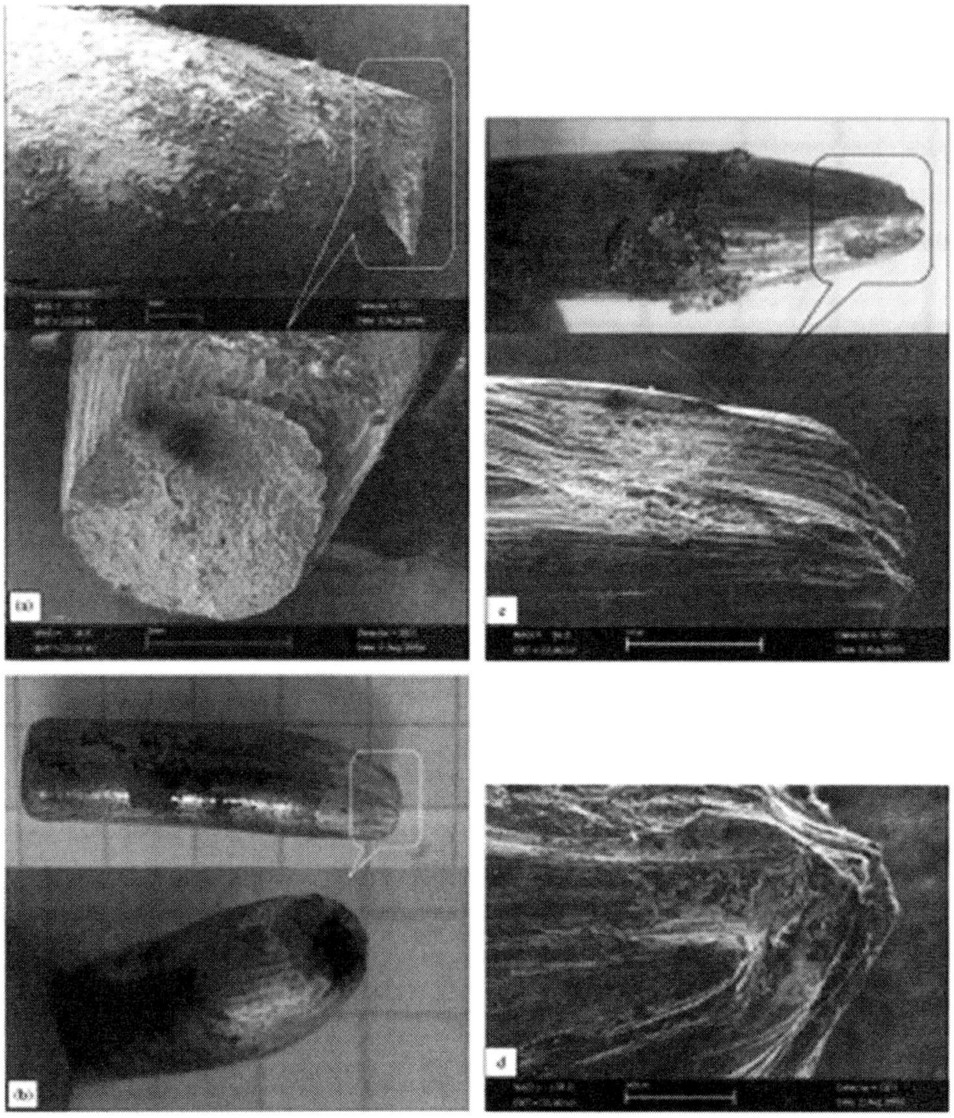

Figure 7. Damage on the projectile tip after impacting: (a) broken AK-47 steel projectile tip, (b) plastic deformation of the tip by bending, (c) crater on the tip nose by breaking, and (d) deep valleys on the tip nose.

The strong friction can lead to the deep grooves on both frictional surfaces. Figures show evidence that the wear mechanisms are predominantly abrasive. Melted zone is observed only on the hole surface. Some matrix material and some SiC particles are adhered to the surface of projectile strongly. Namely, the wear mechanisms are different on both hole and projectile surfaces. While the wear mechanisms are melting, adhesive and abrasive on hole surface, they are predominantly abrasive on the projectile surface. As the strong friction caused by reinforcement particles leads to the deceleration of the projectile in the armours, swelling at the surrounding area of the scratched grooves by particles also takes place. The deep grooves are swelled by plastic yielding to their edges or around the particles buried in the grooves [11].

IOP Conf. Series: Materials Science and Engineering **295** (2018) 012046 doi:10.1088/1757-899X/295/1/012046

4. Conclusions

1. The MMCs can prevent the penetration of a projectile through their body by consuming the projectile's kinetic energy. The energy is consumed through the damage such as cracking, petalling, and some matrix deformation.

2. On the composite targets manufactured with relatively good toughness, some petals are observed at entrance and exit sides of the hole generated by the passing projectile.

3. When the projectile meets reinforcement particles in the composite body, the strong friction is generated between the projectile and hole surface. The velocity of the projectile is decreased suddenly at the inlet of the hole. Some reinforcement particles scratch the projectile surface by burying and the copper alloy jacket of the projectile is left in the hole by heavy deformation (swelling). This section is the largest section of the hole.

4. All composite types show a different deformation mechanism and different deformation rates by the fast punching effects of projectile. In the case of laminate composites, the laminates are fractured and deformed in the direction of projectile movement.

5. If the projectile pierces through and leaves the composite some petalling can be observed at the exit side of the hole. If a strong friction force is generated by the reinforcement particles, the projectile cannot pass through the composite and is kept inside composite.

6. While the wear mechanisms are different on the hole surface such as melting, adhesion and abrasion, on the projectile surface, wear mechanism is predominantly abrasive with some swelling.

7. The projectile nose is broken or deformed plastically when it impacts the composite target. If the composite has hard reinforcement particles such as Al2O3, the friction between projectile tip and hole surface becomes greater and the projectile surface is scratched by ploughing. Plastic damage or breaking of the tip nose leads to increasing the resistance to the projectile movement through the target hole

References

1. Preet, M.S. and Lewandowski, J.J., Effect of Heat Treatment and Reinforcement Size on Reinforcement Fracture During Tension Testing of SiCp Discontinuously Reinforced Aluminium Alloy,Metallurgical Transaction A, 24A: 2531–2543,1993.
2. Marchand A, Duffy J, Christman TA, Suresh S. An experimental study of the dynamics mechanical properties of an Al–SiCw composite. Eng Fract Mech 30(3):295–315, 1988.
3. Vasudevan AK, Richmond O, Zok F, Embury D. The influence of hydrostatic pressure on the ductility of Al–SiC composites. Mater Sci En, ;A107:63–9, 1989.
4. Perng CC, Hwang JR, Doong JL. High strain rate tensile properties of an (Al2O3particles)–(Al alloy 6061-T6) metal matrix composite. Mater Sci Eng A171:213–21, 1993.
5. Manoharan, M. and Lewandowski, J.J., Effect of Reinforcement Size and Matrix Microstructure on the Fracture Properties of an Aluminium Metal Matrix Composite,Material Science and Engineering, A150: 179–186, 1992.
6. Karamıs, M. B., Cerıt, A.A., And Naır, F. , Mutual Action Between MMCS Structure and Projectile After Ballistic Impact, Journal of Composıtematerıals, Vol. 42, No. 23,2008.

7. Karamis, M.B., Cerit, A.A. and Nair, F., Surface Characteristics of Projectiles after Frictional Interaction with Metal Matrix Composites under Ballistic Condition, Wear, 261(7–8): 738–745, 2006.
8. Karamis, M.B., Nair, F. and Tasdemirci, A., Analyses of Metallurgical Behaviour of Al-SiCp Composites after Ballistic Impacts, Composite Structures, 64(2): 219–226, 2004.
9. Z. Rosenberg, Z. Surujon, Y. Yeshurun, Y. Ashuach, E. Dekel, Ricochet of 0.3 AP projectile from inclined polymeric plates, Int. J. Impact Eng. 31, 221–233, 2005.
10. Z. Fawaz, W. Zheng, K. Behdinan, Numeric simulation of normal and obligue ballistic impact on ceramic composite armours, Compos. Struct. 63 (2004) 387–395.
11. Karamis, M.B., Tribology at high-velocity impact, Tribology International 40, 98–104, 2007.